BIOINFORMATICS

Databases, Tools, Algorithms

ORPITA BOSU
Centre for Research in Cognitive Systems
National Institute of Informational Technology
New Delhi

SIMMINDER KAUR THUKRAL
Department of Biotechnology & Bioinformatics
National Institute of Informational Technology
New Delhi

YMCA Library Building, Jai Singh Road, New Delhi 110001

Oxford University Press is a department of the University of Oxford.
It furthers the University's objective of excellence in research, scholarship,
and education by publishing worldwide in

Oxford New York
Auckland Cape Town Dar es Salaam Hong Kong Karachi
Kuala Lumpur Madrid Melbourne Mexico City Nairobi
New Delhi Shanghai Taipei Toronto

With offices in
Argentina Austria Brazil Chile Czech Republic France Greece
Guatemala Hungary Italy Japan Poland Portugal Singapore
South Korea Switzerland Thailand Turkey Ukraine Vietnam

Oxford is a registered trade mark of Oxford University Press
in the UK and in certain other countries.

Published in India
by Oxford University Press

First published 2007

ISBN-13: 978-0-19-567683-9
ISBN-10: 0-19-567683-1

Typeset in Times Roman
by Archetype, New Delhi 110 063
Printed in India by Radha Press, Delhi 110031
and published by Oxford University Press
YMCA Library Building, Jai Singh Road, New Delhi 110001

With an introduction by

Sugata Mitra

Professor
School of Communication, Education and Language Studies
Joseph Cowan Building
St. Thomas Street
Newcastle University
Newcastle upon Tyne
NE1 7RU
UK

The genetic code is a language. It is the first language of non-human origin that we have encountered; it is a truly alien system. Ironically, it has come to us not from outer space, but from within ourselves.

Professor Sugata Mitra

To my grandparents
Late Shri Sachindra Nath Bosu
and
Late Smt. Labanya Bosu

Orpita Bosu

To the memory of Daddy, Col. Madan Singh Thukral,
who always gave much more than he received
and
Mamma, Upinder Kaur Thukral,
whose love and inspiration has an eloquence of its own

Simminder Kaur Thukral

Preface

It was in the seventeenth century that biologists started dealing with problems of information management. For nearly three centuries, early biologists were preoccupied with cataloguing and comparing species of living things. By the middle of the seventeenth century, John Ray introduced the concept of distinct species of animals and plants and developed guidelines based on anatomical features for distinguishing conclusively among species. In 1730, Carolus Linnaeus established the basis for the modern taxonomic naming system of kingdoms, classes, genera, and species. Taxonomy was the first informatics problem for biologists. The University of Arizona's Tree of Life project and NCBI's Taxonomy Database are two examples of online taxonomy projects.

Developments in laboratory technology facilitated collection of data at a rate that was faster than data interpretation. Biologists started facing difficulties in analysing and interpreting this data overload. Collecting and cataloguing information about individual genes (approx. 30,000) in human DNA and determining the sequence of three billion chemical bases that make up the human DNA became the second informatics problem for biologists. In 1990, Human Genome Project was initiated as a major bioinformatics solution to solve this problem, which labelled the 21st century as the era of genomes.

Bioinformatics, as the name suggests, is an interdisciplinary field involving majorly biology, computer science, and information technology. Chemistry and mathematics also play important roles. Bioinformatics deals with methods for manipulating biological data to solve problems in biology. A fine-tuned insight into the structure of DNA, RNA, and protein is not only helpful but also required. It is widely recognized that India has expertise in software and a good pool of qualified and experienced biologists. But it needs to combine the expertise of software/IT engineers and experienced biologists to create trained professionals. Students from varied disciplines such as IT, engineering, pharmaceutics, and life sciences are keen to pursue bioinformatics. The need for a lucid book that adopts a holistic approach to the subject, to suit the requirements of both engineers and life scientists, has been felt for long.

About the Book

This book is intended to serve as an introduction to bioinformatics. The 'do-it-along' approach of this book and in-between notes will be beneficial at the beginner level. In keeping with this approach, it does not assume any background knowledge of the subject. The appendices at the end of the book provide the required basic information on biological and computational resources required for easy and effective understanding of bioinformatics tools. The book comes with chapter-end exercises for the students to test their understanding.

The data required to solve some of the tasks and chapter-end exercises is provided in the CD-ROM that comes with this book. A icon has been put alongside such tasks and exercises. The CD-ROM also contains coloured images of the figures given in the text as well as 3D models.

It is expected that knowledge in bioinformatics presented here will enable readers to go beyond a mere push-button approach of using bioinformatics tools. It will hopefully help the students to generate and interpret data with ease and speed.

Content and Coverage

The book describes the methods used to store, retrieve, and derive data from databases using bioinfromatics tools. Bioinformatics, as we all know, is an applied domain and needs intensive hands-on practice. With this conviction, this book leads the students to concept through practice.

The book has been divided into four parts. Each part discusses a distinct aspect of bioinformatics.

Part A gives an introduction and also throws some light on the present bioinformatics scene in India. This part emphasizes the need to develop qualified and trained human resources to cater to the demands of the fast developing bioinformatics market.

Part B introduces databases and spans Chapters 1 to 3. It describes the various types and categories of databases as well as the various file formats in which data is stored in these databases. This part describes the methods used to store, access, and retrieve information from various data databases. Chapters 1, 2, and 3 discuss sequences databases, structure databases, and other databases, respectively, in detail.

Part C introduces the various tools used by the bioinformatics community. This part comprises Chapters 4, 5, 6, and 7. Chapter 4 discusses the various tools required for submission of data to various databases. Chapter 5 covers data analysis tools in detail and underlines their utility. Prediction tools have been elaborately dealt with in Chapter 6. Chapter 7 gives an account of the tools and concepts used in modelling.

Part D deals with algorithms in the field of bioinformatics. There are two separate chapters discussing two different aspects of algorithms. While Chapter 8 describes the algorithms dealing with data analysis, Chapter 9 discusses algorithms used for prediction purposes.

Additionally, there are three appendices that make the book a self-contained, compact resource material for beginners in bioinformatics. These appendices provide some very useful information needed by the students during the course their study.

Acknowledgements

I am grateful to Professor Sugata Mitra for introducing me to this field. I am thankful to my organization, supervisor, and co-author without whose help it would not have been possible to write this book. I must acknowledge Jawaharlal Nehru University, Delhi, and Professor Jayaram and his team from Supercomputing Facility, IIT Delhi, for inspiring me. I acknowledge Oxford University Press for their support and kind patience in publishing this book. Over and above, I would like to acknowledge the support and encouragement from my dear ones, which helped me to complete this book.

Orpita Bosu

I would like to thank Professor Sugata Mitra (Newcastle University, UK) and Professor Jayram (IIT Delhi) for introducing me to the field of bioinformatics. I am indebted to Professor Sugata Mitra and Orpita Bosu without whose belief in the worth of this collaborative enterprise this book would not have been possible. I would also like to thank Udai Singh and Meenakshi Goel (NIIT Ltd) and Professors Sunil Khanna, Parimal Mandake, and M.P. Kapoor (TNI) for generously extending the encouragement and resources required to complete this mammoth task. I do not know how to thank Mom and Dad for the core values that they helped me inculcate, of which this work is a small reflection. To Professor Prashant Mishra (my husband), Nishant (my son), and little Anushka (my daughter) I am thankful, for their love, patience, understanding, and encouragement, which enabled this book to proceed even when sunny days and other enticing distractions beckoned.

Simminder K Thukral

Contents

PART A

Bioinformatics: The Information Technology of Living Things **3**

A.1 Bioinformatics: When and Why 8
A.2 What is Bioinformatics? 8
A.3 In This Book 9
A.4 Bioinformatics: Applications and Research 11
A.5 Present Bioinformatics Scenario in India 12
Conclusion *17*

PART B

Databases **21**

B.1 Characteristics of Bioinformatics Databases 22
B.2 Categories of Bioinformatics Databases 22
B.3 Navigating Databases 35
B.4 Information Retrieval Systems 37
Conclusion *42*
Exercises *42*

1. Sequence Databases **44**

1.1 Nucleotide Sequence Databases 45
1.2 Secondary Nucleotide Sequence Databases 58
1.3 Protein Sequence Databases 60
1.4 Secondary and Specialized Protein Sequence Databases 75
1.5 Information Retrieval System: Entrez 81
1.6 Information Retrieval System: SRS 93
Conclusion *100*
Exercises *100*

2. Structure Databases **106**
2.1 Structure File Formats 107
2.2 Protein Structure Database Collaboration 108
2.3 PDB 108
2.4 MMDB 117
2.5 CATH 119
2.6 FSSP 127
2.7 DALI 128
2.8 SCOP 131
Conclusion *141*
Exercises *142*

3. Other Databases **145**
3.1 Enzyme Databases 145
3.2 MEROPS 147
3.3 BRENDA 167
3.4 Pathway Databases: CAZy 174
3.5 Disease Databases 177
3.6 Literature Databases 193
3.7 Other Specialized Databases 197
Conclusion *234*
Exercises *235*

PART C

Tools **239**
C.1 Need for Tools 241
C.2 Knowledge Discovery 242
C.3 Industry Trends 244
C.4 Data-mining Tools 245
Conclusion *250*

4. Data Submission Tools **251**
4.1 Nucleotide Sequence Submission Tools 252
4.2 Protein Submission Tools 267
4.3 tbl2asn (Command Line Tool for GenBank) 278
Conclusion *279*
Exercises *280*

5. Data Analysis Tools **283**
5.1 Tools for Nucleotide Sequence Analysis 283
5.2 Tools for Protein Sequence Analysis 324
Conclusion *349*
Exercises *350*

6. Prediction Tools **354**
- **6.1** Phylogenetic Trees and Phylogenetic Analysis 354
- **6.2** Gene Prediction 372
- **6.3** Protein Structure and Function Prediction 387
- *Conclusion* *397*
- *Exercises* *397*

7. Modelling Tools **400**
- **7.1** Tools for 2D Protein Modelling 400
- **7.2** Tools for 3D Protein Modelling 414
- *Conclusion* *425*
- *Exercises* *425*

PART D

Algorithms **429**
- **D.1** Classification of Algorithms 433
- **D.2** Implementing Algorithms 440
- **D.3** Biological Algorithms 441
- **D.4** Bioinformatics Tasks and Corresponding Algorithms 442
- **D.5** Algorithms and Bioinformatics Software 443

8. Data Analysis Algorithms **444**
- **8.1** Sequence Comparison Algorithms 444
- **8.2** Substitution Matrices Algorithms 452
- **8.3** Sequence Alignment Optimal Algorithms 458
- *Conclusion* *472*
- *Exercises* *472*

9. Prediction Algorithms **476**
- **9.1** Gene Prediction Algorithm 477
- **9.2** Phylogenetic Prediction Algorithm 482
- **9.3** Protein Structure Prediction 491
- *Conclusion* *501*
- *Exercises* *501*

Appendix A1: Biology for Bioinformatics **503**

Appendix A2: PERL for Bioinformatics **552**

Appendix A3: LINUX for Bioinformatics **562**

Glossary of Bioinformatics Terms **569**

References **580**

Index **582**

PART A
Bioinformatics: The Information Technology of Living Things

A

Bioinformatics: The Information Technology of Living Things

An Introduction by Prof. Sugata Mitra

One of the important properties of living things is their ability to reproduce. Living things are able to make copies of themselves: not perfect copies, but close enough. The subtle differences between parents and offspring are the reason for evolution and biodiversity. If we could understand this copying process, we would know a lot about ourselves, perhaps even about the purpose of our existence.

The process of copying anything always involves a transfer of information. For example, in a photocopier, a paper document is scanned by a beam of light and the patterns of black and white in the reflected light are stored. This is the information that is then used to direct ink to the right places on a blank piece of paper so that a copy of the original document is produced. However, even in the best photocopiers, there are always subtle differences between the original document and the copy, all caused by tiny random errors of movement or ink application. Transformation of information from one form to another always has some error associated with it, just as the transformation of energy from one form to another will always have some waste. Indeed, energy and information are connected through these 'errors' and the concept of entropy in thermodynamics ties the two together. It is this connection between information and energy that makes life possible on earth. Life, as we know it on earth, is dependent on processes that convert information into chemical energy and eventually into the chemistry of macromolecules. This was unknown until the discovery of deoxyribonucleic acid (DNA).

DNA is a single, huge molecule, sometimes up to two metres long. It consists of two long strings of smaller molecules, called nucleotide, wound up against each other in a structure, now famous, called the *double helix*. The two strings of the double helix are connected to each other, periodically, by a set of four molecules or bases—adenine, guanine, cytosine, and thymine. The structure of DNA is like a ladder, twisted around to form a helix shown in Fig. A.1. The steps of this ladder are formed of pairs of molecules-adenine with thymine, or guanine with cytosine (AT or GC in short) (see Fig. A.2). There are no other connections allowed. Of course, an AT connection can also be seen as a TA connection, depending on which way it is looked at. The same holds for GC and CG. DNA molecules exist inside every cell nucleus in a living organism. A two-metre long molecule of DNA is coiled tightly and packed into the nucleus of every human cell.

If all the AT and GC pairs (called base pairs) in a section of DNA were to be opened, the unmatched base pairs could be thought of as a string of characters, for example, AAAGTTCTTCTTCAATTA. This, of course, would be matched on the other string by its complements, that is, TTTCAAGAAGAAGTTAAT. Such strings of 'alphabets' contain the procedure for assembling a certain sets of proteins from the necessary amino acids after several intermediate processes. These proteins eventually make up most of the organism. In other words, the sequences of base pairs in DNA are the 'raw materials' as well as 'instructions' for building and maintaining the organism. Sequences of base pairs, often millions of alphabets in number, are grouped together into genes, much as the English alphabet are grouped together into words, sentences, and paragraphs in order to make sense. When a gene inside a DNA molecule needs to be activated, the DNA molecule in a cell nucleus uncoils and unfurls to just the right extent to expose that gene. At this time, another molecule, RNA, is formed.

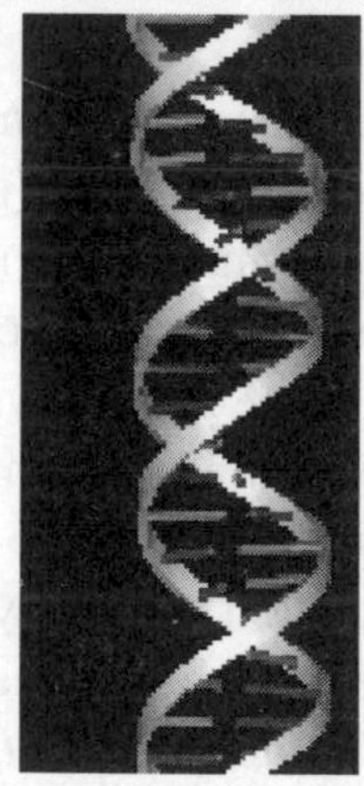

Fig. A.1 DNA double helix

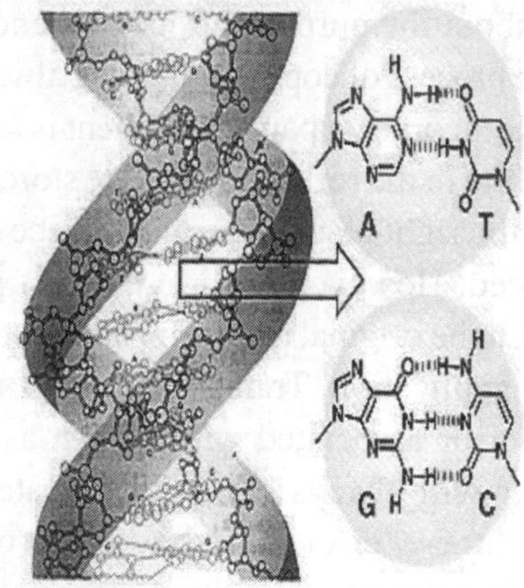

Fig. A.2 Zoomed view of double helixed DNA with side chains

Ribonucleic acid, or RNA, is thought to be even more fundamental to life than is DNA. A simple understanding of the structure of RNA would be to imagine a DNA double helix that has been partly separated into a section with the broken ends of the base pairs sticking out of them. Free peptides and the A, T, G, C molecules floating in the nucleus are attracted to this open section of the DNA. Here they form an RNA molecule. For example, if the open section of the DNA contained the sequence CCG, the corresponding section of the RNA that is formed would contain the sequence GGC. In effect, the RNA that is formed would be a 'print' of the open DNA section. Such RNA is called messenger RNA or mRNA. However, RNA and DNA differ in an important respect. RNA does not contain the base thymine (T). Instead, it contains another base called uracil, coded as U. This is because uracil, though similar to thymine, takes less energy to bind and is, therefore, a more efficient form of copying, provided the copy does not need to last for long. This is indeed the case, as the mRNA has a very short-lived and specific purpose.

Once mRNA is formed, the open section of the DNA molecule closes, the gene is no longer exposed, and its copy, in the form of the mRNA, travels out of the cell nucleus into the cytoplasm. Here it encounters the ribosome, a relatively large structure made mostly of rRNA (another form of RNA) and some proteins to hold the mass together. Ribosome is capable of a form of biological computation. It 'reads' the base sequences on an mRNA molecule, and for every three bases it produces an amino acid from the cytoplasm (for example, for three bases CGC, it produces arginine).

Note

Amino acids are compounds that have an amine group at one end and a carboxyl group at other end. Proteins are formed by the polymerization of amino acids where the carboxyl and amine groups of adjacent amino acids bond together, releasing a molecule of water. There are 20 standard amino acids that are active in human beings. Each of these amino acids is related to one or more codons. You can refer Table 1 for a list of 20 amino acids and their related information.

A set of three consecutive bases in RNA is called a *codon*. In effect the ribosome is a process control computer that takes codons as the input and produces amino acids as the output. The amino acids are brought to the ribosome by another form of RNA, called transfer RNA or tRNA. These are relatively short RNA molecules that float freely in the cytoplasm with an amino acid attached to one end. When a ribosome detects a codon in an mRNA, it attracts that tRNA molecule which is 'carrying' that amino acid that the codon is referring to. Once close to the ribosome, the tRNA releases the amino acid. As a result, a sequence of codons results in a sequence of amino acids, which join together to form a polypeptide chain or a protein. Eventually, masses of intricately folded proteins form the gross structures such as skin and hair of living organisms. While it is usual to consider the human brain as a powerful central computer that controls all our actions and functions, it is interesting to note that our life processes are far more dependent, at a fundamental level, on the information processing that takes place in hundreds of millions of ribosomes. Life, in this context, appears as an immense distributed computing system.

Table 1 List of 20 amino acids and their related information

Amino acid	Abbreviation	Side chain	Hydrophobic	Polar	Charged	Small	Tiny	Aromatic or aliphatic	van der Waals volume	Codon	Occurrence in proteins (%)
Alanine	(Ala, A)	$-CH_3$	X	-	-	X	X	-	67	GCU GCC GCA GCG	7.8
Cysteine	(Cys, C)	$-CH_2SH$	X	-	-	X	-	-	86	UGU UGC	1.9
Aspartate	(Asp, D)	$-CH_2COOH$	-	X	Negative	X	-	-	91	GAU GAC	5.3
Glutamate	(Glu, E)	$-CH_2CH_2COOH$	-	X	Negative	-	-	-	109	GAA GAG	6.3
Phenylalanine	(Phe, F)	$-CH_2C_6H_5$	X	-	-	-	-	Aromatic	135	UUU UUC	3.9
Glycine	(Gly, G)	$-H$	X	-	-	X	X	-	48	GGU GGC GGA GGG	7.2
Histidine	(His, H)	$-CH_2-C_3H_3N_2$	-	X	Positive	-	-	Aromatic	118	CAU CAC	2.3
Isoleucine	(Ile, I)	$-CH(CH_3)CH_2CH_3$	X	-	-	-	-	Aliphatic	124	AUU AUC AUA	5.3
Lysine	(Lys, K)	$-(CH_2)_4NH_2$	-	X	Positive	-	-	-	135	AAA AAG	5.9
Leucine	(Leu, L)	$-CH_2CH(CH_3)_2$	X	-	-	-	-	Aliphatic	124	UUA UUG CUU CUC CUA CUG	9.1

(contd)

Table 1 (contd)

Amino acid	Abbreviation	Side chain	Hydrophobic	Polar	Charged	Small	Tiny	Aromatic or aliphatic	van der Waals volume	Codon	Occurrence in proteins (%)
Methionine	(Met, M)	$–CH_2CH_2SCH_3$	X	-	-	-	-	-	124	AUG	2.3
Asparagine	(Asn, N)	$–CH_2CONH_2$	-	X	-	X	-	-	96	AAU AAC	4.3
Proline	(Pro, P)	$–CH_2CH_2CH_2$-	X	-	-	X	-	-	90	CCU CCC CCA CCG	5.2
Glutamine	(Gln, Q)	$–CH_2CH_2CONH_2$	-	X	-	-	-	-	114	CAA CAG	4.2
Arginine	(Arg, R)	$–(CH_2)3NH–C(NH)NH_2$	-	X	Positive	-	-	-	148	CGU CGC CGA CGG AGA AGG	5.1
Serine	(Ser, S)	$–CH_2OH$	-	X	-	X	X	-	73	UCU UCC UCA UCG AGU AGC	6.8
Threonine	(Thr, T)	$–CH(OH)CH_3$	X	X	-	X	-	-	93	ACU ACC ACA ACG	5.9
Valine	(Val, V)	$–CH(CH_3)_2$	X	-	-	X	-	Aliphatic	105	GUU GUC GUA GUG	6.6
Tryptophan	(Trp, W)	$–CH_2C_8H_6N$	X	-	-	-	-	Aromatic	163	UGG	1.4
Tyrosine	(Tyr, Y)	$-CH_2–C_6H_4OH$	X	X	-	-	-	Aromatic	141	UAU UAC	3.2

A somewhat lyrical (and not too accurate!) description of the process of life

You have a small cut on your finger. The skin is broken and so are some minor blood vessels. There is some bleeding. The blood clots quickly and the bleeding stops in a few minutes. Clotting is a physical property of blood useful for cuts, although it can be deadly when it happens internally. Your bleeding has stopped but the wound has to heal. The cells surrounding the cut begin to react. The DNA in their nuclei unfold and twist to expose those genes that have the instructions for building new skin. Floating molecules fit themselves to the open genes to form strands of mRNA that float out of the nuclei into the cytoplasm of the cells. Here they are scanned by floating ribosomes that process the codons in the mRNA and attract other floating tRNA from the cytoplasm. The tRNA bring with them the amino acids that will polymerize to form proteins. Wisps of proteins twist and fold to form structures that eventually join to form new cells—cells of skin. In a few days, the wound heals, as though it had never been there at all.

A.1 Bioinformatics: When and Why

It was in the 17th century that biologist started dealing with problems of information management. Early biologists were preoccupied with cataloguing and comparing species of living things. By the middle of the 17th century, John Ray introduced the concept of distinct species of animals and plants and developed guidelines based on anatomical features for distinguishing conclusively between species. In 1730, Carolus Linnaeus established the basis for the modern taxonomic naming system of kingdoms, classes, genera, and species. Taxonomy was the first informatics problem in biology. The University of Arizona's Tree of Life project and NCBI's taxonomy database are two examples of online taxonomy projects.

From Mark S. Boguski's article in *Trends Guide to Bioinformatics* Elsevier, Trends Supplement 1998, p1, '.... So bioinformatics has, in fact, been in existence for more than 30 years and is now middle-aged.'

Growth in laboratory technology facilitated collection of data at a rate that was faster than the rate of data interpretation. Biologists reached a similar information overload and started facing a lot of difficulties in the field of data analysis and interpretation. Collecting and cataloguing information about individual genes (approx. 30,000) in human DNA and determining the sequence of three billion chemical bases that made up the human DNA became the second informatics problem in biology.

In 1990, Human Genome Project was initiated as a prominent bioinformatics solution to the problem, and this labelled the 21st century as the era of genomes.

A.2 What is Bioinformatics?

Fredij Tekaia at the Institute Pasteur offers the definition of bioinformatics as 'The mathematical, statistical and computing methods that aim to solve biological problems using DNA and amino acid sequences and related information.'

From *Introduction to Bioinformatics* by T.K. Attwood and D.J. Parry-Smith (published by Prentice Hall in 1999): 'The term bioinformatics is used to encompass almost all computer applications in biological sciences…'

As per Cynthia Gibas, Bioinformatics is the intersection of information technology and biology. It was in 1730 that Carolus Linnaeus established the basis for the modern taxonomic naming system of kingdoms, classes, genera, and species. Taxonomy was the first informatics problem in biology.

Bioinformatics is a highly interdisciplinary field of biology, relying on basic principles from computer science, biology, physics, chemistry, and mathematics (statistics). The subject facilitates biological information to be handled by the use of computers. Or to be more precise, bioinformatics is the subject that involves the use of techniques from all these subjects to deal with problems of biology, understand biological processes, and find methods and solutions with information technology to solve biological problems.

Harold Morowitz believes that 'computers are to biology, what mathematics is to physics.'

The following four statements sum up the use of IT in bioinformatics:

1. Database management is used to store, retrieve, analyse, and/or predict the huge biological data.
2. Software development is used for implementing algorithms and developing applications and tools for insilico experiments.
3. CAD 4 multimedia are used for developing user interfaces, static/dynamic web pages, and graphic representations of data for prediction and visualization.
4. Next generation operating systems, networking, and software can facilitate seamless data transfer and cxccution of tools.

A.3 In This Book

As mentioned before, bioinformatics deals with databases. Among many varieties of databases, sequence databases are records of the sequence of bases found in DNA, sequence of amino acids, and protein sequences. While such databases are essential for storing the vast information collected from labs, their real use lies in the interpretation of the stored data. Searching a base sequence database for genes forms an important task in bioinformatics. A secion of DNA may be a part of a gene, or it may contain many genes. Software applications are used to determine where genes lie in the sequence databases of DNA with searching techniques. Once a gene is identified, the amino acid and protein sequence that it generates can be visualized and modelled. How proteins fold is an important, unsolved problem. In order to do all of this efficiently, new algorithms need to be developed and new software or tools written. The creation of tools is an important and emerging area in bioinformatics. While consolidating the learnings of a new subject such as bioinformatics what emerged as broad sections (called parts in this book) are databases, tools, and algorithms. Each part begins with a part-introducting chapter, followed by chapters comprising that part. The content and coverage of these parts is as follows.

Part A: Introduction This is an introductory part of the whole book and does not have its own introduction chapter or related chapters unlike other parts.

Part B: Databases This part talks about the characteristics and categories of Bioinformatics databases wherein the technical design aspect is given more importance than other database categories. This will not only help to understand any new database you come across, it will also help you to design your own mirror database in case you need to create one, with a different structure of an already existing public database. You may also create a subset of an existing database and extend that as your own private database by adding data from your own experiments. This part also explains how you can navigate in a database and retrieve information that you need. This part has three main chapters after part introduction, namely, Sequence Databases, Structure Databases, and Other Databases.

Part C: Tools The need for tools, their role in knowledge discovery, and various types of bioinformatics tools are discussed in this part. All these tools are software applications developed specifically for the databases to perform specific tasks. The chapters in this part are Data Submission Tools, Data Analysis Tools, Prediction Tools, and Modelling Tools. Each chapter discusses systematically one or two samples of each type of tools. The purpose of the tool, the set attributes, desired input, and expected output with a broad analysis is also discussed in the chapters. Customized tools can be developed to suit specific needs of research using specialized programming languages such as BioCORBA, BioJava, BioPerl, BioPython, BioXML, CellML, GEML (Gene Expression Markup Language), SBML (Systems Biology Markup Language). However it is a separate area of study altogether and is not covered in this book.

Part D: Algorithms This part discusses major algorithms that are used in bioinformatics tools for various bioinformatics tasks. These algorithms are categorized based on the methods of algorithm design as well as on their implementation. Understanding these algorithms will help one to know the method adopted to perform a task as well as help decide which tool to select to get the specific input. The chapters included in this part are Data Analysis Algorithms and Prediction Algorithms. You may learn Part D at your own pace and you do not need Internet connectivity to study and understand the content in this part.

Each chapter starts with chapter objectives and comprises general content, analogies, tasks, notes, examples, and exercises. While all of these are self-explanatory, tasks are step-by-step procedures that you can follow to explore a database or tool to get their respective outputs. This feature actually strengthens ones confidence by giving step-by-step route to explore or solve a problem without leaving any chance for getting defocused. The respective output ensures that you are going in the right direction. Thus, it is recommended that while you go through a chapter, read the concepts first and then sit at your computer (with internet connectivity) to explore the database and tools.

A.4 Bioinformatics: Applications and Research

With adequate data and right tools, it is possible to explore a number of new areas in biology. Chief among these are given below.

Biodiversity In addition to species and physical measures, bioinformatics provides a genetic measure of biodiversity. Eventually, it is the genetic diversity in a biosphere that will provide the correct measure of its extent.

Analysis of gene expression A physical or chemical change in a living system is not caused by a single gene but by the combined effect of many genes. Understanding the action of many genes on a single condition will, one day, provide a genetic basis for disease and change control.

Analysis of gene regulation Regulation is the chain of events, starting with an extracellular event (such as temperature change) and leading to a change in the activity of proteins. The analysis of what promotes and regulates the activity of genes and proteins forms a part of this study.

Comparative genomics By comparing the genes of different organisms, it is possible to trace evolutionary pathways by which one organism could have evolved into another. Such studies can not only throw new light on evolution, but also provide evidence for the migration of species, thereby bringing new evidence to historical and anthopological studies.

Molecular medicine More emphasis needs to be given on tracing the fundamental cause of diseases rather than treating symptoms. Specific diagnostic tests and faster generation of test reports can improve the situation. It is believed that gene testing and thereby through gene therapy may even make replacement of defective genes feasible.

Microbial genomics The genomes of bacteria can help throw light on energy sources, environmental monitoring to detect pollutants, find disease-producing properties of genes, and improve industrial efficiency.

Risk assessment More research on human genome can help to assess individual risk on exposure to toxic elements as resistance to external agents varies from person to person. It can also help to reduce the likelihood of heritable mutations.

Bioarchaeology, anthropology, evolution, and human migration Understanding human and other genomes will help to understand evolution, inheritance, traits, and disease carriers. The study of the genome comparison across organisms can help to understand similar genes with associated disease.

DNA forensics (identification) DNA profile of an individual, called DNA fingerprints, can help in identifying criminals, establishing family relationships, protecting rare wildlife species, and matching organ donors.

Agriculture, livestock breeding, and bioprocessing Genome research on plants can provide nutritious, disease-resistant, pesticide-free crops. Even alternate use of crops can be found, e.g., tobacco has been found to produce bacterial enzymes that break down explosives such as TNT and dinitroglycerin.

Bioinformatics research is a multidisciplinary and vast area. It ranges from drawing a mathematical or physical model of a biological system to implementing data analysis algorithms to develop databases and web tools to access them. The large number of informatics tools and data resources already available or still being developed need to be fully integrated and coordinated. Drug design is one of the major areas of research that is gaining importance in academics as well as pharmaceutical companies. In this direction, a major challenge is to integrate genome and genome-related databases to design common interfaces, implementing 'minimonolithic' databases containing subsets of relevant data extracted from a set of larger public databases. Major scope lies in the improvement of laboratory-systems integration and information-management systems to promote large-scale genomics and other biology programs in academia and industry.

To sum up, the sky is the limit, and bioinformaticians have ample scope for research in specialized fields within bioinformatics, e.g., informatics, cheminformatics, proteomics, genomics, etc. It is not only biology that stands to benefit from bioinformatics. The genetic code is a language. It is the first language of non-human origin that we have encountered; it is a truly alien system. Ironically, it has come to us not from outer space, but from within ourselves. Understanding the structure and function of this language is vital to our understanding of this language, of communication and, eventually, of sentience.

As discussed RNA protects itself by replacing the uracil base with thymine and twisting itself into the DNA molecule. It then copies itself (mRNA), reads itself (rRNA), and assembles itself (tRNA) to perpetuate a cycle of self-referential, self-organized, reproduction. The evolution of life on earth, looked at from this perspective, appears as an incredibly efficient attempt by one molecule, RNA, to survive. The key to this immense process of survival lies in the connections between information science and biology. You will learn more about the present bioinformatics scenario in India in the following section.

A.5 Present Bioinformatics Scenario in India

The following subsections discuss the present bioinformatics scenario in India as also the role of various organization in this field.

A.5.1 Government Organizations

In 1986, the Department of Biotechnology (DBT) launched the Biotechnology Information System (BTIS), during the 7th five-year plan. It is a nationwide network with ten distributed information centres (DICs) and 48 distributed information subcentres (sub-DICs). Its mission was to establish India as a leader in bioinformatics. In the 8th five-year plan, BTISnet was established—a distributed database and network infrastructure. It comprised the above as well as six national facilities for high-end interactive graphics and molecular modelling. All these centres were connected through satellites and terrestrial links under two major network service providers—NICNET and ERNET. They also use X.25 links of DoT/VSNL for international access.

These DICs and sub-DICs have been given the task of providing discipline-oriented information to all institutions interested in relevant fields. The information catered to the major areas of research. A list of various disciplines with respective DICs responsible for the same is given below:

1. Genetic engineering: Jawaharlal Nehru University, New Delhi; Indian Institute of Science, Bangalore; Madurai Kamraj University, Madurai; Bose Institute, Calcutta.
2. Protein modelling and protein engineering: Institute of Microbial Technology, Chandigarh.
3. Plants tissue culture, photosynthesis, and plant molecular biology: Indian Agricultural Research Institute, New Delhi.
4. Animal cell culture and virology: University of Pune, Pune.
5. Oncogenes, reproduction physiology, cell transformation, nucleic acid and protein sequences: Centre for Cellular and Molecular Biology, Hyderabad.
6. Immunology: National Institute of Immunology, New Delhi.
7. Neuro-informatics: National Brain Research Centre, Delhi.
8. Terra flops supercomputing facility: Supercomputing facility for Bioinformatics and Computational Biology, at IIT Delhi.
9. Biotechnology and other related fields: Department of Biotechnology, Government of India, New Delhi.

A.5.2 Private Organizations

There are a lot of international biotech companies such as Strand Genomics, Jubilant Biosys, etc., with their offices in India, with plans to harness Indian manpower. There are also Indian based companies such as Biocon India, Informatics Pvt. Ltd, Advance Biochemical Lab, working as contract research organizations for companies abroad as well as companies involved in customized software development work for India. There are pure Information Technology (IT) companies heading towards bioinformatics by collaborations with government organizations. Few such companies are TCS, Nicholas Piramal, Satyam, with collaborative ventures with the Centre for DNA Fingerprinting and Diagnostics (CDFD), Centre for Biochemical Technology (CBT), and Center for Cellular and Molecular Biology (CCMB), respectively. Spectramind eServices is one of the IT enabled service (ITES) companies where professionals are given to analyse biological literature and create databases. GE call centre also has plans on similar lines. Other companies such as Avestha Gengraine, Mahindra-BT, and DSQ software also ventured in the field with a people-intensive business model for the bioinformatics service sector.

Almost all Indian pharmaceutical companies are into bioinformatics research and Development. Some of them are Ranbaxy, Dr. Reddy's Lab, Dabur, Smith Klein Beecham, Pfizer, Cipla, Zydus Cadila, Wockhardt, Astra Zeneca, and East India Pharmaceuticals, Pharma and pharma-related companies in India have a unique and exciting opportunity of growing their annual revenue from $5.5 bn in 2000 to $25 bn in 2010 see Table 2. Of this, $1.5–2.0 bn can accrue from IT-related new horizon areas including bioinformatics, genomics/proteomics, data management for contract research, and remote sales and marketing. And $20 bn from the growth in the current core business and the rest $3 bn from emergence of a new horizon of non-IT innovation. However, success in IT-related fields (especially informatics) will be difficult and will require a focus on biological knowledge and innovation and not just investment.

Table 2 Projected IT-related revenue for year 2010 from pharma and pharma-related companies

Total IT-related revenue	$1.5–2.0 bn
Informatics research	~$0.7 bn
Development (data management)	~$0.5–0.6 bn
Sales and marketing	~$0.3–0.7 bn

Trends that can create new and exciting opportunities for India, especially through the proliferation of IT-related technologies, are the following.

Dramatic technologies Global bio-pharmacos are keen to foman alliance with Indian companies, especially for genomics, proteomics, and bioinformatics. Companies in India can provide contract research services in bioinformatics and cheminformatics.

Global harmonization There is a need to reduce costs across the business system, globally. To meet this need, outsourced IT services are likely to be better supported, provided significant quality and confidentiality can be ensured.

Cost containment Global pharmacos are finding it acceptable to source from non-domestic markets. Their growing demand for low-cost R&D, especially IT-related work can be off-shored from India.

India, thus, becomes the preferred destination for pharma and bioinformatics companies for their network with global academic and biotech community. Thus annual profit and market capital of IT-related Indian pharma companies have the opportunity to grow five times, i.e., from $1 bn to $5 bn and ten times, i.e., $15–20 bn to $150–200 bn, respectively. As pharmaceutical research is in the midst of a paradigm shift towards gene-based drug discovery where proteins are tailor made, companies are entering into alliances to access emerging research technologies.

A.5.3 Informatics Research

In research, India can focus on several options with the main focus on informatics. The strengths on which India can compete is skill, infrastructure, availability of research sample, and cost competitiveness. As said by a director of a premier research institution in India, 'Biotech is one of the areas where India can compete...To create a biotech hub in India is a dream but it is one worth pursuing'. Different research areas in the descending order of Indian strengths are informatics, synthetic and medicinal chemistry, structure-based drug design, positional cloning, and expression profiling and proteomics. Of these, informatics and structure-based drug design can give a higher profitability.

Looking from the IT perspective, research in informatics has broadly three categories.

Research enterprise application service providers (ASPs) They provide a user-friendly interface and access to bioinformatics and cheminformatics data (public and proprietary) as well as analytical solutions. Such companies require data acquisition assets along with bioinformatics capabilities. In India, bioinformatics skills are available at (1/3) rd cost. Other than US and Western Europe, Ireland is one of the potential competitors of India. India's global share in this technology is about 15%.

Software application developer and service providers They provide analytical solutions to discrete stages of the discovery process. They require extensive IT and bioinformatics capabilities. India's superiority in mathematics and statistics allows informatician to think at different levels. India's global share compared to the emerging competitors such as Hungary, other than US and Western Europe, is 15%.

Database content providers These companies provide access to proprietary and public data on genomics, proteomics, and combinatorial chemistry. They require extensive IT capabilities. The genetic diversity in India is of tremendous advantage for the generation of genomic databases. With the US and Western Europe as dominating players, India's other emerging competitors are Iceland, Ireland, and China, with only 5% global share. India can capture $400–500 million IT-related data management opportunity for clinical trials.

Table 3 shows the likely Indian market capital of informatics for year 2010.

Table 3 Global versus Indian market capital of information for year 2010

Informatics research areas	Global market (in million dollars) Year 2010	Indian market share In %	Indian market share (in million dollars)
Research enterprise application service providers	500	15%	75
Software application developer and service providers	4100	15%	615
Database content providers	900	5%	45
Total	**5500**	**13.36%**	**735**

A.5.4 Data Management

Data management is rapidly evolving into an IT-driven paperless activity. Earlier data management was done using software installed on dedicated servers, which were batch processed. This resulted in delay in query processing and report generation. Now it involves browser-based programs which are machine and operating system independent. Also data processing is in real time, enabling receiving and resolving queries in real time. The key activities are purely IT, starting from the development of the database to data capturing, analysis, and monitoring. To support all these, the key technology enablers are electronic data capture, point-of-care data collection, point-of-entry queries, data warehousing, and 'real-time' data monitoring. India can leverage its IT skills and emerging track record to participate in this opportunity. Database service is an easier and quicker revenue opportunity. It involves annotation of gene and protein sequences for large database providers and pharmacos and curation of scientific literature on an outsourced basis.

A.5.5 Academic Scenario in India

India is gearing up to improve the academic scenario in the field of bioinformatics. DBT has started one-year Advance Diploma in Bioinformatics in five Indian universities—Jawaharlal Nehru University, Calcutta University, Pune University, Madurai Kamraj University, and Pondicherry University. The admission to the course is through All India Entrance Examination, conducted by the respective centres. The Union Government is planning to set NBI under the DBT in the 10th five year plan, which will be on the lines of the National Centre for Bioinformatics under the National Institute of Health (NIH) of USA. In the 10th plan, there is a substantial increase in the allocation for bioinformatics, i.e., nearly 1.2 billion rupees as against a total of about 300 million rupees during the 9th plan. The proposed NBI (the location of the centre is yet to be decided) is planned to regulate bioinformatics research in India. It would have separate wings for various bioinformatics activities such as software and database development, human resources, genomics, proteomics, and services. Indian Institute of Technology (IIT) Delhi conducted certificate courses in supercomputing facility and plans to offer M.Tech in bioinformatics.

Other than what we have from the DBT charter, there are other universities as well with their own courses and resources development planning. Bharathiar University, Coimbatore, offers a two-year MSc course in bioinformatics. Karnataka Government in association with ICICI has set up the Institute of Bioinformatics and Applied Biotechnology (IBAB) at Bangalore offering one-year diploma course. These courses are not DBT authorized.

Bioinformatics Institute of India (BII), the only institute providing a distance learning programme (DLP) on bioinformatics in the country, plans to invest 150 million rupees to open 500 centres by the year 2003. This is to fulfil the global biotech market share that is expected to cross US$ 500 billion by the year 2010.

A.5.6 IT Contribution: Skills

Presently institutes providing bioinformatics as primary degree course or offering it as an elective subject are providing students with basic knowledge of molecular biology, protein (bio) chemistry, and evolutionary biology for computer science students. On the contrary, biology students need to cover programming, optimization, cluster analysis, C, C++, Java, Perl, Fortran, CGI scripts, Linux, RDBMS such as Oracle/ Sybase, Maths including statistical techniques and calculus. There are a host of new languages tailored for bioinformatics need, e.g., BioCORBA, BioJava, BioPerl, BioPython, BioXML, Cell Modelling (CellML), gene expression markup language (GEML), systems biology markup language (SBML), etc.

The job profiles for non-biology bioinformatics science graduates are the same as that of IT sectors. They are system analyst, system engineer, software engineer, scientific programmer, application analyst, application programmer, database administrator, database designer, database programmer, database developer, database services, network administrator, technical support, marketing, etc. Bioinformatics tools to be handled by non-IT bioinformatics science graduates are production and data submission tools, data mining tools, research tools, analysing tools, annotation tools, map integration tools, visualizing tools, etc.

Conclusion

The discussions so far highlight the tremendous scope in the field of bioinformatics. Low cost Indian IT and biology workforce are two of the key success factors to achieve the target. Consultant scientists with a network of global scientific alliances and scientific advisory board is likely required. It is widely recognized that India has expertise in software and a good pool of qualified and experienced biologists. In this context, India needs to get its act together by way of helping create a workforce that is capable of integrating the expertise of software developers and biologists to achieve the objective.

PART B
Databases

Chapter 1: Sequence Databases

Chapter 2: Structure Databases

Chapter 3: Other Databases

B

Databases

Introduction

Unlike biology, and especially after the completion of the Human Genome Project, bioinformatics has turned into a data-rich science. The goal of this data collection and analysis is to unravel the wealth of biological information hidden in the jumble of DNA, RNA sequences, structures, literature, and other biological data. It would enable mankind to enter the realm of customized medicines, prevent and cure diseases, reap environmental benefits, produce high-yield-but-low-cost crops, and just 'play God'.

The accumulation of data and its analysis accentuated the need for storing, retrieving, updating, and communicating large datasets. The goal of this section is to introduce you to the databases used for bioinformatics available today. There are many specialized databases that have mushroomed over an incredibly short period of time. The obvious examples are the nucleotide sequences, the protein sequences, and the three-dimensional (3D) structural data produced by x-ray crystallography, nuclear magnetic resonance (NMR), and those pertaining to metabolic pathways and gene expression data (micro arrays). Most of the databases in the field of bioinformatics are freely accessible from the Internet.

The chapters in this part deal with a variety of different kinds of bioinformatics databases such as taxonomy, genome, sequence, structure, disease, literature, and many other specialized databases. In this part, we will try understand the design of these databases. The database design offers an organized mechanism for storing, managing, and retrieving information. Major considerations for a database design will be the storage and availability of data at one place as well as fast multi-user data access; data is structured and standardized; data from dissimilar sources may be interconnected and used jointly; data is amenable to verification; data may be used directly by many different application programs (tools), including programs whose purpose differs from those for which the original data was compiled. But before you start to use these databases you need to

- understand the characteristics of databases,
- understand the storage system of a database,
- learn about information retrieval systems and how to use them,
- understand how to manipulate and update a database, and
- understand various categories of databases in bioinformatics.

B.1 Characteristics of Bioinformatics Databases

Biological data and databases have some common characteristics. All of them by and large have the following features.

Complex data type The data of molecular biology databases can be text-based sequences, blobs and images of cells and tissues, three-dimensional molecular structures as well as complex data structured biochemical pathways. All of these have linkages between each other.

Hierarchical data organization These databases hold data ranging from molecules, molecular pathways, cells, tissues, to organisms and populations.

Heterogeneous content Most databases in bioinformatics are heterogeneous in their content. General differences, such as the database size and location, and syntactic differences, such as the storage format or the access method, are common. Databases may even have differences that are semantic, e.g., a gene is a locus associated with a particular disease, a specific region of an organism's DNA as well as a sequence that gets expressed to a particular kind of protein.

Dynamic nature The data content of a database is constanlly. This leads to a constant change in the database schema.

Accessibity All these databases critically need the Internet access as well as the search/ browsing facility. Along with this, the flexibility to support external analysis tools and federation is a basic need. NCBI and ExPASY are federated databases with smaller databases within, having either the same or different data models, or schemas that are linked with each other. Sometimes they also share common vocabulary such as the HUGO Gene Nomenclature project and NIH's MeSH.

Quality Unfortunately, it is difficult to specify integrity constraints in databases. There is always a need to curate data. Also as constant work is going on in this field, there is a need for updating data constantly.

B.2 Categories of Bioinformatics Databases

Analogy

In daily life, we all maintain databases of various kinds. The most common one is databases of contacts—official or personal. Almost all of us record contact details of each person we know either in our diary or in an address book. Few of us (those organized lots) also maintain databases of books and CDs. Many housewives maintain a database of materials required to help them update the stock to cook in kitchen.

Bioinformatics databases available today can be categorized on the basis of

- data type,
- maintainer status,
- technical design,
- data source,
- data access, and/or
- any other parameter.

B.2.1 Categories of Databases: Type of Data

Database categories on the basis of the data content are the following:

- Taxonomy databases
- Genome databases (organism specific)
  - Genomics databases (non-vertebrate)
  - Human and other vertebrate genomes
- Sequence database
  - Nucleotide sequence databases
  - Protein sequence databases
  - RNA sequence databases
- Structure databases
- Proteomic databases
- Micro-array databases
- Chemical databases
- Expression databases
- Enzyme databases
- Pathway databases (metabolic and signalling pathways)
- Disease databases (human genes and diseases)
- Literature databases
- Other molecular biology databases

B.2.2 Categories of Databases: Maintainer Status

The proprietors and maintainers of databases are predominantly agencies funded by the government which enable free data access to support research and work in the subject area. These may be federal bodies, such as NCBI, EBI, SIB maintaining a pool of databases hosted independently on their own servers. Also, some of them might join hands so to enable autoupdation of each other's databases when any of their databases get populated/ updated. Besides, the 'linked status' affords easy inter-operability. This provides the users with more information and ease of query. Of late, there has been an increase in commercialization of databases. Examples of the companies with proprietary rights of core data are Incyte and Celera. Some representative examples are discussed below.

(a) NCBI (http://www.ncbi.nih.gov/Entrez/)

The National Center for Biotechnology Information (NCBI) is a division of the National Library of Medicine (NLM) at the National Institutes of Health (NIH), a federal agency of the US government. The Entrez site for PubMed, nucleotide, genome, protein and structure databases is being maintained in the USA. The mission of the NCBI is to ensure that the growing body of information from molecular biology and genome research is placed in the public domain and is accessible freely to all facets of the scientific community in ways that promote scientific progress.

(b) EBI (http://www.ebi.ac.uk/)

The European Bioinformatics Institute (EBI) is a non-profit academic organization in Hinxton, Cambridge, UK. It is a site for nucleotide and protein sequences, protein structure and signature, seqs (SRS), and literature database. It is a centre for research and services in bioinformatics. The institute manages databases of biological data including nucleic acids, protein sequences, and macromolecular structures. It maintains the European Molecular Biology Laboratory (EMBL) database which represents over 185,000 organisms. It also created more than 200 computational biology tools. Currently the EBI's services receive around one million requests per day. EBI is a large public institutes funded by the UK government.

(c) SIB (http://www.isb-sib.ch/)

The Switzerland Institute of Bioinformatics (SIB) is an quasi-academic, non-profit foundation (established on March 30, 1998) whose mission is to promote research, development of databanks and computer technologies, teaching, and service activities in the field of bioinformatics, both individually and with international collaborators. Mainly, protein databases are maintained at this site. The ExPASy (expert protein analysis system) proteomics server of the SIB is dedicated to the analysis of protein sequences and structures as well as 2-D PAGE.

B.2.3 Categories of Databases: Technical Design

To a computer scientist, developing a biological database might prove to be a daunting task. Most fields are hard to define, and there always is a need to create new ones. Assigning an object to a particular field is almost never final, and there are numerous exceptions to almost every rule. There is a lot of connectivity between different objects, and this, too, is subject to change. Although all the data submitted is checked for integrity and obvious errors omitted by the data library staff, the quality of the data is the responsibility of the submitter. As a consequence, there are many errors in the database; many sequence entries are either mislabelled, contaminated, incompletely or erroneously annotated, or contain sequencing errors. In addition, the database is very redundant, in the sense that the same sequence from the same organism may be included many times, simply reflecting the redundancy of the original scientific reports.

As you all know, bioinformatics databases are publicly available and are designed, developed, and maintained by different organizations located across countries in the world. So it is obvious that these databases must have been developed using different database management systems, e.g., DB2, Informix, MySQL, Sybase, Oracle, etc., or stored as spreadsheets, flat files, and simple text files on different hardware platforms. If you look at the data organization of each of them, you can group them into four basic types of data models as follows.

(a) Flat file

A flat-file database is the simplest database model. In this model, all the information is stored in text files, containing records in the one-record-per-line format. Each record represents one entry having fields of a fixed width or fields separated by delimiters such as white space, tabs, commas, or other characters. Extra formatting may be needed to distinguish data from a delimiter where the character used as delimiters may be a part of the data. Thus a flat file is a database that exists in a single file consisting of data and delimiters in the form of rows and columns, with no relationships between records and fields except sharing the table structure. This type of data is said to be 'flat' in contrast to models such as a relational database. A common example of a flat-file database is a list of contacts comprising names, addresses, and phone numbers. This model supports simple text files that can be accessed easily by simple programs to handle a query. Flat files are indexed for easier search. Let us find a gene with ID G13163, i.e., search only 1 record. The search will be slow if you follow the simple strategy of reading each block from a gene table of the query database. It takes 10–100 sec to read a table size of 40 MB. The search will be fast if you build an index that ideally occupies 10% of the size of the table. So it will take 0.003 to 0.030 sec to read a table size of 40 MB (see Fig. B.1).

The FASTA format (free text) and EMBL format (tag, value pairs) are two common file formats supported by flat files and indexed flat files, respectively.

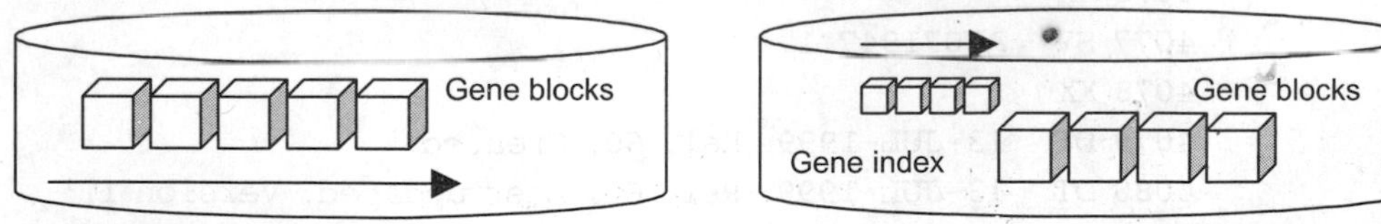

Fig. B.1 Search in non-index and index file

Example B.1: FASTA (free text: flat file) file format

```
>gi|1040960|gb|U35641.1|MMU35641 Mus musculus Brca1 mRNA, complete
cds
GGCACGAGGATCCAGCACCTCTCTTGGGGCTTCTCCGTCCTCGGCGCTTGGAA
GTACGGATCTTTTTTCTCGGAGAAAAGTTCACTGGAACTGGAAGAAATGGATTTATC
```

```
TGCCGTCCAAATTCAAGAAGTACAAAATGTCCTTCATGCTATGCAGAAAATCTTAG
AGTGTCCGATCTGTTTGGAACTGATCAAAGAACCTGTTTCCACAAAGTGTGACCAC
ATATTTTGCAAATTTTGTATGCTGAAACTTCTTAACCAGAAGAAAGGGCCTTCACAAT
GTCCTTTGTGTAAGAATGAGATAACCAAAAGGAGCCTACAGGGAAGCACAAGG
TTTAGTCAGCTTGCTGAAGAGCTGCTGAGAATAATGGCTGCTTTTGAGCTTGACAC
GGGAATGCAGCTTACAAATGGTTTAGTTTTTCAAAAAAGAGAAATAATTCTTGTGAG
CGTTTGAATGAGGAGGCGTCGATCATCCAGAGCGTGGGCTACCGGAACCGTGT
CAGAAGGCTTCCCCAGGTCGAACCTGGAAATGCCACCTTGAAGGACAGCCTA
GGTGTCCAGCTGTCTAACCTTGGAATCGTGAGATCAGTGAAGAAAAACAGGC
```

Example B.2: EMBL (tag, value pairs: indexed flat file) file format

```
1 ID   HSECTXT01 standard; DNA; HUM; 5579 BP.
2 XX
3 AC   U34367;
4 XX
5 SV   U34367.1
6 XX
7 DT   24-JAN-1996 (Rel. 46, Created)
8 DT   02-JUL-1999 (Rel. 60, Last updated, Version 7)
9 XX
10 DE  Human protein tyrosine kinase TEC (tec) gene, partial cds, and
11 DE  tyrosine kinase TXK (txk) gene, exon 1.
12 XX
13 KW  .
14 XX
    .
    .
4073 ID   AF071947  standard; DNA; ROD; 1381 BP.
4074 XX
4075 AC   AF071947;
4076 XX
4077 SV   AF071947.1
4078 XX
4079 DT   13-JUL-1999 (Rel. 60, Created)
4080 DT   13-JUL-1999 (Rel. 60, Last updated, Version 1)
4081 XX
4082 DE   Mus musculus protein tyrosine kinase Txk (Txk) gene, exon 1 and
4083 DE   partial cds.
.
.
N lines
........
```

Here *N* represents the number of fields for each entry.

Example B.3: Corresponding index file of Example B.2

```
U34367.1     1
AF071947.1  4073
.
.
.
M entries
```

Here *M* represents the number of fields for each entry.

(b) XML

This is a hierarchical and semi-structured model that has text-based files in the form of extensible markup language (XML). It supports XQuery as the query language. The nested data structures in XML files can be exchanged across the Internet. It has implicit, flexible schema. The schema is usually defined in the document type definition (DTD) that defines valid tags, valid value types used for format, and data validation. XSLT defines how to translate documents into other formats.

Example B.4: XML file format

```
<?xml version="1.0"?>
<!DOCTYPE GBSeq PUBLIC "-//NCBI//NCBI GBSeq/EN"
"http://www.ncbi.nlm.nih.gov/dtd/NCBI_GBSeq.dtd">
<GBSet>
<GBSeq>
<GBSeq_locus>MMU35641</GBSeq_locus>
<GBSeq_length>5538</GBSeq_length>
<GBSeq_strandedness value="not-set">0</GBSeq_strandedness>
<GBSeq_moltype value="mrna">5</GBSeq_moltype>
<GBSeq_topology value="linear">1</GBSeq_topology>
<GBSeq_division>ROD</GBSeq_division>
<GBSeq_update-date>18-OCT-1996</GBSeq_update-date>
<GBSeq_create-date>25-OCT-1995</GBSeq_create-date>
<GBSeq_definition>Mus musculus Brca1 mRNA, complete
cds</GBSeq_definition>
<GBSeq_primary-accession>U35641</GBSeq_primary-accession>
<GBSeq_accession-version>U35641.1</GBSeq_accession-version>
```

(c) Relational model

It is a highly structured model with a predefined schema. It has tables with rows and columns. The DBMS that supports this model is known as the relational database management system (RDBMS). SQL is one of the popular, freely accessible query languages with basic RDBMS features. Other RDBMS are Oracle, DB2, MySQL, PostgreSQL, and Sybase. RDBMSs have a relational data model where data is stored in a table, called a relation. Each table is named and contains rows, called *tuples* or *records*, and columns, called *attributes* or *fields*. Columns are also named. A *key* is a combination of attributes that identify each record. RDBMSs have a more structured form of storing data.

Example B.5: A table of a relational database file

Table: Gene

ID	Name	Alias	Organism	Sequence	...
G13163	Trypsinogen B	TRY5	Homo sapiens	atccgggatat	...
G47113	Trypsin 1	TRY1	Homo sapiens	gggatatgtga	...
...	...	...	...	...	

In this table, the primary key is ID and ID data G13163 and G47113 are unique.

A database can have many tables and each table can be linked with another table through a common field, called a *foreign key*. It is one of the attributes of the table, but a key of another table that acts like a logical pointer or link between two tables. The column name need not be the same when used in different tables, but the data under that field should be common. Foreign keys are automatically maintained during an update process.

Example B.6: The link between two tables of a relational database file

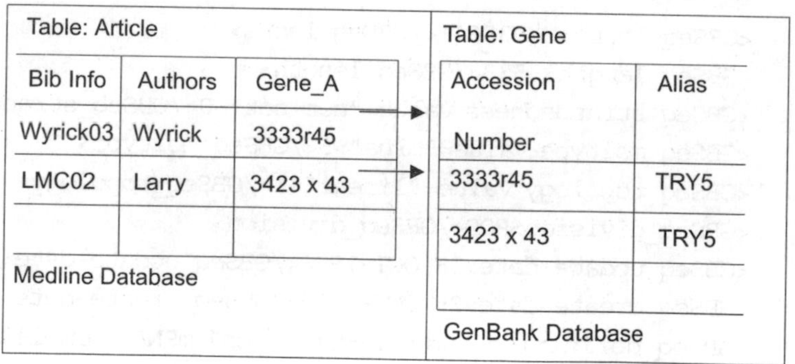

In this table, the foreign key of the Medline database is Gene_A, which is mapped to Accession Number of the GenBank database.

Various RDBMSs available in the market are the following: BerkeleyDB used for public domain databases. It supports only storage but provides no query language; MySQL is a free database also used for public domain with limited SQL implementation and functionality; Interbase, Postgres, Oracle, and Sybase are commercial RDBMSs with full functionality and applications supporting query for data management.

(d) Object-oriented/Object Relational Model

A database with an object-oriented model is known as an object-oriented database management system (OODBMS) and that with object relational model is known as an object relational database management system (ORDBMS). Such databases have objects and classes with structured as well as abstract data types. Classes correspond to data types and objects correspond to instances of classes. Classes may or may not inherit the attributes of super class and can have some specific attributes of their own. These databases are quite popular in molecular biology as the attributes of the data model have much similarity with the organism model such as classes, objects, inheritance, etc.

Example B.6: Using RDBMS as well as OODBMS

Using RDBMS:

```
CREATE TABLE eukaryote (
id text,
desc text,
intronsno int,
geneno int,
);
```

```
CREATE TABLE prokaryote (
id text,
desc text,
geneno int,
);
```

Using OODBMS:

```
CREATE TABLE organism (
id text,
desc text,
geneno int,
);
```

```
CREATE TABLE eukaryote (
intronsno int)
INHERITS (organism);
```

In early days these databases had small datasets, so standard database features were not needed. Also, the involvement information technology was less. But there was a definite need for data exchange and query-based applications to be executed upon these datasets, e.g., a query generated by a user from India to find information about TRY5 from data located in a server in Japan. The results of the query were required to be converted into various standard text file formats such as ASN.1, FASTA, HTML, XML. Also, tools may be file-type specific, e.g., BLAST works easily on the text files. Further, it is easier to develop applications to read data from a text file. Thus, owing the ease of use, molecular biologist used text files. Bioinformatics databases, whether DBMSs or text-based datasets, need to be accessed through various information retrieval systems. Most of these databases are linked to each other. So there was a need for an international standard.

(e) ASN.1

In early 1980s, ASN.1 (or abstract syntax notation one) was developed and adopted by the ISO (International Organization for Standardization). It is a notation used for representing data to communicate information across various platforms. One can also understand it as a language that is used to define information (i.e., database structure and data) in such a way that it can be ported over the internet. Thus it is an application protocol (communication protocol used in the application layer of the network architecture based on the OSI model) to represent data format. It has two basic kinds of syntax; *abstract syntax* for defining information and *transfer syntax* for communicating information. In NCBI it is used to store, send, and receive data. It has a structured flat-file format. It also supports multiple data types. The complex data types of the molecular biology databases (sequences, structures, genetic maps, etc.) are built using small, simple data in the nested form in a hierarchical manner. It is reasonably stable and widely used. The format of ASN.1 being simple, data submission is also easy. The flat-file formats of GenBank and EMBL are not the true format of the database. These files are basically reports generated by the computer converting the ASN.1 file into a format that can be easily read.

The text-based files, however, have some definite disadvantages. Only one person can add or modify data at one point of time. There is neither any provision for data validity check nor any scope for redundancy. For multiple accesses, either the data file needs to be split or a special application needs to be developed. There is no facility to recover data if computer crashes. Such files cannot support queries without additional data structures and/or search tools, e.g., BLAST. Many systems also have constraints for low file-size limits.

Example B.7: ASN.1 file

```
Seq-entry ::= set {
  level 1 ,
  class nuc-prot ,
  descr {
    title "Mus musculus Brca1 mRNA, and translated products" ,
    source {
      org {
        taxname "Mus musculus" ,
         db {
            {
             db "taxon" ,
              tag
                id 10090 } } ,
        orgname {
          name
            binomial {
              genus "Mus" ,
             species "musculus" } , ...
```

With increase in sophistication of database models, there has been an increase in the sophistication of physical storage mechanisms. The best part is that, based on the need, flat files can be stored in a relational database and a highly structured dataset can be stored in a flat file. Here we studied five different kinds of database organization, but broadly they are of three kinds, flat files, hierarchical files, and relational files. Let us look at the advantages and disadvantages of these databases.

Structure	**Advantages**	**Disadvantages**
Flat files	• Fast data retrieval • Simple structure and easy to program	• Difficult to process multiple values of a data item • Adding new data categories requires reprogramming • Slow data retrieval without the key
Hierarchical files	• Addition and deletion of records is easy • Fast data retrieval through higher level records • Multiple associations with like records in different files	• Pointer path restricts access • Each association requires repetitive data in other records • Pointers require large amounts of computer storage
Relational files	• Easy access and minimal technical training for users • Flexibility for unforeseen inquiries • Easy modification and addition of new relationships, data, and records • Physical storage of data can be changed without affecting relationships between records	• New relations can require considerable processing • Sequential access is slow • Method of storage impacts processing time • Prone to logical mistakes due to flexibility of relationships between records

Let us now have a quick glance at the various databases, their description, format, and data type.

Database name	**Description of data**	**Data storage format**	**Data type**
GenBank	DNA/RNA sequence	Text file/ASN.1	Text, numeric
OMIM	Disease, phenotypes, and genotypes	Text file/ASN.1	Text file
GDB	Genetic map	Relational/MySQL	Text, numeric
AceDB	Sequence and variants	Object-oriented	Text, numeric, complex data

(contd)

(Table contd)

Database name	Description of data	Data storage format	Data type
Medline	Literature	ASN.1	Text
NCBI databases	Sequence, structure, literature, etc.	ASN.1	Text, numeric
PDB	Structure	Oracle	3D images
BLAST	Sequence, analysis	FASTA	Text, numeric
ClustalW	Sequence, analysis	FASTA	Text, numeric
KEGG	Metabolic Pathways	HTML text, binary images	Text, images
Microarray	Microarray Data	RDBMS, Excel	Text, images

(f) Data format interchangeability

To use various application programs available, we need to cater to different formats of data. Not only this, each format should be flexible enough to be changed to another format so that those data can be used by relevant applications. The challenge is to represent complex nested data models to abstract hierarchical structures of biological systems and processes.

Example B.8: Data model represented in two different formats

Data model				
GBSeq				
locus	length	strandedness	moltype	topology
MMU35641	5538	0	5	1

RDBMS representation

```
CREATE TABLE GBSeq
(
 locus varchar(20) NOT NULL,
  length int(10) NOT NULL,
  strandness int(10) CONSTRAINT cstrand CHECK (strandness IN ('not-set',
'single-
stranded', 'double-stranded', 'mixed-stranded')),
  moltype int(10) NOT NULL, CONSTRAINT cmol CHECK(moltype IN ('nucleic-
acid',
'dna', 'rna', .......)),
  ...............
  PRIMARY KEY(locus),
);
INSERT TUPLE INTO GBSeq
('MMU35641', 5538, NULL, 5, 1);
..........
```

XML representation

```
<!ELEMENT GBSeq (
          GBSeq_locus ,
          GBSeq_length ,
             GBSeq_strandedness? ,
```

```
      GBSeq_moltype?
      .........................)>

<!ELEMENT GBSeq_locus (#PCDATA)>
<!ELEMENT GBSeq_length ( %INTEGER; )>
<!ELEMENT GBSeq_strandedness ( %INTEGER; )>
<!ATTLIST GBSeq_strandedness value (
            not-set |
           single-stranded |
           double-stranded |
           mixed-stranded )  #IMPLIED >
<!ELEMENT GBSeq_moltype ( %INTEGER; )>
<!ATTLIST GBSeq_moltype value (
           nucleic-acid |
            dna |
            rna |
            trna |
            rrna |
            mrna |
            urna |
            snrna |
            snorna |
           peptide )  #IMPLIED >
............
<GBSeq>
  <GBSeq_locus>MMU35641</GBSeq_locus>
  <GBSeq_length>5538</GBSeq_length>
  <GBSeq_strandedness value="not-set">0</GBSeq_strandedness>
  <GBSeq_moltype value="mrna">5</GBSeq_moltype>
............
```

(g) Data integration

The ability to access any data across multiple disparate databases and apply computational tools for data analysis is the need of the hour. Whether the data is developed in any DBMS or stored in any format, the federated database approach implements data integration where the user queries and other bioinformatics applications access the raw data through the following layers (see Fig. B.2):

- The *first layer* is either SQL or JDBC/ODBC application protocol interfaces (API). These are software interfaces for queries and applications that can interact with the *kernel*.

- The middle layer, or the kernel, is a high-level query language to transform, manipulate, and integrate data that facilitates data exchange and data integration. It interacts with interfaces of both the applications and the core data.
- The third layer is called *wrappers*. These are software interfaces of the data sources, e.g., flat files, XML files, RDBMS, or OODBMS/ORDBMS files implemented as wrapper class files.

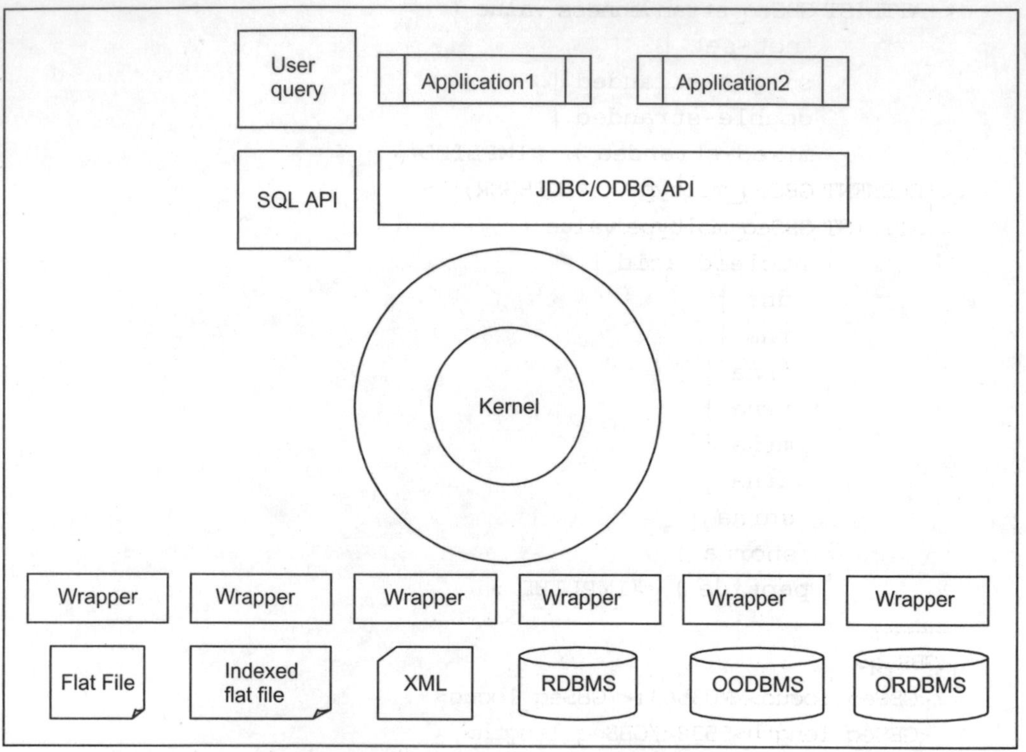

Fig. B.2 Data Integration of various databases

B.2.4 Categories of Databases: Data Source

Data source databases are mostly populated from two basic sources, either from experimental, publication, and patent data forming the primary databases or from already existing primary databases forming secondary databases. There are portals with aggregates of databases. Following are the examples of databases from the perspective of data source.

Primary databases

Nucleic acid sequences—Genbank, EMBL, and DDBJ
Protein sequences—Swiss-Prot (curated), TrEMBL (predicted)
Protein structures—PDB (NMR and x-ray crystallography coordinates)

Secondary databases
Data derived in the analysis of primary databases
Genomic—TiGR human gene index
Proteomic—PROSITE organizes proteins into families and domains
Portals
Entry to a federation of databases—NCBI, Expasy, etc.
Literature—PubMed

B.2.5 Categories of Databases: Data Access

Databases can have various kinds of access status, as mentioned below:

- Publicly available with no restrictions, such as databases of NCBI and EBI, Swiss-Prot
- Available with copyright
- Browsing only, accessible but not downloadable
- Academic, but not freely available
- Proprietary, commercial with access on payment, e.g., databases of Celera
- Ad-hoc query facility, e.g., Boolean queries
- Unrestricted SQL queries against underlying DBMS

B.2.6 Categories of Databases: Others

There are various other parameters to classify databases. They are as follows.

1. *Completeness*: Entries of the databases can be complete or incomplete.
2. *Annotation*: Databases either can hold raw data (i.e., not annotated), or can have the analysis of the data (i.e., as annotated data).
3. *Curation*: When annotation is established, the databases are known as curated, e.g., RefSeq, Swiss-Prot. There are annotated databases that are not curated.
4. *Data submission source*: The source of database population can be from experiments, patents, publication, or from primary databases.
5. *Data submission method*: Data can be submitted manually or data submission can be automated.
6. *Exchange/publication technologies*: Technologies used for data exchange are also many. They can be through FTP, HTML, CORBA, or XML.

B.3 Navigating Databases

Before you learn to navigate through a database, you must have some preliminary information as to how these databases are organized. All databases, whether in bioinformatics or otherwise, use a system where an entry can be identified and traced. The entry can be called by these names:

- Identifier
- GI number
- Version number

- Accession code (or number)
- Nucleic acid identifier

Identifier It is a unique integer which identifies a particular sequence. You will notice that this number will change every time the sequence changes. Conventionally, the identifier is a string of letters and digits which generally is interpretable in some meaningful way by a scientist and is usually an abbreviation of the full protein or gene name (e.g., KRAF_HUMAN is the entry name for the Raf-1 oncogene from *Homo sapiens*). A qualified identifier is one that conforms to the NCBI standard FASTA identifier (NSID) syntax. An unqualified identifier is just a bare word, lacking any indication of its database domain or name space in which it was assigned. For example, gb|AF218085.2|, sp|P18646|, and pir|A00008| are GenBank ACCESSION.VERSION, Swiss-Prot ACCESSION and PIR ACCESSION numbers, respectively.

The identifier serves three main purposes.

1. An identifier is assigned to all sequences processed, including International Nucleotide Sequence Database Collaboration nucleotide sequences of DDBJ/EMBL/GenBank and protein sequences from Swiss-Prot, PIR, and many others. This number provides a unique sequence identifier, which is independent of the database source.
2. When a sequence is modified (even by a single base pair), a new identifier is assigned to it. However, its accession number remains unchanged. Thus the preferred method of citing sequences is with the accession version number (e.g., U49845.1). The accession identifies a particular sequence record and the version number tracks changes to the sequence itself.
3. The identifier is stable and retrievable. NCBI keeps the last version of every GI number and includes this history in the record.

There are four types of sequence identification numbers. If a sequence changes in any way, it receives a new GI number, and the version number is incremented by 1. The two types of sequence identification numbers, GI and VERSION, have different formats and were implemented at different points in time. They run in parallel to each other.

GI numbers It stands for GenInfo identifier. It is a series of digits that are assigned consecutively by NCBI to each sequence it processes. It bears no resemblance to the accession number of the sequence record. It is not consistent across different databases.

- The nucleotide sequence GI number is shown in the VERSION field of the database record.
- The protein sequence GI number is shown in the CDS/db_xref field of a nucleotide database record and the VERSION field of a protein database record.

Version numbers It consists of the accession number followed by a dot and a version number. It is consistent across databases.

- The nucleotide sequence version contains two letters followed by six digits, a dot, and a version number (or for older nucleotide sequence records, the format is one letter followed by five digits, a dot, and a version number).

- The protein sequence version contains three letters followed by five digits, a dot, and a version number.

For example, the RefSeq record for the *Homo sapiens* cystic fibrosis transmembrane conductance regulator (cftr) has the following numbers:

accession number NM_000492. The record contains

- one nucleotide sequence with sequence identifiers GI: 6995995, VERSION: NM_000492.2
- one amino acid translation with sequence identifers GI: 6995996, VERSION: NP_000483.2

Accession number It is the unique identifier assigned to the entire sequence record when the record is submitted to GenBank. The GenBank accession number is a combination of letters and numbers which are usually in the format of one letter followed by five digits (e.g., M12345) or two letters followed by six digits (e.g., AC123456). The accession number for a particular record does not change even if the author submits a request to change some of the information in the record. It is often called the primary key for the entry. The accession code, once issued, must always be possible to find again, even after many/major changes have been made to the entry. Sometimes there is a need to merge two entries into one, and then the new entry will have both accession codes, where one will be the primary and the other the secondary accession code. When an entry is split into two, both the new entries will get new accession codes, but will also have the old accession code as secondary codes.

Nucleic acid identifier It is a number assigned to each version of an entry. While the identifier and the accession number never change, a new NI number is allocated each time the sequence is modified, however minor the change might be.

B.4 Information Retrieval Systems

Analogy

From a general store counter you can buy products of various kinds. All that you need to do is to mention the product code, the required quantity you want to purchase, and pay the amount. You get the product as an output.

At a railway portal you can look for the train schedule, check for ticket availability, buy train tickets, or check PNR status. To do any of these, you need to click the right option and give the desired input.

Accesses to databases is made possible through various retrieval systems. These systems are nothing but portals developed by various organizations independently or jointly to provide an interface for the users to retrieve information from their databases. These portals not only have archives of various kinds of databases, but also provide links to databases of other portals. Hence these systems are also called *gateways*. The interfaces of the portals are generally user friendly and not only give access to many databases but also provide tools for various operations. We will learn about online tools in Part C. Here we will introduce you to some of such prominent and commonly used systems.

- ENTREZ, NCBI with nucleotide, protein, Genbank, EMBL, DDBJ, and others databases
- The Sequence Retrieval System (SRS), EBI with nucleotide, protein, other databases
- Sequence search and retrieval tool, The Protein Information Resource (PIR), with protein sequence databases
- Expert Protein Analysis System (ExPASy), SIB with Swiss-Prot and Others
- Ensemble, Sanger Institute with organisms databases

These systems have links to all kinds of databases, however you will find that each one is commonly used for one kind of database. For example, Ensemble is used to access genome databases, ENTREZ for nucleotide sequence databases (GenBank), disease databases (OMIM), and ExPASy for protein databases. We will explore some of these systems while learning different databases in the chapters in this Part. For a query submitted to the retrieval system, there are various display formats for the results corresponding to the query. These results can be from single or multiple databases. Some common databases are mentioned below:

Display format	**Description**
Summary	It is the default display. Accession number is linked with a brief description.
Brief	Accession number is linked with abbreviated description.
ASN.1	It displays the data in Abstract Syntax Notation 1 format.
FASTA	It displays the data in FASTA format with the definition line and then the sequence.

When an entry from the result displayed is selected, the information retrieval system also allows various views, depending on the option selected by the user, of the data. These options are Brief, Summary, ASN.1, GenBank/GenPept, FASTA, XML, INSDSeq, Graph, links, etc.

B.4.1 Storage System of a Database

Analogy

Consider the database of your contacts. It will have entries or records of all your contacts. Each record will typically have information such as name, date of birth, mobile number, office and/or residence phone number, address. Each piece of such information from a record is called an attribute or a field. You cannot store your friend's contact details with the details of the books from your personal library. But yes, you can store contact details of all people you know in one database by recording additional information about the contact. You can add an attribute called contact type and mention friends, relatives, or co-workers for each entry.

A database is meant to store related information. You cannot store a protein sequence and a protein structure in one database. Yes, definitely you can link them. For a particular type of database, if you want to look up all related information stored for each entry, you need to look at one entry in detail. Unfortunately, the bioinformatics DBMS architecture follows no standard. Till date there is no defined design methodology that can cater to all kinds of data of the molecular biology database. Apart from the federated databases, there is no common nomenclature used. We will explore some of the databases and their entries to know relevant information stored as attributes in the chapters in this part. Databases typically support the following operations:

- Retrieval
- Insertion
- Updating
- Deletion

B.4.2 Retrieving Information from a Database

Analogy

Consider the same contact database as mentioned earlier. Each record will have information such as name, date of birth, mobile number, office and residence phone numbers, address. From this database if you wish to you can retrieve information about only those contacts with birthdays falling in February. You can also retrieve contact details of your friends from a specific location as well as birthdays falling in February. But you cannot view contacts with anniversaries in February. This is simple because the database does not have related information stored in a different attribute that can support such a query.

In order to search and select data from databascs, you can cither download data from web pages or use scripts to download data using Perl/BioPerl. An Internet protocol called *file transfer protocol* (FTP) can be used to download entire databases, e.g., ftp.ncbi.nih.gov

To view a record or a set of records from a database that matches with the query or holds true for a given condition, you need to mention a keyword or search word specific to an attribute of that database. If the keyword matches with the data for the mentioned attribute, it will give an output or else will report a message like 'Record Not Found'. Usually to find out any information it takes a lot of clicking.

For a search based on some condition or for multiple searches, the result of a query operates based on the functioning of the *set theory* represented as a Venn diagram mentioned below. When you submit a search for all proteins that are common to human and mouse, the search narrows down to three steps: search all records for 'Human Beings', say set A; search all records for 'Mouse', say set B; search all 'Protein' records, say set C. Final search is $A \cap B \cap C$. Similarly when you submit a search for all proteins from both human and mouse, search result will be $A \cup B \cup C$.

Wild card characters such as ? and * can be used while writing a search statement. For example, in the search box of genome database you can type bacteria * and click **Go**.

(a) Search rules and syntax/Use of Boolean operators

- **AND:** To AND two search terms from either the same field or different fields gives all entries that contain BOTH terms.
- **OR:** To OR two search terms from either the same field or different fields gives all entries that contain EITHER term.
- **NOT:** To NOT two search terms from either the same field or different fields gives all entries that contain search term 1 but not search term 2.
- Search terms used with Boolean operators should be in lowercase.
- Boolean operators AND, OR, and NOT must be entered in uppercase.
- The sequence of Boolean operations is from left to right.
- The order of operation can be changed by using parentheses. The terms inside the parentheses are processed first as a unit and then the result is processed with the rest of the terms.

(b) Query language

SQL or structured (english) query language is used in almost 99% of relational databases. It has three major versions, which were made available in 1980, 1992, and 2003.

(c) SQL used for data query language

```
SYNTAX: SELECT <some columns> FROM <some tables> WHERE <condition is true>
```

Simple condition:

```
SELECT ID FROM Gene WHERE Organism = 'Homo sapiens';
```

More complicated conditions:

```
SELECT * FROM Gene WHERE Organism = `Homo sapiens' AND ID = 'G13163';
SELECT gi, description FROM Sequences WHERE locus = 'MMTECH13';
```

The 'WHERE' clause is optional:

```
SELECT gi, description FROM Sequences;
SELECT * FROM Sequences;
SELECT * FROM Sequences WHERE locus IS NULL;
```

SQL-Joins

- Tables are typically joined on foreign keys
- Which authors have written articles related to TRY5?

```
SELECT Author FROM Gene, Articles WHERE ID = Gene_ID AND Alias = 'G13163';
```

Table: Article

Bib Info	Authors	Gene_ID
Wyrick03	Wyrick	G13163
LMC02	Larry	G13163
....	....	....

Medline Database

Table: Gene

ID	Name	Alias
G13163	Trypsinogen B	TRY5
G47113	Trypsin 1	TRY1
....	....	....

GenBank Database

A view is a table constructed by a query that is used to implement access control. It may or may not be 'materialized' and updateable.

```
CREATE VIEW HumanGenes
AS SELECT *
   FROM Genes
  WHERE g.species = 'human';
GRANT SELECT ON HumanGenes TO bob;
```

Creation and population of database Database creation, i.e., the provision of database design and creation of database structure is the sole discretion of the owner. Data population is done either manually or automatically. Scientists gather data experimentally and populate the database. Publishing mechanisms are limited. Data can be also added from published papers and patents. Data is append-only. Usually when data is submitted, it is imported to many sites or rather it gets updated to the related databases either immediately or at a periodic interval.

SQL can create tables or drop tables

```
CREATE TABLE Sequences
(
 gi int(10) NOT NULL,
 accession     varchar(15) NOT NULL,
 locusvarchar(20),
 description   varchar(200),

 PRIMARY KEY(gi),
 UNIQUE(accession),
 INDEX(accession)
);

CREATE TABLE Sequences
(
 gi   int(10),
 accession     varchar(15),
 locusvarchar(20),
 description   varchar(200)
);

GRANT SELECT ON Sequences TO fred;

GRANT SELECT ON Sequences TO fred
WITH GRANT OPTION;
```

```
DROP TABLE Sequences;
CREATE TABLE Gene (
ID numeric(7) primary key,
Name varchar(50),
Alias varchar(20),
Organism varchar(30),
Sequence BLOB);
```

SQL can insert data

```
INSERT TUPLE INTO Gene
('G13163', 'trypsinogen B', 'TRY5', 'Homo sapiens', 'tcctcacagg...');
INSERT INTO Sequences VALUES
(14728383, 'XM_003323.4', NULL, 'Homo sapiens', mRNA');

INSERT INTO Sequences VALUES
(5453027, 'AF071947.1', 'MMTECH13',
'Mus musculus protein tyrosine kinase Txk (Txk) gene, exon 1 and partial
cds');
```

Manipulation and updating of databases Datasets of public databases cannot be manipulated, i.e., updated or deleted, unless you have write permission.

SQL can update data

```
UPDATE Sequences SET locus = 'CLOCK' WHERE gi = 14728383;
SQL can delete data
DELETE FROM Sequences WHERE gi = 14728383;
```

Conclusion

Research efforts in this field include sequence alignment, gene finding, genome assembly, protein structure alignment, protein structure prediction, prediction of gene expression, and protein–protein interactions. Along with the databases of various kinds, the modelling of evolution has led to the creation of tools for various activities. You will get to learn more about that in Part C.

EXERCISES

Exercise B.1 Find all nucleotide sequences of mouse with D-loop annotations. In the nucleotide database, use the following expression:

```
D-loop[FKEY] AND "Mus musculus" [ORGN]
```

Exercise B.2 Find drosophila population studies published in the journal *Nature* written by Wang W. In the PopSet database, use the following expression:

```
"Drosophila" [Organism] AND "Science" [Journal] AND "Wang W" [Author]
```

Exercise B.3 Find all drosophila protein sequences with lengths between 50 and 60 amino acids that were entered into the database since 2000 till date. Exclude data submitted by third party. In the protein database, use the following expression:

```
"Drosophila"[Organism] AND 000050[SLEN] : 000060[SLEN] AND 2000[MDAT]
AND (1900[MDAT] : 3000[MDAT] NOT srcdb_tpa_ddbj/embl/genbank[PROP]) AND
gene_in_genomic[PROP] AND ("2000/01/01"[PDAT] : "2005/02/28"[PDAT])
```

Exercise B.4 Find the DNA sequence of mouse expressing myoglobin protein: In the nucleotide database use the following expression:

```
mouse[Organism] AND myoglobin[Protein Name]
```

1 Sequence Databases

Learning Objectives

- To describe the various kinds of sequence databases
- To learn about the organization of databases
- To list the attributes of each database
- To browse databases using various information retrieval systems
- To access data using Boolean algebra
- To analyse sequence data

Introduction

'I have an unknown DNA sequence and I would like to analyse it for the existence of known consensus sequences.' This has been the most intriguing query that continues to arise in the minds of hundreds of thousands of molecular biologists and bioinformaticians. Current genome projects continue to generate billions of base pairs of data at an explosive rate. Ever since its inception in 1982, GenBank has doubled in size in about every 14 months. Table 1.1 shows the growth of the sequence data at GenBank and EMBL collectively. You may visit http://www.ncbi.nlm.nih.gov/Genbank/genbankstats.html for latest statistics.

Given this unknown DNA sequence, how are encoded genes determined and positioned? Translating from all translational start codons to all chain-terminating stop codons, every frame provides a list of open reading frames (ORFs). But which of them, if any, actually code for proteins? If one is dealing with eukaryotic genomic DNA, then exons and introns complicate the matter further. Like nucleic acid sequences, there is a huge amount of information that is available today on protein sequences as well. Information on proteins ranges from their sequences, structures, domains, motifs, and fingerprints. This data can be best stored, accessed, and utilized in the form of databases. Databases that deal with sequences are called sequence databases. These are primarily of two types:

- DNA sequences made up of nucleotides or bases
- Protein sequences made up of amino acids

Sequence databases are designed differently to meet generalized or specific requirements. Sequences when stored in these databases need to follow the design of the database model. Thus the same sequence when stored in a different database may differ in format.

Table 1.1 Number of sequences and base pairs as an indicator of the growth of GenBank and EMBL databases from 1982–2005, revised: March 7, 2006.

Year	GenBank/EMBL Data Base Pairs	Sequences
1982	680,338	606
1983	2,274,029	2,427
1992	101,008,486	78,608
1997	1,160,300,687	1,765,847
2000	11,101,066,288	10,106,023
2001	15,849,921,438	14,976,310
2002	28,507,990,166	22,318,883
2003	36,553,368,485	30,968,418
2004	44,575,745,176	40,604,319
2005	56,037,734,462	52,016,762

1.1 Nucleotide Sequence Databases

Nucleotide sequence databases are data repositories that accept nucleic acid sequence data and make it freely available to the public. The data in these repositories is heterogeneous with respect to the source of material (e.g., genomic, cDNA), quality (e.g., finished, expressed sequence tags), annotation (un-annotated, partial, or complete), and the intended completeness of the sequence relative to its biological target (e.g., complete or partial gene or a genome).

The pace of human genomic sequencing has outstripped the capacity of sequencing centres to annotate and validate the sequence, which is usually done prior to submitting it to the archival databases. Multiple third-party groups have stepped in and are currently annotating the human sequence with a combination of computational and experimental methods. Their analytic tools, data models, and visualization methods are diverse.

Much of the relevant and useful information available today is housed in secondary databases. These databases have information that is either curated or annotated. DDBJ, EMBL, and TrEMBL are examples of primary databases, and Swiss-Prot, Prosite, and PDB are examples of secondary databases.

Primary versus secondary databases

- Primary sequence → primary database
- Secondary structure → motif → secondary database (or pattern and profile databases)

1.1.1 International Nucleotide Sequence Database Collaboration

Instead of getting all information in one database, bioinformatics researchers often have to check multiple databases to get information about a particular region on a gene of interest, download the data in several different formats, and perform a manual integration in order to get a complete picture. Clearly, this is cumbersome, and a convenient one-stop database would surely make the whole process easier and help researchers save their effort and time. The ***I****nternational* ***N****ucleotide* ***S****equence* ***D****atabase* ***C****ollaboration* (*INSD*) consists of *nucleotide sequence databases* from three groups. They are

- DNA Databank of Japan (DDBJ) of National Institute of Genetics, Japan
- GenBank of National Center for Biotechnology Information (NCBI), USA
- The European Molecular Biology Labs (EMBL) Data Library of European Bioinformatics Institute, UK

Databases EMBL, GenBank, and DDBJ are the *three primary nucleotide sequence databases*. The scientific community today often refer to the combination of these three public databases simply as GenBank. These databases include sequences submitted directly by scientists and genome sequencing groups and sequences taken from literature and patents. Automated procedures are provided for submissions from large-scale sequencing projects and data from various patent offices such as the European Patent Office. There are various data submission tools such as Webin, BankIt, and Sequin. You will learn more about these when we come to the chapter on data submission tools. Data includes **T**hird-**P**arty **A**nnotation (TPA) and alignments. There is comparatively little error checking, which leads to a fair amount of redundancy.

These three databases exchange and update data on a daily basis to achieve optimal synchronization (Fig. 1.1). The data is exchanged using a common format for data distribution. In spite of the fact that these databases have common entries that are mirrored among each other, the entries are stored in different formats. Thus the data exchanged among the databases are the same, but their representation in each of them is different. Access to these sequence databases is provided via ftp and several WWW interfaces. The collaboration allows you to either refer to a single database or to a collaborated data library. For example, you can collectively access all the sequences in GenBank and EMBL, by using the GenEMBL data library. You can also access data in separate groups, by using the smaller sets of categories such as primate, bacterial, etc., as logical names. Sequence databases are split into many different divisions. For example, sequence data stored as RDBMS uses the concept of tables with rows and column as divisions, whereas the same data when stored as a flat file uses delimiters for data segregations. The main purpose of the divisions is to help effective searching of the most relevant sequences only. These divisions may contain certain taxonomic categories, individual species, and special classes of loci, which offers several obvious advantages:

- Minimize search time.
- Obtain less noisy results.
- Allows access to single species categories, which tend to be indicated by the first letters of the scientific name, e.g., 'em:hs*' specifies all human (*Homo sapiens*) entries in the EMBL database, and 'gb:rn*' specifies all rat (*Rattus norvegicus*) entries in Genbank.
- Allows access to specific taxonomic categories within a database. For instance, 'em_ba:*' refers to EMBL bacterial sequences, and 'gb_ro:*' refers to GenBank rodent sequences.

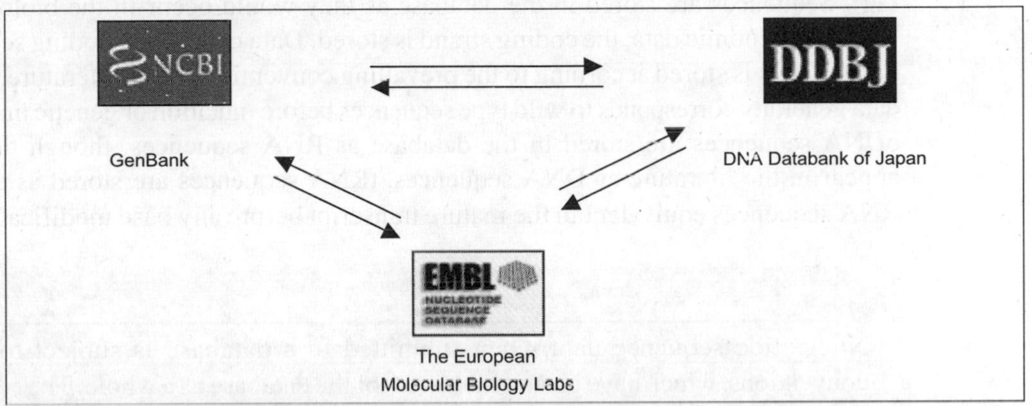

Fig. 1.1 Exchange of data among DDBJ, EMBL, and NCBI

DDBJ, EMBL, and GenBank adhere to documented guidelines to regulate the content and syntax of the database entries. Details about the database policies issued can be obtained from the European Bioinformatics Institute website http://www.ebi.ac.uk/ Likewise, other institutes also have their own database policies, which are available in their respective websites.

1.1.2 EMBL www.ebi.ac.uk/embl/

The ***E****uropean* ***M****olecular* ***B****iology* ***L****aboratory* (EMBL) is a nucleotide sequence database maintained by the ***E****uropean* ***B****ioinformatics* ***I****nstitute* (EBI). The current database release (Release 81, Dec. 2004), complete with release notes and user manual, are available from the EBI servers. Databases provided at the EBI include the EMBL nucleotide sequence database, the protein databases Swiss-Prot, TrEMBL, and UniProt, InterPro, the macromolecular structure database (E-MSD), the gene expression database (ArrayExpress), and the Ensembl automatic genome annotation database.

(a) EMBL entry

The entries in the EMBL database are structured in a format that is equally friendly to both humans and computer programs. The explanations, descriptions, classifications, and other comments are in standard english. Symbols and formatting have been specially designed to help understand the entries. Wherever possible, familiar symbols generally used in molecular biology laboratories have been used. However, care has been taken to organize the database systematically so as to allow reading, identification, and manipulation of data with ease with the help of computer programs.

The EMBL nucleotide sequence database (also known as EMBL-Bank) is divided into sections that reflect taxonomic as well as non-taxonomic divisions. Taxonomic divisions include invertebrates, mammals, organelles, bacteriophages, plants, prokaryotes, rodents, unclassified viruses, and other vertebrates. Non-taxonomic sequences come from patents, htg (**h**igh-**t**hroughput **g**enomic sequences), htc (**h**igh-**t**hroughput **c**DNA sequencing), gss (**g**enome **s**urvey **s**equence), wgs (**w**hole **g**enome **s**hotgun), and est (**e**xpressed **s**equence **t**ag). Sequences are stored in the database as they would occur in the biological state. Thus, for genomic data, the coding strand is stored. Data containing coding sequences on both strands is stored according to the prevailing conventions in the literature. The stored data generally corresponds to wild type sequences before mutation or genetic manipulation. cDNA sequences are stored in the database as RNA sequences, though they usually appear in the literature as DNA sequences. tRNA sequences are stored as unmodified RNA sequences equivalent to the mature transcript before any base modification occurs.

> **Note**
>
> Nucleotide sequence data when submitted to a database is subject to certain conventions, which have been established for the database as a whole. The sequences are always listed in the 5′ to 3′ direction, regardless of the published order. Bases are numbered sequentially, beginning with 1 at the 5' end of the sequence.

(b) Sample EMBL format

```
ID    U87107    standard; DNA; SYN; 8840 BP.
XX
AC    U87107;
XX
SV    U87107.1
XX
DT    15-OCT-1997 (Rel. 52, Created)
DT    15-OCT-1997 (Rel. 52, Last updated, Version 4)
XX
DE    Cloning vector pAL-F insertion sequence IS1 galactokinase (galK),
DE    aminoglycoside 3′-phosphotransferase (kn), beta-galactosidase (lacZ),
      small
```

```
DE     ribosomal protein and beta-lactamase (Ap) genes, complete cds.
XX
KW
XX
OS     Cloning vector pAL-F
OC     artificial sequence; vectors.
XX
RN     [1]
RP     1-8840
RA     Ahmed A., Podemski L.;
RT     'Use of ordered deletions in genome sequencing';
RL     Gene 197:367-373(1997).
XX
RN     [2]
RP     1-8840
RA     Ahmed A.;
RT     ;
RL     Submitted (27-JAN-1997) to the EMBL/GenBank/DDBJ databases.
RL     Biological Sciences, University of Alberta, Edmonton, Alberta, T6G
       2E9,
RL     Canada
XX
DR     REMTREMBL; AAC53713; AAC53713.
DR     REMTREMBL; AAC53714; AAC53714.
DR     REMTREMBL; AAC53715; AAC53715.
DR     REMTREMBL; AAC53716; AAC53716.
DR     REMTREMBL; AAC53717; AAC53717.
XX
FH     Key          Location/Qualifiers
FH
FT               source    1..8840
FT               /db_xref='taxon:56954'
FT               /organism='Cloning vector pAL-F'
FT               /insertion_seq='IS1'
FT               /specific_host='Escherichia coli'
FT     CDS       complement(933..2081)
FT               /codon_start=1
FT               /db_xref='REMTREMBL:AAC53713'
FT               /transl_table=11
FT               /gene='galK'
FT               /product='galactokinase'
FT               /protein_id='AAC53713.1'
FT
/translation='MSLKEKTQSLFANAFGYPATHTIQAPGRVNLIGEHTDYNDGFVLP
```

```
FT
.............................................................................FT
LEQGDLKRMGELMAESHASMRDDFEITVPQIDTLVEIVKAVIGDKGGVRMTGGGFGGCI
FT     VALIPEELVPAVQQRVAEQYEAKTGIKETFYVCKPSQGAGQC'
FT   CDS              complement(7087..7947)
FT                    /codon_start=1 FT    /
db_xref='REMTREMBL:AAC53717'
FT                   /db_xref='REMTREMBL:AAC53717'
FT                    /transl_table=11
FT                     /gene='Ap'
FT                   /function='ampicillin resistance'
FT                   /product='beta-lactamase'
FT                   /protein_id='AAC53717.1'
/translation='MSIQHFRVALIPFFAAFCLPVFAHPETLVKVKDAEDQLGARVGYI
FT
ELDLNSGKILESFRPEERFPMMSTFKVLLCGAVLSRVDAGQEQLGRRIHYSQNDLVEYS
FT
.........................................................................FT
LIKHW'
XX
SQ Sequence 8840 BP; 2068 A; 2288 C; 2319 G; 2165 T; 0 other;
   caattactgc aatgccctcg taattaagtg aatttacaat atcgtcctgt tcggagggaa
60
  gaacgcggga tgttcattct tcatcacttt taattgatgt atatgctctc ttttctgacg 120
..........................................................................
gacagctgat agaaacagaa gccactggag cacctcaaaa acaccatcat acactaaatc
8820
agtaagttgg cagcatcacc
8840
//
```

Note
Each entry begins with an identification line (ID) and ends with a terminator line (//). However, all entries do not carry all the line types mentioned above.

(c) Abbreviation key

Each entry in the database is composed of lines. Different types of lines, each with its own format, are used to record various types of data, which make up the entry. In general, fixed-format items for an entry are kept to a minimum (e.g., ID, AC, DT). The rest of the items of that entry have a more syntax-oriented structure adopted for the lines, which is different for different entries (e.g., number of CDS in feature table 'FT' and number of sequence data lines in 'SQ'). Note that each line in an entry begins with a two-character line code. This code indicates the type of information contained in the line. The line codes along with their corresponding descriptions and the accepted number of line codes per entry are listed in Tables 1.2–1.4.

Table 1.2 Line codes that have fixed number of entries

Line code	Description	Number of entries
ID	Identification	begins each entry; 1 per entry
SQ	Sequence header	1 per entry
DT	Date	2 per entry

Table 1.3 Line codes that do not have fixed number of entries

Line code	Description	Number of entries
AC	Accession number	>=1 per entry
DE	Description	>=1 per entry
KW	Keyword	>=1 per entry
OS	Organism species	>=1 per entry
OC	Organism classification	>=1 per entry
RN	Reference number	>=1 per entry
RP	Reference positions	>=1 per entry
RA	Reference author(s)	>=1 per entry
RT	Reference title	>=1 per entry
RL	Reference location	>=1 per entry
XX	Spacer line	many per entry

Table 1.4 Line codes that have may not appear in all entries

Line code	Description	Number of entries
OG	Organelle	0 or 1 per entry
RC	Reference comment	>=0 per entry
DR	Database cross-reference	>=0 per entry
FH	feature table header	0 or 2 per entry
FT	feature table data	>=0 per entry
CC	comments or notes	>=0 per entry

The GenBank home page at http://www.ncbi.nlm.nih.gov/Genbank/ lists the meaning of these and all other abbreviations used.

Note

Although the nucleotide sequence data represented in GenBank, EMBL, and DDBJ differ slightly due to differences in sequence formats, each nucleotide sequence is identified with the same accession number in all the three databases. Therefore, you can connect to these three databases and essentially get the same nucleotide sequence using the same accession number.

TASK Searching and retrieving from EMBL database

Step 1 Locate the EMBL nucleotide database on the Internet.
Step 2 Search EMBL ID AF 193508.
Step 3 Try to answer the following questions:

(a) From which organism is your entry?
(b) What is the molecule type and sequence length?
(c) Give its entry division?
(d) What is the UnitProt/TrEMBL ID of the corresponding protein sequence entry?
(e) What is its NCBI Tax ID?
(f) Using SMART, gather the abstract about this ATPase. What family does it belong to?
(g) What is the CDS of this entry (the start and stop)?
(h) Using the links to Medline, find the abstract of one article relating to this entry. On the pages with the article abstract, try to follow the link 'Related Articles by NCBI'. How many related articles are identified? Check article titles to see if these articles are really relevant.
(i) How can the EMBL database be searched using the ID AF 193508 to find this entry directly (i.e., without finding all ATPase-related Candida genes)?

Note

Before you continue, it would be good if you learnt how to save data from a web page in a format so that it is easy to use later on. With the entry page for AF 193508 in your browser, click on the button marked 'Save' on the left side, and then choose to save as a browser window (which is default). Thereafter, check what happens when changing the 'Save with view' option in the 'ASCII text/table' part under 'Save As'.

1.1.3 GenBank (www.ncbi.nlm.nih.gov/Genbank/)

The GenBank nucleotide database is maintained by the National Center for Biotechnology Information (NCBI), which is part of the National Institute of Health (NIH, a federal agency of the US government). It can be accessed and searched through the Entrez gateway at NCBI. You can also download the entire database as flat files. This database is huge and consists of approximately 44,575,745,176 bases in 40,604,319 sequence records as of February 2005.

(a) GenBank entry

Each entry in GenBank contains information on the sequence, the scientific name and taxonomy of the source organism, a features that identify coding regions, transcription units, mutation sites, and repeats. All additional information is captured in 'feature table'. Protein translations for coding regions are also a part of the feature table. Links to the Medline unique identifier for all published sequences are also provided. The format specifications for GenBank files and other related information can be found in the GenBank release notes, gbrel.txt, on the GenBank website at ftp://ncbi.nlm.nih.gov/genbank/gbrel.txt There are many programs that can be used to search the GenBank database, e.g., GCG package. You will find it useful to remember that the EMBL ID and GenBank accession numbers are unique. It is possible to search these databases and others (e.g., protein databases) quoting just these unique identifiers. Any other information is stored along with the sequence. Each piece of information is written on its own line, with a code defining the line, e.g., DE, description; OS, organism species; AC, accession number, etc. Some information is self-explanatory from the content. Sometimes, a user-friendly graphical report is made available, making it even easier to read the results of a search. Any relevant biological information is usually described in the feature table (FT).

(b) Sample GenBank format

An example of what an entry looks like is given for the *Listeria ivanovii* sod gene.

```
LOCUS       LISOD                  756 bp    DNA    linear   BCT 30-JUN-1993
DEFINITION  Listeria ivanovii sod gene for superoxide dismutase.
ACCESSION   X64011 S78972
VERSION     X64011.1  GI:44010
KEYWORDS    sod gene; superoxide dismutase
SOURCE      Listeria ivanovii
  ORGANISM  Listeria ivanovii
            Bacteria; Firmicutes; Bacillales; Listeriaceae; Listeria
REFERENCE   1  (bases 1 to 756)
  AUTHORS   Haas,A. and Goebel,W.
  TITLE     Cloning of a superoxide dismutase gene from Listeria ivanovii by
            functional complementation in Escherichia coli and
             characterization
            of the gene product
  JOURNAL   Mol. Gen. Genet. 231 (2), 313-322 (1992)
  MEDLINE   92140371
REFERENCE   2  (bases 1 to 756)
  AUTHORS   Kreft,J.
  TITLE     Direct Submission
  JOURNAL   Submitted (21-APR-1992) J. Kreft, Institut f. Mikrobiologie,
            Universitaet Wuerzburg,Biozentrum Am Hubland, 8700 Wuerzburg,FRG
FEATURES             Location/Qualifiers
     source          1..756
                     /organism='Listeria ivanovii'
                     /strain='ATCC 19119'
                     /db_xref='taxon:1638'
                     /mol_type='genomic DNA'
     RBS             95..100
                     /gene='sod'
     gene            95..746
                     /gene='sod'
     CDS             109..717
```

```
                     /gene='sod'
                     /EC_number='1.15.1.1'
                     /codon_start=1
                     /transl_table=11
                     /product='superoxide dismutase'
                     /db_xref='GI:44011'
                     /protein_id='CAA45406.1'
                     /db_xref='Swiss-Prot:P28763'
/translation='MTYELPKLPYTYDALEPNFDKETMEIHYTKHHNIYVTKLNEAVS
                     GHAELASKPGEELVANLDSVPEEIRGAVRNHGGGHANHTLFWSSLSPNGGGAPTGNLK
                     AAIESEFGTFDEFKEKFNAAAAARFGSGWAWLVVNNGKLEIVSTANQDSPLSEGKTPV
                     LGLDVWEHAYYLKFQNRRPEYIDTFWNVINWDERNKRFDAAK'
     terminator      723..746
                     /gene='sod'
ORIGIN
        1 cgttatttaa ggtgttacat agttctatgg aaatagggtc tataccttc gccttacaat
       61 gtaatttctt ..........
//
```

(c) Abbreviation key

A GenBank entry is composed of lines, each line having its own format. The lines record the various types of data, which make up the entry. Fixed-format items for an entry are kept to a minimum such as LOCUS, DEFINITION, etc. The abbreviation key for a typical entry is given below:

- LOCUS : Short name for this sequence (maximum of 32 characters)
- DEFINITION : Definition of sequence (maximum of 80 characters)
- ACCESSION : accession number of the entry
- VERSION : Version of the entry
- DBSOURCE : Shows the source, the dates of creation and last modification of the database entry
- KEYWORDS : Keywords for the entry
- AUTHORS : Authors for the work
- TITLE : Title of the publication
- JOURNAL : Journal reference for the entry
- MEDLINE : Medline ID
- COMMENT : Lines of comments
- SOURCE ORGANISM : The organism from which the sequence was derived
- ORGANISM : Full name of organism (maximum of 80 characters)
- AUTHORS : Authors of this sequence (maximum of 80 characters)
- ACCESSION : ID number for this sequence (maximum of 80 characters)
- FEATURES : Features of the sequence
- ORIGIN : Beginning of sequence data
- // : End of sequence data

TASK Searching and retrieving from GenBank database

Step 1 Let us say we want to study *Homo sapiens* pancreatic lipase.

Step 2 Go to GenBank at www.ncbi.nlm.nih.gov/Genbank/

Step 3 Observe the GenBank flat-file format. Pay special attention to the Header, Feature Table, and the Nucleotide Sequence Section.

Step 4 Open the Related Sequence Link.

Step 5 Explore the HPR database for the colipase pancreatic (Fig. 1.2).

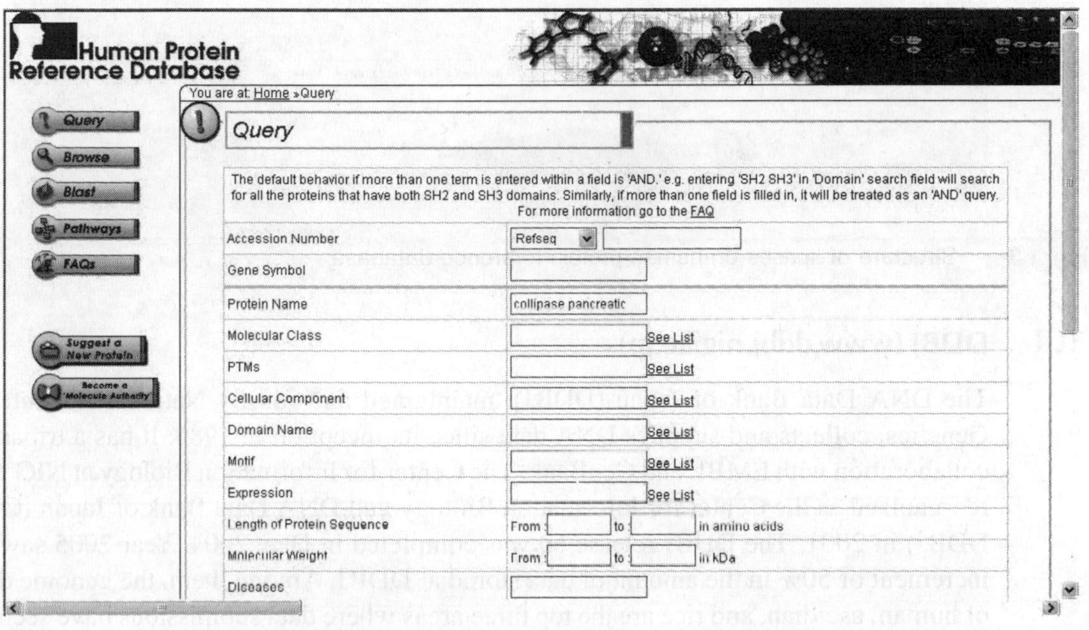

Fig. 1.2 View of human protein reference database

Step 6 Prepare a report on all information you gather from here. Organize the report, taking care of the following heads:

- Nature of the molecule—molecular class and function, mol. wt, gene map locus, site of expression (primary).
- Description of the motif.
- DNA (and the ORF) and protein sequences.
- Interactions of colipase lipase *in vitro*, the JORNAL, PMID (Result: PMID: 11278590 [PubMed—indexed for Medline])
- OMIM ID (Result: 120105)
- Number of PTMs

Step 7 Check your result of the structure of lipases with Fig. 1.3.

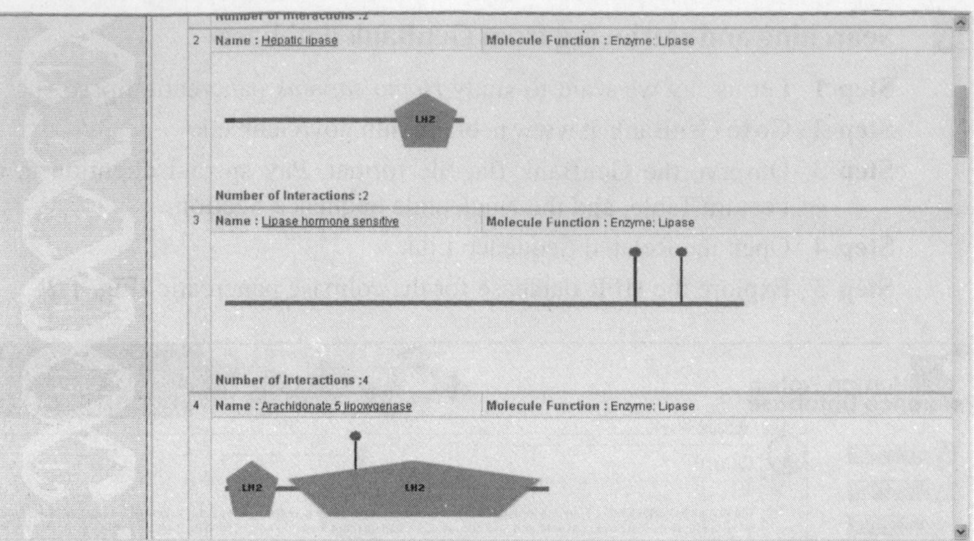

Fig. 1.3 Structure of lipases on human protein reference database

1.1.4 DDBJ (www.ddbj.nig.ac.jp)

The DNA Data Bank of Japan (DDBJ), maintained and run by National Institute of Genetics, collects and supplies DNA data since its inception in 1986. It has a tripartite collaboration with EMBL and GenBank. The Center for Information Biology at NIG was reorganized as the Center for Information Biology and DNA Data Bank of Japan (CIB-DDBJ) in 2001. The DDBJ release 60 was completed in Dec. 2004. Year 2005 saw an increment of 50% in the amount of data stored at DDBJ. Among them, the genome data of human, ascidian, and rice are the top three areas where data submissions have seen an increase. The DNA data can be accessed by online retrieval systems by the registered users with the help of different retrieval software or through distribution on magnetic tapes. One can search for entries by the accession number, FASTA/BLAST, keywords, and regular expressions. SAKURA is a nucleotide sequence data submission system operating accessible through the WWW server and has been functional since 1995. To help researchers, DDBJ has recently implemented a simple object access protocol (SOAP) server and web services at http://www.xml.nig.ac.jp

(a) DDBJ entry

Like the GenBank and EMBL databases entries, each DDBJ entry contains not only information about the sequence, the scientific name and taxonomy of the source organism, but also features that identify coding regions, transcription units, mutation sites, and repeats. This additional information is captured in a feature table. More details about the feature tables of EMBL, GenBank, and DDBJ can be found at http://www.ncbi.nlm.nih.gov/collab/FT/#2

(b) Sample DDBJ format

Its format is the same as the GenBank format. Refer Section 1.1.3(b) for detail.

(c) Abbreviation key

The abbreviation key is the same as in the GenBank. Refer Section 1.1.3(c) for detail.

TASK Searching and retrieving from DDBJ database

Step 1 Locate the DDBJ nucleotide database on the Internet.

Step 2 Search with the keywords human growth factor through the SRS (left link).

Step 3 On the left hand side click Standard Query Form to get the standard query form as displayed in Fig. 1.4.

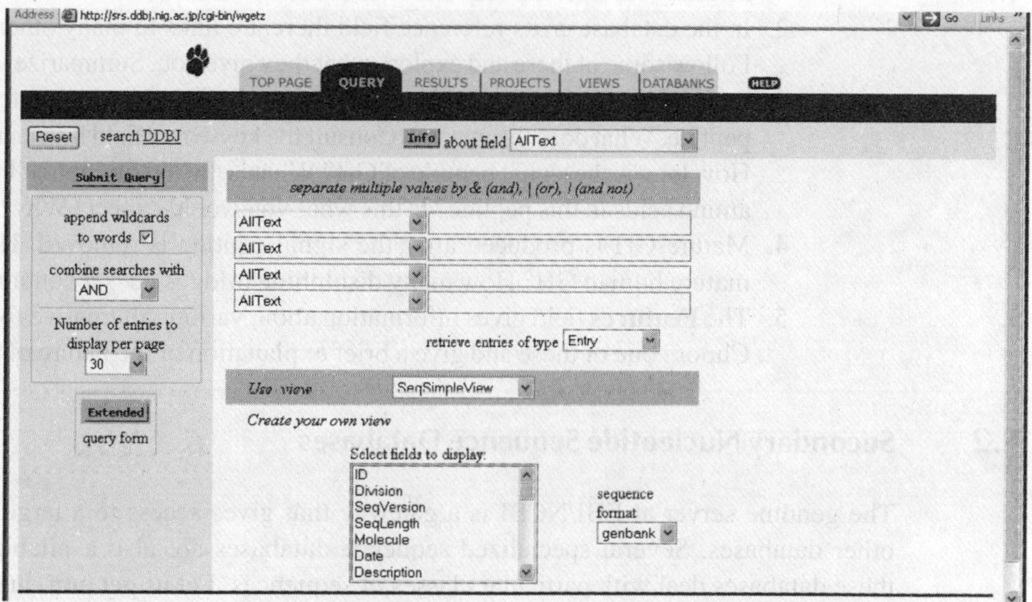

Fig. 1.4 Interface of standard query form at DDBJ

Step 4 Examine this form. The four bars labelled **All Text** allow you to select individual fields within the database to search, using search terms entered in the adjacent boxes. There are various parameters that can be changed which we will leave as the default for now (a more complex form is available via the **Extended Query Form** choice).

Step 5 Change the top bar to **Accession Number** and enter P01241 (the accession number for human GH) in the adjacent box. Click the **Search** button on the top right of the screen. After a short delay, while the database is being searched, a new page appears indicating that a single item has been found.

Step 6 The column headed **Description** indicates that this is human growth hormone (note that somatotropin is an alternative name for the growth hormone). Click on the entry under **UniProt**. The next page is the full entry for human GH. The sequence (actually the sequence of the precursor of human GH) is at the bottom of the page, in one letter code. *You should know the one-letter code for amino acids. If you do not know it yet, now is the time*! The page also gives you a large amount of information about the protein, with appropriate links to explore further. Spend a few minutes exploring this page.

Step 7 Compile a report about the GH sequence.

1. In the **References** field more than 30 references are given. Why are there so many for this one sequence? Explore the links to Medline and PubMed.
2. In the database cross-reference field there are links to many other databases. Follow some of these and explore what they give you. Summarize your search.
3. In the **Features** field there is a reference to **SIGNAL**, which refers to signal peptide. What does this mean? (Consult the keywords field if you are not sure.) How long is the signal peptide of GH? What is the distribution of hydrophobic amino acids in this peptide? Is this what you would expect? Why?
4. Mature GH is produced after the signal peptide is removed. How long is mature human GH? How many disulphide bridges does it contain?
5. The **Features** field gives information about various alternative forms of GH. Choose one of these and give a brief explanation of the nature of the variant.

1.2 Secondary Nucleotide Sequence Databases

The genome server at EBI/NCBI is a gateway that gives access to a large number of other databases. Several specialized sequence databases are also available. Some of these databases deal with particular classes of sequences. Let us get introduced to some of these.

(a) UniGene (http://www.ncbi.nlm.nih.gov/UniGene/index.html)

NCBI, which hosts GenBank and BLAST, has created the UniGene database. This database has records with unique gene clusters. Each UniGene cluster contains sequences that represent a unique gene, as well as related information such as the tissue types in which the gene has been expressed and map location. Since only 2%–3% of human genomic DNA actually encodes proteins, it makes sense to focus sequencing efforts on these 'expressed' regions first. This is the basis for a number of **E**xpressed **S**equence **T**ag (EST) projects. Therefore, in addition to sequences of well-characterized genes, the database is equally well populated with EST sequences from Genbank. These ESTs have been grouped into relevant clusters by sequence similarity. This is a sequence-cluster database that has brought together similar sequences, which *may* have been derived from the same gene.

(b) STACK (http://ziggy.sanbi.ac.za/stack/)

The South African National Bioinformatics Institute (SANBI) at the University of the Western-Cape has created an *EST consensus database* known as **S**equence **T**ag **A**lignment and **C**onsensus **K**nowledgebase (STACK). All available human EST sequences in GenBank are clustered, aligned, and made into *joined consensus records*. Thus *STACK consensus sequences* are longer than TIGR or UniGene clusters. It is organized into 15 tissue-based categories and one disease category.

(c) Ribosomal database project (http://rdp.cme.msu.edu/index.jsp)

This project provides ribosome related data services, including online data analysis information, rRNA derived phylogenetic trees, aligned and annotated rRNA sequences, to the scientific community. This database is updated on a monthly basis. Release 9 (8 Feb. 2005) contains 124,165 aligned and annotated bacterial small-subunit rRNA sequences. Each update consists of *new 16S rRNA sequences* that have appeared in GenBank in the previous month. It has four analysis tools. These are hierarchy browser, RDP classifier, sequence match, and probe match.

(d) HIV Sequence database (http://hiv-web.lanl.gov/content/index)

This database collects, curates, and annotates HIV and SIV sequence data and provides various tools for analysing this data. It contains information about HIV genetic sequences, immunological epitopes, drug-resistance-associated mutations, and vaccine trials. Like the ribosomal database project, it contains several analysis tools. The HIV protease database, run by the National Cancer Institute, USA, is an archive of experimentally determined three-dimensional (3D) structures for HIV-1, HIV-2, and SIV.

(e) Eukaryotic promoter database (http://www.epd.isb-sib.ch/)

The **E**ukaryotic **P**romoter **D**atabase (EPD) was designed and developed at the Weizmann Institute of Science in Rehovot (Israel) and is currently maintained at ISREC in Epalinges/ Lausanne (Switzerland). EPD is a specialized annotation database that uses the EMBL data library. It is an annotated, non-redundant collection of eukaryotic POL II promoters, for which the transcription start site has been determined experimentally. All information is either directly extracted from scientific literature or, starting from release 73, compiled by a new in silico primer extension method.

(f) REBASE (http://rebase.neb.com/rebase/rebase.html)

It is a database for restriction enzymes and restriction enzyme sites. This database has information about restriction enzymes, methylases, the microorganisms from which they have been isolated, recognition sequences, cleavage sites, methylation specificity, the commercial availability of the enzymes, and their references. REBASE is *used as the default* in the restriction mapping programs such as GCG, MAP, MAPPLOT, MAPSORT, etc. It can be also reached using SRS; however, it is not available for BLAST.

1.3 Protein Sequence Databases

Most of the available protein-related information that can be gleaned today arises out of one protein database or the other. There are two categories of universal protein databases that exist today: one that stores simple archives of sequence data, and the other that stores annotated data. In the annotated databases, additional information is added to the sequence record. Databases act as 'universal' databases having entries from all the species and specialized data about specific families or groups of proteins, or about the proteins of a specific organism.

1.3.1 International Protein Sequence Database Collaboration

The EBI/SIB Swiss-Prot and the TrEMBL databases and the PIR protein sequence database (PIR-PSD) were created as separate protein databases. Each one of them differed in its protein sequence coverage and annotation. In 2002, EBI, SIB, and PIR came together as the UniProt (universal protein) consortium (Fig. 1.5). It is a central repository of protein sequences and functions created by integrating the information contained in Swiss-Prot, TrEMBL, and PIR. The consortium provides accurate, consistent, and rich sequence and functional annotation. It is at present the world's most comprehensive database of information on proteins. The main objective behind the formation of the consortium was to facilitate biological research by maintaining a high-quality database with cross references in the form of pointers to information in other databases and freely accessible querying interfaces.

European Bioinformatics Institute

Swiss Institute of Bioinformatics

Georgetown University

Fig. 1.5 Uniprot consortium, the central repository of protein data

There are two basic methods for searching from protein sequence databases. These search methods are by using the following:

(1) Particular keyword label (e.g., 'cytochrome c'). PIR allows a variety of keyword searches.
(2) Search engines to hunt for sequences that are similar.

Analogy

You can search a phonebook for a number (sequence) associated with a particular name (protein), or you can search a phonebook to determine the names of all the people who had phone number starting with 2658 (pattern or domain).
UniProt is comprises three components. These are discussed below.

(a) UniProt Knowledgebase (UniProtKB)

It is the central database of protein sequences with accurate information on sequences and functional annotation. It acts as a central access point for not only protein information, but also their functions, classifications, and cross references. It consists of two sections:

- A section containing manually annotated records with information extracted from literature and curator-evaluated computational analysis named 'Swiss-Prot'.
- A section with computationally analysed records that await full manual annotation named 'TrEMBL'.

The database contains data from a wide variety of organisms (more than 6,000 species).

(b) UniProt reference clusters (UniRef)

These combine closely related sequences into a single record to speed up searches. These non-redundant clustered sets of sequences from the UniProt knowledgebase include splice variants and isoforms and UniParc records, in order to obtain complete coverage of sequence space.

Note

UniRef90 and UniRef50 are built by clustering UniRef100 sequences with 11 or more residues using the CD-HIT algorithm (Li, W., Jaroszewski, L., and Godzik, A., *Bioinformatics*, 17: 282-83, 2001) such that each cluster is composed of sequences that have at least 90% or 50% sequence identity, respectively, to the representative sequence. UniRef90 and UniRef50 also reduce the database size by ~40% and 65%, thus increasing the speed of sequence searches.

(c) UniProt archive (UniParc)

It is a comprehensive repository consisting only of unique identifiers and sequences. Protein sequences are integrated on a daily basis from many different publicly accessible sources such as UniProt Consortium databases, Swiss-Prot, TrEMBL, PIR-PSD, EMBL-Bank/DDBJ/GenBank, EnsEMBL, PDB, and others. While a protein sequence may exist in multiple databases, UniParc stores each unique sequence only once and assigns it a unique UniParc identifier. Cross references back to the source databases are provided. These include source accession numbers, sequence versions, and status (active or obsolete). UniParc also provides the history about a sequence. These databases can be accessed and searched through the SRS system at ExPASy, or one can download the entire database in flat-file format.

TASK

Step 1 Go to UniProt at http://www.ebi.uniprot.org/index.shtml

Step 2 Copy and paste the following human amino acid sequence into the query sequence window. You will not need to specify advanced options, but you should check the program and choose the database.

```
PSYTVTVATG SQWFAGTDDY IYLSLVGSAG CSEKHLLDKP
FYNDFERGAV DSYDVTVDEELGEIQLVRIE KRKYWLNDDW
YLKYITLKTP HGDYIEFPCY RWITGDVEVV LRDGRAKLAR
DDQIHILKQH RRKELETRQK QYRWMEWNPG FPLSIDAKCH
KDLPRDIQFD SEKGVDFVLN YSKAMENLFI NRFMHMFQSS
WNDFADFEKI FVKISNTISE RVMNHWQEDL MFGYQFLNGC
NPVLIRRCTE LPEKLPVTTE MVECSLERQL SLEQEVQQGN
IFIVDFELLD GIDANKTDPC TLQFLAAPIC LLYKNLANKI
VPIAIQLNQI PGDENPIFLP SDAKYDWLLA KIWVRSSDFH
VHQTITHLLR THLVSEVFGI AMYRQLPAVH PIFKLLVAHV
RFTIAINTKA REQLICECGL FDKANATGGG GHVQMVQRAM
KDLTYASLCF PEAIKARGME SKEDIPYYFY RDDGLLVWEA
IRTFTAEVVD IYYEGDQVVE EDPELQDFVN DVYVYGMRGR
KSSGFPKSVK SREQLSEYLT VVIFTASAQH AAVNFGQYDW
CSWIPNAPPT MRAPPPTAKG VVTIEQIVDT LPDRGRSCWHA
LGAVWALSQF QENELFLGMY PEEHFIEKPV KEAMARFRKN
LEAIVSVIAE RNKKKQLPYY YLSPDRIPNS VAI
```

Step 3 Run the search and identify the protein. Use the link provided to see the report. If no links to structural databases are displayed, you can just repeat the search selecting a structure database (pdb).

Step 4 Prepare a report about the following:

(a) What is the name (ID) of the entry?
(b) What is the primary accession number?
(c) How many amino acid residues does the protein contain?
(d) What is the most common name of the protein?
(e) What does it 'do'?
(f) What is the gene called?
(g) Is there a structure model available for this protein? If so, what was the year and who was the first author. If not, are there any homologous or similar proteins in the structural databases (yes/no)? 'Yes', Of what protein(s)?
(h) What is a leukotriene?

1.3.2 Swiss-Prot and TrEMBL (www.expasy.ch/sprot/)

Swiss-Prot is a protein sequence database that was started in 1986. It is maintained collectively by the Swiss Institute of Bioinformatics and the EMBL Data Library. For standardization purposes, Swiss-Prot follows the format of EMBL nucleotide sequence database as closely as possible. Data redundancy is low, and, as discussed earlier, it is integrated with other databases.

(a) Swiss-Prot entry

Sequence entries in Swiss-Prot are composed of different line types, each with their own format. The database gives to its user a high level of annotations. These annotations include the description of the function of a protein, its domains structure, post-translational modifications, variants, etc.

TrEMBL (translation of EMBL) is a computer-annotated supplement of Swiss-Prot that contains all the translations of EMBL nucleotide sequence entries that is not yet integrated in Swiss-Prot. It was created to accommodate the deluge of sequence information and the time-consuming curating process. TrEMBL has two major divisions: SP-TrEMBL and REM-TrEMBL. SP-TrEMBL contains the entries, which will finally merge into Swiss-Prot. REM-TrEMBL (REMaining TrEMBL) contains the entries that will not get included in Swiss-Prot. It has entries that are synthetic, truncated, pseudogenes, patented, small fragments or immunoglobulins and T-cell receptors.

(b) Sample Swiss-Prot format

Swiss-Prot: P43021

NiceProt—a user-friendly view of this Swiss-Prot entry

```
ID  NODAL_MOUSE   STANDARD;    PRT;   354 AA.
AC  P43021;
DT  01-NOV-1995 (Rel. 32, Created)
DT  01-NOV-1995 (Rel. 32, Last sequence update)
DT  01-FEB-2005 (Rel. 46, Last annotation update)
DE  Nodal precursor.
GN  Name=Nodal;
OS  Mus musculus (Mouse).
OC  Eukaryota; Metazoa; Chordata; Craniata; Vertebrata; Euteleostomi;
OC  Mammalia; Eutheria; Rodentia; Sciurognathi; Muridae; Murinae; Mus.
OX  NCBI_TaxID=10090;
RN   [1]
RP  NUCLEOTIDE SEQUENCE.
RX  MEDLINE=93156841; PubMed=8429908 [NCBI, ExPASy, EBI, Israel, Japan];
DOI =10.1038/361543a0;
RA  Zhou X., Sasaki H., Lowe L., Hogan B.L., Kuehn M.R.;
RT  'Nodal is a novel TGF-beta-like gene expressed in the mouse node

RT  during gastrulation.';
RL  Nature 361:543-547(1993).
CC  -!- FUNCTION: Seems essential for mesoderm formation and subsequent
```

```
CC   organization of axial structures in early mouse development.
CC   -!- SUBUNIT: Homodimer; disulfide-linked (By similarity).
CC   -!- DEVELOPMENTAL STAGE: Expressed in the node during gastrulation.
CC   Expression is first detected in primitive streak-stage embryos at
CC   about the time of mesoderm formation. It then becomes highly
CC   localized in the node at the anterior of the primitive streak.
CC   -!- SIMILARITY: Belongs to the TGF-beta family.
CC   ----------------------------------------------------------------------
          ----------
DR   EMBL; X70514; CAA49914.1; -. [EMBL/GenBank/DDBJ] [CoDingSequence]
DR   PIR; S29718; S29718.
DR   HSSP; P08476; 1NYS. [HSSP ENTRY / SWISS-3DIMAGE / PDB]
DR   Ensembl; ENSMUSG00000037171; Mus musculus. [EMBL / GenBank / DDBJ]
     [CoDingSequence]
DR   MGD; MGI:97359; Nodal.
DR   GeneLynx; Nodal.
DR   Ensembl; P43021. [Entry / Contig view]
DR   SOURCE; Nodal.
DR   GO; GO:0005615; C:extracellular space; TAS.
DR   GO; GO:0045165; P:cell fate commitment; IMP.
DR   GO; GO:0007368; P:determination of left/right symmetry; IDA.
DR   InterPro; IPR002400; GF_cysknot.
DR   InterPro; IPR001839; TGFb.
DR   InterPro; Graphical view of domain structure.
DR   Pfam; PF00019; TGF_beta; 1.
DR   Pfam; Graphical view of domain structure.
DR   PRINTS; PR00438; GFCYSKNOT.
DR   ProDom; PD000357; TGFb; 1.
DR   ProDom [Domain structure / List of seq. sharing at least 1 domain
          ]
DR   SMART; SM00204; TGFB; 1.
DR   PROSITE; PS00250; TGF_BETA_1; 1.
DR   CMR; P43021.
DR   HOVERGEN [Family / Alignment / Tree]
DR   BLOCKS; P43021.
DR   ProtoNet; P43021.
DR   ProtoMap; P43021.
DR   PRESAGE; P43021.
DR   DIP; P43021.
DR   ModBase; P43021.
DR   SMR; P43021.
DR   SWISS-2DPAGE; GET REGION ON 2D PAGE.
KW   Cytokine; Developmental protein; Glycoprotein; Growth factor; Signal.
```

```
FT   SIGNAL          1     26       Potential.
FT   PROPEP         27    244       Potential.
FT   CHAIN         245    354       Nodal.
FT   DISULFID      254    320       By similarity.
FT   DISULFID      283    351       By similarity.
FT   DISULFID      287    353       By similarity.
FT   DISULFID      319    319       Interchain (By similarity)
FT   CARBOHYD       73     73       N-linked (GlcNAc...) (Potential).
SQ   SEQUENCE   354 AA;  40448 MW;  4CC4021780815356 CRC64;
     MSAHSLRILL LQACWALLHP RAPTAAALPL WTRGQPSSPS
       PLAYMLSLYR DPLPRADIIR
SLQAQDVDVT GQNWTFTFDF SFLSQEEDLV WADVRLQLPG PMDIPTEGPL TIDIFHQAKG
DPERDPADCL ERIWMETFTV IPSQVTFASG STVLEVTKPL SKWLKDPRAL EKQVSSRAEK
CWHQPYTPPV PVASTNVLML YSNRPQEQRQ LGGATLLWEA ESSWRAQEGQ LSVERGGWGR
RQRRHHLPDR SQLCRRVKFQ VDFNLIGWGS WIIYPKQYNA YRCEGECPNP VGEEFHPTNH
AYIQSLLKRY QPHRVPSTCC APVKTKPLSM LYVDNGRVLL EHHKDMIVEE CGCL
//
```

Format differences between Swiss-Prot and EMBL databases

Swiss-Prot and EMBL databases make a deliberate effort to structure their database designs as close as possible to each other. However, certain subtle differences are there. The general structure of an entry is identical in both databases. The data classes (the second item on the ID line) are the same, but Swiss-Prot does not have 'BACKBONE', 'UNREVIEWED', and 'UNANNOTATED' data classes. Differences between the EMBL and Swiss-Prot database ID line formats are:·

- The entry name can be up to 10 characters long (instead of 9 in EMBL) and can begin with a numerical character in Swiss-Prot;
- EMBL entry ID lines have an additional three-letter taxonomic division 'token' inserted between the data class and the molecule type
- The molecule type is listed as 'PRT' rather than 'DNA' or 'RNA' in case of Swiss-Prot
- The length of the molecule is followed by 'AA' (amino acid) in Swiss-Prot instead of 'BP' (base pairs) in EMBL.

  Other differences between the two database formats include the AC line (ACcession number) and DT line format (two DT lines per entry in EMBL in contrast to three in Swiss-Prot). Some other minor differences also occur in the organism, organelle, and comments fields.

TASK

Step 1 Go to the expert protein analysis system (ExPASy) at http://au.expasy.org/
Step 2 Under **Databases** choose **Swiss-Prot and TrEMBL—protein knowledgebase**
Step 3 Search for urokinase and get 15 hits.
Step 4 Take a closer look at the chicken urokinase entry.
Step 5 Try to get a NiceProt view of the same entry.

Step 6 There is a reference that describes the 'chicken urokinase-type plasminogen activator gene'. Look at the entry and figure out who is the author of this entry. Search the bibliographic database PubMed to find how many articles have been published by the same author.

Step 7 Do the following:

(a) Prepare a brief report on the type of enzyme. Focus on the EC number. Try to decipher the activity of the enzyme.

(b) What is the length and molecular weight of this compound?

(c) Go to the 'Keywords' entry and make sure that there is information about them in the report.

(d) Go to MotifScan and search for similarity with other domains.

1.3.3 PIR-PSD (pir.georgetown.edu)

The Protein Information Resource (PIR) is a division of the National Biomedical Research Foundation (NBRF) in the US. PIR-PSD (protein sequence database) is a function-annotated database of PIR, MIPS, and JIPID at NBRF, Georgetown University, USA. Release 80.00 is the final release for the PIR-International Protein Sequence Database (PIR-PSD), the world's first database of classified and functionally annotated protein sequences that grew out of the *Atlas of Protein Sequence and Structure* edited by Margaret Dayhoff. However, it is generally believed that it does not reach the level of completeness in the entry annotation as much as does Swiss-Prot. Although Swiss-Prot and PIR overlap extensively, there are still many sequences which can be found in only one of them. Like Swiss-Prot, this database can be searched for entries or sequence-similarity searches. The PIR-PSD flat files are available in the XML format with the associated DTD file from release 66.0 on (September 2000) in addition to the original NBRF and CODATA formats. The PSD sequence file is distributed in the FASTA format. The database can also be downloaded as a set of files.

(a) PIR-PSD entry

PSD data is collated from sequences in GenBank, EMBL, and DDBJ translations, published literature, and direct submission to PIR International. These sequences are then assigned an accession number and are entered in both PIR-PSD for further merging, annotation, and classification and PIR Archive for retrieval of originally submitted/reported sequences. The accession numbers appear in two places, individually in the accession block nested within the reference blocks from which they are derived, and collectively in a single accession line after the date information in each entry. This annotated database employs a controlled vocabulary for most annotations and adopts standard nomenclature in an attempt to promote annotation quality and database inter-operability. PIR is partitioned into four sections as listed below:

- PIR1: Fully annotated and merged with other entries
- PIR2: Not all entries are merged, classified, and annotated
- PIR3 (temporary buffer): Not all entries are merged, classified, or annotated
- PIR4: Not naturally occurring or expressed sequences

PIR maintains several databases that belong to one of the four sections. The databases are mentioned below:

- PIR-PSD: The main protein sequence database
- iProClass: Classification of proteins according to their structure and function
- ASDB: **A**nnotation and **S**imilarity **D**ata**B**ase; each entry is linked to a list of similar sequences
- NRL_3D: A database of sequences and annotations of proteins of known structures deposited in the PDB
- ALN: A database of protein sequences
- RESID: A database of covalent protein structure modifications

(b) Sample PIR format

The PIR entry for the urokinase from Gallus gallus (NREF no. NF00046864) is shown in Fig. 1.6.

Nf00046864	iProClass View Submit Bibliography XML View	**Last Updated: 22-Mar-2004**
Protein	Urokinase-type plasminogen activator precursor (EC 3.4.21.73) (uPA) (U-plasminogen activator)	
Name	**Gallus gallus** (chicken) *NCBI Taxon ID*: 9031	
Taxonomy	Lineage: cellular organisms; Eukaryota; Fungi/Metazoa group; Metazoa; Eumetazoa; Bilateria; Coelomata; Deuterostomia; Chordata; Craniata; Vertebrata; Gnathostomata; Teleostomi; Euteleostomi; Sarcopterygii; Tetrapoda; Amniota; Sauropsida; Sauria; Archosauria; Aves; Neognathae; Galliformes; Phasianidae; Phasianinae; Gallus	
Source Organism	Gallus gallus (Taxon ID: 9031)	
Bibliography	▶View Bibliography information. ▶ Submit Bibliography. PubMed: PMID:2295632	

Sequence Database

Database	Protein ID	Accession	Taxon ID	Protein Name
PIR	A35005	A35005	9031	u-plasminogen activator (EC 3.4.21.73) precursor *ALT_NAMES*:uPA
SwissProt	UROK_CHICK	P15120	9031	Urokinase-type plasminogen activator precursor (EC 3.4.21.73) (uPA) (U-plasminogen activator)
GenPept	g212857	AAA49130.1	9031	plasminogen activator
GenPept	g212859	AAA49131.1	9031	plasminogen activator precursor (E.C.3.4.21.31)
RefSeq	g45382101	NP_990774	9031	plasminogen activator

Protein MKLIIFLTVTLCTLVTGLDSVYIRQYYKLSHKHRPQHRECQCLNGGTCITYRFFSQIKRC

Fig. 1.6 PIR entry for the urokinase from Gallus gallus

TASK

Step 1 From the Internet browser log onto PIR at Georgetown.

Step 2 In the **Text search protein databases** feed in the keywords rhodopsin and Haloarcula japonica strain TR-1.

Step 3 Observe the result page closely. PIR gives our designation for the sequence (NRef, IPro, Uniprot, and PIR-PSD). What do these numbers indicate?

Step 4 Prepare a report on the following:

(a) What is the protein ID and its accession number in the TrEMBL database?

(b) Give the sequence length of the rhodopsin protein from *Haloarcula japonica* strain TR-1.

(c) Write a note on cruxrhodopsin protein in Haloarcula and compare it with that of *Halobacterium salinarium* R1M1.

(d) Click on the **iProClass** designation to get a page on the 'Summary Report of ProClass Entry: Q9YGHB7 HALJP'.

(e) Retrieve the Nucleic Acids Symp Ser. 1997;(37):111-12 paper on PubMed.

(f) Go to the UniProt entry and get the dates of submission and annotation.

(g) Read about the database cross references and additional information provided on this page.

1.3.4 Protein Data Bank (http://www.rcsb.org/pdb/index.html)

The Protein Data Bank (PDB) is managed by the Research Collaboratory for Structural Bioinformatics (RCSB). It is a non-profit consortium of three organizations: Rutgers, The State University of New Jersey; the San Diego Supercomputer Center at the University of California, San Diego; and the National Institute of Standards and Technology. The European Bioinformatics Institute Macromolecular Structure Database group (UK) and the Institute for Protein Research at Osaka University (Japan) are the international participants in data deposition and processing. The RCSB also works closely with the EBI and the NCBI to improve the level and consistency of annotation for each structure.

(a) PDB entry

PDB is a database of the experimentally determined three-dimensional structures of biological macromolecules. The archive contains atomic coordinates, bibliographic citations, primary and secondary structure information, as well as experimental data from X-ray crystallography and NMR experiments. The database supports several data formats for representing structures, sequences, and graphical displays. Here we will deal with only sequences and consider structures in a later chapter. Protein sequences are currently available for display and can be downloaded in the FASTA format. A FASTA formatted file consists of a header line, beginning with a greater than (>) symbol followed by a sequence of one-letter codes.

(b) Sample PDB format

A typical PDB entry is as shown in Fig. 1.7. It will have both sequence and structure information. But in this chapter, we will look at the sequence portion only.

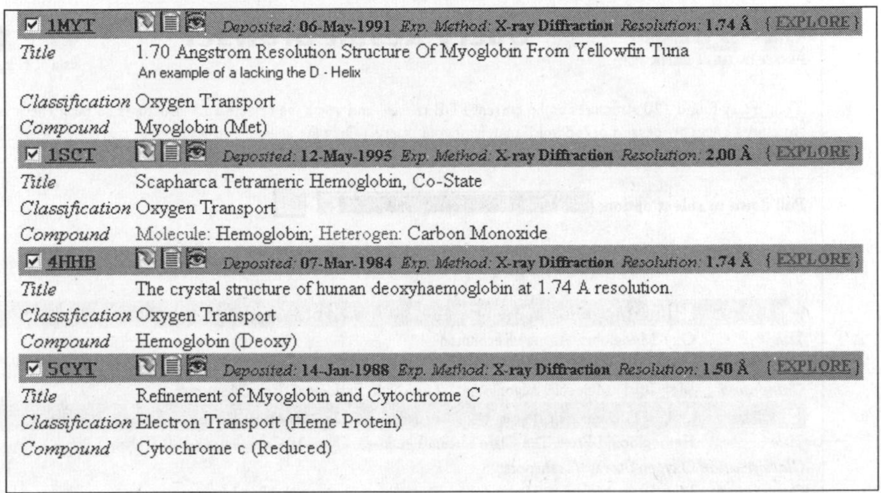

Fig. 1.7 Sequence and structure information in a typical PDB file

TASK

Step 1 To search from the PDB database, type this url http://www.rcsb.org/pdb/index.html in your internet browser. You will get a search page as shown in Fig. 1.8.

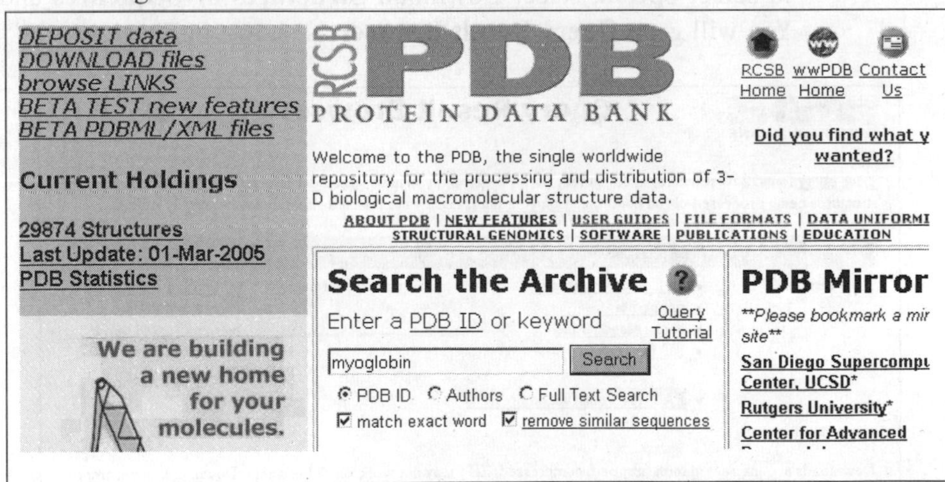

Fig. 1.8 Search and query page of PDB on http://www.rcsb.org/pdb/index.html

Step 2 Type myoglobin in the search box, select **PDB ID**, check **match exact words** and **remove similar sequences**, and click **Search**. You will get a result page as shown in Fig. 1.9.

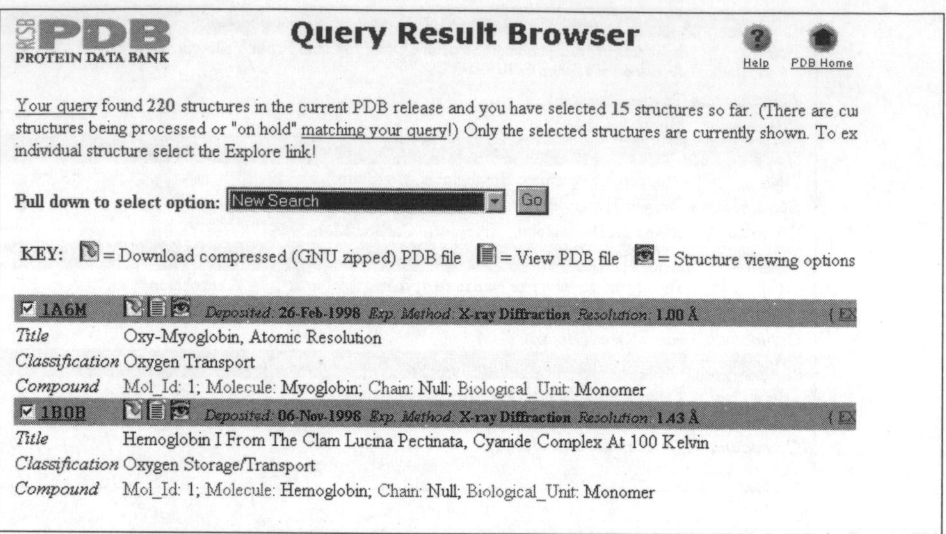

Fig. 1.9 Result page of PDB on http://www.rcsb.org/pdb/index.html for myoglobin

Step 3 Under **Pull down to select option**, select **Deselect All Selected Structures** instead of **New Search** and Click **Go**.

Step 4 Select **4HHB** by scrolling down and clicking the checkbox. Under **Pull down to select option**, select **Download Structures or Sequences** and click Go. You will get a **Query Result Browser** as shown in Fig. 1.10.

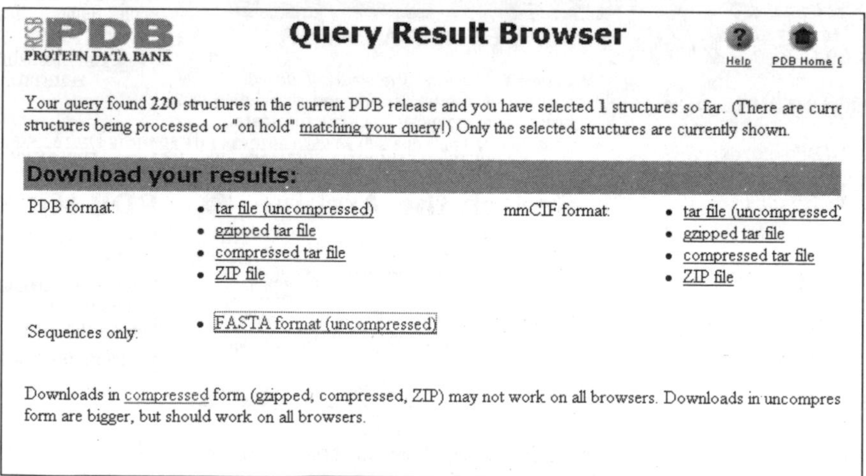

Fig. 1.10 Result page of PDB for myoglobin displaying various means of accessing the data

Step 5 Click FASTA format (uncompressed). You will get the result as shown in Fig. 1.11.

```
>4HHB:A HEMOGLOBIN (DEOXY) - CHAIN A
VLSPADKTNVKAAWGKVGAHAGEYGAEALERMFLSFPTTKTYFPHFDLSHGSAQVKGHGK
KVADALTNAVAHVDDMPNALSALSDLHAHKLRVDPVNFKLLSHCLLVTLAAHLPAEFTPA
VHASLDKFLASVSTVLTSKYR
>4HHB:B HEMOGLOBIN (DEOXY) - CHAIN B
VHLTPEEKSAVTALWGKVNVDEVGGEALGRLLVVYPWTQRFFESFGDLSTPDAVMGNPKV
KAHGKKVLGAFSDGLAHLDNLKGTFATLSELHCDKLHVDPENFRLLGNVLVCVLAHHFGK
EFTPPVQAAYQKVVAGVANALAHKYH
>4HHB:C HEMOGLOBIN (DEOXY) - CHAIN C
VLSPADKTNVKAAWGKVGAHAGEYGAEALERMFLSFPTTKTYFPHFDLSHGSAQVKGHGK
KVADALTNAVAHVDDMPNALSALSDLHAHKLRVDPVNFKLLSHCLLVTLAAHLPAEFTPA
VHASLDKFLASVSTVLTSKYR
>4HHB:D HEMOGLOBIN (DEOXY) - CHAIN D
VHLTPEEKSAVTALWGKVNVDEVGGEALGRLLVVYPWTQRFFESFGDLSTPDAVMGNPKV
KAHGKKVLGAFSDGLAHLDNLKGTFATLSELHCDKLHVDPENFRLLGNVLVCVLAHHFGK
EFTPPVQAAYQKVVAGVANALAHKYH
```

Fig. 1.11 Result in FASTA format

Step 6 You can also do the following after Step 2. Click **Explore** for the entry **4HHB**, as shown in Fig. 1.12.

☑ 1MYT *Deposited:* **06-May-1991** *Exp. Method:* **X-ray Diffraction** *Resolution:* **1.74 Å** { EXPLORE }
Title 1.70 Angstrom Resolution Structure Of Myoglobin From Yellowfin Tuna
An example of a lacking the D - Helix
Classification Oxygen Transport
Compound Myoglobin (Met)

☑ 1SCT *Deposited:* **12-May-1995** *Exp. Method:* **X-ray Diffraction** *Resolution:* **2.00 Å** { EXPLORE }
Title Scapharca Tetrameric Hemoglobin, Co-State
Classification Oxygen Transport
Compound Molecule: Hemoglobin; Heterogen: Carbon Monoxide

☑ 4HHB *Deposited:* **07-Mar-1984** *Exp. Method:* **X-ray Diffraction** *Resolution:* **1.74 Å** { EXPLORE }
Title The crystal structure of human deoxyhaemoglobin at 1.74 A resolution.
Classification Oxygen Transport
Compound Hemoglobin (Deoxy)

☑ 5CYT *Deposited:* **14-Jan-1988** *Exp. Method:* **X-ray Diffraction** *Resolution:* **1.50 Å** { EXPLORE }
Title Refinement of Myoglobin and Cytochrome C
Classification Electron Transport (Heme Protein)
Compound Cytochrome c (Reduced)

Fig. 1.12 Result page of PDB for myoglobin displaying various matches

You will get a result page as shown in Fig. 1.13.

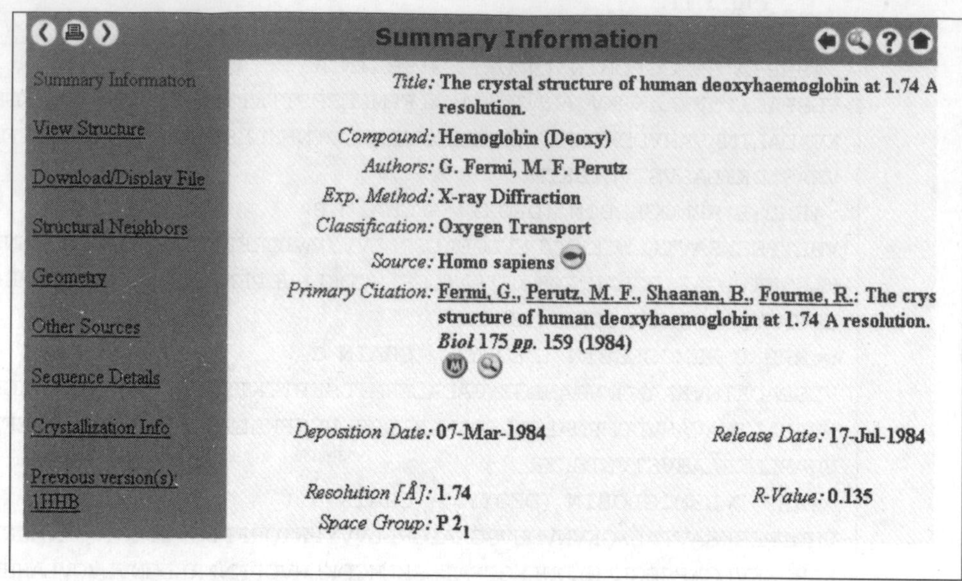

Fig. 1.13 Summary information page of PDB for myoglobin

Step 7 Click **Sequence Details**. Here sequences of a single chain or all chains can be downloaded in the FASTA format. The sequence information does not include covalent residue modifications. You will get the result as shown in Fig. 1.14.

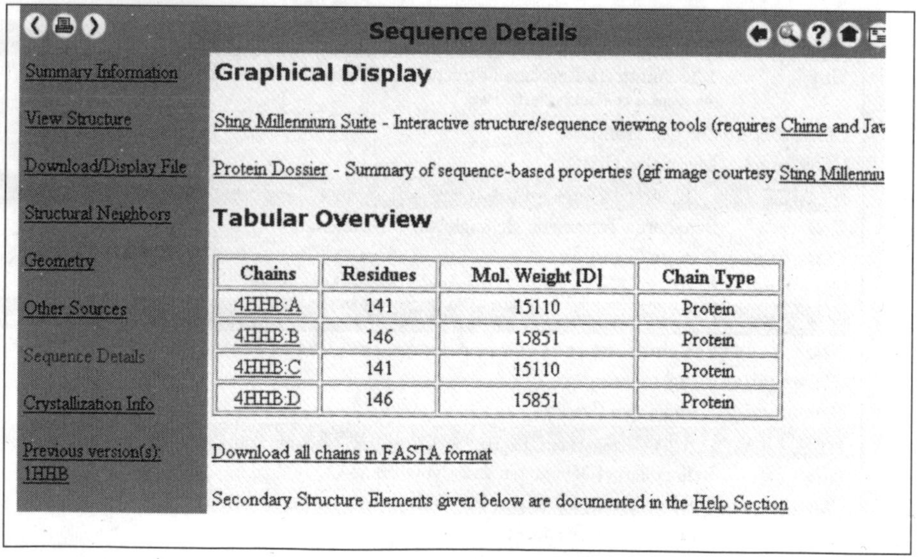

Sequence Details

Summary Information
View Structure
Download/Display File
Structural Neighbors
Geometry
Other Sources
Sequence Details
Crystallization Info
Previous version(s): 1HHB

Graphical Display

Sting Millennium Suite - Interactive structure/sequence viewing tools (requires Chime and Jav

Protein Dossier - Summary of sequence-based properties (gif image courtesy Sting Millenniu

Tabular Overview

Chains	Residues	Mol. Weight [D]	Chain Type
4HHB:A	141	15110	Protein
4HHB:B	146	15851	Protein
4HHB:C	141	15110	Protein
4HHB:D	146	15851	Protein

Download all chains in FASTA format

Secondary Structure Elements given below are documented in the Help Section

Fig. 1.14 Sequence details offering graphical, tabular, and FASTA format page of PDB for myoglobin

Step 8 To see the summary information, click **Summary Information**. You will get a result page as shown in Fig. 1.15.

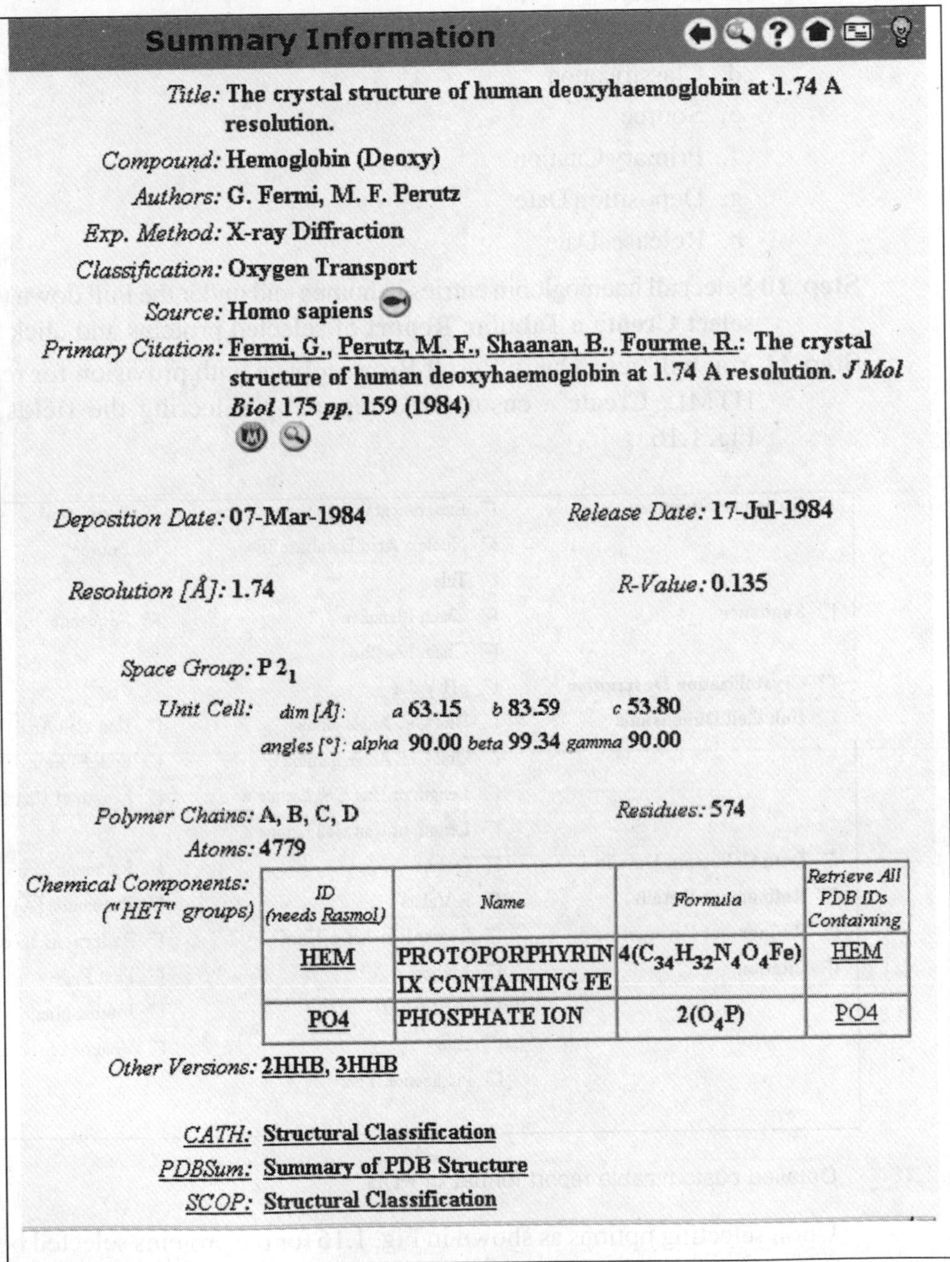

Fig. 1.15 Detailed summary information page of PDB for myoglobin

Step 9 Prepare a report on the explanation of summary information for the following:

a. Compound
b. Authors
c. Exp. Method
d. Classification
e. Source
f. Primary Citation
g. Deposition Date
h. Release Date

Step 10 Select all haemoglobin entries in human and under the Pull down to select option, select **Create a Tabular Report** of selected proteins and click **Go**.

Step 11 You will get a Query Result Browser page with provision for report in Text or HTML. Create a customized report by selecting the fields, as shown in Fig. 1.16.

☐ **Structure Summary**	☐ Experimental Technique	☐ Raw Data
	☑ Nucleic Acid Database ID	☐ Source
	☐ Title	
☐ **Sequence**	☑ Chain Identifier	☑ Sequence
	☑ Chain Length	
☐ **Crystallization Description**	☐ pH Value	
☐ **Unit Cell Dimensions**	☐ Unit Cell Angle alpha	☐ Unit Cell Angle beta
	☐ Unit Cell Angle gamma	☐ Space Group
	☐ Length of Unit Cell Lattice a	☐ Length of Unit Cell Lattice b
	☐ Length of Unit Cell Lattice c	
☐ **Data Collection Details**	☐ Device	☐ Radiation Type
☐ **Refinement Details**	☐ R Value	☐ Resolution [Å]
☐ **Refinement Parameters**	☐ Lower Resolution Limit	☐ Reflections used for Refinement
☐ **Citation**	☐ Authors	☐ First Page
	☐ PubMed ID	☐ Journal Name
	☐ Title	☐ Volume No
	☐ Publication Year	

Fig. 1.16 Detailed customizable report format of PDB

Upon selecting options as shown in Fig. 1.16 for the proteins selected before, you will get a result page as shown in Fig. 1.17.

'PDB ID', 'Nucleic Acid Database ID', 'Chain Identifier', 'Sequence', 'Chain Length'

'4HHB','PDB4HHB','A','141'

'4HHB','PDB4HHB','B','146'

'4HHB','PDB4HHB','C','141'

'4HHB','PDB4HHB','D','146'

Fig. 1.17 Result page in text format

Step 12 The ligand tabular report includes covalently or non-covalently bonded organic or inorganic molecules. They may also include single atoms or ions (e.g., Ca, Cl). In PDB files, all ligands are described in the HETATM records. A sample ligand report is shown in Fig. 1.18.

'PDB ID', 'Ligand name', 'Ligand formula', 'Ligand ID'

'4HHB', 'PHOSPHATE ION', '2(O4 P1 3-)', 'PO4'

'4HHB', 'PROTOPORPHYRIN IX CONTAINING FE', '4(C34 H32 N4 O4 FE1)',

'HEM'

Fig. 1.18 Result page of ligand report in text format

1.4 Secondary and Specialized Protein Sequence Databases

Secondary protein sequence databases have become reliable tools for working out relationships between novel sequences and thus getting some idea about their function. These databases have evolved by using signature-recognition methods to address different sequence analysis problems, resulting in rather different and independent databases. To perform a comprehensive analysis, a user therefore has to know attributes associated with these databases. Attributes such as their performance, family coverage, search outputs, access and links to other databases, etc., can make a large difference in the choice of database that a researcher may use for a specific purpose.

In addition to the secondary databases, there are many specialized protein sequence databases that are available today. These vary not only in their data types, but also in their size; some of them are quite small, while others are wider in their scope and larger in size. Some of specialized protein sequence databases will be described here in an attempt to familiarize you with them. The most commonly used secondary protein databases are PROSITE, PRINTS, and Pfam. Each of these is unique and is employed for special diagnostic and analytical purposes owing to the different strengths and weaknesses.

(a) GOA (http://www.ebi.ac.uk/GOA/index.html)

GO or Gene Ontology is an international consortium of scientists at the EBI. The goal of the consortium has been to form a dynamic yet controlled vocabulary that can be applied to all organisms. This is quite a challenge especially when data about a gene or a protein is still accumulating and changing. The Gene Ontology Consortium developed three separate sections to describe gene products: molecular function, biological process, and cellular component. These allow annotation of molecular characteristics across species. Each vocabulary is structured as directed acyclic graphs. Here any term may have more than one parent as well as zero or more children. This has been found to work better than the traditional hierarchical graph. Currently the GO vocabulary consists of more than 17,000 terms. The Gene Ontology Annotation (GOA) project is run in parallel with GO. Controlled vocabulary to a non-redundant set of proteins is described in the UniProt/Swiss-Prot and Ensembl databases and it collectively provides complete proteomes.

(b) MEROPS (http://merops.sanger.ac.uk/)

This database is a resource for catalogue and structure-based classification of peptidases. It contains a set of files, termed PepCards. Each of these cards provides information on a single peptidase. The information contains classification and nomenclature and gives hypertext links to the relevant entries in other databases. The peptidases are classified into families on the basis of statistically significant similarities between the protein sequences in the part termed the 'peptidase unit' which is most directly responsible for activity. Similarly, families with common evolutionary origin (FamCards) are expected to have similar tertiary folds and are grouped into clans (ClanCards). Each FamCard provides links to other databases for sequence motifs and secondary and tertiary structures showing the distribution of the family across the major taxonomic kingdoms.

(c) GPCRDb (http://www.gpcr.org/7tm/)

This database contains information not only on the sequences but also on other data relevant to G-protein coupled receptors (GPCRs). GPCRs is a very large protein family and generally forms a part of the signalling systems in animals. Much of the information is taken from EMBL. This database is primary put together with a view to ease the analysis (e.g., multiple alignments, classification into subfamilies). More information on this is available on http://www.gpcr.org/7tm/multali/multali.html

(d) YPD (http://www.proteome.com/YPDhome.html)

The yeast protein database (YPD) is a database for the proteins of *Saccharomyces cerevisiae*. YPD is the first annotated database for the complete proteome of any organism. YPD houses experimentally determined properties of each protein, e.g., molecular weight, isoelectric point, subcellular localization, post-translational modifications, and extensive annotations from the yeast literature. Annotations are drawn from approximately 3500 yeast papers and abstracts. More than 8700 yeast papers are cited in the reference lists. YPD complements the sequence databases GenBank, PIR-International, and Swiss-Prot. YPD interacts with the major genome databases for *S. cerevisiae*, namely, Martinsreid Institute for Protein Sequences (MIPS) and Saccharomyces Genome Database (SGD).

(e) ENZYME (http://au.expasy.org/enzyme/)

This database is an annotated extension of the Enzyme Commission (EC) and acts as a repository of information pertaining to the nomenclature of enzymes. It contains data for each type of characterized enzyme and its corresponding EC number, recommended name, alternative names (if any), catalytic activity, cofactors (if any), and links to Swiss-Prot protein sequence entrie(s). The data also links to the human disease(s) associated with a deficiency of the enzyme.

(f) CATH (http://cathwww.biochem.ucl.ac.uk)

The CATH database provides a hierarchical domain classification of protein structures in the Brookhaven Protein Databank. These cluster proteins are at four major levels: class (C), architecture (A), topology (T), and homologous superfamily (H). Only crystal structures that are solved to a resolution better than 3.0 Å together with NMR structures are kept. This databse contains multidomain proteins, which are further classified using a consensus procedure. The procedure is based on three independent algorithms for domain recognition DETECTIVE (Swindells 1995), PUU (Holm & Sander 1994), and DOMAK (Siddiqui & Barton 1995).

Class (C) Class is determined according to the secondary structure, composition, and packing within the structure.

Architecture (A) Architecture is used to describe the overall shape of the domain structure as determined by the orientations of the secondary structures.

Topology (T) Depending on the overall shape and connectivity of the secondary structures, proteins are grouped into fold families. This is done using the structure comparison algorithm SSAP. 7

Homologous superfamily (H) This level groups together protein domains that share a common ancestor and are homologous. Similarities are identified by sequence comparisons and are followed by structure comparison using SSAP.

(g) PROSITE (http://us.expasy.org/prosite/)

This database has entries of many protein families. The families are further segregated and grouped by sequence domains or motifs. Proteins or protein domains belonging to a particular family generally share functional attributes and are derived from a common ancestor. This forms the basis of classification.

While studying protein sequence families it is generally observed that some regions have been conserved during evolution. These regions are generally important for a protein to either perform a function and/or maintain its three-dimensional structure. Analysis of these and other domains within several groups of similar sequences has led to the derivation of signatures for a protein family or domain. This signature distinguishes its members from all other unrelated proteins. Such signatures can be used to assign a newly sequenced protein to a specific family of proteins and also predict its possible function. These biologically significant regions or residues that form a signature include

- enzyme catalytic sites
- prostethic group attachment sites (heme, pyridoxal-phosphate, biotin, etc.)
- amino acids involved in binding a metal ion
- cysteines involved in disulfide bonds
- regions involved in binding a molecule (ADP/ATP, GDP/GTP, calcium, DNA, etc.) or another protein

These patterns also facilitate computational tools to rapidly and reliably identify to which family of proteins the new sequence belongs. PROSITE also has several computational tools such as ScanProsite, MotifScan, InterProScan, ps_scan, pftools, and PRATT. You will learn about them in coming chapters.

(h) PRINTS (http://umber.sbs.man.ac.uk/dbbrowser/PRINTS/)

This database of proteins has a slightly different approach when compared to PROSITE. This approach to pattern recognition is termed as 'fingerprinting'. Usually within a sequence alignment it is possible to find more than one motifs that characterize the aligned family. These should be all used to develop a family signature. The advantage of this approach is that it gives a greater chance of identifying a distant relative, whether or not all parts of the signature are matched. This makes fingerprinting a very powerful diagnostic technique since it also has the ability to tolerate mismatches, both at the level of residues within individual motifs and at the level of motifs within the fingerprint as a whole. Release 36.0 of PRINTS contains 1800 entries, encoding 10,931 individual motifs. Overall, the database is still relatively small, largely because the detailed annotation of fingerprints is very time consuming. The database has two types of fingerprints, simple and composite. Depending on their complexity, simple fingerprints are mostly single motifs, while composite fingerprints encode multiple motifs.

The fingerprinting method used here is based on the fact that in any protein family, only parts of a sequence are common and these usually are key functional regions or core structural elements of the folds. Multiple sequence alignment (using SOMAP, XALIGN, or VISTAS manual alignment programs) is the first step in fingerprinting. The initial alignment is performed with only a few sequences, since the method itself is designed to add to the alignment with each database scan. Once a motif, or set of motifs, has been identified, the conserved regions are excised in the form of local alignments. These motifs may occur immediately adjacent to one another or may be separated by any distance along the length of the alignment. Several other database scans are made with each aligned motif, resulting in a set of hit lists, one for each motif. The hit lists are then analysed to determine which sequences in the database have matched with all elements of the fingerprint, and which have matched only with part of it. Only those sequences that match with all elements are regarded as true matches. This process is repeated until convergence, the point at which the true set remains constant between successive scans. The final aligned motifs from this iterative procedure constitute the refined fingerprint that is entered into the PRINTS database.

Difference between a motif and a fingerprint

A motif is usually defined as any conserved element of a sequence alignment. It is formed as a result of a local alignment corresponding to a region whose function or structure is known. The pattern is conserved, and hence can be used for prediction of such a structural/functional region in any other protein sequence. A fingerprint, on the other hand, is a set of motifs used to predict the occurrence of similar motifs, either in an individual sequence or in a database. Database searches with such aligned motifs are actually frequency scans.

(i) InterPro (http://www.ebi.ac.uk/interpro/)

While browsing through databases you will realize that these secondary databases do not use a common format and nomenclature. This acts as a major hindrance especially when the searches are to be automated. Therefore, the UniProt/Swiss-Prot group at the EBI has developed the integrated resource of protein domains and functional sites, more commonly known as Interpro. This database is an integration of the PROSITE, PRINTS, Pfam, and ProDom databases. InterPro is a searchable database, providing information on sequence function and annotation. It addresses problems and presents an integrated view of commonly used pattern databases and provides an intuitive interface for text- and sequence-based searches. It is maintained using a relational database system. The core entries are released in a single ASCII (text) flat file, which is written in XML. The overall data flow—from the individual data provider, through the DBMS, out to the flat file, and on to the user—is fairly complex.

The web interface allows text-based searches using SRS and sequence-based searches using the software provided by the consortium members. InterPro today also includes SMART, PIR SuperFamilies, and TIGRfams protein family databases.

(j) SWISS-2DPAGE (http://au.expasy.org/ch2d/)

SWISS-2DPAGE contains data on proteins identified on various 2DPAGE reference maps. It is possible to locate these proteins on the 2DPAGE maps or display the region of a

2DPAGE map where one might expect to find a protein from Swiss-Prot. Each SWISS-2DPAGE entry contains textual data on one protein, mapping procedures, physiological and pathological information, isoelectric point, molecular weight, amino acid composition, peptide masses, and bibliographical references. It is cross linked to other federated 2D databases such as COMPLUYEAST-2DPAGE, ECO2DBASE, HSC-2DPAGE, YPD) and to Swiss-Prot, which in turn is linked to other molecular databases (EMBL, Genbank, PROSITE, OMIM, etc). The format of protein entries in SWISS-2DPAGE is similar to that used in Swiss-Prot.

(k) Clusters of orthologous groups (COG)

(http://www.ncbi.nlm.nih.gov/Class/NAWBIS/Modules/Genomes2Other/genomes41.html)

This database is maintained at the NCBI and contains clusters of orthologous groups of proteins (COGs) that are formed by comparing protein sequences from complete genomes. The database has 44 complete genomes from 30 major phylogenetic lineages. Each COG consists of individual orthologous proteins or orthologous sets of paralogs from at least three lineages. Since orthologs typically have the same function even though they evolve it is easy to extrapolate information from one COG member to an entire COG. This relation automatically yields a number of functional predictions for poorly characterized genomes. The COGs has been constructed by applying the criterion of consistency of genome-specific best hits to the results of an exhaustive comparison of all protein sequences from these genomes. The COG database is accompanied by the COGNITOR program that is used to fit new proteins into the COGs and can be applied to functional and phylogenetic annotation of newly sequenced genomes.

Difference between an ortholog(ue) and a paralog(ue)

Orthologs are genes in different species that evolved from a common ancestral gene by speciation. Normally, orthologs retain the same function in the course of evolution. Identification of orthologs is critical for reliable prediction of gene function in newly sequenced genomes. *Paralogs* are genes related by duplication within a genome. Paralogs evolve new functions, even if these are related to the original one.

(l) HGVbase (http://hgvbase.cgb.ki.se)

This database was formerly known as HGBASE. Human Genome Variation Database (HGVbase) is stored locally in a MySQL relational database running on a digital UNIX server. It is a non-redundant database of genomic variation data of all types, mostly comprising single-nucleotide polymorphisms (SNPs or snips). The database has entries that include neutral polymorphisms as well as disease-related mutations. Online search tools and downloads are freely available in a variety of formats such as XML, Fasta, SRS, SQL, and tagged-text file formats. Each entry is presented in the context of its surrounding sequence and many records are related to neighbouring human genes and affected features therein.

Since SNPs are amenable to stability over generations, population allele frequencies are included. Records are assigned unique and permanent HGVbase identification numbers to facilitate cross-database referencing, publication, and similar use by the community. The database also contains information about SNP-related meetings, genotyping methods, SNP databases, and analysis software. Online search tools facilitate data interrogation of all information fields in HGVbase, comprising five major categories: (i) sequence location, (ii) predicted functional importance, (iii) validation status, (iv) allele frequencies in populations, and (v) data sources.

Why are SNPs important?

SNPs are DNA sequence variations that occur when a single nucleotide (A, T, C, or G) in the genome sequence is altered. For example, an SNP might change the DNA sequence AAGGCTAA to ATGGCTAA. For a variation to be considered an SNP, it must occur in at least 1% of the population. Ninety per cent of all human genetic variation occurs because of SNPs. This translates into every 100 to 300 bases for the three-billion-base human genome. Of these variations, two of every three SNPs involve the replacement of cytosine (C) with thymine (T). SNPs can occur in both coding (gene) and non-coding regions of the genome. Many SNPs have no effect on cell function, but scientists believe others could predispose people to disease or influence their response to a drug. This is why SNPs are of great importance to biomedical researchers and pharmaceutical diagnostics. SNPs are evolutionarily stable, so SNP maps will help in identifying the multiple genes associated with such complex diseases as cancer, diabetes, vascular disease, etc.

1.5 Information Retrieval System: Entrez

Entrez was developed by the National Centre for Biotechnology Information, a component of the United States National Library of Medicine. It is an integrated text-based search and retrieval system of databases for Nucleotide, Protein, Structure, Taxonomy, Genome, Expression, and Chemical. Each category has one or more databases that support specific purposes. Here we will look at the common ones and leave the readers to explore the rest either during web-based exercises or later when required as part of their project or job. Go to http://www.ncbi.nlm.nih.gov/Entrez/ to access the Global Query Cross-database Search System for further exploring.

1.5.1 Entrez Global Query Cross-database Search System

It is a multi-database search page of all databases that are under Entrez retrieval system. The search page is shown in Fig 1.19.

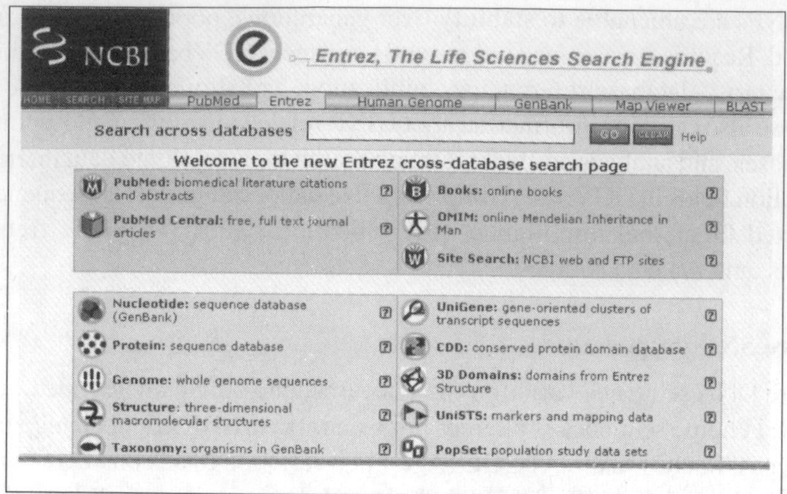

Fig. 1.19 Interface of Entrez search page

It allows users to search databases in two different ways:

1. The user can enter one or more search terms in the search page and click **Go**. The query runs simultaneously against a set of Entrez databases, or to be more precise it searches all the fields of all the tables of all the databases. The hits for each database are displayed on the results page along with each database name as shown in Fig. 1.20.

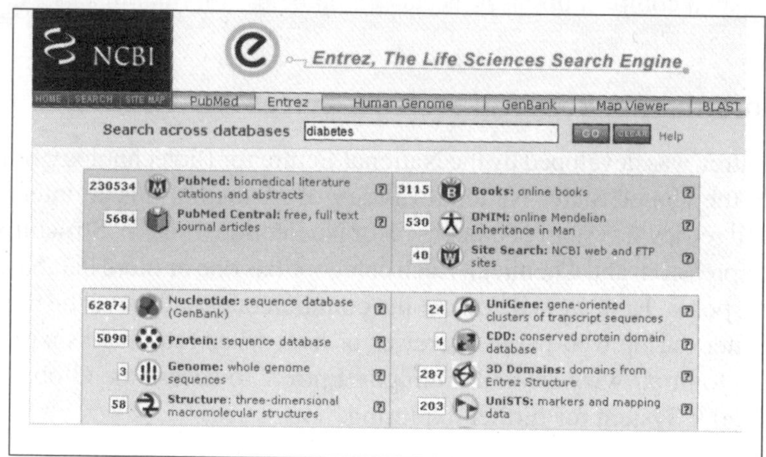

Fig. 1.20 Interface of Entrez hits for each database displaying the Results Page along with each database name

2. The user can click on any database to see the result of that particular database, as shown in Fig. 1.21.

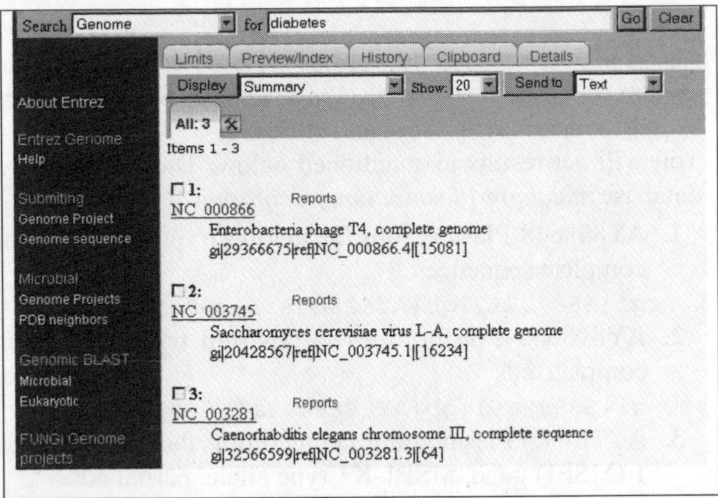

Fig. 1.21 Interface of NCBI hits displaying the results

The user can also go directly to an individual database's search page from the Entrez search page, as shown in Fig. 1.22. In this case, instead of entering a keyword in the search box you need to click on any database of your choice.

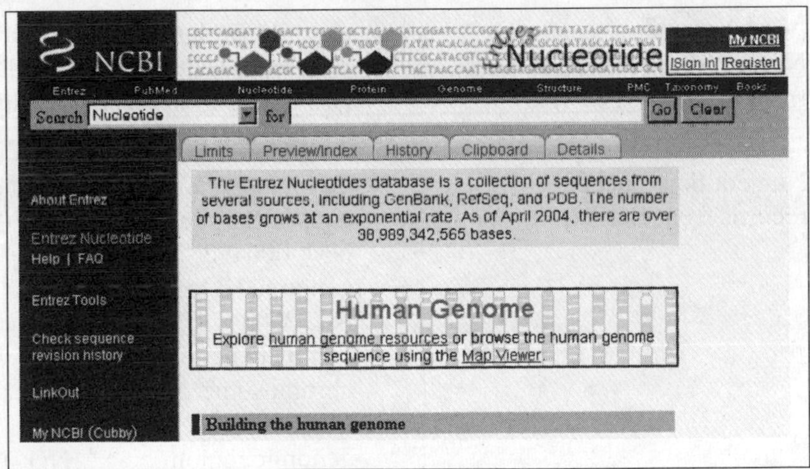

Fig. 1.22 Interface of NCBI that allows user to select a database through its dropdown box

1.5.2 Information Storage System in a Nucleotide Database

In Part 2, *Databases*, you learnt about the storage system of a database. Let us learn about the organization of nucleotide database under NCBI.

TASK

Step 1 Let us browse through the various sections of the Entrez information retrieval system for DNA sequence databases. Select **Nucleotide**, enter the search term as plasmodium falciparum, and click **Go**.

You will get results as mentioned below. These are entries of the nucleotide database *matching Plasmodium falciparum*

1. AY898648 Plasmodium falciparum tryptophan-rich antigen pseudogene, complete sequence
`gi|58892717|gb|AY898648.1|[58892717]`
2. AY898647 Plasmodium falciparum tryptophan-rich antigen 3 gene, complete cds
`gi|58892691|gb|AY898647.1|[58892691]`
3. AY714586 Plasmodium falciparum isolate 404 merozoite surface protein 1 (MSP1) gene, MSP1-K1-type allele, partial cds
`gi|56805411|gb|AY714586.1|[56805411]`

You will note that each entry has an accession number followed by the description of the sequence (genome sequence/complete sequence/complete cds/partial cds). The entry also provides information about sequence identification numbers, e.g., GI number, GB number, and emb.

More about accession number

Type of Record	Sample Accession Format
• Nucleotide sequence records from GenBank/EMBL/DDBJ	• One letter followed by five digits, or two letters followed by six digits, e.g. U12345, AY123456
• Nucleotide sequence records from RefSeq	• Two letters, an underscore and six digits, e.g., mRNA records (NM_*): NM_000492 genomic DNA contigs (NT_*) : NT_000347 complete genome or chromosome (NC_*): NT_000907 genomic region (NG_*): NG000019
• RefSeq model (predicted)	• Two letters (XM, XP, or XR), an underscore, and six digits,

Step 2 Let us select one entry from the result page shown in Step 1. For example, the following entry is selected

1. AY898648 *Plasmodium falciparum* tryptophan-rich antigen pseudogene, complete sequence

`gi|58892717|gb|AY898648.1|[58892717]`

You will get result in the GenBank format that you have learnt earlier. You can also select other display options like XML, ASN.1, FASTA, etc.

Step 3 Go to the FASTA format, locate the GI number and find out the corresponding tag in the XML format. You will find `<Seq-id_gi>58892717</Seq-id_gi>`. Similarly, locate the same information in the ASN.1 format. The tag associated with the GI number in the ASN.1 format is as shown in Fig. 1.23.

```
Seq-entry ::= seq {
  id {
    genbank {
      accession "AY898648" ,
        version 1 },
    gi 58892717 } ,............
```

Fig. 1.23 ASN.1 format

Step 4 Find out tags associated with the sequence data in XML and compare your result with the format given in Fig. 1.24.

```
<Seq-data_iupacna>
<IUPACna>ATGGAATTGAATACTTATAACAATATTAATTCTACAGTTAAAATATTTCA
AACTACATCAACTAGTGCATTAAATCATGACAAACAAATAGTGGAGAAAA
TAAATATGACCATGTTGTCCAACTTTTTATATATTAACTATTATTATTTTAAAATATGTG
TTTAAAAGTGTGAATAAAGTAAGATAAAATGAAATGTAATAAAAAATATATATGTGT
ATTTATTATT
ATTTTGTACATATATATATATATATATAT..................................
CTTGAAGAATTATATGATATAGAGCATAATGGAAAGCTAGCTGTTTTGTTTGAATATTAA</
IUPACna>
</Seq-data_iupacna>
```

Fig. 1.24 Sequence data XML format

Step 5 Notice the difference in the sequence data in ASN.1. The whole four-letter sequence is stored in a hexadecimal form. Also note that the 2543 bp sequence is reduced to 1274 hexadecimal numbers as shown in Fig. 1.25.

```
........................
seq-data
ncbi2na
'3A0F831F30433C3DC4BC033F4071341CB93C0D384040CBA200C0CE853BED41F
FF333C1CF3CFF00CEEFC02EE0C0B08C0380EC3000CCCEECFCF3CFFB13333333333333BBB1333FF7
FFB3C3BFD50EF0330FFCA33C200A29A000009CC281C033BC31C382F8E73280828E30EB83190301
308E7144000E0463F40CC422F20DF1383EFC080C81303B0235000218D33000C28828200E003F0B
E0E0813E6482002000301303 92D30200A00A3003023CF302104F0230232E30200232630232E302
B02B02802326302B02302326302B32326302B32326302B02326303B02302326303B02304302302
326303B02304302302326302002100B0230213210A00000200840200 2030F71082B1A080063C32
0028B2203F434A0E00030083A3A0ACC8C05C38E33D40FCF2C12F0C080F8AC5D2A00BD083F820A0
C83EFE0828B00F1F0280C4B3E8208680E030380C80FBFF8005033C78E08002820000 1272C803E8
F2E130300EBE31A03D300F3F480C8018C38213B44F4F208030603F380CDC9423B8070400B0A78C
30CED20B0C40C0E2007CC3423B83B03343803F030C3D03303FBF043F0B00310FDD08D7F13053F0
B803603DE0823F0300BF0B9339030690A3B0300E8E84C070333381B00F024E83C3A0000C48C483
A810FCE73E80080000203D32B0003808FA820EBF504333CC3E9C9F3BC307E8208FF3938C2EA0FF
AF03E800208FA802F8C8F78BA14040E80C03D020FCE2049200CBF0040043F01BE8B33A0370C0E8
3AFC0E8E8000883A0FF82B8F03BA303433D67B100E0040F80E9083CD4C42080D83AF23EAC02002
02F33870C043E3E0EA113A020800201EBF800CCF0FC3C20A3EBC8D30D4E800FA02C4083C03BF41
44602E081F820F338CC8930E809C9EFFBF833C0'H } ,
........................
```

Fig. 1.25 Sequence data in ASN.1 format

1.5.3 Information Retrieval from a Database

Referring back to Step 1, the search term should correspond to the data matching with the data of any of the attributes of the database mentioned. To fine-tune the search you can use multiple keywords corresponding to multiple attributes of the database. Now let us explore a database and limit the search to a specific field. The **Limit page** of nucleotide database is as shown in Fig. 1.26.

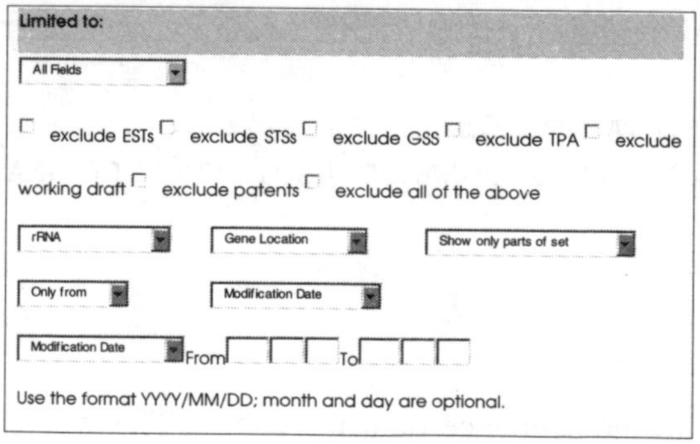

Limited to:
All Fields
exclude ESTs exclude STSs exclude GSS exclude TPA exclude working draft exclude patents exclude all of the above
rRNA | Gene Location | Show only parts of set
Only from | Modification Date
Modification Date From To
Use the format YYYY/MM/DD; month and day are optional.

Fig. 1.26 Limit page

As the contents of each Entrez database differ, the limits available for each database will also differ.

1. For the nucleotide database the search limit can be restricted to one of 'All Fields' of the database or a combination of more than one field of the same database. Through the format of the nucleotide database has many fields as mentioned before, you can put a limit on the fields mentioned below to refine the search: Accession, Author, EC/RN number, feature key, filter, gene name, issue, journal, keyword, modification date, organism, page number, primary accession, properties, protein name, publication date, SEQID string, sequence name, substance name, textword, title, volume
2. As shown in the above **Limited to** page, you can exclude ESTs, STSs, GSS, TPA, working draft, patents, or exclude all of these.
3. You can also select options from the following pull-down menu:
   - Genomic DNA/RNA, mRNA, rRNA from **Molecule**.
   - Genomic DNA/RNA, mitochondrion, chloroplast from **Gene Location.**
   - Show only master of the set, Show only parts of the set from **Segmented Sequence**.
   - RefSeq, GenBank, EMBL, DDBJ, PDB from Only from.
   - 30 days–10 years from Modification Date.
   - Specify **From** and **To** dates for **Modification Date** or **Publication Date**.

TASK

Let us take the example of database 'Nucleotide' and limit the search for organism 'homo sapiens'. Also restrict the search to Field: Organism, Publication Date from 2005/02/1 to 2005/02/05, check **exclude all of the above, Genomic DNA/RNA from Molecule and Gene Location, GenBank, Show only master of set** as in the search page shown in Fig. 1.27.

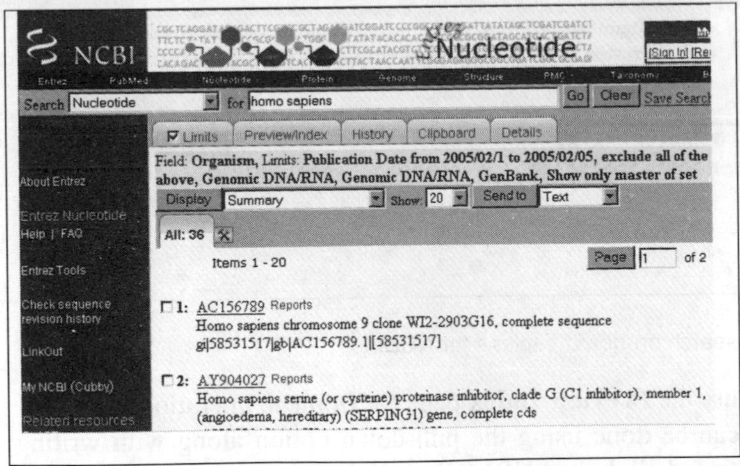

Fig. 1.27 Search page for NCBI for 'homo sapiens'

(a) Limit your search

Let us search the Nucleotide database to find out the common sequences between Homo sapiens and mouse. What all we need to do is the following:

Step 1 Select **Nucleotide** under **Search.**

Step 2 Select Limits and within that, select **Organism** under **Limited to**, check boxes for **exclude TPA** and **exclude Patents**, select **Genomic DNA/RNA** under **Gene Location**, select **show only master of set** under **Segmented Sequence**, **GenBank** under **only from.**

(b) Preview/Index

Step 3 Select **Preview/Index** and within that, select **Organism** under **All Fields**, type homo sapiens in the textbox and click the **AND** button, then type mouse in the textbox and click **Preview**. The preview/index page is displayed as shown in Fig. 1.28 and will list 16 entries.

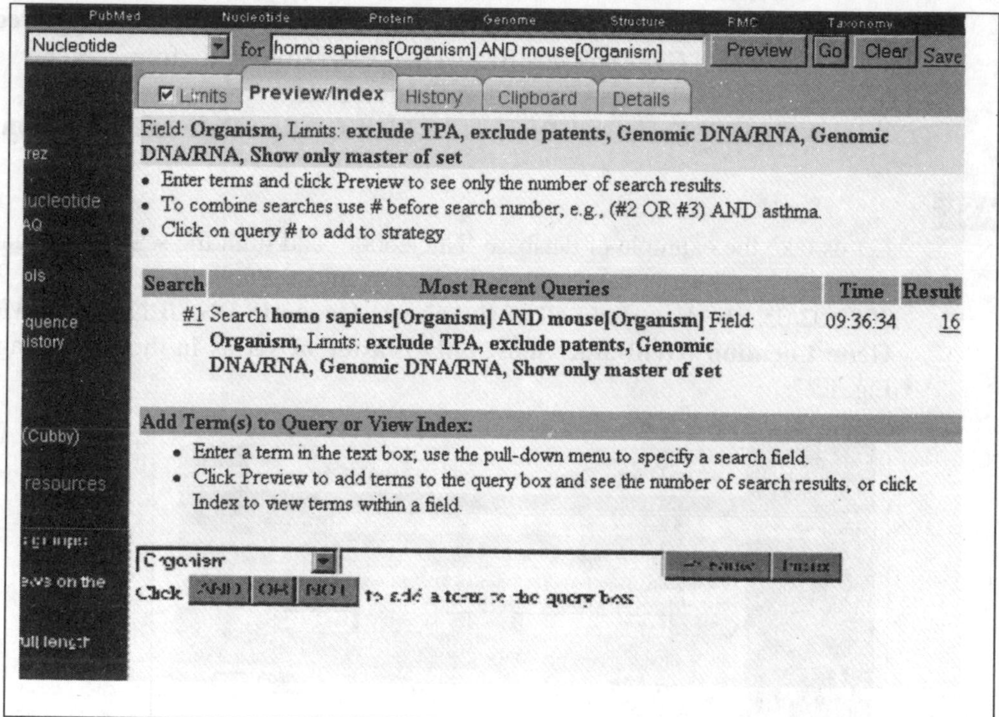

Fig. 1.28 Refining the search on homo sapiens for NCBI

- You can use the **Preview/Index** page for Boolean operations:
  - This can be done using the pull-down option along with writing text in the textbox and clicking **AND**, **OR**, or **NOT** buttons.
  - You can see the index under each search term by clicking the **Index** button.

  - You can see the result by clicking the **Preview** button.
- You can also process a complex search statement directly:
  - This can be done by writing and executing a complex search statement directly in the query box.
  - Use this syntax to perform a search: term [field] OPERATOR term [field].
  - Where term(s) is the search term, the field(s) is the Search Fields and qualifiers and the OPERATOR(s) is the Boolean Operators.

(c) History

All your search queries are stored even after eight hours of inactivity. The queries can be viewed in the History option. Search queries are numbered as #1, #2, and so on. You can operate on these queries by clicking on their numbers. You can also combine queries by adding search query numbers like #1 AND #3.

(d) Clipboard

When you shoot a query and click **Go**, Entrez displays the view where accession numbers, reports, and links are hyperlinked. For your analysis you may need to view the result in text format or save it to the clipboard or file for your later use. This can be done by selecting **Summary** in **Display** option, setting a number ranging from 5 to 500 for selecting the number of entries per page, and finally selecting **Clipboard** under **Send to** option. Clipboard will hold a maximum of 500 items. Like **History**, **Clipboard** items will be lost after eight hours of inactivity. Like you can save your result in clipboard, you can also save your result in a file where you need to provide the file name of your choice.

(e) Detail

To see the Entrez translation of Boolean search click the **Details** button and check your result with Fig. 1.29.

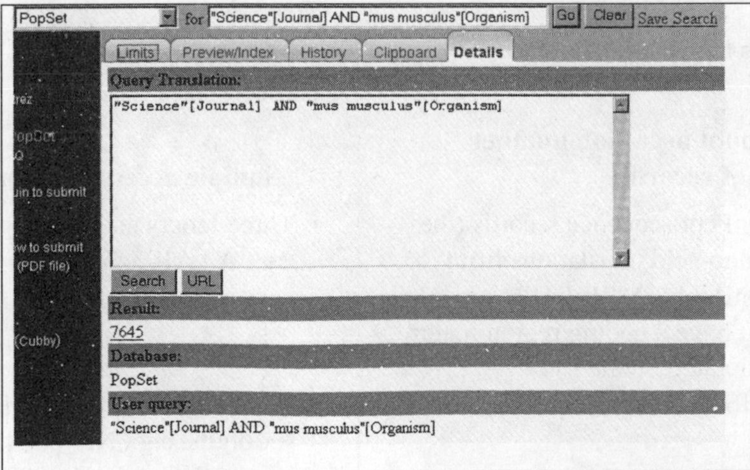

Fig. 1.29 Refining a search using Boolean operators

1.5.4 Information Storage System in Protein Database

You have already learnt how to access data from the Nucleotide database using Entrez as well as look at the different data divisions.

TASK

Step 1 Let us browse various sections of the Entrez information retrieval system for amino acid sequence databases. Click **Protein**, enter the search term as 'plasmodium falciparum', and click **Go**. You will get results as shown in Fig. 1.30 giving entries of the protein database matching *Plasmodium falciparum*. Each entry will have the accession number followed by the description of the amino acid sequence. It will also provide sequence identification numbers, e.g., gi number, gb/emb/dbj/sp number.

1: CAF25316
NBP2b protein [Plasmodium falciparum]
gi|56409776|emb|CAF25316.1|[56409776]

2: CAF25315
NBP2b protein [Plasmodium falciparum]
gi|56409774|emb|CAF25315.1|[56409774]

3: BAD83636
valyl tRNA synthetase [Plasmodium falciparum]
gi|57157213|dbj|BAD83636.1|[57157213]

Fig. 1.30 Entries from protein database

More about accession number

Type of record	Sample accession format
• GenPept sequence records (the amino acid translations from GenBank/EMBL/DDBJ records that have a coding region feature annotated on them)	• Three letters and five digits, e.g., AAA12345
• RefSeq protein sequence records	• Two letters (NP), an underscore and six digits, e.g., NP_000483 genomic region (NG_*): NG000019

Step 2 When one of the entries is selected, you will get the result in the GenPept format similar to the GenBank format. You can also select other display options such as XML, ASN.1, FASTA, etc.

Step 3 Go to the XML format and locate the tag for the following information. Also observe the differences in the display and storage formats. You can see a part of a sample entry in the XML format as shown in Fig. 1.31.

LOCUS CAF25316: <Textseq-id_accession>CAF25316</Textseq-id_accession>
AUTHORS Cortes,A:

```
<Author>
        <Author_name>
              <Person-id>
                    <Person-id_name>
                          <Name-std>
<Name-std_last>
                                Cortes
                          </Name-std_last>
                          <Name-std_initials>
                                A.
                                </Name-std_initials>
                          </Name-std>
                                </Person-id_name>
                    </Person-id>
              </Author_name>
</Author>
```

JOURNAL Submitted (07-FEB-2004):

```
              <Date-std>
                    <Date-std_year>2004</Date-
std_year>
                    <Date-std_month>2</Date-std_month>
                          <Date-std_day>7</Date-std_day>
              </Date-std>
```

Fig. 1.31 Result in XML format

1.5.5 Information Retrieval from Protein Database

Referring back to Step 1, the search term should correspond to the data matching with the data of any of the attributes of the database mentioned above. To fine-tune the search, you can use multiple keywords corresponding to multiple attributes of the database.

(a) Limit your search

Now let us explore a database and limit the search for a specific field. The limit page of the Protein database is the same as that of the Nucleotide database. However, the Protein database will provide you with only two excluded fields. They are TPA and patents.

TASK

Let us search the Protein database to find out the common proteins between *Homo sapiens* and mouse.

Step 1 Select **Protein** under **Search**.

Step 2 Select **Limits** and within that, select **Organism** under **Limited to**, check boxes for **exclude TPA** and **exclude Patents**, select **Genomic DNA/RNA** under **Gene Location**, select **show only master of set** under **Segmented Sequence**, **Swiss-Prot** under **only from**.

Step 3 Select **Preview/Index** and within that select **Organism** under **All Fields**, type homo sapiens in the textbox and click the **AND** button, then type mouse in the textbox and click **Preview**. The Preview/Index page is displayed as shown in Fig. 1.32.

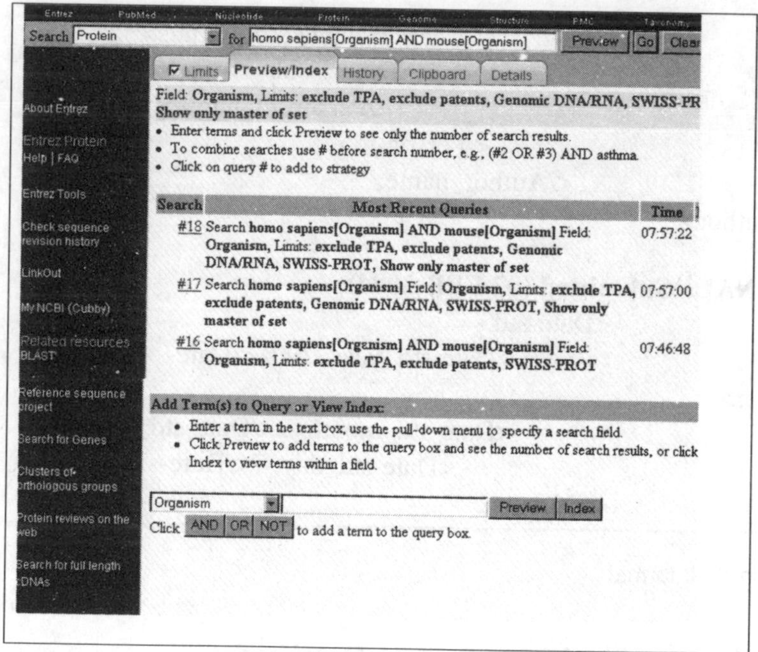

Fig. 1.32 Preview/Index page of NCBI

Step 4 Try **Preview/Index**, **History**, **Clipboard**, and **Detail** options by clicking each of them. You will find them similar to that in the Nucleotide database.

1.6 Information Retrieval System: SRS

By now you have an idea about Entrez, which is an information retrieval system. What do you expect to learn more from SRS, which is another information retrieval system? Yes, to start with, you need to know the full form of SRS and the url with the help of which you can access the system. Before learning to use SRS, why not give a try to retrieve information about SRS itself. You have search engines at your fingertips if you are connected to the Internet. Let us learn this system with the help of a 'Task'. In between you will get a handholding whenever it is required. It will be worthwile to do the following:

- Look at the functionalities of SRS that you are already familiar with in Entrez.
- Look at the specialties in SRS that are not there in Entrez.
- Try to find out the shortcomings in SRS, if any, that are not there in Entrez.

TASK

Step 1 Locate the link to SRS using the search feature in Google. Find out the full form of SRS.

Step 2 The top most portion of the SRS page is the Navigation Bar that contains the following tabs:

Quick Searches, **Select Databanks**, **Query Form**, **Tools**, **Results**, **Projects**, **Custom Views**, **Information**

Start Page and **Quick Search** pages are virtually identical. The only difference is that the SRS start page provides an option to start a permanent project, whereas the SRS quick search page gives the information about the current session of a temporary project. Both the pages provide two kinds of searches:

- Quick text search as shown in Fig. 1.33.

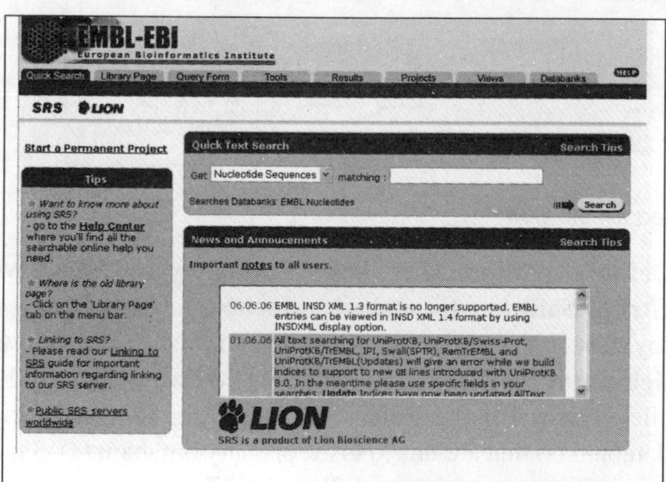

Fig. 1.33 SRS with quick text search

- List search as shown in the Fig. 1.34.

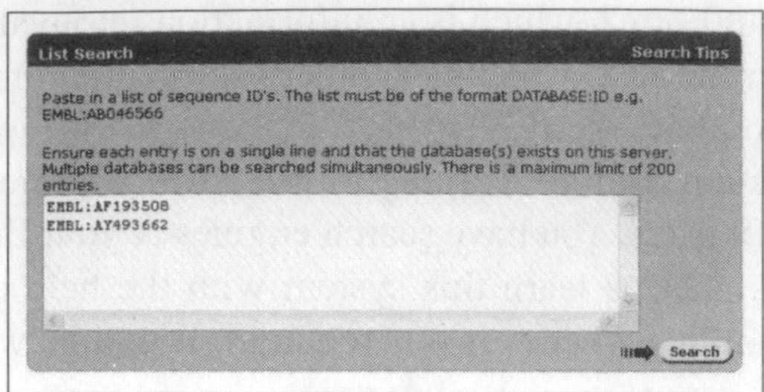

Fig. 1.34 SRS with list search

Step 3 Perform a **Quick Text Search**. This search will perform searches in respective databanks based on the selection of the drop-down get menu.

(a) Get **Nucleotide Sequences** matching **Tuberculosis**. Try **Quick Text Search** for all the options of the **Get** menu and note down the databank that is looked at under each option.

(b) Select few sequences, apply **Options to Selected Results only** and view results using **EMBLSeqSimpleView**. Show 5 results per page in a printer-friendly view. Check your result with Fig. 1.35.

EMBL EMBL (Coding Sequences) EMBL (Contig)	Primary Accession (Links to SVA)	Accession List	Description	Sequence Length
☐ EMBL:DD235846	DD235846	DD235846	Method Of Screening For Gamma-Glutamylcysteine Synthetase Inhibitors.	1299
☐ EMBL:DD250120	DD250120	DD250120	A Process for Determining Mycobacterium tuberculosis.	195
☐ EMBL:DD250121	DD250121	DD250121	A Process for Determining Mycobacterium tuberculosis.	195
☐ EMBL:DD250110	DD250110	DD250110	A Process for Determining Mycobacterium tuberculosis.	20
☐ EMBL:DD250111	DD250111	DD250111	A Process for Determining Mycobacterium tuberculosis.	20

Fig. 1.35 EMBLSeq simple view

Step 4 Define a view for EMBL, link to Taxonomy as well as OMIM (you may try with **Ctrl** or **Shift** key for multiple selections). Create a new view with view name **myview** so that the result is displayed in **list mode**. Show fields from **common fields only**. Select the datafields that you want displayed in your view by using checkboxes for the fields corresponding to your primary databank(s) and databank(s) that are linked to the primary databank(s). Save your view. Compare your view page with Figs 1.36 and 1.37.

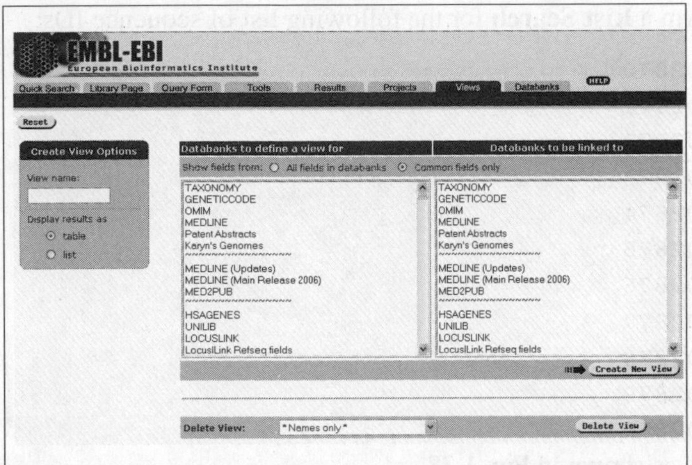

Fig. 1.36 View page 1

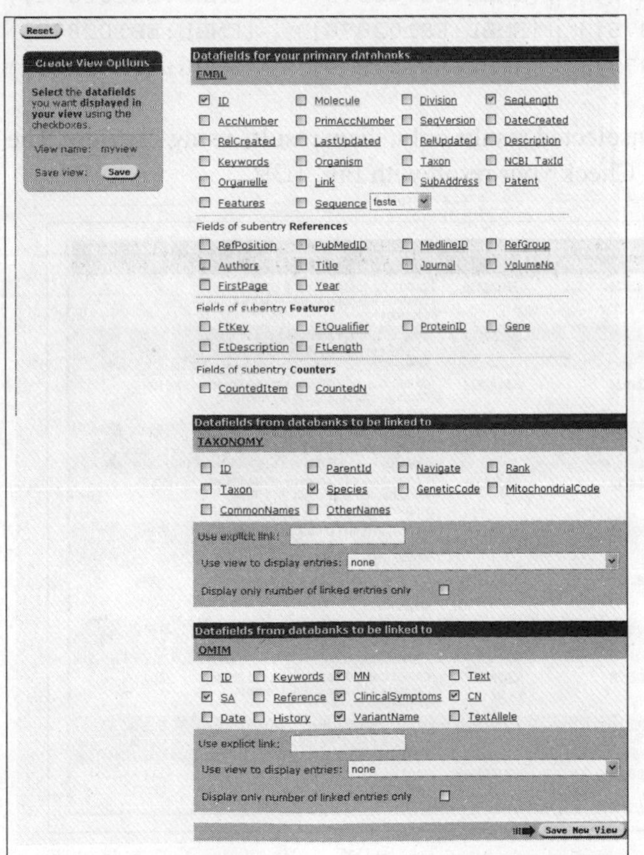

Fig. 1.37 View page 2

Step 5 Perform a **List Search** for the following list of sequence IDs:

```
EMBL:EB102870
EMBL:EB102871
EMBL:EB102872
EMBL:EB102873
EMBL:EB102874
EMBL:EB102875
EMBL:EB102876
EMBL:EB102877
EMBL:EB102878
EMBL:EB102879
EMBL:EB102880
```

The result is as shown in Fig. 1.38.

`Query '((((((((((([EMBL:EB102870] | [EMBL:EB102871]) | [EMBL:EB102872]) | [EMBL:EB102873]) | [EMBL:EB102874]) | [EMBL:EB102875]) | [EMBL:EB102876]) | [EMBL:EB102877]) | [EMBL:EB102878]) | [EMBL:EB102879]) | [EMBL:EB102880])'` found 11 entries

Apply options to unselected results only; view results using **myview**. The view need not be printer friendly. Check your result with Fig. 1.39.

EMBL	Primary Accession (Links to SVA)	Accession List	Description	Sequence Length
☑ EMBL:EB102870	EB102870	EB102870	CA001 Candida albicans cDNA library Candida albicans cDNA 3', mRNA sequence.	427
☐ EMBL:EB102871	EB102871	EB102871	CA002 Candida albicans cDNA library Candida albicans cDNA 3', mRNA sequence.	159
☐ EMBL:EB102872	EB102872	EB102872	CA003 Candida albicans cDNA library Candida albicans cDNA 3', mRNA sequence.	199
☐ EMBL:EB102873	EB102873	EB102873	CA004 Candida albicans cDNA library Candida albicans cDNA 3', mRNA sequence.	92
☑ EMBL:EB102874	EB102874	EB102874	CA005 Candida albicans cDNA library Candida albicans cDNA 3', mRNA sequence.	252
☑ EMBL:EB102875	EB102875	EB102875	CA006 Candida albicans cDNA library Candida albicans cDNA 3', mRNA sequence.	374
☑ EMBL:EB102876	EB102876	EB102876	CA007 Candida albicans cDNA library Candida albicans cDNA 3', mRNA sequence.	476
☐ EMBL:EB102877	EB102877	EB102877	CA008 Candida albicans cDNA library Candida albicans cDNA 3', mRNA sequence.	116
☑ EMBL:EB102878	EB102878	EB102878	CA009 Candida albicans cDNA library Candida albicans cDNA 3', mRNA sequence.	281
☑ EMBL:EB102879	EB102879	EB102879	CA010 Candida albicans cDNA library Candida albicans cDNA 3', mRNA sequence.	213
☐ EMBL:EB102880	EB102880	EB102880	CA011 Candida albicans cDNA library Candida albicans cDNA 3', mRNA sequence.	118

Fig. 1.38 Result page of list search

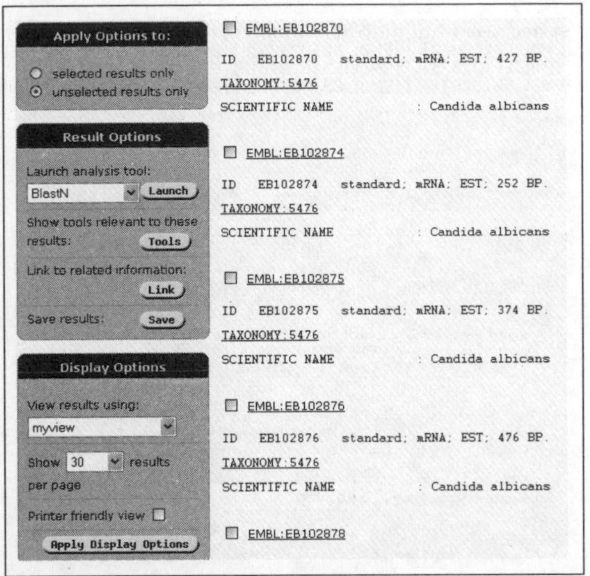

Fig. 1.39 Result page of own created view

Step 6 Result

(a) Check your **Result History**, select queries of importance, give comments as in Fig. 1.40.

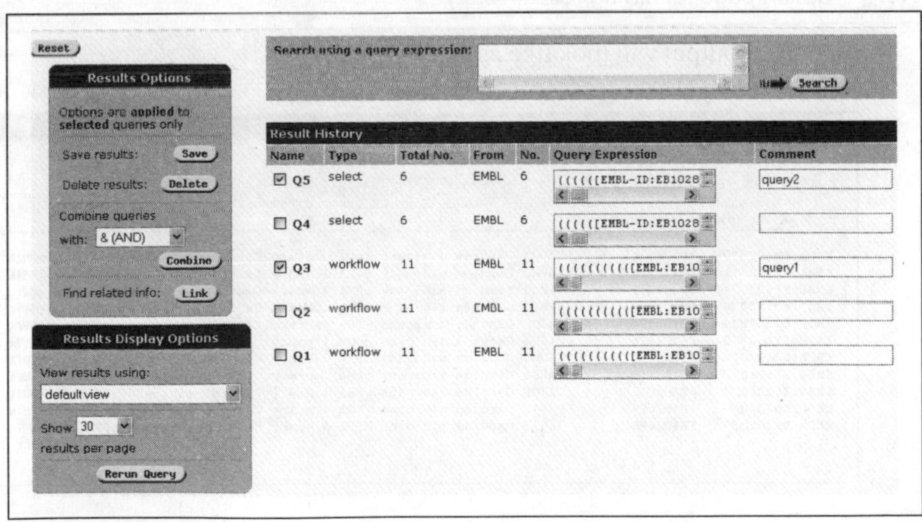

Fig. 1.40 Result history page

(b) Save your result as ASCII text/table with **Output to Browser Window/ HTML** as shown in Fig. 1.41.

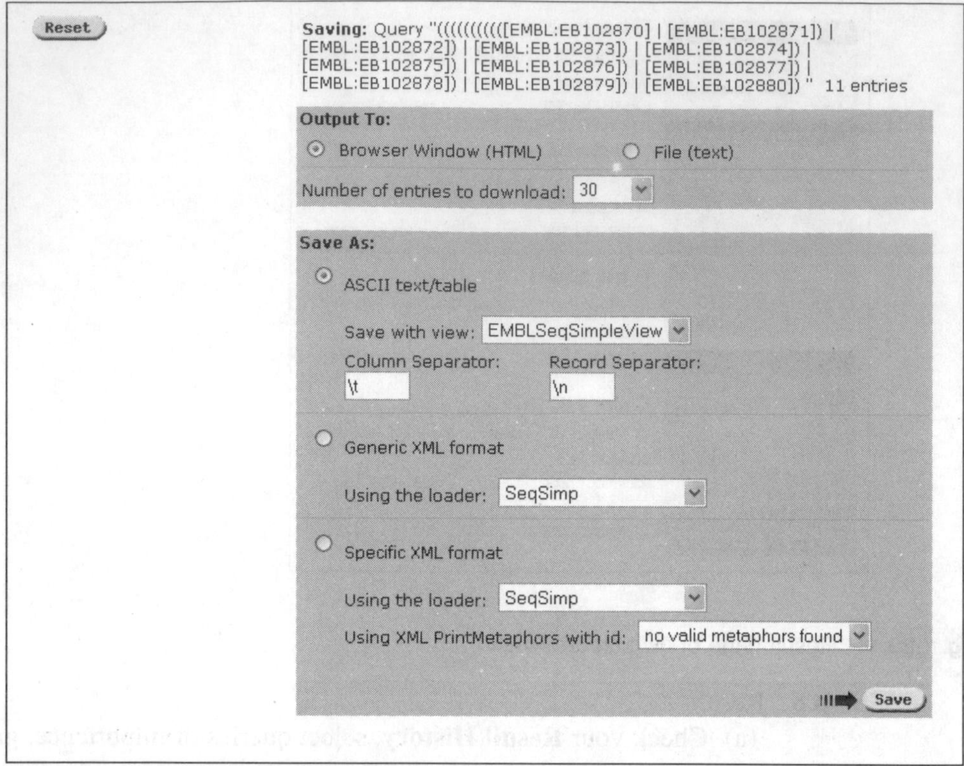

Fig. 1.41 Option page for saving result history

(c) Your output will look like as given in Fig. 1.42.

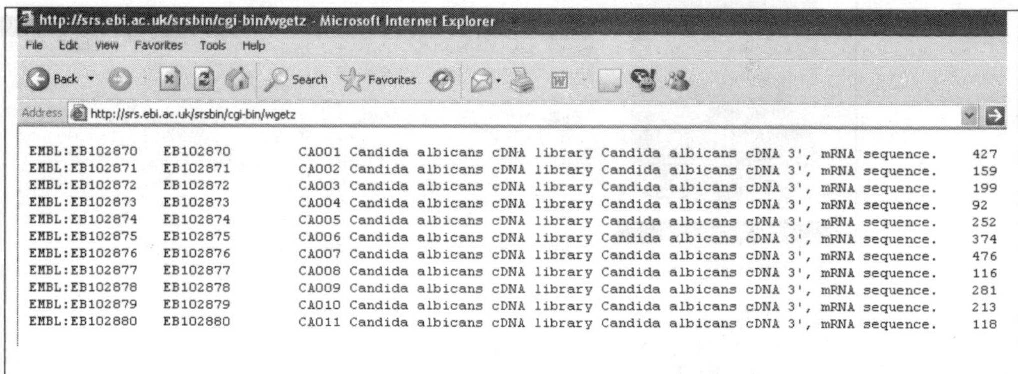

Fig. 1.42 Output to browser window (HTML)

(d) You may also save your result as a file with the default file name.

Step 7 Project

(a) Look at the contents of your temporary project as in Fig. 1.43.

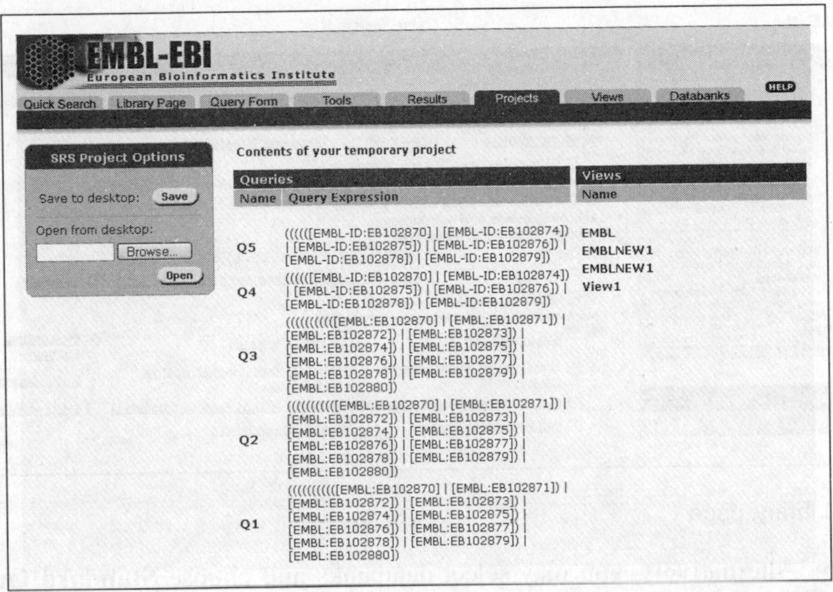

Fig. 1.43 Project page

(b) Save the project in the default name using **SRS Project Option**. You can also browse to open a project file from your local hard disk which was saved earlier.

Permanent project

- Keep track of the work performed during the current project.
- Save a project to a file.
- Share a project with others (by saving it).
- Open a previously saved project.
- See a list of queries or views in this or another project.
- See the current login/account name.
- See the name of the current project.
- Rename a project.
- Delete a project.
- Create/start a new project.
- Change to a different project.

Step 8 Click the **Databanks** tab to get information about various databanks.

Step 9 Library page

(a) Check the databanks of your choice for performing a **Quick Search**. The library page can be compared with Fig. 1.44.

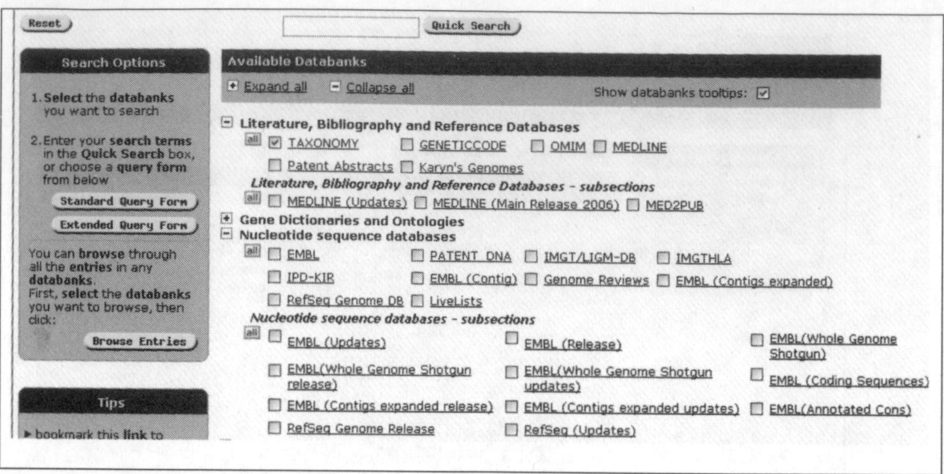

Fig. 1.44 Library page

(b) Alternatively, you may select databanks and choose **Standard Query Form** or **Extended Query Form**.

(c) You already got exposed to SRS Query Page while learning DDBJ. You may try a couple of queries for *Candida albicans*.

(d) Can you perform a canned query on your own?

Conclusion

From the genes themselves to the structural or functional properties of the predicted proteins, most of the biological features discovered through sequencing projects are inferred by the computational analysis of sequences. All these results have a great impact on the biologists worldwide who are working in this domain, offering them new data (and tools), but with few clues for enabling them to fully exploit these. In order to evaluate the biological relevance of the available methods and data as well as to make the relevant choices with respect to particular biological problems, an understanding of what is inside the programs and databases is needed. The computational representation and analysis of biological sequence data are based on theoretical (mathematical or biological) models and make use of algorithms, leading to methods whose efficiency, accuracy, sensitivity, and specificity may vary greatly from one program to another depending on their parameters.

EXERCISES

Exercise 1.1 A media company wishes to develop a database that will keep track of information of all the TV serials and programmes that have been aired by it. It wants to track all the actors that have taken part in the serials, the serial directors, their anchors, and choreographers. The lists of the queries that the database must be able to answer are the following:

- Is the same actor part of more than one serials?
- Has the same director directed more than one serial?
- What are the ratings of the serials?
- How many times has the same serial got aired in the last five years?
- What commercials are most fit to be aired along with theses serials?

Develop a model/design for the required database. Assign attributes and entities. Enlist which attributes can serve as primary keys to the database.

Exercise 1.2 Consider the information given in the box that follows. It pertains to the sample entry from GenBank. List all those attributes that you think are useful for:

- experimental measurements
- updating and maintaining the database
- conducting a search
- literature indexing
- crosslinking to other databases

```
LOCUS       LISOD      756 bp    DNA      linear   BCT 30-JUN-1993
DEFINITION  Listeria ivanovii sod gene for superoxide dismutase.
ACCESSION   X64011 S78972
VERSION     X64011.1  GI:44010
KEYWORDS    sod gene; superoxide dismutase.
SOURCE      Listeria ivanovii
  ORGANISM  Listeria ivanovii
            Bacteria; Firmicutes; Bacillales; Listeriaceae; Listeria.
REFERENCE   1  (bases 1 to 756)
  AUTHORS   Haas,A. and Goebel,W.
  TITLE     Cloning of a superoxide dismutase gene from Listeria ivanovii by
            functional complementation in Escherichia coli and characterization
            of the gene product
  JOURNAL   Mol. Gen. Genet. 231 (2), 313-322 (1992)
  MEDLINE   92140371
REFERENCE   2  (bases 1 to 756)
  AUTHORS   Kreft,J.
  TITLE     Direct Submission
  JOURNAL   Submitted (21-APR-1992) J. Kreft, Institut f. Mikrobiologie,
            Universitaet Wuerzburg, Biozentrum Am Hubland, 8700 Wuerzburg,
```

```
FRG
  FEATURES              Location/Qualifiers
     source             1..756
                     /organism='Listeria ivanovii'
                     /strain='ATCC 19119'
                     /db_xref='taxon:1638'
                     /mol_type='genomic DNA'
     RBS                 95..100
                      /gene='sod'
     gene                 95..746
                      /gene='sod'
     CDS                 109..717
                      /gene='sod'
                     /EC_number='1.15.1.1'
                      /codon_start=1
                      /transl_table=11
                     /product='superoxide dismutase'
                     /db_xref='GI:44011'
                     /protein_id='CAA45406.1'
                     /db_xref='Swiss-Prot:P28763'
\translation='MTYELPKLPYTYDALEPNFDKETMEIHYTKHHNIYVTKLNEAVS
GHAELASKPGEELVANLDSVPEEIRGAVRNHGGGHANHTLFWSSLSPNGGGAPTGNLK
AAIESEFGTFDEFKEKFNAAAAARFGSGWAWLVVNNGKLEIVSTANQDSPLSEGKTPV
                  LGLDVWEHAYYLKFQNRRPEYIDTFWNVINWDERNKRFDAAK'
   terminator          723..746
                     /gene='sod'
ORIGIN
      1 cgttatttaa ggtgttacat agttctatgg aaatagggtc tataccttc gccttacaat
     61 gtaatttctt ..........
//
```

Exercise 1.3 Go to Entrez. Glean data about 'Human Christmas Factor' from the six different databases that can be accessed from:

- PubMed—A public version of Medline presented by NCBI, NLM.
- Protein—Protein sequence
- Nucleotide—DNA/mRNA sequence
- Structure—Information about three-dimensional structure, mainly proteins.

- Genomes—A database focused on all organisms where we know the complete genome sequence. Physical maps.
- Taxonomy—Classification of organisms. Search an organism and find its closest relatives
- OMIM-Online Mendelian inheritance in man

Exercise 1.4 Go to the NCBI site and enter one of the keywords or phrases from the following list:

- adhesin
- tubulin
- poliovirus
- MHC
- ribulose bisphosphate carboxylase

Use the **All Fields** pull-down menu to specify a field. Use Boolean operators **AND, OR, NOT** to refine your search. Note that Boolean operators must be in uppercase.

Exercise 1.5 Go to the GenBank and enter one of the keywords or phrases from Exercise 1.4. Find out the accession number of the keyword of your interest. Retrieve the sequence of your interest. Choose which type of display you want, and review the type of information available. Display and save the report of your interest in the FASTA format.

Exercise 1.6 Again using any of the keywords given in Exercise 1.4, try and retrieve a protein sequence matching a nucleotide sequence and vice versa. Save MMDB and PDB accession numbers of the proteins as well.

Exercise 1.7 Use the Sequence Retrieval System (SRS) to obtain the entry in the EMBL database for the sequence of the *Amoeba histolytica*. Generate a map of unique restriction sites.

Exercise 1.8 Open PubMed database and look for information on **blood and screening**. How many items are there? Choose the type of blood in the search box and choose five blood donors. Click the **Send to Search** box with **AND** operator. To retrieve information on screening, choose mandatory testing and click the **Send to Search** box again using the **AND** operator. Limit your search and look at the options.

Exercise 1.9 Copy the following sequence:

aaaactgcga ctgcgcggcg tgagctcgct gagacttcct ggacgggga caggctgtgg

Paste this information into the query sequence window of GenBank. Select the primate and a human database and search. Note the E value of your search and justify the match to the reported sequences.

Exercise 1.10 Compare the PIR and Swiss-Prot entries, after their retreival, for the tyrosine kinase I.

Exercise 1.11 Use ExPaSy (http://www.expasy.ch/sprot/) to obtain the sequence of mouse rhodopsin kinase. Find out the following information:

- What is its molecular mass?
- What residues make up the catalytic domain?
- It has a C-terminal modification: What is it?
- Use ProtScale to determine its amino acid composition.

Exercise 1.12 Get the PDB files for the two proteins from Exercise 1.4. State their PDB file names. Look at the protein structures for the two proteins. Identify, in both structures, those residues that align with a gap in the other protein and those residues between which a gap is found, i.e., if this was your alignment: XYYXZXZ or YXZXZ. Then identify in the structure the residues shown in bold type. State the residue names and numbers of these residues, as found in the PDB structure, and state what type of the secondary structure these residues are found in.

Exercise 1.13 Use the PDB mirror at the EBI to retrieve the structure of the actin–myosin complex. View it with Chime. Note: Search PDB with OCA. Use the controls on the PDB Chime page to view the different representations of the protein. Click on the ligand bound into the cleft in actin to find out what it is.

Exercise 1.14 You are given a protein accession number P78504.
(a) Search for its chromosome location and the the gene coding for it.
(b) Does the gene have a GDB nomenclature committee approved symbol? If so, what is it?
(c) What are its other family members?
(d) Does the protein have any specific domains?
(e) Is there a disease associated with the gene?
(***Hint:*** Start the search in SRS or Entrez with the accession number and select Swiss-Prot and SPTrEMBL sequence libraries).

Exercise 1.15 Go to Entrez and search with the keyword all 'photosystem' related sequences in the Nucleotide database (use wildcard "*"). Do the following:
(a) Count the number of spinach sequences in the nucleotide, protein, structure, and genome databanks?
(b) Display the FASTA view of the Protein entry AAD02267.
(c) Display the graphics view of the Nucleotide entry corresponding to it.
(d) Search for all spinach proteins with a molecular weight range from 50,000 to 50,050 daltons (use the field range format '050000:050050[MOLWT]').

Exercise 1.16 Perform a database search using the following partial amino acid sequence of a protein from *Saccharomyces cerevisiae*: VAENIIQ HATHNST.
(a) Determine what is the likely function of this protein.
(b) Determine what is the predicted isolectric point (pI) .
(c) Determine what is the chromosome and gene location.
(d) Which genes are located upstream and downstream?

(e) Does this protein have a sequence motif that belongs to a certain protein group (family)? If so then what is its protein group and sequence motif.
(***Hint:*** Use ExPaSy to conduct the search)

Exercise 1.17 Retrieve a PDB file on cyclin-dependent protein kinase CDK6. Use the following link: http://bip.weizmann.ac.il/oca-bin/ocamain/. The results page contains the summary of the PDB entry in its upper part and various links with additional information about the protein in the lower part. Go through the page and determine the following for **1JOW**:

(a) When was this structure deposited to PDB (**date** field)?
(b) What is a primary reference for this structure?
(c) What method was this structure determined by?
(d) What is the resolution of the structure?
(e) How many polypeptide chains appear in the file (**compounds** field)?
(f) In order to save the file locally in your computer, go to the **Data retrieval** section and save the complete structure in your computer.

Exercise 1.18 Open the file that you have downloaded and saved on your computer in Exercise 17(f). Open the file in any text editor and take a look:

(a) What organism was this kinase extracted from (**SOURCE** records)?
(b) What types of secondary structures does it contain (**HELIX**, **SHEET** records)?
(c) Look at the coordinate part (**ATOM** records). Can you identify x, y, z coordinates for every atom?
(d) How many chains are there in the crystal?
(e) What can you say about the mobility of these residues when compared to other residues in the PDB file?

Exercise 1.19 Go to NCBI's site at http://www.ncbi.nlm.nih.gov/structure. Use this tool to search for 'Protein Kinase'. Do you get similar results as in Exercise 1.17? Perform three independent searches and use **History** to combine them [((#1 AND #2) NOT #3)]

Exercise 1.20 Troponin C and calmodulin are two Ca^{2+} binding proteins (P02585 and P02593). Go to http://www.expasy.ch/tools/sim-prot.html to determine whether their sequences are related.

2 Structure Databases

Learning Objectives

- To describe the various kinds of structural databases
- To learn about the organization of these databases
- To list the attributes of each of these databases
- To use any information retrieval system to obtain data
- To analyse structural data

Introduction

The journey from a string of amino acids to a functionally relevant protein is not easy, and predicting its structure and function is next to impossible. The problem calls for the prediction of the tertiary structure of a protein from its amino acid sequence. This is highly desirable for the following reasons.

- The structure of a protein can be directly correlated to its function as a biological macromolecule.
- Structure prediction using experimental techniques such as X-ray crystallography are time consuming and expensive in comparison to the structure determination from a primary sequence.
- Many proteins such as membrane proteins do not lend themselves to the structure solution.

Figure 2.1 illustrates the protein prediction process from a primary sequence. But before we get into the realm of structure prediction, let us learn about the databases that house useful data on amino acid sequences. These databases are very useful for protein structure prediction. The Protein Data Bank (PDB) is the main primary database used toady for 3D structures of biological macromolecules. SCOP (Structural Classification of

Proteins) is another database that classifies protein 3D structures in a hierarchical scheme of structural classes. Its data is based on the primary database PDB, but it holds additional information on analysis, organization, and classification of structures into folds, families, and super families. It is regularly updated when new entries are deposited in the PDB. CATH (class, architecture, topology, homologous superfamily) is another common structure database. Like SCOP, this utilizes a hierarchical classification of protein domain structures. These domains are further clustered at four major structural levels. MMDB is a structure database associated with the NCBI website. In this chapter we will look at each of these databases.

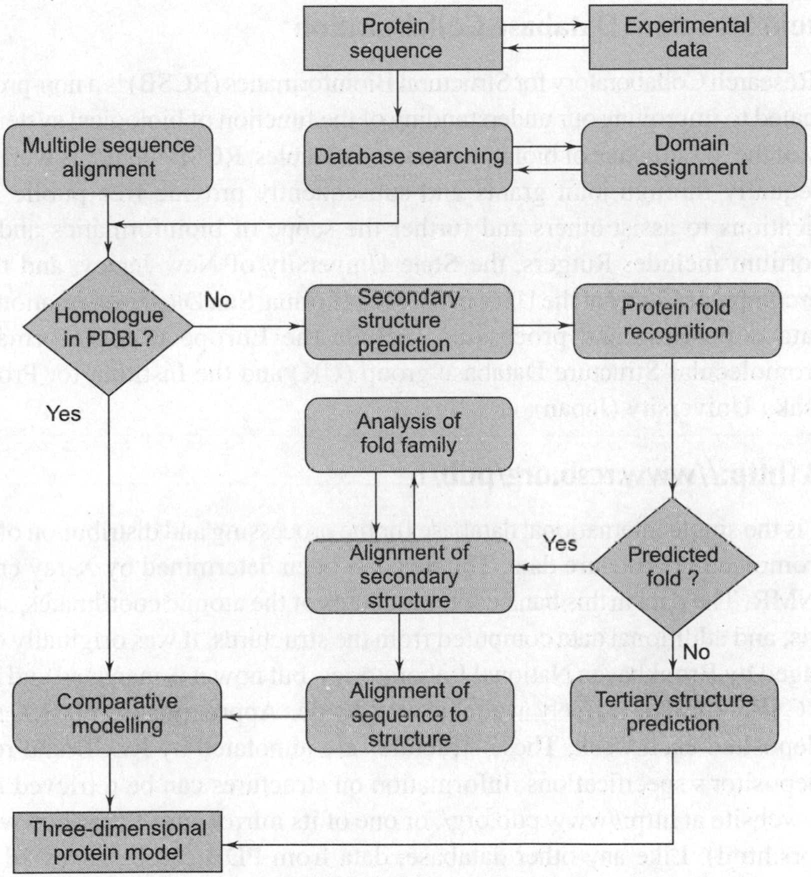

Fig. 2.1 Protein prediction process from primary sequence

2.1 Structure File Formats

The structure database records contain the analysis and interpretation of practical data derived from X-ray crystallography and nuclear magnetic resonance (NMR) spectroscopy and not from non-experimental computational three-dimensional modelling procedures.

There are two basic approaches to store bond data. The first approach is the *chemistry rule approach* which is the basis of the PDB format. It does not require a residue dictionary and lacks complete information of bonds. It just maintains a table of bond lengths and bond types for every conceivable pair of bonded atoms. Thus different software interprets the bonding differently, depending on the algorithm, distance tolerance, and the method of exception handling. The second approach is the *explicit bonding approach* used in MMDB. It records explicit bonding information in the form of a residue dictionary. It also stores information related to bonding rule exceptions. Here any software developed to read MMDB data would interpret the data consistently.

2.2 Protein Structure Database Collaboration

The Research Collaboratory for Structural Bioinformatics (RCSB) is a non-profit consortium dedicated to improving our understanding of the function of biological systems through the study of the 3D structure of biological macromolecules. RCSB members work cooperatively and equally through joint grants and subsequently provide free public resources and publications to assist others and further the scope of bioinformatics and biology. The consortium includes Rutgers, the State University of New Jersey, and the San Diego Supercomputer Center at the University of California, San Diego. International participants in data deposition and processing include the European Bioinformatics Institute Macromolecular Structure Database group (UK) and the Institute for Protein Research at Osaka University (Japan).

2.3 PDB (http://www.rcsb.org/pdb/)

PDB is the single international database for the processing and distribution of 3D biological macromolecular structure data. The data has been determined by X-ray crystallography and NMR. The data in this bank contains details of the atomic coordinates, some structural details, and additional data computed from the structures. It was originally developed and managed by Brookhaven National Laboratories, but now it is managed and maintained by the RCSB. The data is freely available worldwide. Approximately 50–100 new structures are deposited each week. These structures are annotated by RCSB and released as per the depositor's specifications. Information on structures can be retrieved from the main PDB website at http://www.pdb.org/, or one of its mirror sites (http://www.rcsb.org/pdb/mirrors.html). Like any other database, data from PDB can be accessed using several search methods (search methods will be discussed later in the chapter). You can also fetch the structure files from the main FTP site at ftp://ftp.rcsb.org/ or from one of its mirrors. The PDB database is a historical archive, so its contents are not uniform and the search may produce incomplete query results. The issue of data management practices is being addressed through a data uniformity project.

- The PDB archive contains macromolecular structure data on proteins, nucleic acids, protein–nucleic acid complexes, and viruses.

- A variety of information associated with each structure is available, including sequence details, atomic coordinates, crystallization conditions, 3D structure neighbours computed using various methods, derived geometric data, structure factors, 3D images, and a variety of links to other resources.

2.3.1 PDB Entry

There are two query engines available on the RCSB home page to look at PDB entries. One of them is SearchLite, which is used for text-based search across the database. The second is SearchFields, which is used to search specific fields within the database. Both the query engines report summary information of the structure as shown in Fig. 2.2.

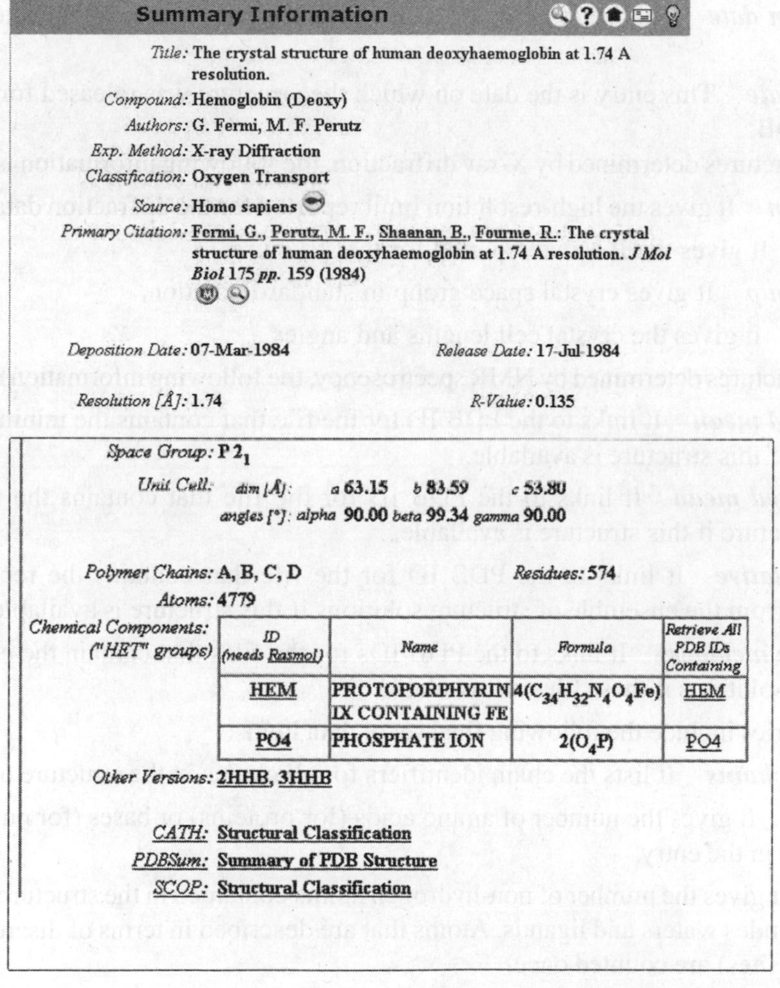

Fig. 2.2 Summary information of a PDB search

Description of entries

Compound This entry may contain one or more fields specifying the type of protein.

Authors This entry contains the names of the authors responsible for the deposition.

Exp. method This entry is the experimental method that was used to determine the structure.

Classification This entry provides a description of the molecule according to its biological function.

Source This entry specifies the biological and/or chemical source of the molecule.

Primary citation This entry provides the primary journal references to the structure and includes a link to Medline.

Deposition date This entry is the date on which the structure was deposited with the PDB.

Release date This entry is the date on which the structure was released for public use by the PDB.

For structures determined by X-ray diffraction, the following information is provided:

Resolution It gives the high-resolution limit reported for the diffraction data.

R-Value It gives the R-value reported for the structure.

Space group It gives crystal space group in standard notation.

Unit cell It gives the crystal cell lengths and angles.

For structures determined by NMR spectroscopy, the following information is provided:

Minimized mean It links to the PDB ID for the file that contains the minimized mean structure if this structure is available.

Regularized mean It links to the PDB ID for the file that contains the regularized mean structure if this structure is available.

Representative It links to the PDB ID for the file that contains the representative structure from the ensemble of structure solutions if this structure is available.

Ensemble members It links to the PDB IDs for the files that contain the ensemble of structure solutions if these files are available.

All entries include the following final set of data items:

Polymer chains It lists the chain identifiers for all chains in the structure entry.

Residues It gives the number of amino acids (for proteins) or bases (for nucleic acids) contained in the entry.

Atoms It gives the number of non-hydrogen atoms contained in the structure entry. This count includes waters and ligands. Atoms that are described in terms of discrete disorder (multiple sites) are counted once.

Chemical component (HET groups) It lists the three-letter codes that identify chemical components (typically, bound ions and ligands) in the structure entry. The chemical component ID has no special significance. The chemical names are either names used commonly or systematic names. The links to the chemical component IDs activate viewing using Rasmol.

Other versions It lists those structures that are either current or previous but not obsolete versions. These structures have replaced the structures that are explored.

2.3.2 Sample PDB format

The structure record of the Protein Data Bank is a unique four-character, alphanumeric code called PDB-D or PDB code. The code uses digits 0 to 9 and the uppercase letters A to Z. Structural data in a PDB entry is maintained as flat files as shown below:

```
load pdb inline
select [HEM]
spacefill
exit
HEADER   OXYGEN TRANSPORT               07-MAR-84  4HHB      4HHB   3
COMPND    HEMOGLOBIN (DEOXY)                   4HHB  4
SOURCE    HUMAN (HOMO SAPIENS)                  4HHB  5
AUTHOR    G.FERMI, M.F.PERUTZ                     4HHB  6
REVDAT  2  15-OCT-89 4HHBA  3       MTRIX     4HHBA 1
REVDAT  1   17-JUL-84 4HHB     0              4HHB   7
SPRSDE 17-JUL-84 4HHB       1HHB                4HHB  8
JRNL  AUTH  G.FERMI, M.F.PERUTZ, B.SHAANAN, R.FOURME  4HHB  9
JRNL     TITL  THE CRYSTAL STRUCTURE OF HUMAN DEOXYHAEMOGLOBIN AT  4HHB  10
JRNL    TITL 2 1.74 ANGSTROMS RESOLUTION                          4HHB  11
JRNL    REF   J. MOL. BIOL.                 V. 175   159 1984    4HHB  12
JRNL    REFN   ASTM JMOBAK UK ISSN 0022-2836               070  4HHB  13
REMARK  1                                                       4HHB  14
REMARK   1 REFERENCE 1                                          4HHB  15
REMARK   1  AUTH   M.F.PERUTZ,S.S.HASNAIN,P.J.DUKE,J.L.SESSLER,   4HHB  16
REMARK   1  AUTH 2 J.E.HAHN                                     4HHB  17
REMARK   1  TITL   STEREOCHEMISTRY OF IRON IN DEOXYHAEMOGLOBIN     4HHB  18
REMARK   1  REF    NATURE                      V. 295   535 1982  4HHB  19
REMARK   1  REFN   ASTM NATUAS  UK ISSN 0028-0836          006  4HHB  20
REMARK   1 REFERENCE 2                                      4HHB  21
REMARK   1  AUTH    G.FERMI,M.F.PERUTZ                          4HHB  22
REMARK   1  REF    HAEMOGLOBIN AND MYOGLOBIN.   V.   2   1981      4HHB  23
REMARK    1  REF   2 ATLAS OF MOLECULAR                         4HHB  24
REMARK    1  REF   3 STRUCTURES IN BIOLOGY                        4HHB  25
REMARK    1  PUBL    OXFORD UNIVERSITY PRESS                         4HHB  26
REMARK    1  REFN          ISBN 0-19-854706-4                986  4HHB  27
...............
REMARK    2              4HHB  69
REMARK    2 RESOLUTION. 1.74 ANGSTROMS.                              4HHB  70
```

```
REMARK 3              4HHB 71
REMARK 3 REFINEMENT. UNRESTRAINED REFINEMENT. THE CONFORMATION OF 4HHB 72
REMARK 3 THE HEME GROUP WAS MODIFIED BEFORE STARTING THE 4HHB 73
REMARK 3 UNRESTRAINED REFINEMENT. THE FINAL R VALUE IS 0.135. 4HHB 74
REMARK 4                                                      4HHB 75
REMARK 4 THE CRYSTALLOGRAPHIC ASYMMETRIC UNIT CONTAINS TWO ALPHA AND 4HHB 76
REMARK 4 TWO BETA CHAINS. ONLY ONE CHAIN OF EACH TYPE IS REPRESENTED 4HHB 77
REMARK 4 HERE.                                                4HHB 78
REMARK 5                                                      4HHB 79
REMARK 5 THE COORDINATES GIVEN HERE ARE IN THE ORTHOGONAL ANGSTROM 4HHB 80
REMARK 5 SYSTEM STANDARD FOR HEMOGLOBINS. THE Y AXIS IS THE 4HHB 81
REMARK 5 (NON CRYSTALLOGRAPHIC) MOLECULAR DIAD AND THE X AXIS IS THE 4HHB 82
REMARK 5 PSEUDO DIAD WHICH RELATES THE ALPHA-1 AND BETA-1 CHAINS. 4HHB 83
REMARK 5 THE TRANSFORMATION GIVEN IN THE *MTRIX* RECORDS BELOW 4HHB 84
REMARK 5 WILL GENERATE COORDINATES FOR THE *C* AND *D* CHAINS FROM 4HHB 85
REMARK 5 THE *A* AND *B* CHAINS RESPECTIVELY.                  4HHB 86
-----------------
REMARK 7                                                      4HHB 102
REMARK 7 STRUCTURE FACTORS FOR HUMAN DEOXYHEMOGLOBIN CORRESPONDING 4HHB 103
REMARK 7 TO ENTRY 2HHB ARE AVAILABLE FROM THE PROTEIN DATA BANK AS A 4HHB 104
REMARK 7 SEPARATE ENTRY.                                      4HHB 105
REMARK 8                                                      4HHBA 2
REMARK 8 CORRECTION. CORRECT FORMAT OF MTRIX RECORDS. 15-OCT-89. 4HHBA 3
SEQRES 1 A  141  VAL LEU SER PRO ALA ASP LYS THR ASN VAL LYS ALA ALA 4HHB 106
SEQRES 2 A  141  TRP GLY LYS VAL GLY ALA HIS ALA GLY GLU TYR GLY ALA 4HHB 107
SEQRES 3 A  141  GLU ALA LEU GLU ARG MET PHE LEU SER PHE PRO THR THR 4HHB 108
SEQRES 4 A  141  LYS THR TYR PHE PRO HIS PHE ASP LEU SER HIS GLY SER 4HHB 109
SEQRES 5 A  141  ALA GLN VAL LYS GLY HIS GLY LYS LYS VAL ALA ASP ALA 4HHB 110
SEQRES 6 A  141  LEU THR ASN ALA VAL ALA HIS VAL ASP ASP MET PRO ASN 4HHB 111
SEQRES 7 A  141  ALA LEU SER ALA LEU SER ASP LEU HIS ALA HIS LYS LEU 4HHB 112
SEQRES 8 A  141  ARG VAL ASP PRO VAL ASN PHE LYS LEU LEU SER HIS CYS 4HHB 113
SEQRES 9 A  141  LEU LEU VAL THR LEU ALA ALA HIS LEU PRO ALA GLU PHE 4HHB 114
SEQRES 10 A  141 THR PRO ALA VAL HIS ALA SER LEU ASP LYS PHE LEU ALA 4HHB 115
SEQRES 11 A  141 SER VAL SER THR VAL LEU THR SER LYS TYR ARG     4HHB 116
SEQRES 1 B  146  VAL HIS LEU THR PRO GLU GLU LYS SER ALA VAL THR ALA 4HHB 117
.................
FTNOTE 1                                                      4HHB 152
FTNOTE 1 PROBABLY PHOSPHATE GROUP.                            4HHB 153
HET    HEM  A   1  43   PROTOPORPHYRIN IX GRP CONTAINS FE(2+)  4HHB 154
HET    HEM  B   1  43   PROTOPORPHYRIN IX GRP CONTAINS FE(2+)  4HHB 155
HET    HEM  C   1  43   PROTOPORPHYRIN IX GRP CONTAINS FE(2+)  4HHB 156
HET    HEM  D   1  43   PROTOPORPHYRIN IX GRP CONTAINS FE(2+)  4HHB 157
HET    PO4      1   1   PHOSPHATE GROUP                        4HHB 158
HET    PO4      2   1   PHOSPHATE GROUP                        4HHB 159
```

```
FORMUL  5  HEM    4(C34 H32 N4 O4 FE1 ++)          4HHB 160
FORMUL  6  PO4   *2(O4 P1)                         4HHB 161
FORMUL 7  HOH   *223(H2 O1)               4HHB 162
HELIX  1  AA SER A   3  GLY A   18  1         4HHB 163
HELIX  2  AB HIS A   20  SER A   35  1          4HHB 164
HELIX  3  AC PHE A   36  TYR A   42  1             4HHB 165
HELIX  4  AD HIS A   50  GLY A   51  1 DEGEN 2 RES HLX RETAIN  HOMOL 4HHB 166
HELIX   5  AE SER A   52  ALA A   71  1              4HHB 167
HELIX   6  AF LEU A   80  ALA A   88  1                    4HHB 168
HELIX   7  AG ASP A   94  HIS A  112  1                    4HHB 169
HELIX   8  AH THR A  118  SER A  138  1                    4HHB 170
HELIX   9  BA THR B    4  VAL B   18  1                    4HHB 171
HELIX  10  BB ASN B   19  VAL B   34  1                    4HHB 172
HELIX  11  BC TYR B   35  PHE B   41  1                    4HHB 173
HELIX  12  BD THR B   50  GLY B   56  1                    4HHB 174
HELIX  13  BE ASN B   57  ALA B   76  1                    4HHB 175
HELIX  14  BF PHE B   85  CYS B   93  1                    4HHB 176
HELIX  15  BG ASP B   99  HIS B  117  1                    4HHB 177
HELIX  16  BH THR B  123  HIS B  143  1                    4HHB 178
HELIX  17  CA SER C    3  GLY C   18  1                    4HHB 179
HELIX  18  CB HIS C   20  SER C   35  1                    4HHB 180
HELIX  19  CC PHE C   36  TYR C   42  1                    4HHB 181
HELIX  20  CD HIS C  50  GLY C   51  1 DEGEN 2 RES HLX RETAIN  HOMOL 4HHB 182
HELIX  21  CE SER C   52  ALA C   71  1                    4HHB 183
HELIX  22  CF LEU C   80  ALA C   88  1                    4HHB 184
HELIX  23  CG ASP C   94  HIS C  112  1                    4HHB 185
HELIX  24  CH THR C  118  SER C  138  1                    4HHB 186
HELIX  25  DA THR D    4  VAL D   18  1                    4HHB 187
HELIX  26  DB ASN D   19  VAL D   34  1                    4HHB 188
HELIX  27  DC TYR D   35  PHE D   41  1                    4HHB 189
HELIX  28  DD THR D   50  GLY D   56  1                    4HHB 190
HELIX  29  DE ASN D   57  ALA D   76  1                    4HHB 191
HELIX  30  DF PHE D   85  CYS D   93  1                    4HHB 192
HELIX  31  DG ASP D   99  HIS D  117  1                    4HHB 193
HELIX  32  DH THR D  123  HIS D  143  1                    4HHB 194
CRYST1 63.150   83.590   53.800  90.00  99.34  90.00 P 21        4  4HHB 195
ORIGX1 .963457    .136613    .230424       16.61000            4HHB 196
ORIGX2      -.158977    .983924    .081383       13.72000      4HHB 197
ORIGX3      -.215598  -.115048    .969683       37.65000       4HHB 198
SCALE1       .015462    .002192    .003698        .26656       4HHB 199
SCALE2      -.001902    .011771    .000974        .16413       4HHB 200
SCALE3      -.001062  -.001721    .018728        .75059        4HHB 201
MTRIX1   1 -1.000000    .000000  -.000000       .00001    1     4HHBA  4
MTRIX2   1  -.000000  1.000000    .000000          .00002    1  4HHBA  5
MTRIX3   1   .000000  -.000000 -1.000000          .00002    1     4HHBA  6
```

```
ATOM      1  N   VAL A   1       6.204  16.869   4.854  7.00 49.05  4HHB 205
ATOM      2  CA  VAL A   1       6.913  17.759   4.607  6.00 43.14  4HHB 206
ATOM      3  C   VAL A   1       8.504  17.378   4.797  6.00 24.80  4HHB 207
ATOM      4  O   VAL A   1       8.805  17.011   5.943  8.00 37.68  4HHB 208
ATOM      5  CB  VAL A   1       6.369  19.044   5.810  6.00 72.12  4HHB 209
ATOM      6  CG1 VAL A   1       7.009  20.127   5.418  6.00 61.79  4HHB 210
ATOM      7  CG2 VAL A   1       5.246  18.533   5.681  6.00 80.12  4HHB 211
ATOM      8  N   LEU A   2       9.096  18.040   3.857  7.00 26.44  4HHB 212
ATOM      9  CA  LEU A   2      10.600  17.889   4.283  6.00 26.32  4HHB 213
ATOM     10  C   LEU A   2      11.265  19.184   5.297  6.00 32.96  4HHB 214
ATOM     11  O   LEU A   2      10.813  20.177   4.647  8.00 31.90  4HHB 215
ATOM     12  CB  LEU A   2      11.099  18.007   2.815  6.00 29.23  4HHB 216
ATOM     13  CG  LEU A   2      11.322  16.956   1.934  6.00 37.71  4HHB 217
ATOM     14  CD1 LEU A   2      11.468  15.596   2.337  6.00 39.10  4HHB 218
ATOM     15  CD2 LEU A   2      11.423  17.268    .300  6.00 37.47  4HHB 219
ATOM     16  N   SER A   3      11.584  18.730   6.148  7.00 28.01  4HHB 220
.................
TER    1070      ARG A 141                                            4HHB1274
HETATM 1071 FE   HEM A   1       8.116   7.403 -15.045 24.00 18.07  4HHB1275
HETATM 1072  CHA HEM A   1       8.585   7.902 -18.282  6.00 16.31  4HHB1276
HETATM 1073  CHB HEM A   1      10.355   9.805 -14.208  6.00 26.27  4HHB1277
HETATM 1074  CHC HEM A   1       8.341   6.363 -11.589  6.00 13.23  4HHB1278
HETATM 1075  CHD HEM A   1       6.988   4.088 -15.744  6.00 14.77  4HHB1279
HETATM 1076  N A HEM A   1       9.397   8.686 -16.211  7.00 16.46  4HHB1280
............
ATOM   1114  N   VAL B   1       9.223 -20.614   1.365  7.00 46.08  4HHB1318
ATOM   1115  CA  VAL B   1       8.694 -20.026   -.123  6.00 70.96  4HHB1319
ATOM   1116  C   VAL B   1       9.668 -21.068  -1.645  6.00 69.74  4HHB1320
ATOM   1117  O   VAL B   1       9.370 -22.612   -.994  8.00 71.82  4HHB1321
ATOM   1118  CB  VAL B   1       9.283 -18.281   -.381  6.00 59.18  4HHB1322
ATOM   1119  CG1 VAL B   1       7.449 -17.518   -.791  6.00 57.89  4HHB1323
ATOM   1120  CG2 VAL B   1      10.416 -18.038    .066  6.00 44.20  4HHB1324
..................
CONECT  650  648  649 1071                                            4HHB4988
CONECT 1071  650 1076 1087 1095                                       4HHB4989
CONECT 1071 1103                                                      4HHB4990
CONECT 1072 1077 1107                                                 4HHB4991
CONECT 1073 1080 1088                                                 4HHB4992
CONECT 1074 1091 1096                                                 4HHB4993
CONECT 1075 1099 1104                                                 4HHB4994
CONECT 1076 1071 1077 1080                                            4HHB4995
CONECT 1077 1072 1076 1078                                            4HHB4996
CONECT 1078 1077 1079 1082                                            4HHB4997
---------------
MASTER     94    2    6   32    0    0    0    9 4779    4  180   46  4HHBA  7
END                                                                   4HHB5169
```

2.3.3 Abbreviation Key

`HEADER:`	Type of protein
`COMPND:`	Chemical compound
`SOURCE:`	Organism name
`EXPDTA:`	In case of NMR data, it gives the number of structures considered
`AUTHOR:`	Author name(s) of the paper
`REVDAT:`	Revised date
`JRNL:`	Journal information
`REMARK:`	Information about the paper in which the protein data was published
`SEQRES:`	Explicit protein sequence using three-letter amino acid code. It also stores non-standard amino acids using arbitrarily chosen three-letter names. To represent a discrete sequence corresponding to multiple chains, it also includes entries for chain identifiers using a single uppercase letter or blank space.
`FTNOTE:`	Footnotes
`FORMUL:`	Formula for the compound
`ATOM:`	Describes the orthogonal coordinate system of an implicit sequence and names of each atom
`CONECT:`	Describes the connectivity between atoms
`MODEL:`	NMR data contains several models corresponding to different conformations
`ENDMDL:`	End of model information
`END:`	End of PDB file

TASK Exploring PDB

Step 1 Using the Internet Explorer go to http://www.rcsb.org/pdb/Welcome.do. You will come across a page as shown in Fig. 2.3.

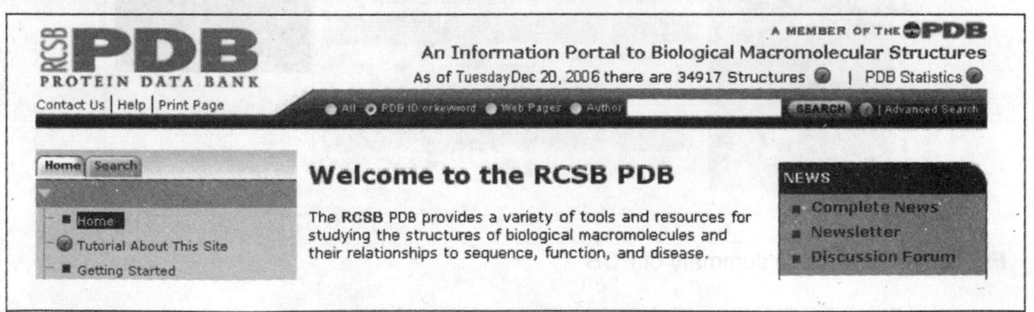

Fig. 2.3 Home page of RCSB PDB

Step 2 Select PDB ID or keyword in the search bar, type 'myoglobin', and click **Go**. You will get a result as shown in Fig. 2.4.

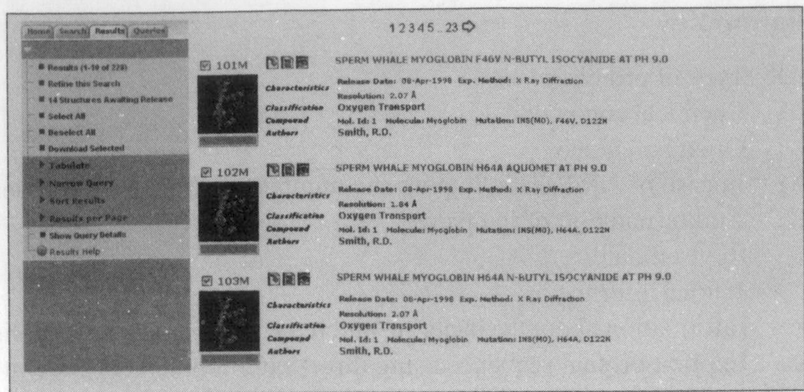

Fig. 2.4 Keyword search result in PDB

Step 3 Look for the following by pointing your mouse pointer at various locations of any entry.

- View structure summary
- Download GNU zipped PDB file
- View PDB file
- Structure visualization with JMOl viewer

Step 4 Click **View Structure Summary**. You will get results as shown in Fig. 2.5.

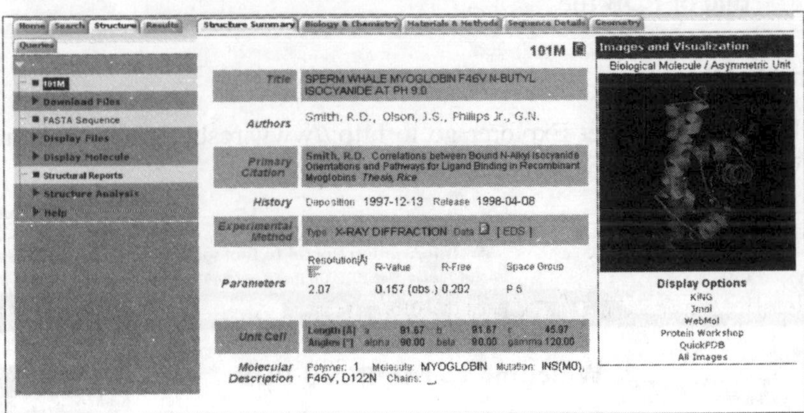

Fig. 2.5 Structure summary of PDB

Step 5 Click the link **Fasta Sequence** in the left column. The result is as shown below:

```
>101M:_|PDBID|CHAIN|SEQUENCE
MVLSEGEWQLVLHVWAKVEADVAGHGQDILIRLFKSHPETLEKFDRVKHLKTEAEMKASEDLKKHGVT
VLTALGAILKKK
GHHEAELKPLAQSHATKHKIPIKYLEFISEAIIHVLHSRHPGNFGADAQGAMNKALELFRKDIAAKYKELG
YQG
```

2.4 MMDB

NCBI's macromolecular 3D structure database is called MMDB or molecular modelling database. It is also known as the NCBI structure division. It was designed to archive structural data from PDB as well as biomolecules generated by electron microscopy (surface models). MMDB is linked to the rest of the NCBI databases.

2.4.1 MMDB Entry

An MMDB entry contains three-dimensional biomolecular or biopolymer structures including proteins and polynucleotides that are experimentally determined from X-ray crystallography and NMR spectroscopy. MMDB 3D structure data is obtained from the Protein Data Bank (PDB) after excluding theoretical models. The records of MMDB are created after sequence validation. The first implicit sequence is derived from the PDB entry. In case of gaps in structure, implicit sequence fragments for a given chain are derived and then aligned with the explicit sequence of the same chain to create a complete chemical graph. An entry also records information regarding secondary structures, domain, citation, and taxonomy. All this information is stored in MMDB in ASN.1 format.

2.4.2 Sample Format

The search field of MMDB is the PDB ID or MMDB ID. MMDBID also offers text-based search that scans through REMARK, AUTHOR, or other bibliographic fields. A text-based search result of myoglobin is as shown in Fig. 2.6.

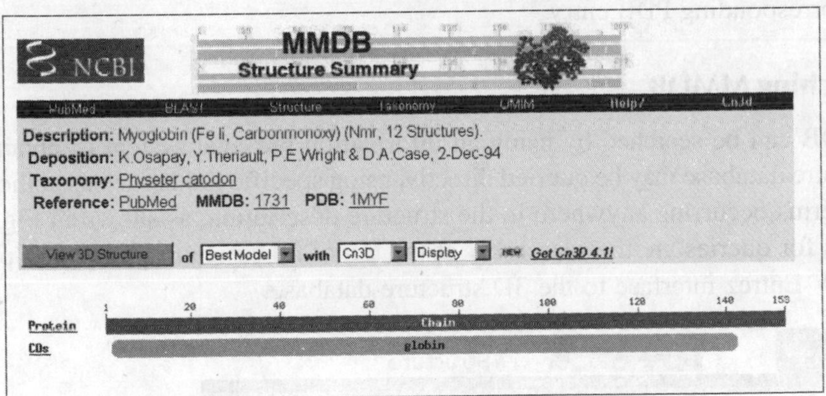

Fig. 2.6 Result page of a text-based search in MMDB

2.4.3 Abbreviation Key

Description information in MMDB comes largely from PDB. It includes the following.

Title This field names and briefly describes the content of an entry.

Deposition This field lists the people who determined the structure as well as the date when the data was deposited in the PDB. The deposition date can be substantially different from the release date (due to the PDB hold policy). The date when structures are available in MMDB may be slightly later than the PDB release date, since MMDB is updated once a month.

Taxonomy This field lists the taxonomic assignments for each chain. If all the chains of the structure have the same assignment, only one link is given. Some chains have no taxonomy assigned. Usually these are short nucleotide chains not specific to any one organism. The assignment name is linked to the corresponding nodes in NCBI's taxonomy database.

Reference This field links to relevant literature references in PubMed. These citations have been taken from PDB and linked to PubMed by citation matching. MMDB records may contain additional citations not linked to PubMed. This may be conveniently viewed via the GenPept sequence report for any chain.

MMDB MMDB identifier can be used to retrieve structure data from MMDB. This unique and stable number is assigned to each new structure when it is entered into MMDB. If the structure data is obsolete, clicking on the MMDB identification following **replaced by** will display the summary of the most current structure. If the entry is deleted from PDB, then the corresponding entry in MMDB will be indicated by **obsolete** following its MMDB ID number. Obsolete structures are archived in MMDB for consistency with original publication and annotation. Only current structures are searched by Entrez.

PDB It is a four-character identifier used in the PDB. Clicking the identifier will lead to the corresponding PDB entry.

TASK Searching MMDB

MMDB can be searched by using an information retrieval system to obtain data. The structure database may be queried directly, using specific fields such as author names, or text terms occurring anywhere in the structure description, as shown in Fig. 2.7. Entry points for queries are the search bar at the top of all structure group web pages or the WWW-Entrez interface to the 3D structure database.

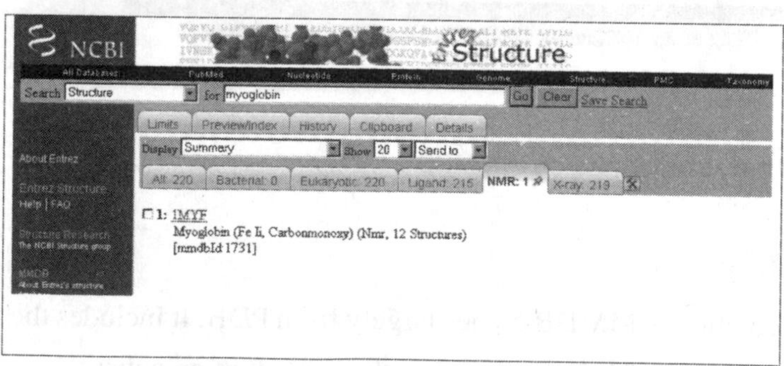

Fig. 2.7 Query page of structure database

You will get result as shown in Fig. 2.8.

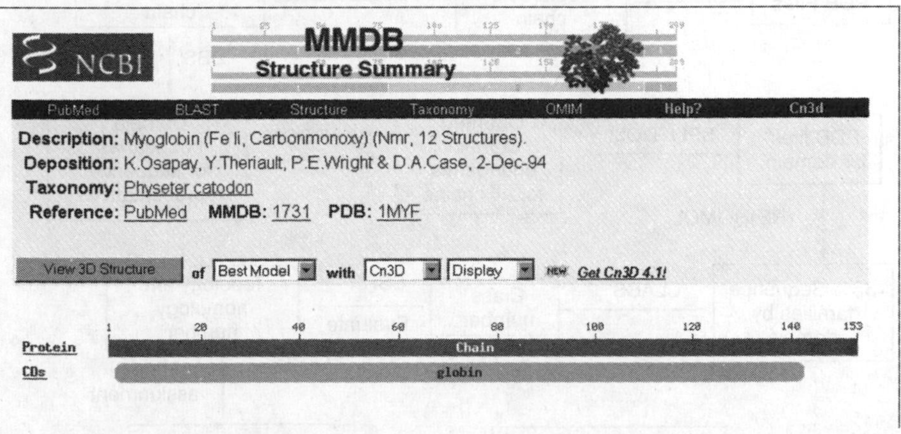

Fig. 2.8 Result page of MMDB search

Alternatively you can use a PDB four-character code or a numerical MMDB ID to retrieve structure summary pages directly, as shown in Fig. 2.9.

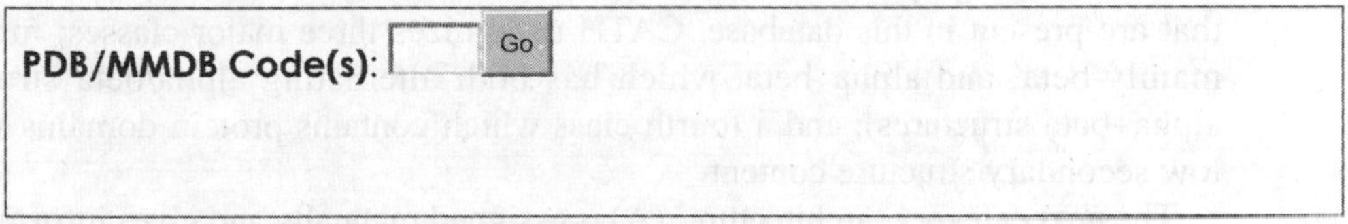

PDB/MMDB Code(s): [ ] Go

Fig. 2.9 Search using PDB or MMDB code

2.5 CATH (http://cathwww.biochem.ucl.ac.uk/latest/)

Like the secondary database that has been described in the previous section, CATH also utilizes the classification of proteins as the basis of its organization. This classification is similar to enzyme classification (EC) that is used to classify enzymes. Like the EC system, CATH is numerically indexed hierarchy of protein domain structures. CATH has four major classes or categories: class (C), architecture (A), topology (T) and homologous superfamily (H), so the name CATH. It currently stores 34287 protein domain structures grouped into 1383 superfamilies and 3285 sequence families.

The basis of classification in CATH is the domain. Domains are arrived at using at least three different methods. Proteins that have significant sequence similarity to a protein already in the database inherit the domain boundaries. For 'new' proteins three different algorithms are used to identify the structural domain boundaries automatically. If no consensus is reached, the domain boundaries are determined manually. Figure 2.10 shows the domain assignment flow sheet of PDB structures.

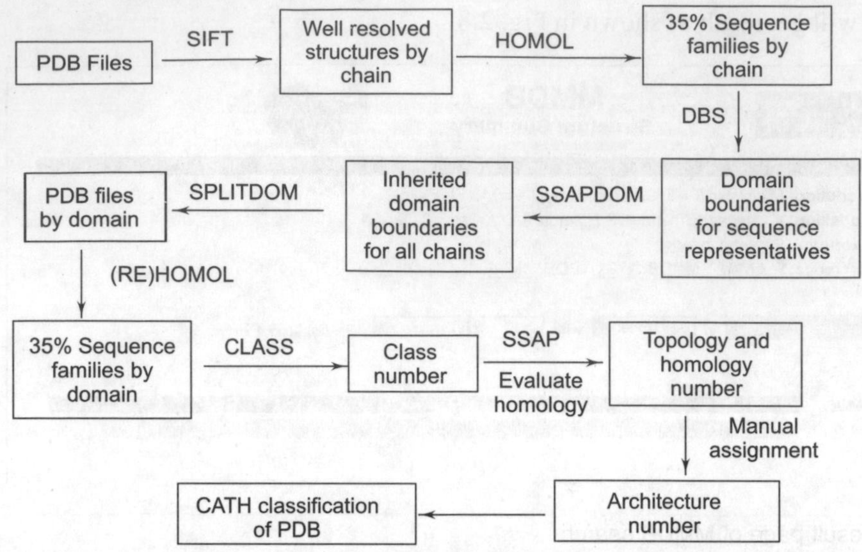

Fig. 2.10 Flow sheet showing domain assignment of PDB structures

The first category 'class' (C) is assigned automatically to more than 90% of the proteins that are present in this database. CATH recognizes three major classes; mainly-alpha, mainly-beta, and alpha–beta, which has both alternating alpha/beta structures and alpha+beta structures), and a fourth class which contains protein domains which have low secondary structure content.

The next category 'architecture' (A) is assigned manually and gives information about the gross orientation of secondary structures. This category describes the overall shape of the domain structure. The overall shape is derived from the orientations of the secondary structures but ignores the connectivity between them.

The next category 'topology' (T) groups the structures according to their toplogical connections. This level of hierarchy is the 'fold' level. Structures are grouped into fold families after considering the overall shape and connectivity of the secondary structures. It uses the SSAP algorithm, which compares protein structures based on the distance plot analysis. This algorithm is relatively insensitive to insertions and deletions in a sequence and is tolerant of the displacement of equivalent substructures between the two molecules being compared. Only those structures that have a SSAP score of 60–80 are assigned to the same T level or fold family.

The last category of this database, 'homologous superfamily' (H), regroups proteins with highly similar structures and functions. Structures are clustered into the same homologous superfamily if they satisfy one of the following criteria:

- Sequence identity >= 35%, 60% of larger structure equivalent to smaller
- SSAP score >= 80.0 and sequence identity >= 20%
  60% of larger structure equivalent to smaller
- SSAP score >= 80.0, 60% of larger structure equivalent to smaller, and domains which have related functions

It is important to understand that the basis for assigning a particular protein in the last two categories is sequence and structure comparisons. This architecture of the database facilitates computerized manipulation and search.

> **Note**
>
> CATH does not store all non-protein models and 'C-alpha only' structures, instead it stores crystal structures (resolution better than 3 Å) along with their NMR structures.

A new category that has been added on to CATH is the sequence or the S Level (shown in Fig. 2.11).

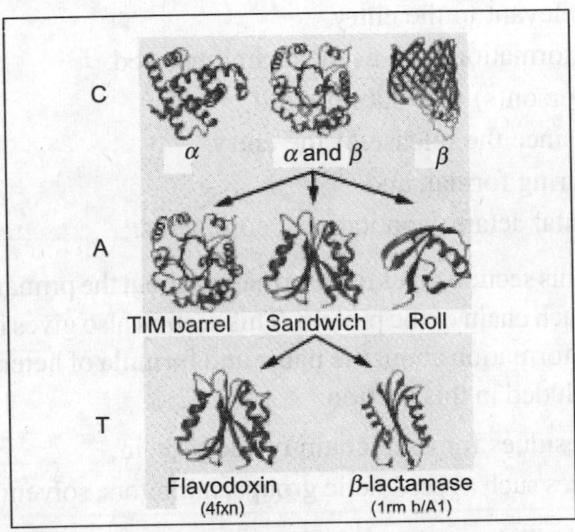

Fig. 2.11 Sequence or S family level of CATH

All the families that evolve out of the last category, that is, homologous superfamily, are further sieved through and grouped into smaller groups based on their domains. Each member of these domains has at least 35% sequence identity and at least 60% of the larger domain should be equivalent to the smaller. This automatically ensures that the members are highly similar in their structure and function.

Each structural domain of a protein is assigned a CATH number, which specifies its position in the hierarchy. However, as the number of PDB files being deposited is increasing, their assignment to CATH is consuming more and more time. To bridge the gap a new method of assignment has been adopted now. This classification protocol identifies structural relatives through matching the 3D template of a given evolutionary family or fold group. Templates are generated by CORA (conserved residue attributes) which is a suite of programs for automatically aligning and analysing protein structural families. CORA plots are now being used to assess a structure to assign it to a family.

2.5.1 Sample CATH Protein File

The structure record of the Protein Data Bank is a unique four-character alphanumeric code called PDB-D or PDB code. CATH houses the same in its database, where information is described as follows:

Title Section This section gives us information on the

- 'HEADER' IDCode field which uniquely identifies the file from the other PDB files, 'TITLE' title for the experiment/analysis,
- 'COMPND' macromolecular contents,
- 'SOURCE' biological/chemical source,
- 'KEYWDS' set of terms relevant to the entry,
- 'EXPDTA' experiment information such as the technique used,
- 'AUTHOR' name of the person(s) who submitted it,
- 'REVDAT' revision data since the release of the entry,
- 'JRNL' name of the featuring formal, and
- 'REMARK' any experimental details, annotations, comments.

Primary structure section This section gives us information about the primary sequence or the sequence of residues for each chain of the protein. This section also gives information about non-standard residues. Information about the name and formula of hetero groups in the macromolecule are also included in this section.

- 'SEQRES' sequence of residues for each chain of the protein;
- 'HET' non-standard residues such as prosthetic groups, inhibitors, solvents, and ions;

Secondary structure section This section contains the data on the helices, sheets, and turns found in protein. Positions of turns, helices and sheets are provided, which are named and numbered.

- `HELIX      1    A ALA A    14   LYS A    28  1`
- `SHEET      1    A 2 GLN A 173    GLN A 175  0`

Connectivity annotation section This section gives information about the existence and location of disulfide bonds and other linkages.

- `SSBOND    1 CYS A    11      CYS A    91`
- `LINK      C    PCA A    1                             N    LEU A    2`
- `LINK      FE    HEM A 300                            NE2 HIS A 169`

Miscellaneous features section This section contains information that is unique to a structure (special features like an active site).

- `SITE   1 APC   5 THR A 170   ASP A 214   THR A 217   LYS A 220`
- `SITE   2 APC   5 ASP A 222`

Crystallographic and coordinate transformation section This section gives information about the geometry of the crystallographic experiment and the coordinate system transformations.

- `CRYST1 48.100 97.200 146.200 90.00 90.00 90.00 P 21 21 21 8`
- `ORIGX1  1.000000  0.000000  0.000000    0.00000`
- `ORIGX3  0.000000  0.000000  1.000000    0.00000`
- `SCALE1  0.020790  0.000000  0.000000    0.00000`
- `MTRIX1 1  0.438990  0.642420 -0.628160    19.31394   1`
- `MTRIX2  1  0.657570  0.246710  0.711850  -23.03639   1`

Coordinate section This section gives only information which pertains to the given atomic coordinates of the structure.

- `HETATM  1  N   PCA A  1 21.053  53.031  41.428 1.00 64.66`
- `ATOM    9 N  LEU A  2  21.768  53.526  44.010  1.00 16.90`

Connectivity section This section gives only information which gives information on chemical connectivity or how the atoms are connected to each other. Information here includes hydrogen bonds, salt bridges, and links.

- `CONECT    1    2    7`
- `CONECT    1    2    7`
- `CONECT    2    1    3    5`

Bookkeeping section This section gives final information about the file itself.

- `MASTER` 252 0 9 26 4 0 12 9 4579 2 174 46
  `END`

This file can be viewed using software, including the 3D structure rendering programs such as Rasmol, Chime, and Cn3D.

2.5.2 CATH Data Formats

CATH supports the following flat-file formats.

CathDomainDescriptionFile Information in this file format is compiled from the following files: CathList, CathDomall, CathDomain Fasta Sequences, CathSegment Fasta Sequences,PdbSum file (in CDDF format), and ChainLimits file (in CDDF format).

CathDomall This file format describes domain boundaries for entries in the CATH database. Only PDB chains that have been split into domains have a CathDomall entry. Note that domains that are composed of whole PDB chains are not found in this file.

CathList CathList contains an entry for each structural entry in CATH. Representative structural domains are selected from CathList based on the numbering scheme.

Cath Names This file contains description of each node in the CATH hierarchy for class, architecture, topology and homologous superfamily levels. It is organized into columns.

Column 1: Node number
Column 2: Representative protein domain (CATH six-character domain name)
Free text: Node description (starting with ':' to indicate description start)
(Note: this is not a single column.)

2.5.3 CATH versus Other Structure Classification Databases

CATH aims at showing how proteins evolve and that although evolution (which occurs by substitutions, insertions and deletions in the amino acid sequence) can be very extensive such that the numbers and orientations of secondary structures vary considerably for a given fold, there is a generally higher structure conservation at the protein core. Here, the protein functions are essentially the same, and structural environments of critical residues are also conserved.

The CATH classification scheme is mostly automated, but some human intervention is required in the process. Various algorithms are used in CATH. It is actually interesting to note that although different classification databases use different methods of classification, they agree on most of the general or higher levels of their hierarchies. CATH emphasizes on clustering proteins that have a similar fold so as to increase the understanding of sequence/structure relationships within protein fold families. CATH aims to improve homology modelling and structure prediction methods. CATH developers try to do this by the algorithm cut off scores they have.

TASK

Now let us try and browse through the CATH database with the intention of retrieving some information.

Step1 Open the CATH search page at http://cathwww.biochem.ucl.ac.uk/latest/index.html

Step 2 There are many options listed there. Choose the 'Browse or search the classification' option. This will lead to a new page as shown in Fig. 2.12.

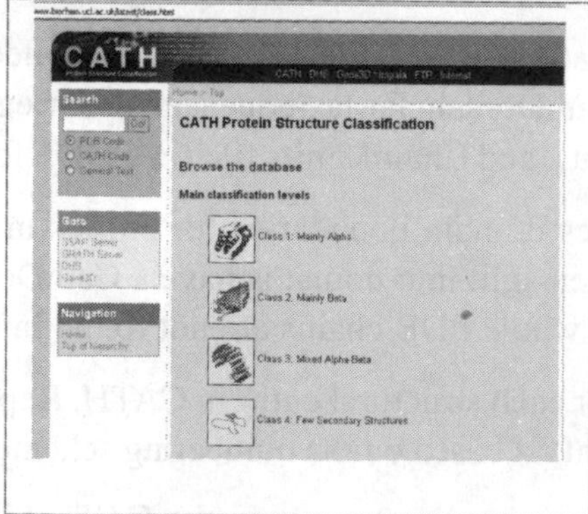

Fig. 2.12 CATH search page

Step 3 In the 'Search' field textbox enter 'urokinase'. Click the **Go** button.
Step 4 The search retrieves 5 hits as shown in Fig. 2.13.

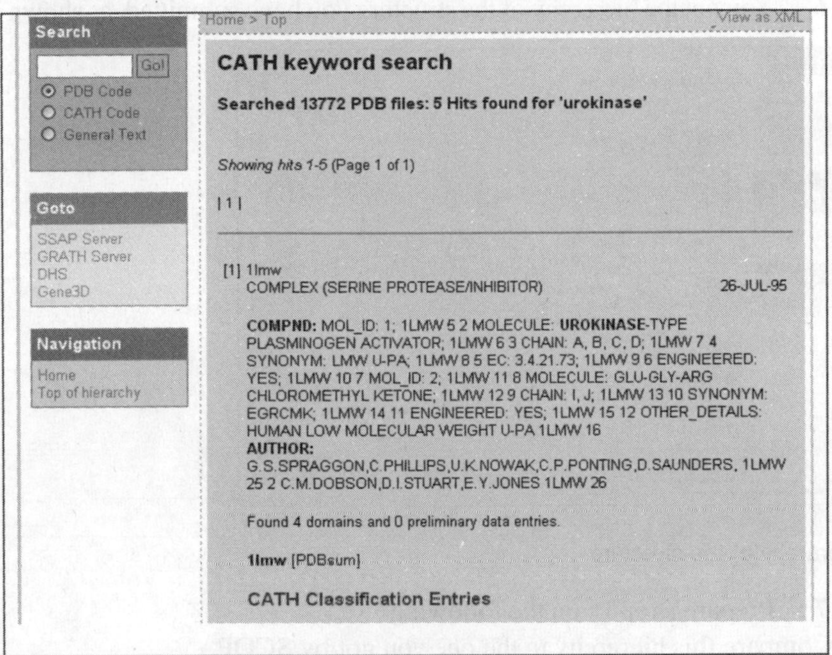

Fig. 2.13 Result page 1 of CATH

Domain	CATH code	Length	Image
1lmwB1	2.40.10.10	128	
1lmwB2	2.40.10.10	120	
1lmwD1	2.40.10.10	128	
1lmwD2	2.40.10.10	120	

[2] 1kdu
PLASMINOGEN ACTIVATION 15-JUL-93

COMPND: PLASMINOGEN ACTIVATOR (**UROKINASE**-TYPE, KRINGLE DOMAIN) 1KDU 3 2 (U-PA K) (NMR, MINIMIZED AVERAGE STRUCTURE) 1KDU 4
AUTHOR: X.LI,A.M.BOKMAN,M.LLINAS,R.A.G.SMITH,C.M.DOBSON 1KDU 7

Found 1 domain and 0 preliminary data entries.

1kdu [PDBsum]

Fig. 2.13A Result page 2 of CATH

Step 5 Choose the complex serine protease/inhibitor 1lmw. Choose the alpha chain domain 1lmwB1.

Step 6 You get the hierarchy of the structure you have submitted, as shown in Fig. 2.14.

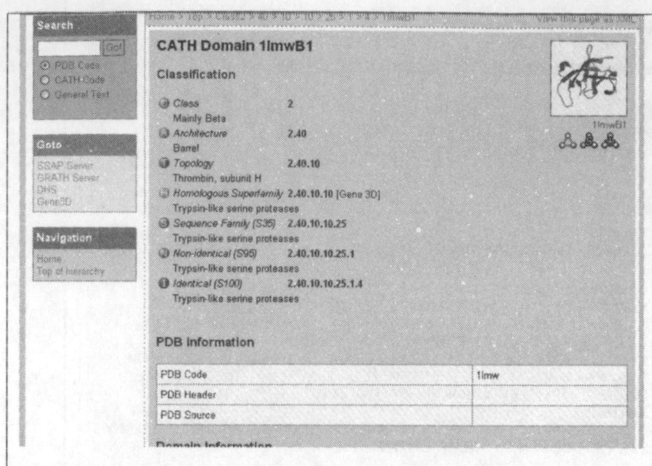

Fig. 2.14 Hierarchy of the structure

Step 7 Prepare a report on the following:

(a) Compare this hierarchy to the one you got by SCOP.

(b) Retrieve the PDB information and the Domain information (sequence).

(c) Click the **PDBsum** link, look at chain A of the protein, and examine the secodary structure topolgy. PDBsum is a comprehensive tool for structural information retrieval.

(d) Study the information provided about the structural relatives.

(e) Comment on the structures displayed on the result page, shown in Fig. 2.15.

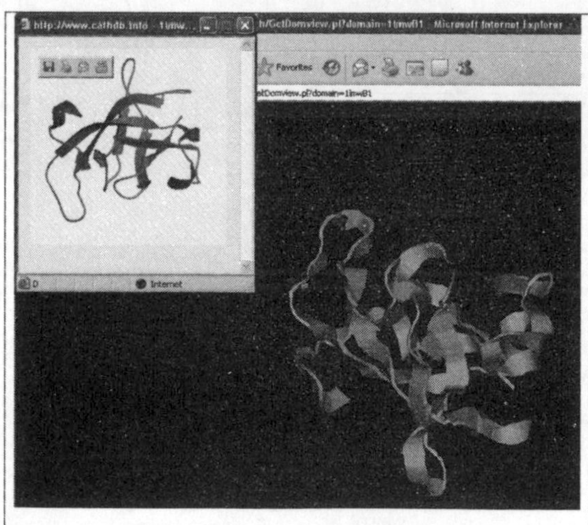

Fig. 2.15 Result page showing secondary structure

2.6 FSSP (http://www2.ebi.ac.uk/dali/fssp/fssp.html)

So far you have studied and learnt about two widely used databases: CATH and SCOP. Both these databases have their own way to compare and classify proteins. The resulting classification schemes are, largely, consistent with each other.

Fold classification based on the structure–structure alignment of proteins and families of structurally similar proteins (FSSP) is a database based on the structural alignment of pairwise combinations of proteins in the Protein Data Bank. Alignments and classification are done automatically and are updated continuously by the DALI search engine. The FSSP database presents a continuously updated structural classification of three-dimensional protein folds. It is derived using an automatic structure comparison program (DALI) for the all-against–all-comparison of over 6000 three-dimensional coordinate sets in the Protein Data Bank. Sequence-related protein families are covered by a representative set of 813 protein chains. Figure 2.16 shows homologous domains having similar structures. It shows a comparison between PH domains of human pleckstrin (a major substrate of protein kinase C in platelets) and dynamin (a large GTPase involved in the scission of nascent vesicles from parent membranes).

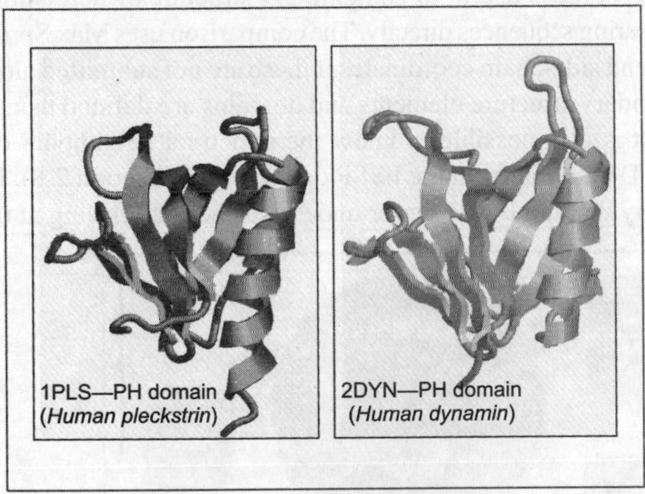

Fig. 2.16 Similar structures of homologous domains

Protein families are known to retain the shape of the fold even when sequences have diverged below the limit of detection of significant similarities at the sequence level unlike that shown in Fig. 2.16. Structural comparisons merge protein families of known 3D structure into structural classes, the members of which may or may not be evolutionarily related. To aid navigation in the database, the 330 protein chains contained in the representative set have been clustered into fold families. A dendrogram of the families was produced by average linkage clustering based on structural similarity scores. Chain length effects were corrected for by transforming the pairwise similarities into statistical significance scores (Z-scores). Families and subfamilies result from truncating the tree at different cut levels of the Z-score. The higher the cut, the larger the resulting number of distinct fold families.

Hierarchical clustering based on structural similarities yields a fold tree that defines 253 fold classes. For each representative protein chain, there is a database entry containing structure–structure alignments with its structural neighbours in the PDB. The database is accessible online through the World Wide Web browsers and by anonymous ftp (file transfer protocol). The overview of fold space and the individual datasets provide a rich source of information for the study of both divergent and convergent aspects of molecular evolution and for defining useful test sets and a standard of truth for assessing the correctness of sequence–sequence or sequence–structure alignments.

2.7 DALI (http://www.ebi.ac.uk/dali/)

The DALI or **D**istance m**A**trix a**LI**gnment server is a network service used to compare three-dimensional protein structures. The query sequence coordinates are compared against those in the Protein Data Bank. A multiple alignment of structural neighbours is the output. The DALI server is useful to compare 3D structures where similarities are not detectable by comparing sequences directly. The comparison uses Max Sprout program to generate backbone and side-chain coordinates if these are not submitted along with the query sequence. Secondary structure elements and domains are defined using the DSSP and PUU programs. It is also possible to know the structural neighbours of a protein already in the Protein Data Bank from the FSSP database (see Section 2.8). Figure 2.17 summarizes the activity that the DALI server undertakes before arriving at the output.

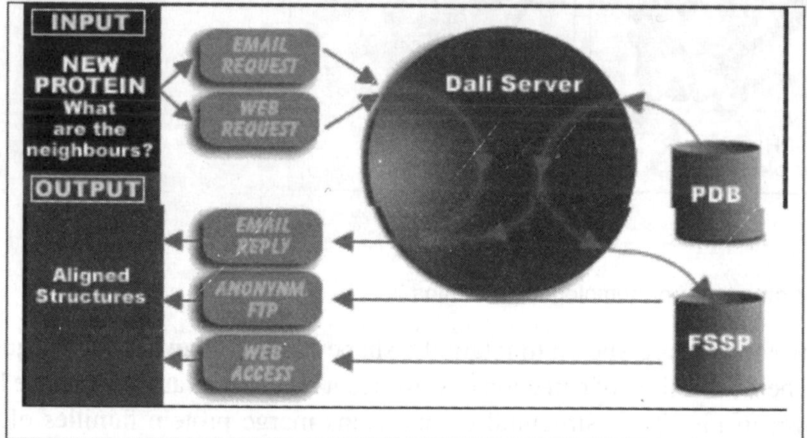

Fig. 2.17 Activity of DALI server

The major difference between the two classification schemes, relevant to our work, is their degree of automation. FSSP is based on a fully automated structure comparison algorithm, DALI, which calculates a structural similarity measure (represented in terms of the Z-score) between pairs of structures of protein chains taken from the PDB. A tree is then constructed by average linkage clustering of the structural similarity score. The tree is cut at DALI Z-score levels 2, 4, 8, 16, 32, and 64. The first level ($Z > 2$) can be used as an operational definition of folds.

FSSP first selects a subset of representative structures from the PDB and then applies the DALI algorithm to calculate the Z-scores for all pairs of representatives. The PDB list includes all chains in PDB90 (a subset of the PDB, where no two chains share more than 90% sequence identity). The domain number is appended as _*n*. Domains are numbered 1, 2, 3, ... Proteins with a domain numbered 0 are not assigned by the structural domain decomposition algorithm but are assigned as single-domain structures by default.

Next, it calculates the Z-scores between each representative and the PDB. Being fully automated, FSSP can be updated fairly often. FSSP was recently extended by a new database, called the DALI database at http://www.bioinfo.biocenter.helsinki.fi:8080/dali.

Table 2.1 DALI structural neighbours in PDB90

Fold index	PDB code	Adda	Browse	Interact	Compound
1.1.1.1.1.1	1w4zB_1	2230	browse	interact	KETOACYL REDUCTASE
1.1.1.1.1.1	____1iy8A_1	223	browse	interact	LEVODIONE REDUCTASE
1.1.1.1.1.1	____1geeA_0		browse	interact	GLUCOSE 1-DEHYDROGENASE
1.1.1.1.1.1	______1edoA_1	223	browse	interact	BETA-KETO ACYL CARRIER PROTEIN REDUCTASE
1.1.1.1.1.1	______1gegE_1	223	browse	interact	ACETOIN REDUCTASE
1.1.1.1.1.1	______1ahiA_1	223	browse	interact	7 ALPHA-HYDROXYSTEROID DEHYDROGENASE
1.1.1.1.1.1	______1ipfA_1	223	browse	interact	TROPINONE REDUCTASE-II
1.1.1.1.1.1	______1ae1B_1	223	browse	interact	TROPINONE REDUCTASE-I
1.1.1.1.1.1	______1vl8B_1	223	browse	interact	GLUCONATE 5-DEHYDROGENASE

The view shown in Table 2.1 can be browsed by clicking on the link to view the list of structural neighbours of the representative and their alignments. In contrast, CATH and SCOP use manual classification at certain levels of their hierarchy, which slows down the classification process and makes it more subjective and error-prone.

TASK

Now let us try to browse through the FSSP database with the intention of retrieving some information about 1c3pA and 1d3vA using Rasmol and SPDV.

Step 1 Open the search page (shown in Fig. 2.18) at http://www2.ebi.ac.uk/dali/fssp/fssp.html OR http://ekhidna.biocenter.helsinki.fi/dali/start

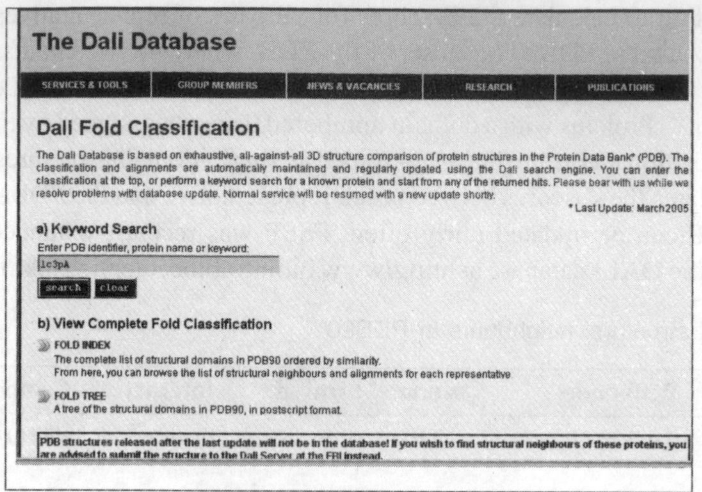

Fig. 2.18 Search page of DALI

Step 2 Under the Keyword Search box enter 1c3pA and click **search**. This will lead to a new page, as shown in Fig. 2.19.

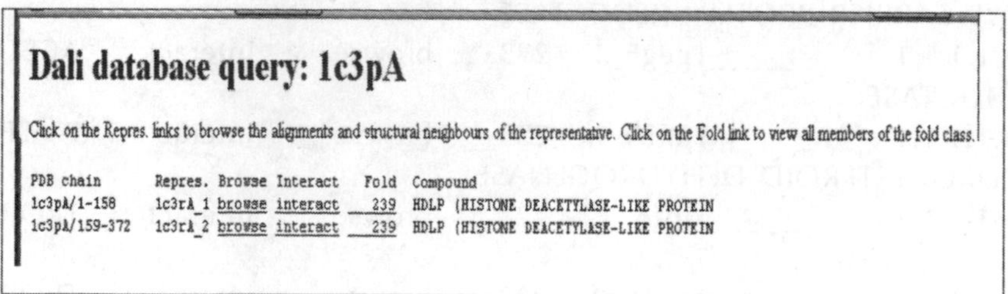

Fig. 2.19 Result page of DALI

Step 3 Select 1d3vA from the list, and use the '3D superposition' button to view the two proteins superposed in Rasmol. Note that the superposition can be viewed with viewers such as Rasmol. Note that the columns in the FSSP family table include the lengths of both proteins, the RMS of the structural alignment, the number of residue pairs in the alignment, and the Z-score of the alignment.

Step 4 Write a small report on the following:

(i) Differences between the RMS values for the best superposition SPDV and the alignment found by DALI.

(ii) The list of structural homologs sorted by the *Z*-score. Emphasize on the use of this strategy rather than sorting the alignments by their geometric similarity.

(iii) Look at the SCOP entry for 1c3pA and 1d3vA. Are these proteins related by evolution?

2.8 SCOP (http://scop.mrc-lmb.cam.ac.uk/scop/)

The SCOP (structural classification of proteins) database was started by the Lab of Molecular Biology, MRC, Cambridge, UK, with a single purpose of classifying protein 3D structures in a hierarchical scheme of structural classes. It is maintained manually, and all protein structures in the PDB are classified. It is updated regularly. This database takes the information present in PDB and adds a layer of information in terms of analysis and/or organization.

This database is a rather large one and houses all proteins of known structure. The organization of the database is such that the proteins are arranged in accordance to their evolutionary, functional, and structural relationships, the basic unit being a protein domain. A domain is considered to be an element of a protein which is self-stabilizing and folds independently of the rest of the protein chain. Domains are not unique to proteins, nor to the products of a gene/gene family, thus are found distributed in a variety of proteins. Often domains carry out a certain specified biological function of the protein, e.g., leucine zipper, calcium-binding domain of calmodulin. These domains arise mainly due to the intrinsic physical and chemical properties of the proteins, which in turn form the basis of protein folding.

Creation of a domain-based database is a very cumbersome and laborious task. In contrast to most of the other database creation processes, such as sequence comparison, clustering, structure comparison, etc., which are routinely performed automatically, there are no automatic methods which are able to define structural domains accurately and consistently in all proteins.

SCOP contains domains present in this database, which are hierarchically classified into species, proteins, families, super families, folds, and classes (Fig. 2.20).

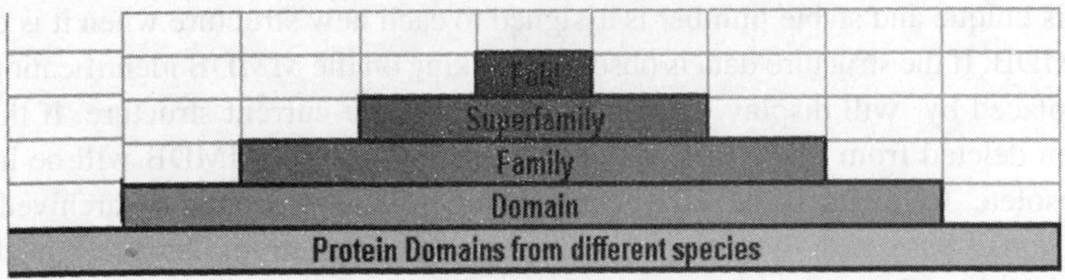

Fig. 2.20 Organization of SCOP database entries

Families in SCOP are organized based on two factors: having residues that are 30% or greater identity and residues having lower identity but greater functional resemblance, e.g., globulins of 15% sequence identity.

Superfamilies consist of collections whose residual identity is not great, but whose functional features suggest a strong ancestral commonality.

A fold is defined in SCOP as a collection of superfamilies having the same secondary structures arranged in the same topological manner. Folds have been classified into five classes: (i) all alpha, (ii) all beta, (iii) alpha or beta, (iv) alpha and beta, and (v) multidomain folds.

2.8.1 SCOP Entry

All SCOP entries have the same format as in PDB files discussed in Section 2.4. Description information in MMDB comes largely from PDB. It includes the following:

Header gives the name(s) and date of an entry.

Title names and briefly describes the content of an entry.

Deposition lists the people who determined the structure and the date when the data was deposited in the PDB. The deposition date can be substantially different from the release date (owing to PDB hold policy). The date when structures are available in MMDB may be slightly later than the PDB release date, since MMDB is updated once a month.

Taxonomy lists the taxonomic assignments for each chain. If all the chains of the structure have the same assignment, only one link is given. Some chains have no taxonomy assigned. Usually these are short nucleotide chains not specific to any one organism. The assignment name is linked to the corresponding nodes in NCBI's taxonomy database.

Reference links to literature references in PubMed. These citations have been taken from PDB and are linked to PubMed by citation matching. MMDB records may contain additional citations not linked to PubMed. This may be conveniently viewed via the GenPept sequence report for any chain.

MMDB gives the identifier which can be used to retrieve structure data from MMDB. This unique and stable number is assigned to each new structure when it is entered into MMDB. If the structure data is obsolete, clicking on the MMDB identification following 'replaced by' will display the summary of the most current structure. If the entry has been deleted from PDB, then the corresponding entry in MMDB will be indicated by 'obsolete' following its MMDB ID number. Obsolete structures are archived in MMDB for consistency with the original publication and annotation. Only current strutures are searched by Entrez.

PDB is a four-character identifier used in the PDB. Clicking the identifier will lead to the corresponding PDB entry.

2.8.2 Search (Keywd Search, Advanced Options)

SCOP can be accessed by using the following methods:

- Keyword search of SCOP entries
- Enter SCOP at the top of the hierarchy
- SCOP parseable files
- All SCOP releases and reclassified entry history
- SCOP domain sequences and PDB-style coordinate files (ASTRAL)

To view the graphic representations of the proteins, your system should conform to the following requirements:

Windows

Hardware	**Software**
Pentium class recommended 64 MB or more recommended 800 × 600 recommended	For Chime 2.6 SP6: • Windows XP with Microsoft Internet Explorer 6.0 and Netscaper Communicator 4.75, 4.79 • Windows 2000 with Microsoft Internet Explorer 6.0, 5.5 SP2, and Netscape Communicator 4.75, 4.79 • Windows NT 4.0 with Microsoft Internet Explorer 6.0, 5.5 SP2, and Netscape Communicator 4.75, 4.79

Macintosh

Hardware	**Software**
Power PC required 32 MB or more recommended 800 × 600 or better recommended	• Mac OS 9.0 with Netscape Communicator 4.75 • Mac OS 8.6 with Netscape Communicator 4.75 • Connectix RamDoubler is recommended if you have less than 32 MB of physical RAM. You need to increase the memory used by Netscape to 15,000 K.

Protein domains can be visualized using either Rasmol or CHIME. MDL® Chime 2.6 SP6 is a plug-in that allows you to correctly register the MIME type for .pdb (Protein Data Bank) files. It will allow you to interactively display 2D and 3D molecules directly in Web pages. You can rotate, reformat, and save the molecules for use in other programs. You can read more about the permitted and non-permitted uses of this software at http://www.mdl.com/mdl/servlet/DownloadServlet.

TASK

Let us first try to search for a known fructokinase using the 'Keyword search of SCOP entries' option.

Step 1 Open the Search page as shown in Fig. 2.21 and type in 'kinase' in the textbox provided. Select the **Search the SCOP database** option and click **Retrieve Information**.

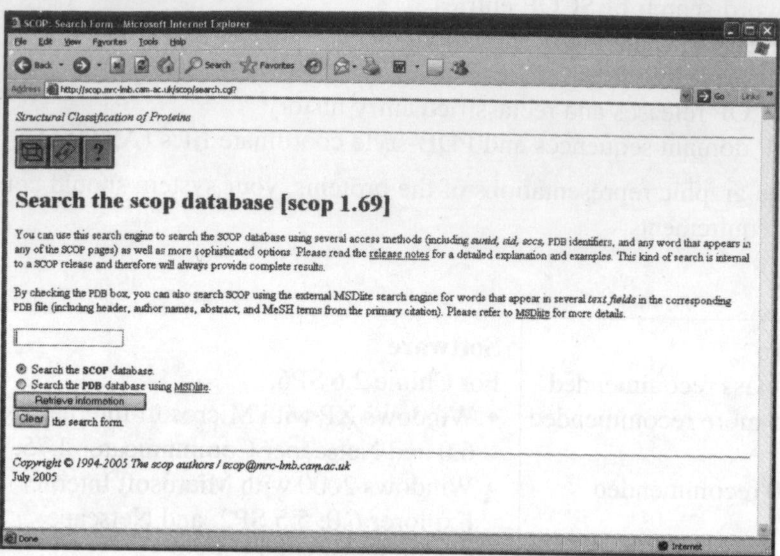

Fig. 2.21

Step 2 The next page will give you a long list of returns (see Fig. 2.22).

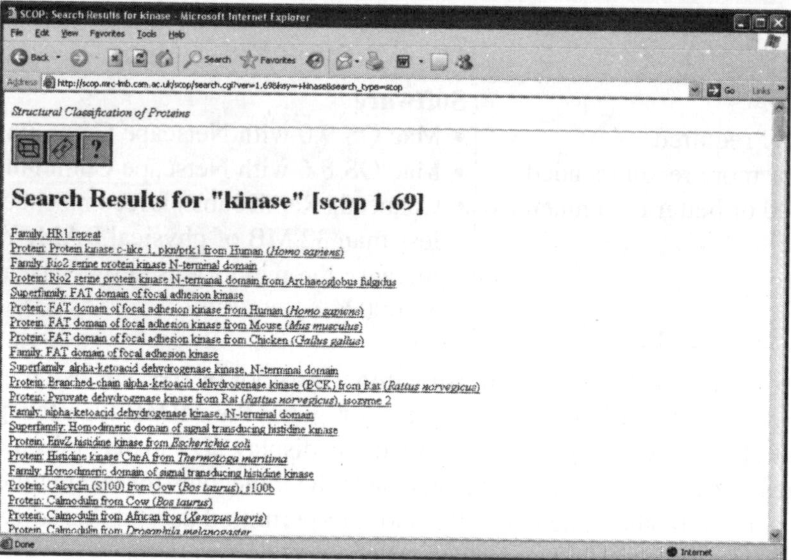

Fig. 2.22

Step 3 Search for a family, that is, **Family: Calmodulin-like**. The result page will look like as shown in Fig. 2.23.

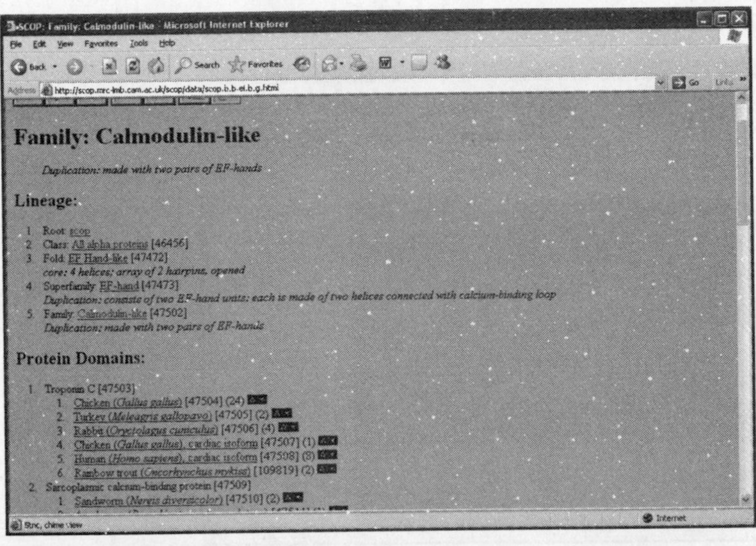

Fig. 2.23

Step 4 Look at how SCOP has organized information:

Root
Class
Fold
Superfamily
Family

Step 5 Now let us try exploring each of these links on the screen.
Click on the link ***scop***. This will give you access to a list of classes (Fig. 2.24).

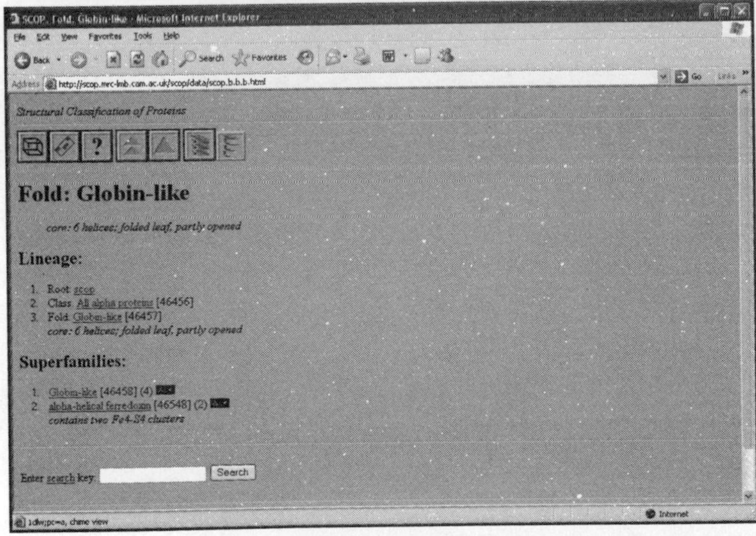

Fig. 2.24

Step 6 Let us explore the first of the classes: ***All alpha proteins [46456]*** (218).

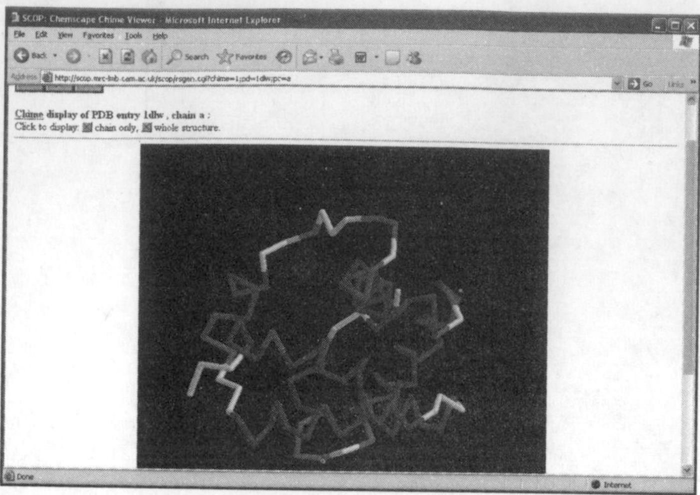

Fig. 2.25

You will note that the chemical structure displayed in Fig. 2.25 can be rotated, reformatted, and saved in various file formats for use in modelling or database software. It also requires Chime as a plug in.

Step 7 Let us explore the fold. There are lots of enzyme subfamilies and families with the general [alpha + beta] fold (Fig. 2.26). Let us retrieve information about ***Fold: Globin-like [46457]*** which has in its core: 6 helices; folded leaf, partly opened.

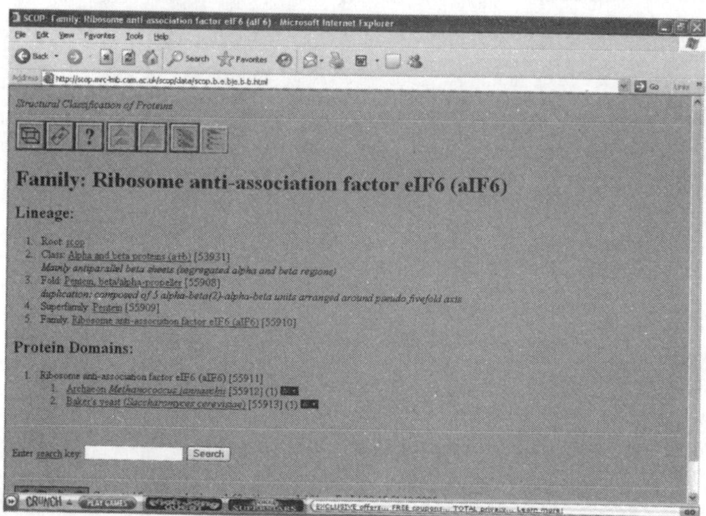

Fig. 2.26

Step 8 Let us try and search for human myoglobin. Type 'myoglobin' in the textbox at the bottom of the interface (Fig. 2.27).

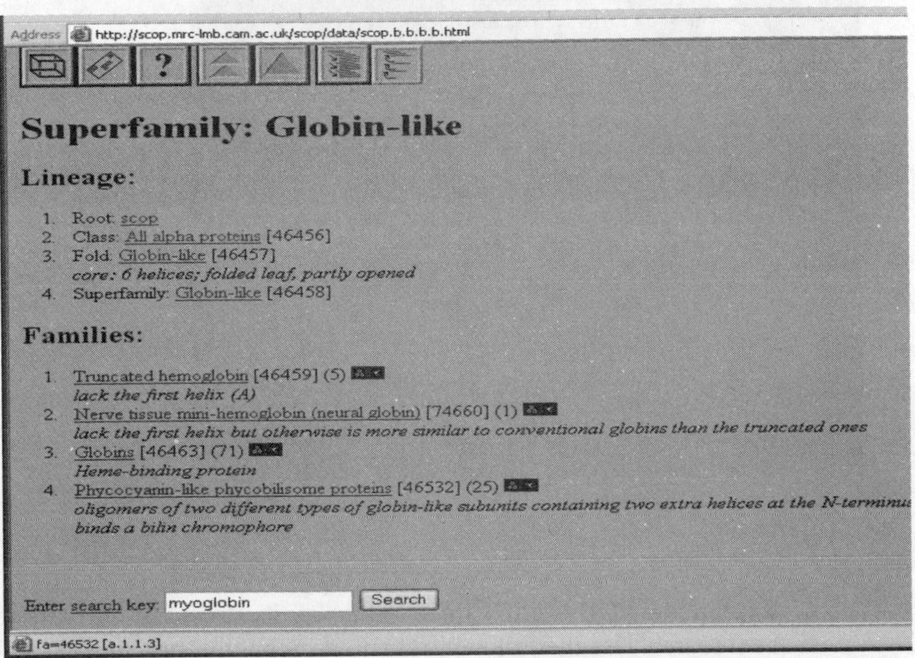

Fig. 2.27

Step 9 The next interface will fetch a list of proteins (Fig. 2.28). Click **Protein: Myoglobin from human (Homo sapiens).**

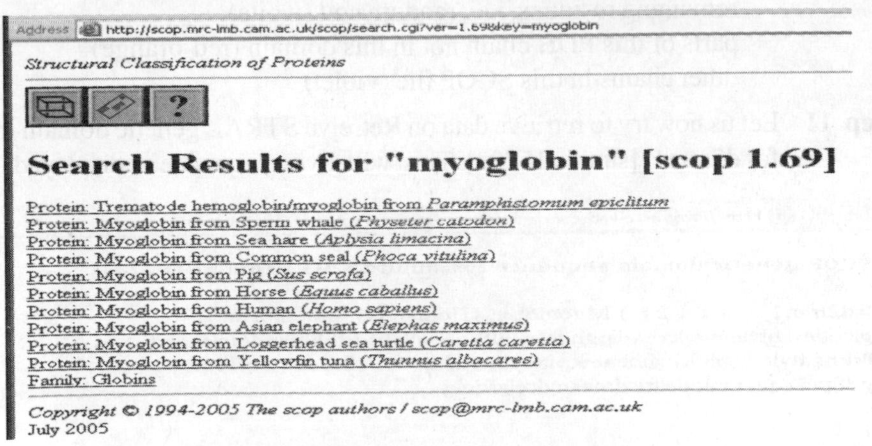

Fig. 2.28

Step 10 The link will display a structure as shown in Fig. 2.29, which shows a PDB entry.

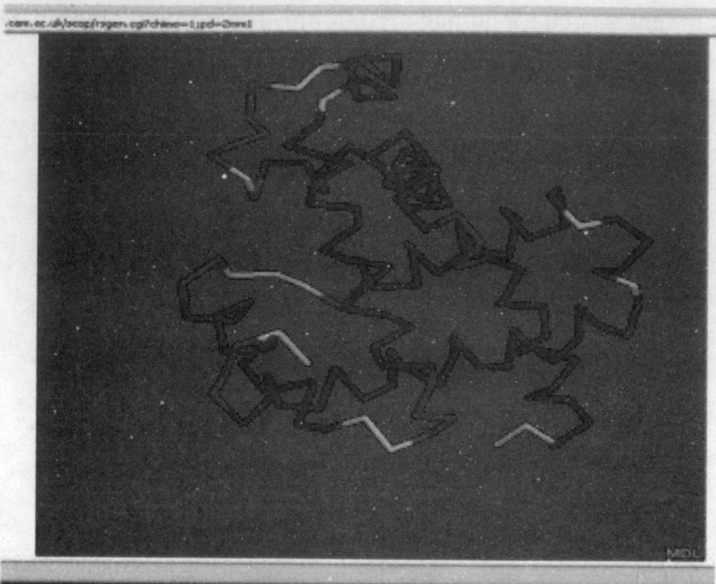

Fig. 2.29

Explore the PDB file in Chime to display its structural attributes or go to the next search. Use the following key to decipher the structure (SCOP will cause a page to be displayed with the molecule coloured):

alpha helix (magenta)
beta strand (yellow)
turn residues (blue)
remaining residues, i.e., random coil (white)
parts of this PDB chain not in this domain (red-orange)
other chains in this SCOP file (violet)

Step 11 Let us now try to retrieve data on Retreive STRAL genetic domain for sequence for d2mm1_[sunid=15203]. This will give a sequence as displayed in Fig. 2.30.

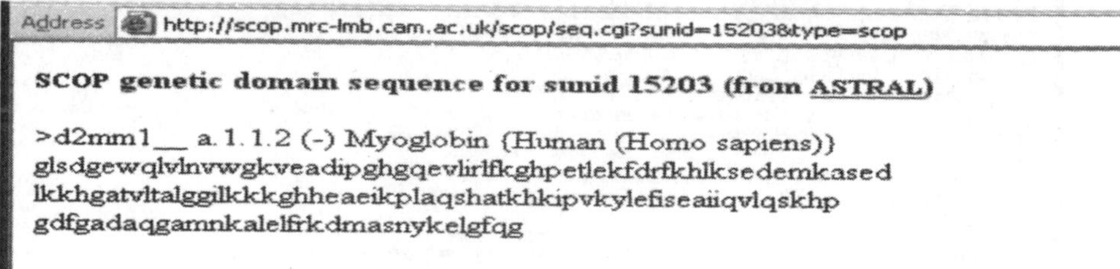

Address http://scop.mrc-lmb.cam.ac.uk/scop/seq.cgi?sunid=15203&type=scop
SCOP genetic domain sequence for sunid 15203 (from ASTRAL)
>d2mm1__ a.1.1.2 (-) Myoglobin {Human (Homo sapiens)}
glsdgewqlvlnvwgkveadipghgqevlirlfkghpetlekfdrfkhlksedemkased
lkkhgatvltalggilkkkghheaeikplaqshatkhkipvkylefiseaiiqvlqskhp
gdfgadaqgamnkalelfrkdmasnykelgfqg

Fig. 2.30

SCOP genetic domain sequences and coordinates, mapping between sequences dromSEQRES and ATOM records in PDB files, and othet SCOP related data can be downloaded from the associated ASTRAL website. The full sets of ASTRAL sequences are available at http://astral.berkeley.edu/

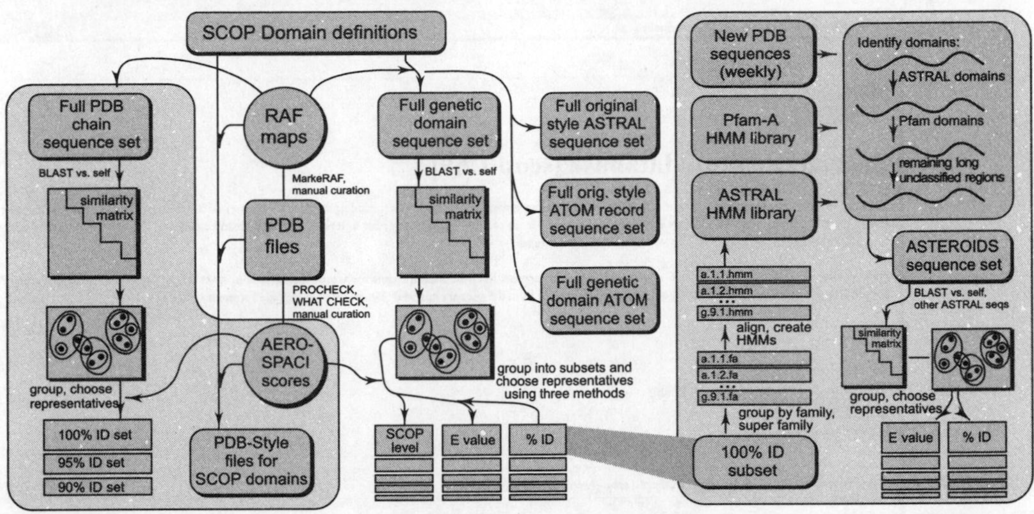

Fig. 2.31 Link of ASTRAL compendium to SCOP and PDB databases

Along with the ASTRAL compendium SCOP can now be used to organize the huge amount of sequence and structural information coming out of various genomics projects, and forms the basis upon which other services, such as Superfamily can be built (Fig. 2.31). In case a SCOP domain includes portions from different PDB chains formed from a single chain precursor, these are listed in the order in which they appear in the original single protein sequence. The purpose is to help develop a common language that can be used to avoid duplication of effort.

2.8.3. Linking SCOP and Parsing SCOP

Each entry in the SCOP hierarchy including levels (i.e., the SCOP domains) and entries corresponding to the protein level are identified through a *sunid*. It is a unique identification number. Another identifier that is used to conduct searches on SCOP is sccs. An sccs is a compact representation of a SCOP domain classification. An sccs identifier includes only the **classs** (alphabetical), **fold**, **superfamily**, and **family** (all numerical) to which each domain belongs to. Sunid and sccs, can be visualized by placing the mouse on the appropriate link (i.e., either in the lineage or in the PDB entry section in a SCOP web page).

TASK

Let us first try to search for a known ribosome anti-association factor domain elF6 (alF6) from *Saccharomyces cerevisiae* using the sccs option. The ribosome anti-association factor domain (PDB entry 1g61, chain A) is d.126.1.1, where d represents the class, 126 the fold 1 the superfamily, and the last 1 the family.

Step 1 Open the SCOP search page as shown in Fig. 2.32 and type in 'd.126.1.1.' in the textbox provided.

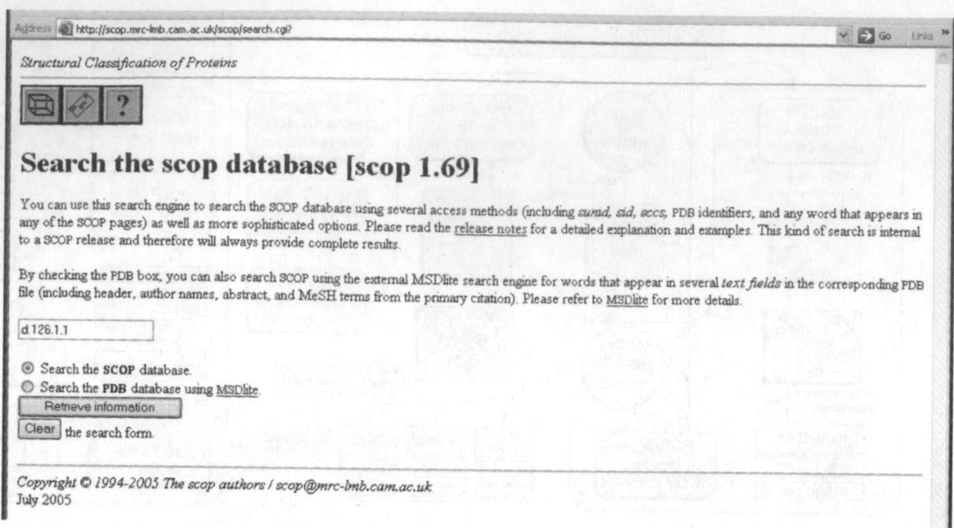

Fig. 2.32

Step 2 The result page will look like As Shown In Fig. 2.33.

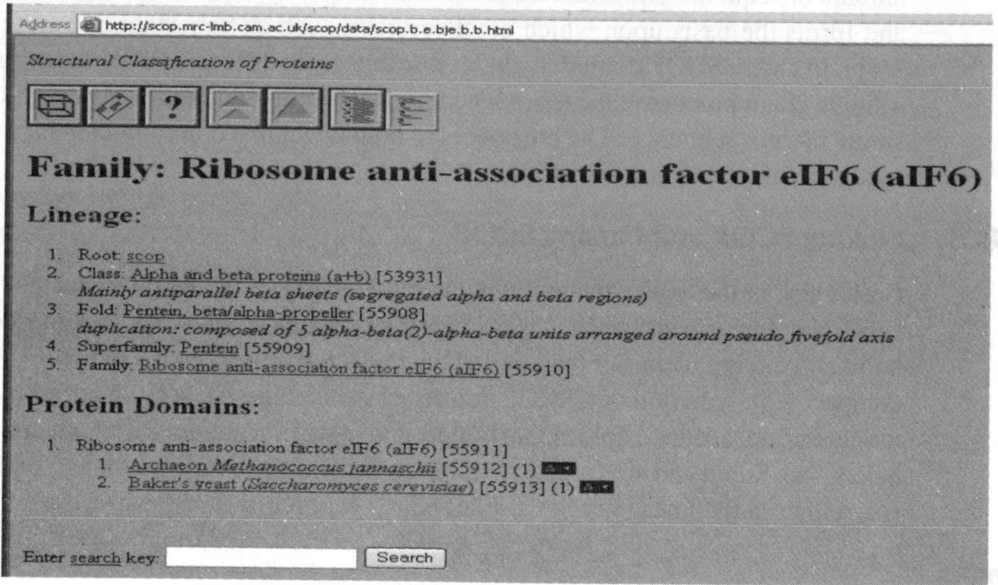

Fig. 2.33

Step 3 From the page displayed in Fig. 2.33, click on the protein domain of your interest: **Baker's yeast (*Saccharomyces cerevisiae*) [55913]**. This will lead to a new page entitled 'Protein: Ribosome anti association factor elF6 (alF6) from baker's yeast *Saccharomyces cerevisiae*

Step 4 Write a report on

- Superfamily pentein
- Beta/alpha-propellar proteins
- Structural similarity between amidino transferase and IF6
- Sequence similarity between amidino transferase and IF6
- The two main difficulties that structural genomics has to face

You can parse SCOP by using the SCOP parser and then save the results to a database and build a tree of SCOP. All the data that is stored in SCOP is now available in three paresable files after the removal of their comments. They replace the older dir.dom.scop.txt and dir.lin.scop.txt. These files have a header that gives the basic information such as release, version, and copyright. These also describe all domains in SCOP and its hierarchy and are suited ideally for computer-based large scale analysis, comparison across releases, and historical summaries.

The file dir.hie.scop.txt was added only after release 1.55. It represents the SCOP hierarchy in terms of sunid. Therefore, each entry has a node in the tree and two more fields. These additional fields correspond to the sunid of the parent of that node (i.e., the node one step up in the tree) and the list of sunids for the children of that node (i.e., the node one step down the tree).

Another file dir.cla.scop.txt contains a description of all domains and their organization in term of sunid and sccs. File dir.des.scop.txt contains details about each node in the hierarchy, including names for proteins, families, superfamilies, folds, and classes.

Conclusion

It is now generally accepted that all proteins have structural similarities with other proteins and, in some of these cases, share a common evolutionary origin. A knowledge of these relationships is crucial to our understanding of the evolution of proteins and their development. It will also play an important role in the analysis of the sequence data that is being produced by worldwide genome projects. Structural domains are considered to be the basic units of protein folding, evolution, function, and design. Thus, it is not surprising that several databases of protein domains have been constructed, which provide a rich source of information for structural and functional prediction/annotation of proteins. One application of these domain databases is threading-based protein structure prediction. Here the best advantage of domain-based threading, compared with protein chain-based threading, is its wider scope of applicability, since a novel protein may share only a domain rather than the entire structure with a known protein structure. Decomposition of multidomain protein structures into individual domains is expected to continue to go up, partly due to the worldwide structural genomics projects.

EXERCISES

Exercise 2.1

(a) Go to the RSCB site.
(b) Find out the molecule of the month.
(c) Note the current statistics of PDB.
(d) Give a keyword search for haemoglobin and look for 'THE STRUCTURE OF HUMAN CARBONMONOXY HAEMOGLOBIN AT 2.7 ANGSTROMS RESOLUTION.
(e) What is the difference between HCO and 2HCO?
(f) What is the common experimental method for deriving the protein structures?
(g) Name a protein ID that is meant for electron transport.
(h) Name a protein ID meant for both oxygen storage and transport.
(i) Is there any structure waiting for release? What is the ID?
(j) In what format can you download the PDB file?
(k) What tool is used to visualize the structure?
(l) Can you view the PDB file without downloading?
(m) View structure summary for 1HGB.
(n) Find out the chemical formula of any ID present in ligands and prosthetic groups from the structural report.
(o) Look in the Geometry tab for bond types where you have maximum and minimum fold deviation score (FDS).

Exercise 2.2

(a) Go to the MMDB page in the NCBI site.
(b) What is the MMDB and GI code corresponding to1HGB?
(c) List down the number of residues for chains A, B, C, and D of protein.
(d) Name the domain family listed along with the chains and describe it in one line.
(e) Give the compound information about Hematin listed as ligand.

Exercise 2.3

(a) Go to the CATH home page.
(b) What are the various CATH data formats?
(c) What are the main classification levels of CATH?
(d) Prepare a summary on CATHSOLID.
(e) What are the various levels under mainly alpha classification?
(f) List various levels under mainly beta.
(g) List various levels under mixed alpha beta.
(h) List is the level under few secondary structures.
(i) Copy and paste COMBS sequence of the domain 1ewfA01.
(j) Copy and paste ATOM sequence of the domain 1ewfA01.
(k) What are the various files you can download from each domain classification page?

Exercise 2.4

(a) Open the DALI search page at http://ekhidna.biocenter.helsinki.fi/dali/start

(b) Under the keyword search box enter '1HGB'.

(c) Select 1dxtB_1 and 1bz1A from the list and compare first 20 alignments for both.

(d) Prepare a report for first 5 alignments of both the representatives for the following:

  (i) raw-score: the sum of weighted similarities of intramolecular distances that DALI maximizes

  (ii) Z-score: normalized score that depends on the size of the structures.

  (iii) %ID: percentage of identical amino acids over all structurally equivalent residues.

  (iv) DALI: number of structurally equivalent residues

  (v) rmsd: root-mean-square deviation of C-alpha atoms in the least-squares superimposition of the structurally equivalent C-alpha atoms.

Exercise 2.5 You have determined the sequence of a gene. Notable features are the absence of a TATA-box and the presence of AGGGA elements in both upstream regions, the presence of several sequences with homology to known regulatory elements (cAMP-, retinoic acid- and interferon-responsive elements) in the gene, and a high G + C-content and several putative Sp1-factor-binding sites in the gene. The corresponding purified protein appeared to be homogeneous by SDS-polyacrylamide gel electrophoresis and shows calcium binding properties. However, the apparent molecular mass of calmodulin-like protein on SDS-polyacrylamide gel electrophoresis was 10 kDa, which was smaller than bovine brain calmodulin (17 kDa). In addition, the expression of this calmodulin-like protein levels are significantly reduced in malignant tumor cells as compared to corresponding normal epithelial cells. Go to SCOP and do a keyword search against this calmodulin-like protein in other vertebrates.

1. Look at the levels in SCOP: root, class, fold, superfamily, family. Within the family look for human calmodulin-like protein.
2. Examine the individual enzyme and obtain structural information about the protein.
3. Obtain information about the superfamily EF-hand. Visualize the protein using Chime.
4. Repeat the steps 1–3 for another protein, Aequorin. It is a photoprotein isolated from luminescent jellyfish (and a variety of other marine organisms).
5. Is there any similarity between the structures of the two proteins?
6. Based on the structural features what inference can you make about the function of the two proteins?

Exercise 2.6 Human JUN protooncogene is assigned to chromosome region 1p31-32, a chromosomal region involved in both translocations and deletions of chromosomes seen in human malignancies. The JUN gene encodes a protein that is structurally and functionally identical to the transcription factor AP-1, which is itself identical to yeast transcriptional activator GCN4.

1. Briefly give one reason why MMDB would yield different structural neighbours from the other classification methods, i.e., CATH and SCOP.
2. Using this result, explain the structure prediction strategy you will adopt for each sequence. Elaborate only on the modelling, secondary structure prediction, and expected accuracy of the predictions.
3. Work out your prediction strategy using Swiss model and PHD.
4. Report the results for secondary structure predictions in a textual format showing the predicted secondary structure for each residue.
5. Can you suggest how the structures of these proteins might eventually be classified in CATH? Does the prediction agree with the classification found in the CATH database?

Exercise 2.7 The primary structure of the cytotoxin alpha-sarcin was determined. Eighteen of the 19 tryptic peptides were purified; the other peptide has only arginine. The complete sequence of 17 of the peptides was determined; the sequence of the remaining peptide was determined in part. Copy the input files named d1de3a.zip. They contain: template structure of d1rgea_ domain from the SCOP database and alignment of d1de3a_ sequence with d1rgea_ sequence. The alignment is saved in two separate files named d1de3a_.txt and d1rgea_.txt. Open the url http://www.cmbi.kun.nl/gv/servers/WIWWWI/ and go to the option Build/check/repair model on the left column. Choose homology modelling link which allows you to build a model on a template structure. Extract the input file at a specific location on your computer. Submit the request for modelling. Compare the model with the real structure of the protein d1de3a_ from SCOP.

Exercise 2.8 Explore the following using Rasmol, MMDB, CATH, and SCOP. Use files in the zipped folder as input files.

- Tim Barrel: 1timA.pdb
- Transmembrane protein 1MSL
- 1hiv.pdb (HIV-1 protease complex with an inhibitor)

Use all display options such as 'Display Cartoons' and 'Colour by Structure', 'Spacefill', and 'Ball & Sticks'.

3 Other Databases

Learning Objectives

- To describe the various kinds of enzyme, pathway, disease, and literature databases
- To learn about the organizations of these databases
- To list the attributes of each of these databases
- To use information retrieval systems to obtain data from these databases

Introduction

Biotechnology organizations (both public and proprietary) maintain a large collection of databases and bioinformatics tools for molecular biology research. A deeper and clear understanding of biotechnological concepts can be achieved with the help of these databases and tools. But how to retrieve the needed information requires an efficient way; this is the fundamental and most important skill required of anyone interested in bioinformatics. In this chapter we will explore some databases which are very specific in nature and store information that has now become essential to conduct meaningful research in bioinformatics. These include databases concerning enzymes, pathways, diseases and literature.

3.1 Enzyme Databases

There is hardly any reaction in a cell that is not controlled by enzymes. Enzymes are responsible for control of every single reaction and thus control cellular metabolism. Enzymes act as catalysts (substances that speed up reactions without actually entering into the reaction). They are used over and over, and a single enzyme molecule may mediate thousands of reactions in a second. One catalase enzyme molecule, for example, can completely break down 5.6 *million* hydrogen peroxide molecules in a *minute*. Yet catalase is considered a relatively slow enzyme! Carbonic anhydrase can break down

36 million carbonic acid molecules in a minute, and so is rightly considered to be one of the fastest enzymes. Enzymes are mostly globular proteins. Each enzyme has a specific structure (which is also known as its native conformation), function, distribution of electrical charges, and surface geometry; its specificity depends on its tertiary structure. The tertiary structure determines the three-dimensional shape.

A typical prokaryotic cell has about 700 enzymes, and there are about 2000 enzymes in an eukaryotic cell. This large number is necessary because enzymes are very particular about the reactions they catalyse; in fact, an enzyme may catalyse only one specific reaction. With that many enzymes, each one differing a little from the rest, it is necessary to store information about them in a specialized database. Such databases contain a very large collection of facts related to enzymes, including reaction specificity, functional parameters, substrates, products, and inhibitors.

Classes of enzymes The International Union of Biochemistry and Molecular Biology (IUBMB) has developed a classification scheme in which enzymes are classified into six major categories. These categories are summarized in Table 3.1.

Table 3.1 IUBMB classification scheme

EC #	Name	Reactions catalysed	Sample reaction	Comments
1	Oxido-reductases	Oxidation-reduction	Lactate + NAD → pyruvate + NADH + H^+	Often involves NAD or FMN
2	Transferases	Transfer a group larger than water	Alanine + a-ketoglutarate → pyruvate + glutamine	Includes kinases
3	Hydrolases	Hydrolysis (elimination or Transfer of H_2O)	Pyrophosphate + H_2O → phosphate + phosphate	
4	Lyases	Lysis making a double bond; (converse: addition across double bond)	Pyruvate + H^+ → acetaldehyde + CO_2	Often called synthases when reverse reaction is emphasized
5	Isomerases	Unimolecular reactions	L-alanine → D-alanine	
6	Ligases	Joining of two substrates	Glutamate + ATP + NH_4+ → glutamine + ADP + Pi	Generally require high-energy phosphate compounds such as ATP

Individual enzymes are numbered in a four-component system, known as the EC system, originally developed by the Enzyme Commission. There are periods (.) separating the levels of organization. Thus pancreatic elastase, a protease (enzyme that cleaves peptide bonds), is a hydrolase (category 3) and its full Enzyme Commission designation is 3.4.21.36:

- Category 3: hydrolases
- Subcategory 3.4: hydrolases acting on peptide bonds (peptidases)
- Sub-subcategory 3.4.21: serine endopeptidases
- Sub-sub-subcategory 3.4.21.36: pancreatic elastase

If it helps you understand the system, think of the EC numbers as being like Internet Protocol (IP) addresses; the same logic applies. If you want to see the details of the EC numbering system, go to http://www.chem.qmw.ac.uk/iubmb/enzyme/. Some of these are ENZYME, BRENDA, KEGG, IntEnz, and PRECISE.

3.2 MEROPS (http://merops.sanger.ac.uk/)

MEROPS is a database of peptidases and their inhibitors. The entries in MEROPS are generated by experts on peptidases/inhibitors and contain detailed functional information about an enzyme. A comprehensive set of references accompanies each entry. The MEROPS database is organized into a hierarchical classification, based on structure, function, and substrate specificity.

This database is a resource for catalogue- and structure-based classification of peptidases. It contains a set of files, termed PepCards. Each of these cards provides information on a single peptidase. The information contains classification and nomenclature as well as hypertext links to the relevant entries in other databases. The peptidases are classified into families on the basis of statistically significant similarities between the protein sequences in the part termed the 'peptidase unit' that is most directly responsible for activity.

Peptidases represent one of the most important and diverse classes of enzymes. These are normally divided into two kinds: endo and exo peptidases. Peptidases have many roles in physiological, pathological, and biotechnology processes. A peptidase can either break proteins down into short peptides or make just one or two cleavages in a specific protein.

(a) History, size, content

> 'The idea of families and clans came from a television documentary on the Whitefronted Bee-eater (*Merops bullockoides*), a common resident along the Zambezi river and its tributaries, which nests in colonies in banks along rivers and streams. These colonies are divided into families and clans each of which occupy a different part of the colony and each has its own discrete area away from the colony where they hunt for flying insects.'
>
> (from http://delphi.phys.univ-tours.fr/Prolysis/merops/history.htm)

The MEROPS team was originally based at the Baraham Institute, but have recently move to the WTSI, where they share an office with the Pfam team. The MEROPS database can be accessed at http://merops.sanger.ac.uk. The project is primarily funded by the MRC.

This database has a structure-based classification of peptidases. These peptidases are classified by catalytic type, clan, and family. To make this process easier, all of these are colour coded. In addition, there are links that give access to database entries for the enzymology, protein and nucleic acid sequences, tertiary structures, and genetics. Viewers such as Rasmol can be used to display tertiary structures.

(b) Classification (by molecular structure and homology, mechanism of catalysis, and reaction catalysed)

Both nomenclature and classification are vital to information handling, storage, and retrieval. A good system serves to highlight important questions and thus prompts new discoveries. Three useful methods of grouping peptidases are currently in use:

1. Grouping by molecular structure and homology,
2. Grouping by the chemical mechanism of catalysis, and
3. Grouping by the details of the reaction catalysed.

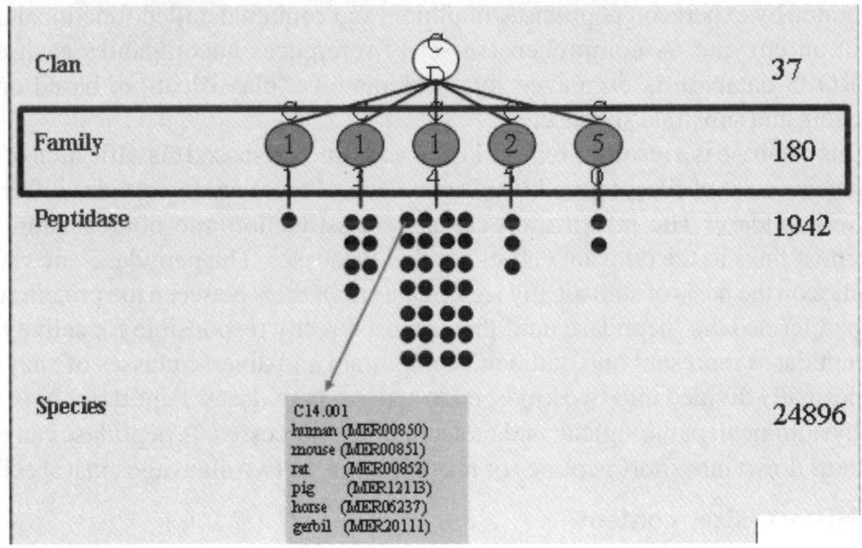

Fig. 3.1 The MEROPS classification hierarchy

The MEROPS classification is a hierarchy (see Fig. 3.1). Starting at the top of the hierarchy, all peptidases having a common ancestor are grouped into a 'clan'. All the peptidases within a clan are expected to have a similar tertiary fold and the position of the catalytic residues therefore is expected to be conserved. In Fig. 3.1 clan CD is one of the thirty-seven clans of peptidases in MEROPS. The first letter of a clan name represents the catalytic type, in this case cysteine, and the second letter is sequentially assigned.

The second level of the hierarchy is 'family'. There are 180 families in MEROPS. All the peptidases within a family have homologous protein sequences. In this example, there are five families within clan CD. In a family name, the first letter represents the catalytic type, and the numbers are sequentially assigned.

The third level of the hierarchy is that of 'peptidase'. One peptidase is distinguished from another on a biochemical basis. The number of peptidases varies greatly from one family to another. Family C11 has only one peptidase, clostripain, whereas there are at least twenty-nine in family C14. In the largest family of peptidases, S1, there are over 400 peptidases. At present we recognize nearly 2000 different peptidases.

A peptidase identifier consists of the family name, a period (.), and a sequential number. The level of peptidase consists of species. Peptidase C14.001 (shown in Fig. 3.1) has species from six mammals, including human. Splice variants, SNPs, alternative translations and sequence variants from subspecies, strains, and isolates are not distinguishable in MEROPS. Essentially, the products of a single gene are given a single identifier, called a MERNUM. There are nearly 25,000 sequences of peptidases and their homologues in MEROPS, about the same number as protein coding genes in an eukaryotic genome. Less than 10,000 of these can be assigned to a peptidase identifier.

Note

While classifying entries in MEROPS, rather than logically start at the top or the bottom of the hierarchy, it is easiest to start at the 'family' level, because you will be familiar with the principles from PFAM.

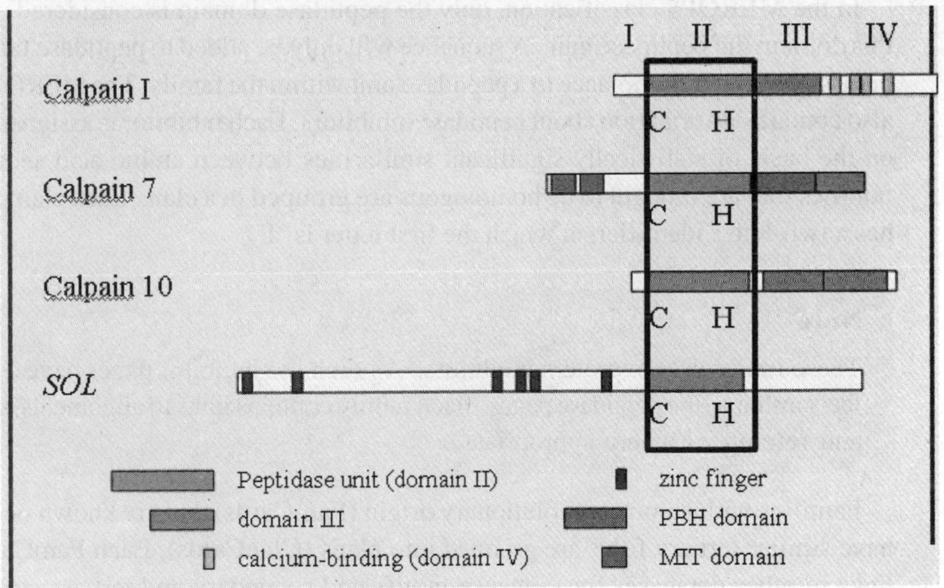

Fig. 3.2 A peptidase unit in MEROPS

Many peptidases are multi-domain proteins, and MEROPS, like PFAM, uses a classification of protein domains. Figure 3.2 shows members of peptidase family C2, which includes the calcium-activated cytoplasmic peptidase calpain I. Calpain I is a heterodimer of a large subunit and a small subunit. The larger subunit carries the active site residues and consists of four domains. Domain I is an activation peptide; domain II is the peptidase domain; the function of domain III is not clear; and domain IV carries EF-hand subdomains, which bind calcium. Other calpains have a variety of domains (mostly of some unknown function) attached to the peptidase domain. One of the most strikingly different calpains is the product of the sol gene in Drosophila, which has a role in the development of the small optic lobe. It has six N-terminal zinc fingers.

A clan can be established when the tertiary structure of a peptidase has been determined. If that structure is unlike that of peptidases from any other clan, a new clan is created. Usually, the first peptidase to have its structure determined becomes the 'type peptidase' for that clan. Another family can be added to that clan if a member has a tertiary fold similar to that type of peptidase.

Out of the 181 families of peptidases, structures are known for only 82 members. Although there are several families that we are unable to assign to a clan, usually because the active site residues are not known, we are able to make some predictions. Where a clan consists of two or more families, we have noticed that the active site residues and their positions are conserved. So we are able to predict to which clan a family belongs by comparing the order of active site residues.

In the MEROPS classification, only the peptidase domain is considered, and we call this domain the peptidase unit. A sequence will only be added to peptidase family C2 if is has a significant resemblance to a peptidase unit within the family. The MEROPS database also contains information about peptidase inhibitors. Each inhibitor is assigned to a family on the basis of statistically significant similarities between amino acid sequences, and families that are thought to be homologous are grouped in a clan. Each clan of inhibitors has a two-letter identifier, in which the first letter is 'I'.

Note

These inhibitors are protein inhibitors. As such the inhibitor pages have a look and feel similar to the peptidase pages. Each family contains links to alignments, structures and references where appropriate.

Families with common evolutionary origin (FamCards) that are known or expected to have similar tertiary folds are grouped into clans (ClanCards). Each FamCard provides links to other databases for sequence motifs and secondary and tertiary structures, and shows the distribution of the family across the major taxonomic kingdoms (see Fig. 3.3).

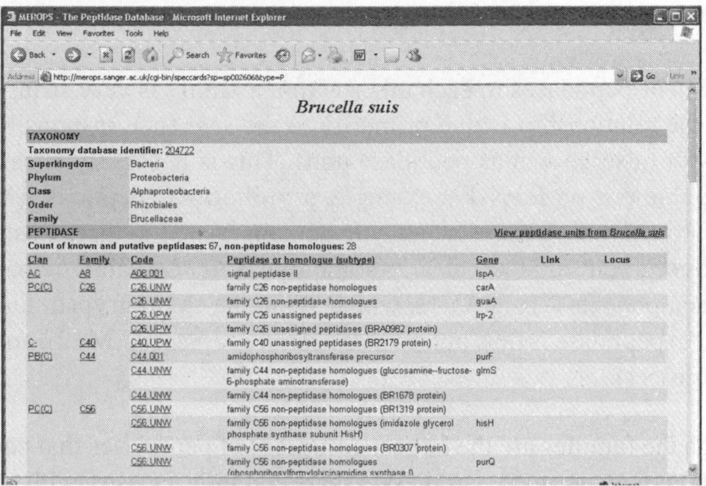

Fig. 3.3

(c) Index of peptidases and inhibitors

The classification of peptidases on the basis of molecular structure and homology is the newest of the three methods. It uses data for amino acid sequences and three-dimensional structures. In 1993, Rawlings and Barrett described a system in which individual peptidases were assigned to families, and the families were grouped in clans. Later this scheme was developed to provide the structure of the *MEROPS* database, and has been extended to include the proteins that inhibit peptidases. The description below relates specifically to the way the classification of individual peptidases and inhibitors by molecular structure and homology is implemented in the *MEROPS* database.

Individual peptidases Each individual peptidase is given a MEROPS identifier that acts much like an accession number. The six-character identifier of the family to which the peptidase belongs consists of an alphabet representing the family, a two-digit family number, a period, and a three-digit number. For example, the identifier of chymotrypsin, the first peptidase in family S1, is S01.001.

Families The families of peptidases are constructed by comparisons of amino acid sequences. Every member of a *MEROPS* family shows a statistically significant relationship in amino acid sequence to at least one other member of the family.

A few of the families in MEROPS contain two or more rather distinct groups of peptidases. These are represented as divergences in a dendrogram. For these, subfamilies are recognized. The naming of the families follows the system introduced by Rawlings and Barrett (1993) in which each family is named with a letter denoting the catalytic type (S, C, T, A, G, M, or U, for serine, cysteine, threonine, aspartic, glutamic, metallo, or unknown), followed by an arbitrarily assigned number.

Note

The relationship of peptidases to each other within a family is very stringent. It is required that the relationship is due primarily to the sequence responsible for its catalytic activity (also known as peptidase unit). This is necessary because some peptidases are chimeric proteins. For example, procollagen C-peptidase (M12.005) is a chimeric protein that contains a catalytic domain related to that of astacin. But it also contains segments that are clearly homologous to non-catalytic parts of the complement components C1r and C1s, which are in the chymotrypsin family. The procollagen endopeptidase is thus placed in the family of astacin (M12), and not that of chymotrypsin (S1).

Clans Although the families are the largest groupings of peptidases that can be proven rigorously to be homologous, many families do share common ancestry with others. These 'sets of families' have proteins which have diverged from a single ancestral protein, but they have diverged so far that their relationship can no longer be proved by comparing the primary structures. These are regrouped and called a clan. The name of each clan is formed from the letter for the catalytic type of the peptidases it contains (as for families), followed by an arbitrary second capital letter. For example, the first clan of cysteine peptidases is clan CA.

Peptidases grouped by the kinds of reaction they catalyse In a sense, all peptidases catalyse the same reaction—hydrolysis of a peptide bond. But they are selective for the position of the peptide bond in the substrate and also for the amino acid residues near the bond that is to be hydrolysed (also known as the 'scissile' bond).

Endopeptidases An endopeptidase hydrolyses internal, alpha-peptide bonds in a polypeptide chain, tending to act away from the N- or C-terminus. Examples of endopeptidases are chymotrypsin (*S01.001*), pepsin (*A01.001*), and papain (*C01.001*). Some endopeptidases act only on substrates smaller than proteins, and these are termed oligopeptidases. thimet oligopeptidase (*M03.001*) is an example.

Omega-peptidases Omega-peptidases are a group of peptidases that have no requirement for a free N- or C-terminus in the substrate. Although they do not have a requirement for a charged terminal group, they act close to one terminus or the other, and are thus distinct from endopeptidases. Some hydrolyse peptide bonds that are not alpha-bonds; that is, they are isopeptide bonds, in which one or both of the amino and carboxyl groups are not directly attached to the alpha-carbon of the parent amino acid. The omega-peptidases are placed in sub-subclass EC 3.4.19 by NC-IUBMB. Some of these are ubiquitinyl hydrolases (e.g., ubiquitinyl hydrolase-L3, *C12.003*), pyroglutamyl peptidases (*C15.001*, *C15.010*, *M01.008*), and gamma-glutamyl hydrolase (*C26.001*).

Exopeptidases Exopeptidases require a free N-terminal amino group, C-terminal carboxyl group, or both. These peptidases hydrolyse a bond not more than three residues from the terminus. They are further subclassified as aminopeptidases, carboxypeptidases, dipeptidyl-peptidases, peptidyl-dipeptidases, tripeptidyl-peptidases, and dipeptidases.

(d) Search (keyword search, advanced options)

Connect to the website of MEROPS (http://merops.sanger.ac.uk/). On that page click the SEARCHES button. The search page of MEROPS is displayed next (as shown in Fig. 3.4).

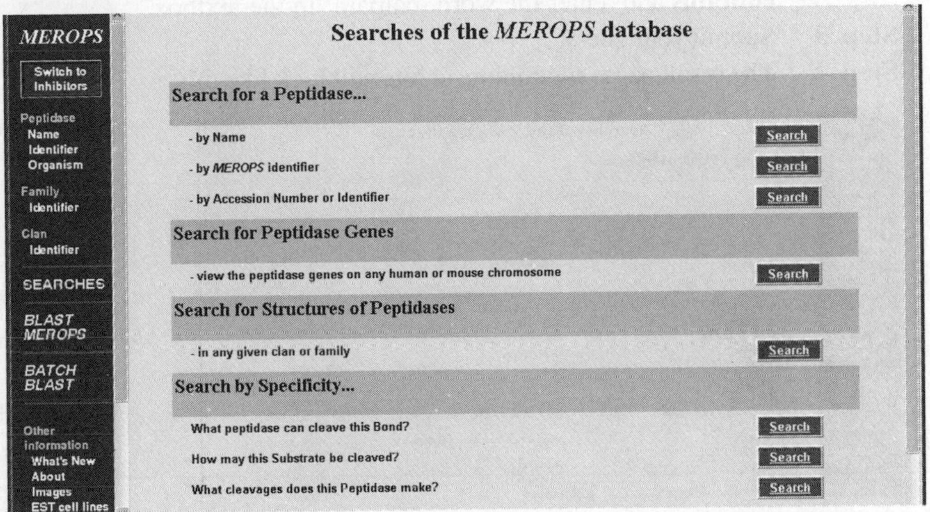

Fig. 3.4 Search page of MEROPS

You can search MEROPS by the following categories:

- Peptidase
  - Name
  - MEROPS identifier
  - Accession number or identifier
- Peptidase genes
  - Human
  - Mouse
- Structures of peptidases
  - In a clan
  - In a family
- Peptidase specificity
  - By cleavage
  - By substrate
  - By bond

Let us look at each of these a little more closely in the following section.

TASK Searching and retrieving from MEROPS

Case (i): Search for a peptidase

Search by name Let us first try to search for a known peptidase using the **by Name** option. Let us look for a well known peptidase calpain I.

Step 1 Open the search page as shown in Fig. 3.4 and click **Search** on the right hand side of by **Name.**

Step 2 The next page gives you three options: **begins with**, **contains**, and **is**. Choose **contains** and enter the word 'calpain' in the textbox.

Step 3 Submit your query.

Step 4 The result page, containing 44 hits will look like this.

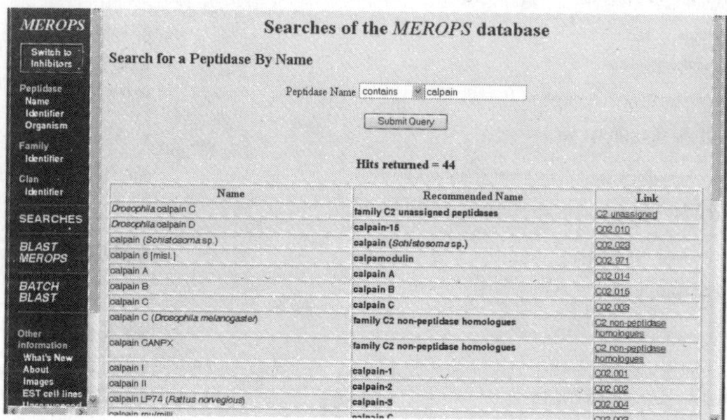

Fig. 3.5

Step 5 Further exploration for Calpain I by clicking **CO2.001** which is the calpain of our interest will lead us to a new page, as shown in Fig. 3.6.

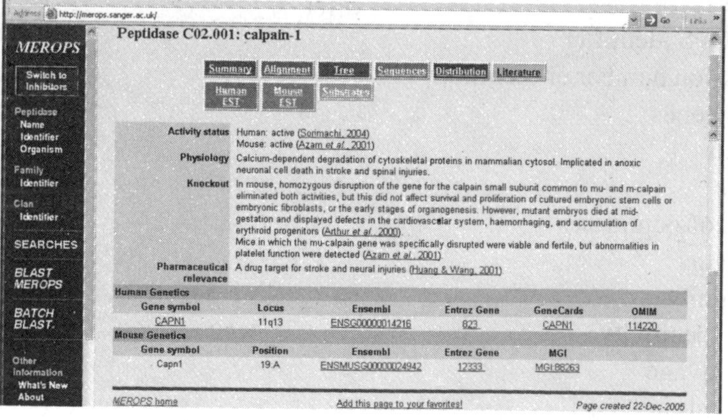

Fig. 3.6

Step 6 Study the summary page carefully and prepare a report on the following subheads:

- Names
- Domains
- MEROPS Classification
- Activity
- Human Genetics
- Mouse Genetics

Step 7 Explore the page further by choosing from the various options such as **Summary**, **Alignment**, **Tree**, **Sequences**, **Distribution**, **Literature**, **Human EST**, **Mouse EST**, and **Substrates**

Search by MEROPS identifier Let us look for the angiotensin-forming enzyme rennin using the MEROPS identifier.

Step 1 Open the search page in the previous exercise (Fig. 3.4) and click **Search** on the right hand side of **by *MEROPS* identifier**. The page that is displayed allows you to enter either the complete MEROPS identifier or just the start of an identifier. Let us enter the identifier 'A01' in the textbox.

Step 2 Click **Submit Query**.

Step 3 The result page, containing 67 hits, will look like as in Fig. 3.7.

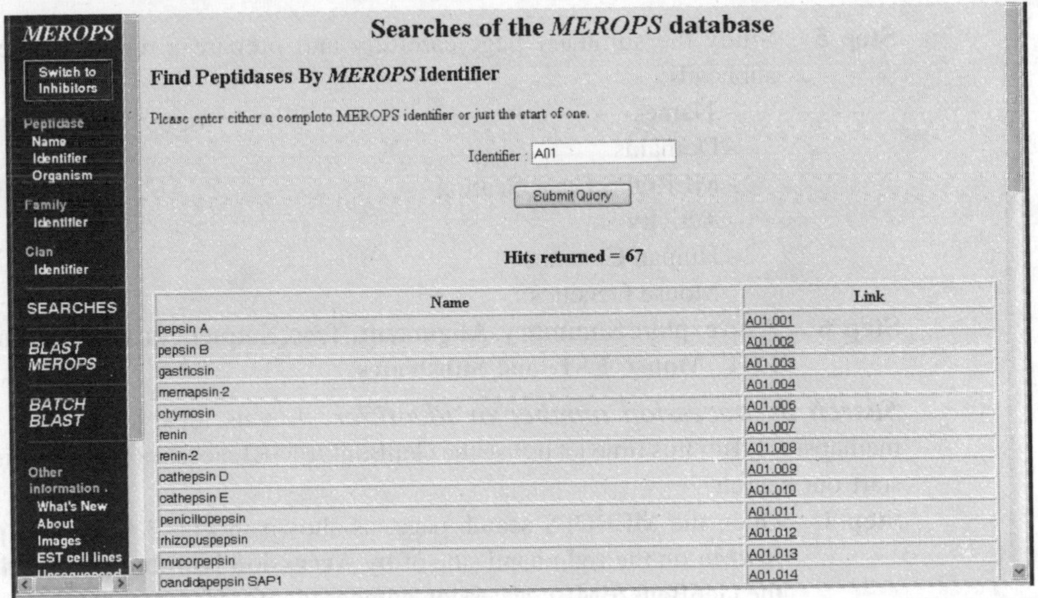

Fig. 3.7

Step 4 Further exploration for renin, by following the link **A01.007**, which is the peptidase of our interest, will lead us to a new page (as displayed in Fig. 3.8).

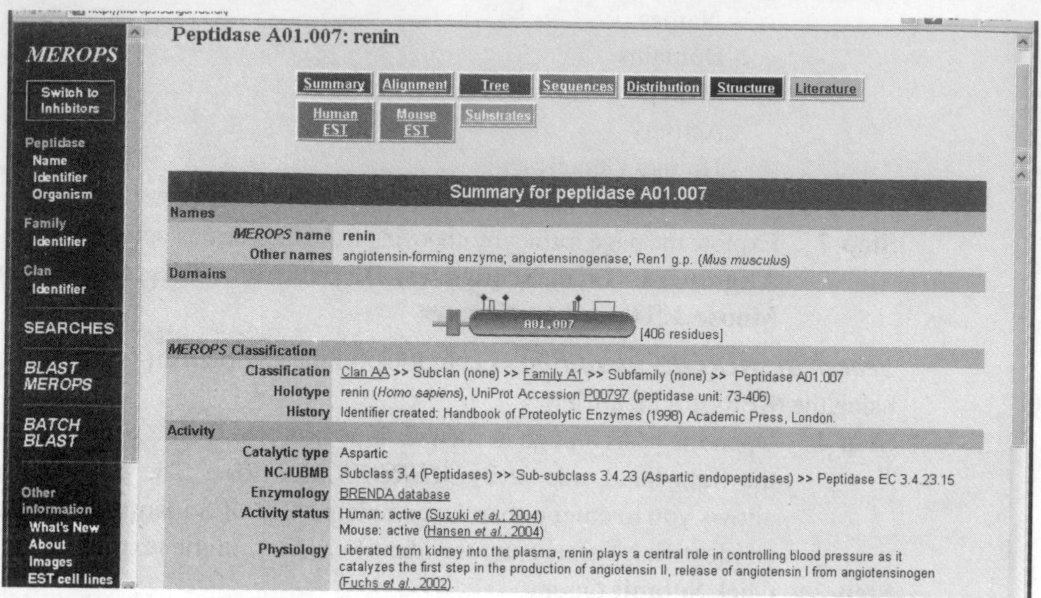

Fig. 3.8

Step 5 Study the summary page carefully and prepare a report on the following subheads:

- Names
- Domains
- MEROPS Classification
- Activity
- Human Genetics
- Mouse Genetics

Step 6 Also explore **Summary**, **Alignment**, **Tree**, **Sequences**, **Distribution**, **Human EST**, **Mouse EST**, and **Substrates**.

Search by accession number or identifier Let us look for another peptidase, memapsin-2. But this time let us use the GenBank/EMBL accession number/identifier to start our search.

Step 1 Open the MEROPS search page, as shown in Fig. 3.4. On that page, click ***Search*** on the right hand side of **by Accession Number or Identifier**. Enter the GenBank/EMBL accession number identifier 'AB032975' in the textbox.

Step 2 Click **Submit Query**.

Step 3 The result page, containing a single hit, is shown in Fig. 3.9.

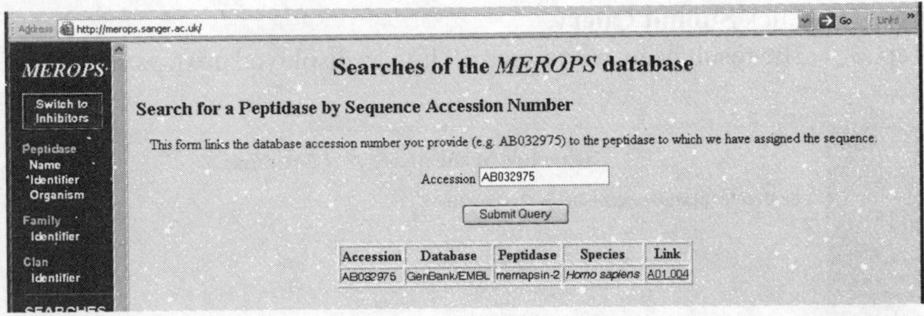

Fig. 3.9

Step 4 Further exploration for memapsin-2 by clicking **A01.004**, which is the peptidase of our interest, will lead to a summary page. This looks like the previous two summary pages.

Step 5 Study the summary page and prepare a report on the following subheads:
- Names
- Domains
- MEROPS Classification
- Activity
- Human Genetics
- Mouse Genetics

Step 6 Also explore **Summary**, **Alignment**, **Tree**, **Sequences**, **Distribution**, **Human EST**, **Mouse EST** and **Substrates**.

Case (ii): Search for peptidase or inhibitor genes

Human or mouse Let us look for the angiotensin-converting enzyme 2 that resides on chromosome number 10 of human beings.

Step 1 Open the MEROPS search page as in the previous exercises. There will be two drop down boxes and a textbox.
Organism: You can choose **Homo sapiens** or **Mus musculus**.
Type: This again gives two choices, that or **Peptidase** and **Inhibitor**.
Chromosome: You need to enter the chromosome number.

The interface appears as shown in Fig. 3.10.

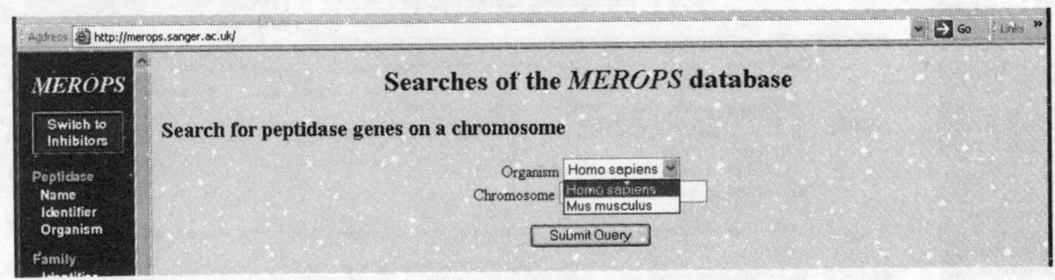

Fig. 3.10

Step 2 Click **Submit Query**.

Step 3 The result page, containing 19 hits, is displayed next, as shown in Fig. 3.11.

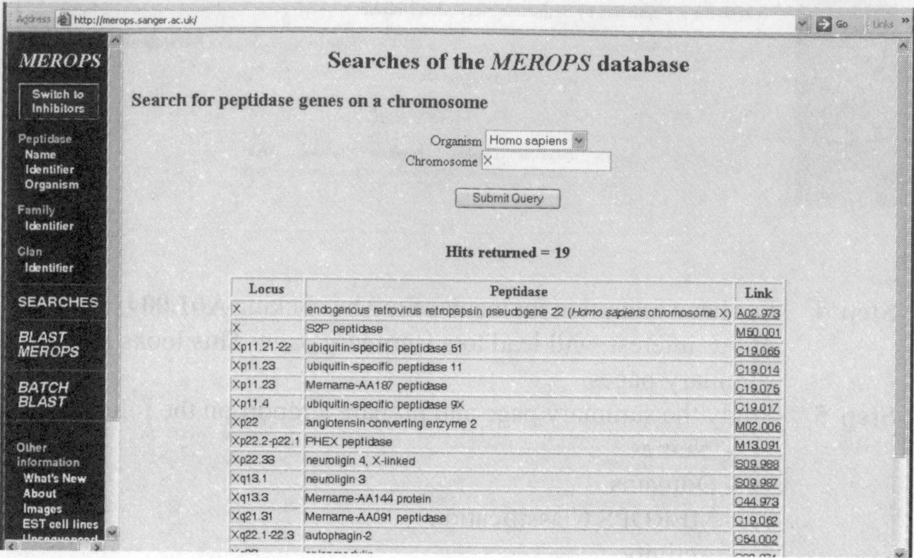

Fig. 3.11

Step 4 Further exploration for angiotensin-converting enzyme 2, by clicking ***M02.006***, which is the peptidase of our interest, will lead to a summary new page, as shown in Fig. 3.12.

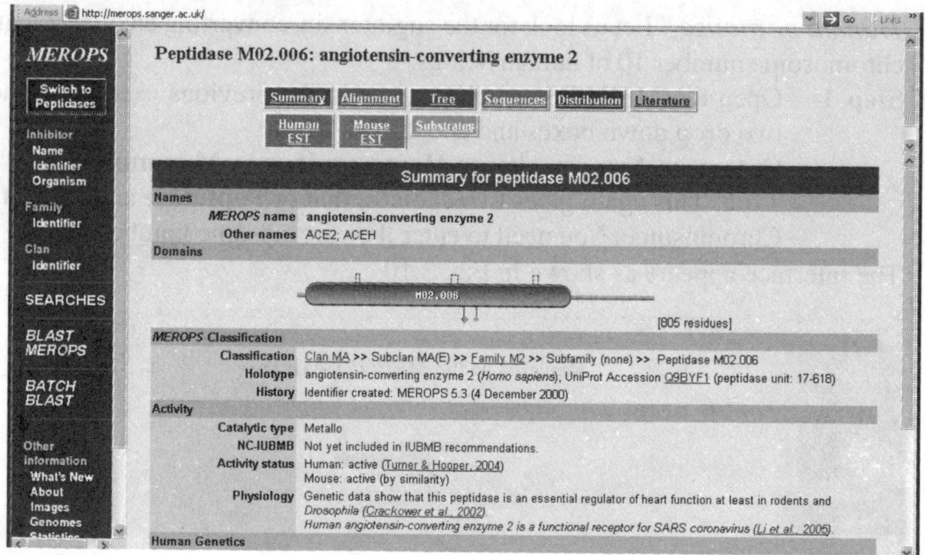

Fig. 3.12

Step 5 Study the page shown in Fig. 3.12 and prepare a report on the following subheads:
- Names
- Domains
- Merops Classification
- Activity
- Human Genetics
- Mouse Genetics

Step 6 Also explore Summary, Alignment, Tree, Sequences, Distribution, Human EST, Mouse EST, and Substrates.

Case (iii): Search for Structures of Peptidases or Inhibitors

Search in a clan or family Let us look for the interleukin 1-beta-converting enzyme belonging to the CD clan.

Step 1 Open the MEROPS search page as in the previous exercise and click **Search** option on the right hand side of any given clan or family. The result page displays a textbox. Enter the clan as 'CD' in the textbox.

Step 2 Click **Submit Query**.

Step 3 The result page, containing 8 hits, looks like as in Fig. 3.13.

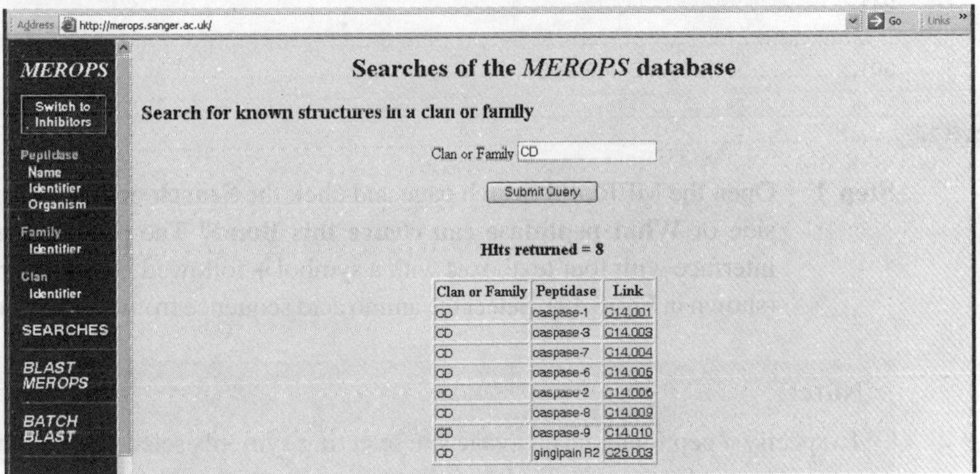

Clan or Family	Peptidase	Link
CD	caspase-1	C14.001
CD	caspase-3	C14.003
CD	caspase-7	C14.004
CD	caspase-6	C14.005
CD	caspase-2	C14.006
CD	caspase-8	C14.009
CD	caspase-9	C14.010
CD	gingipain R2	C25.003

Fig. 3.13

Step 4 Further exploration for caspase-1 by ***C14.001***, which is the peptidase of our interest, will lead us to a new page. Which is the summary page.

Step 5 Study the summary page carefully and prepare a report on the following subheads:
- Names
- Domains

- MEROPS Classification
- Activity
- Human Genetics
- Mouse Genetics

Step 6 Also explore **Summary**, **Alignment**, **Tree**, **Sequences**, **Distribution**, **Human EST**, **Mouse EST**, and **Substrates**.

Case (iv): Search for Peptidase Specificity

Search by cleavage Let us look for the peptidase that can cleave the bond in luteinizing hormone releasing hormone. The enzyme is 334 amino acids long and has the following sequence:

1 GKVKVGVNGF

....

61

....

121

....

181

....

241

....

301......................SWYDNEFGYS...

Fig. 3.14

Step 1 Open the MEROPS Search page and click the **Search** option on the right hand side of **What peptidase can cleave this Bond?** The page will give you an interface with four textboxes with a symbol + followed by four more textboxes (shown in Fig. 3.14). Select the amino acid sequence from the drop down menus.

Note

To specify a peptide bond cleavage site selecting symbols select the amino acids in one or more of the textboxes located on each side of the bond to be cleaved. The bond to be cleaved is denoted by the symbol +. The default settings assume endopeptidase cleavages, so positions that contain unspecified amino acids may be left blank. For an exopeptidase, one or two free termini must be indicated with TER.

Step 2 Enter the highlighted sequence, namely, SWY using their corresponding amino acid three letter codes in the textboxes. Click **Submit Query**.

Step 3 The result page, containing 3 hits, will appear as shown in Fig. 3.15.

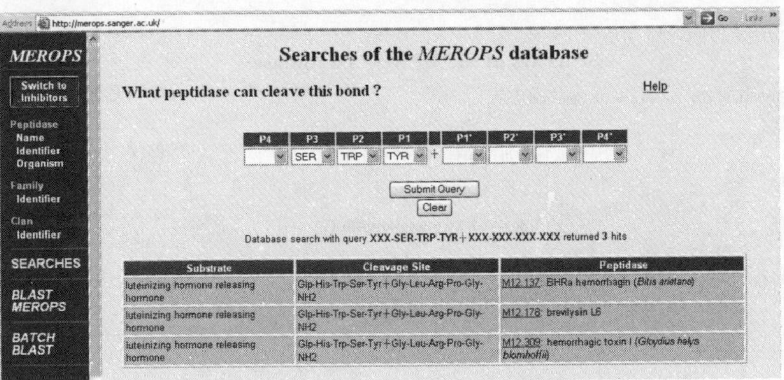

Substrate	Cleavage Site	Peptidase
luteinizing hormone releasing hormone	Glp-His-Trp-Ser-Tyr+Gly-Leu-Arg-Pro-Gly-NH2	M12.137: BHRa hemorrhagin (*Bitis arietans*)
luteinizing hormone releasing hormone	Glp-His-Trp-Ser-Tyr+Gly-Leu-Arg-Pro-Gly-NH2	M12.178: brevilysin L6
luteinizing hormone releasing hormone	Glp-His-Trp-Ser-Tyr+Gly-Leu-Arg-Pro-Gly-NH2	M12.309: hemorrhagic toxin I (*Gloydius halys blomhoffii*)

Fig. 3.15

Step 4 Further exploration for peptidase ***M12.309*: hemorrhagic toxin I (*Gloydius halys blomhoffii*)** by clicking on the link will lead to the summary.

Step 5 Study the summary page carefully and prepare a report on the following subheads:

- Names
- Domains
- MEROPS Classification
- Activity
- Human Genetics
- Mouse Genetics

Step 6 Also explore that the top of the page gives you some links: 'Summary', 'Literature' and 'Substrates'.

Search by substrate

Step 1 Open the MEROPS search page as in the previous exercise and click **Search** on the right hand side of **How may this Substrate be cleaved?** The page will give you an interface as shown in Fig. 3.16. Scroll and choose the substrate to be cleaved. You can also choose the option of ordering the result by either **MEROPS Identifier** or **Cleavage Site.**

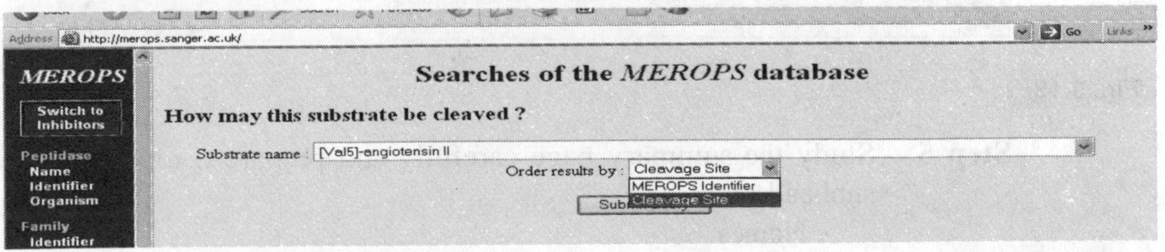

Fig. 3.16

Step 2 Click **Submit Query**.

Step 3 The result page, containing a single hit, will look like as shown in Fig. 3.17.

Address http://merops.sanger.ac.uk/

MEROPS | Switch to Inhibitors | Peptidase Name Identifier Organism | Family Identifier | Clan Identifier | SEARCHES | BLAST MEROPS

Searches of the *MEROPS* database

How may this substrate be cleaved ?

Substrate name : [des-Val-Tyr]-angiotensin IV

Order results by : MEROPS Identifier

Submit Query

'[des-Val-Tyr]-angiotensin IV'

Hits returned = 1

Cleavage Site	Peptidase
Ile ┼ His-Pro-Phe	M01.018: aminopeptidase PILS

Fig. 3.17

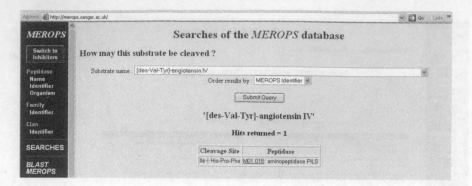

Step 4 Exploring further by clicking ***M01.018*** will lead us to the summary page, as shown in Fig. 3.18.

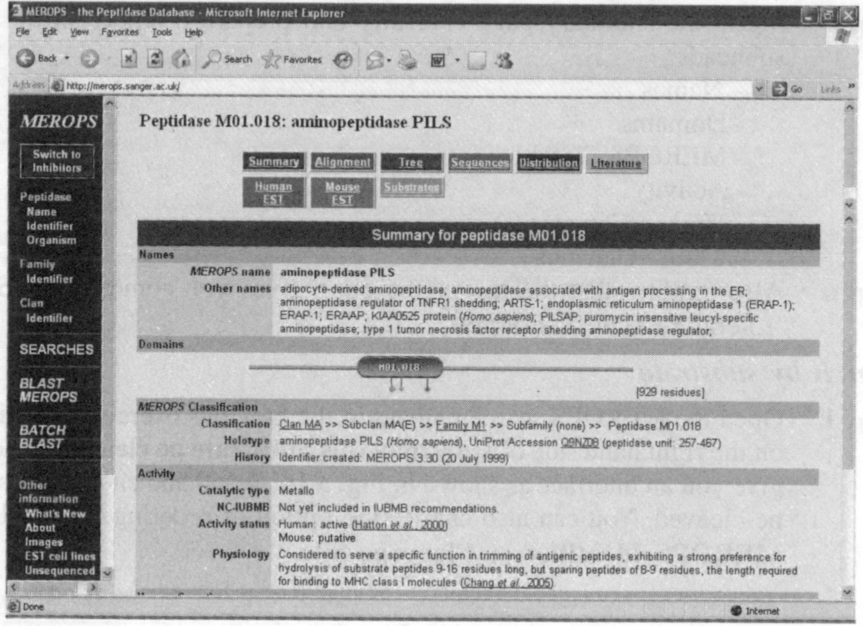

Fig. 3.18

Step 5 Study the summary page carefully and prepare a report on the following subheads:

- Names
- Domains
- MEROPS Classification
- Activity
- Human Genetics
- Mouse Genetics

Step 6 Also explore **Summary**, **Alignment**, **Tree**, **Sequences**, **Distribution**, **Human EST**, **Mouse EST**, and **Substrates**.

Step 7 Add to the report subheads on the following: formation of Angiotensin IV, its cloning, gene function, and the family of zinc metallopeptidases.

Search by bond

Step 1 Open the MEROPS search page as in the previous exercise and click *Search* on the right hand side of **What cleavages does this Peptidase make?** The result page will give you an interface with a scrolling textbox. Choose the substrate to be cleaved as shown in Fig. 3.19.

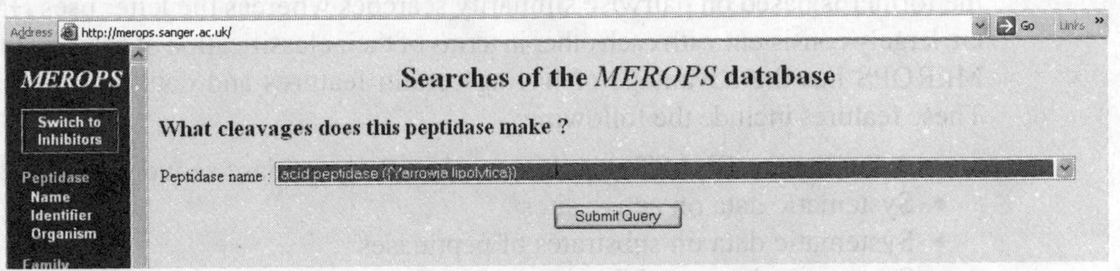

Fig. 3.19

Step 2 Click **Submit Query**.

Step 3 The result page, containing two hits, will look like as shown in Fig. 3.20.

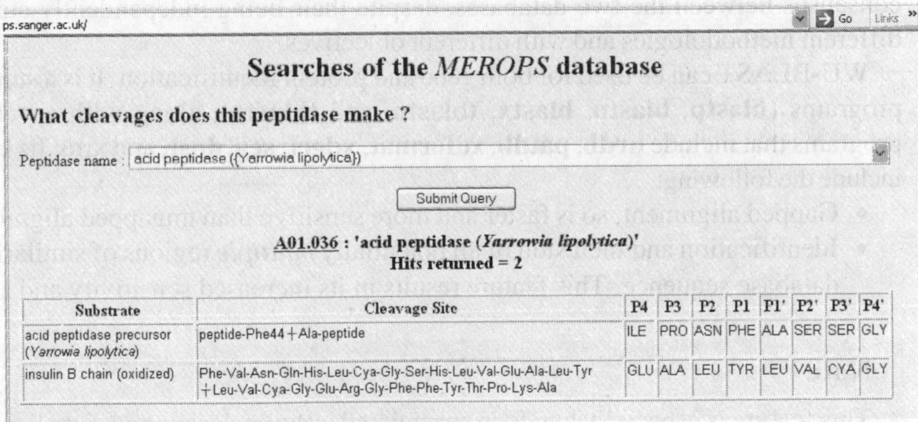

ps.sanger.ac.uk/

Searches of the *MEROPS* database

What cleavages does this peptidase make ?

Peptidase name : acid peptidase ({Yarrowia lipolytica})

Submit Query

A01.036 : 'acid peptidase (*Yarrowia lipolytica*)'
Hits returned = 2

Substrate	Cleavage Site	P4	P3	P2	P1	P1'	P2'	P3'	P4'
acid peptidase precursor (*Yarrowia lipolytica*)	peptide-Phe44 ┼ Ala-peptide	ILE	PRO	ASN	PHE	ALA	SER	SER	GLY
insulin B chain (oxidized)	Phe-Val-Asn-Gln-His-Leu-Cya-Gly-Ser-His-Leu-Val-Glu-Ala-Leu-Tyr ┼ Leu-Val-Cya-Gly-Glu-Arg-Gly-Phe-Phe-Tyr-Thr-Pro-Lys-Ala	GLU	ALA	LEU	TYR	LEU	VAL	CYA	GLY

Fig. 3.20

Step 4 Exploring further by clicking on ***A01.036*** will lead to a new page, which gives the summary for A01.036.

Step 5 Study the summary page carefully and prepare a report on the following subheads:

- Names
- Domains

- MEROPS Classification
- Activity

Step 6 Also explore **Summary**, **Alignment**, **Sequences**, **Distribution**, **Literature**, and **Substrates**.

(e) BLAST MEROPS

MEROPS also offers a BLAST search using the software Washington University BLAST (WU BLAST) version 2.0. It is interesting to compare the sensitivity in the use of BLAST through a database that is comprehensive such as Pfam, and a specialized one such as MEROPS. While both MEROPS and Pfam are successful at identifying family members, the former is based on pairwise similarity searches whereas the letter uses HMMs. Both are largely consistent with each other in terms of their classification of proteins into families. MEROPS has the advantage of having certain features and data not present in Pfam. These features include the following:

- A facility for BLAST searching against the peptidase sequence database
- Systematic data on active sites
- Systematic data on substrates of peptidases
- Systematic data on inhibitors
- A very comprehensive literature database
- MEROPS uses a hierarchical classification whereas Pfam uses a flat classification.

Having noted the differences in the two databases, overall, there is a high degree of consensus between the two databases, despite their being independently curated using different methodologies and with different objectives.

WU-BLAST can be used for both gene and protein identification. It is a suite of search programs (**blastp**, **blastn**, **blastx**, **tblastn**, and **tblastx**), along with several support programs that include **nrdb**, **patdb**, **xdformat**, **xdget**, **seg**, **dust,** and **xnu**. Its key features include the following:

- Gapped alignment, so is faster and more sensitive than ungapped alignments.
- Identification and inclusion of all potentially *multiple* regions of similarity for each database sequence. This feature results in its increased sensitivity and selectivity.

Note

This feature is especially handy in specialized tasks such as search for all exons in a multi-exon gene sequence. Here it identifies not just the longest or best-matching exon but all complete or partial copies of a repetitive element in a genomic sequence. Multiple discrete domains of similarity between sequences are reported besides the highest-scoring one.

- Availability of Sum statistics (and used by default) in all search modes to evaluate the joint probability of multiple regions of similarity. This feature affords a combination of heuristics and statistics. Therefore WU BLAST is often more sensitive/selective than full dynamic programming approach of Smith and Waterman which searches and evaluates the significance of only the highest scoring alignment with each database sequence.

- Availability of Poisson statistics as an option to Karlin–Altschul sum statistics in all search modes.
- Availability of the **postsw** option. Selecting this option results in full Smith–Waterman alignment of sequences and re-ranking of the database matches accordingly prior to output. This option is recommended for conducting search because Smith–Waterman results are combined with the initial BLAST results and redundancy is removed.
- The output is available in several formats such as a table and XML.
- Has the extended database format (**XDF**) which can be used for easier, yet accurate, management of protein and nucleotide sequence databases typically larger than 16 Mbp in length. Thus the WU-BLAST has storage and indexed retrieval of individual sequences up to 1 Gbp (billion base pairs).

Now let us try and use WU-BLAST using MEROPS.

Step 1 Open the MEROPS search page and click **BLAST MEROPS** from the left panel. The BLAST interface that is displayed (see Fig. 3.21).

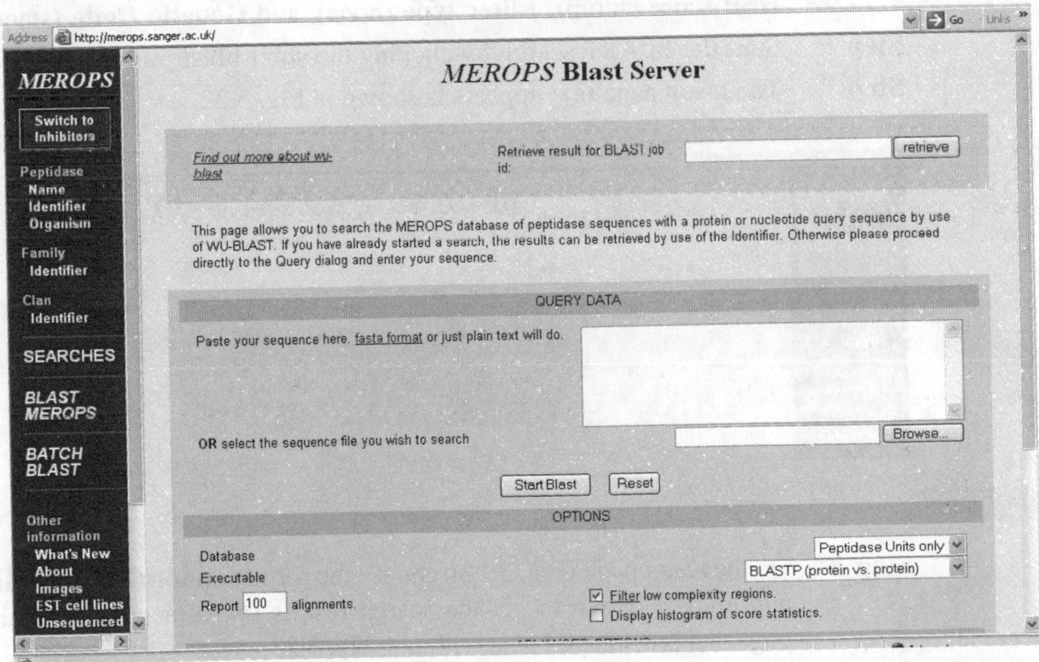

Fig. 3.21

Step 2 Let us take the following protein sequence to conduct a search.

MSTAVLENPGLGRKLSDFGQETSYIEDNCNQNGAISLIFSLKEE
VGALAKVLRLFEENDVNLTHIESRPSRLKKDEYEFFTHLDKRSLPALTNIIKILRHDI
GATVHELSRDKKKDTVPWFPRTIQELDRFANQILSYGAELDADHPGFKDPVYRARRKQ
FADIAYNYRHGQPIPRVEYMEEEKKTWGTVFKTLKSLYKTHACYEYNHIFPLLEKYCG
FHEDNIPQLEDVSQFLQTCTGFRLRPVAGLLSSRDFLGGLAFRVFHCTQYIRHGSKPM
YTPEPDICHELLGHVPLFSDRSFAQFSQEIGLASLGAPDEYIEKLATIYWFTVEFGLC
KQGDSIKAYGAGLLSSFGELQYCLSEKPKLLPLELEKTAIQNYTVTEFQPLYYVAESF
NDAKEKVRNFAATIPRPFSVRYDPYTQRIEVLDNTQQLKILADSINSEIGILCSALQKIK

Step 3 Paste or write the sequence in the textbox below **QUERY DATA**.

Step 4 Under **OPTIONS**, choose the following options: Peptidase Units only, BLASTP (protein vs. protein), **Filter low complexity regions**, and report desired number of alignments as 100.

Step 5 Under **ADVANCED OPTIONS**, choose the following options: **Matrix** (blosum62), **Descriptions** (50), **Sort results by** (pvalue), **Expect** (E) (.01), **HSP score** (sump), **Filter type** (none), and **Genetic Code** (standard).

Step 6 Start the BLAST search by clicking the **start blast** button.

Step 7 The result page that appears is shown in Fig. 3.22.

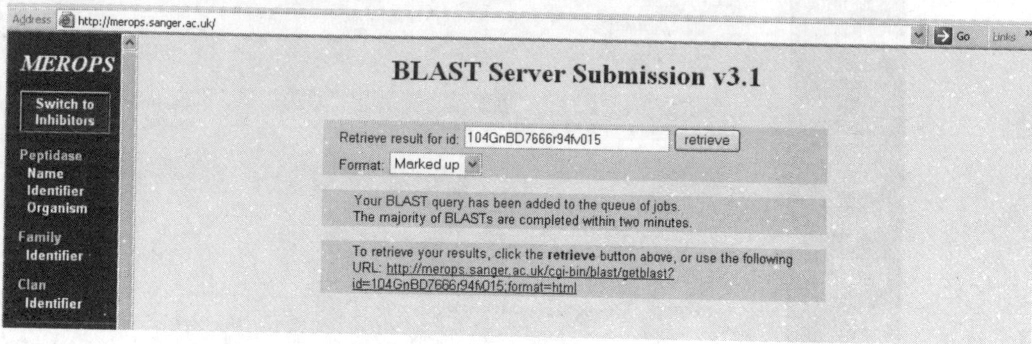

Fig. 3.22

Step 8 On the result page (Fig. 3.22), choose the format option (**Plain** or **Marked up**) and click **retrieve** to get the search results for your chosen sequence. The page that is displayed next will look Fig. 3.23.

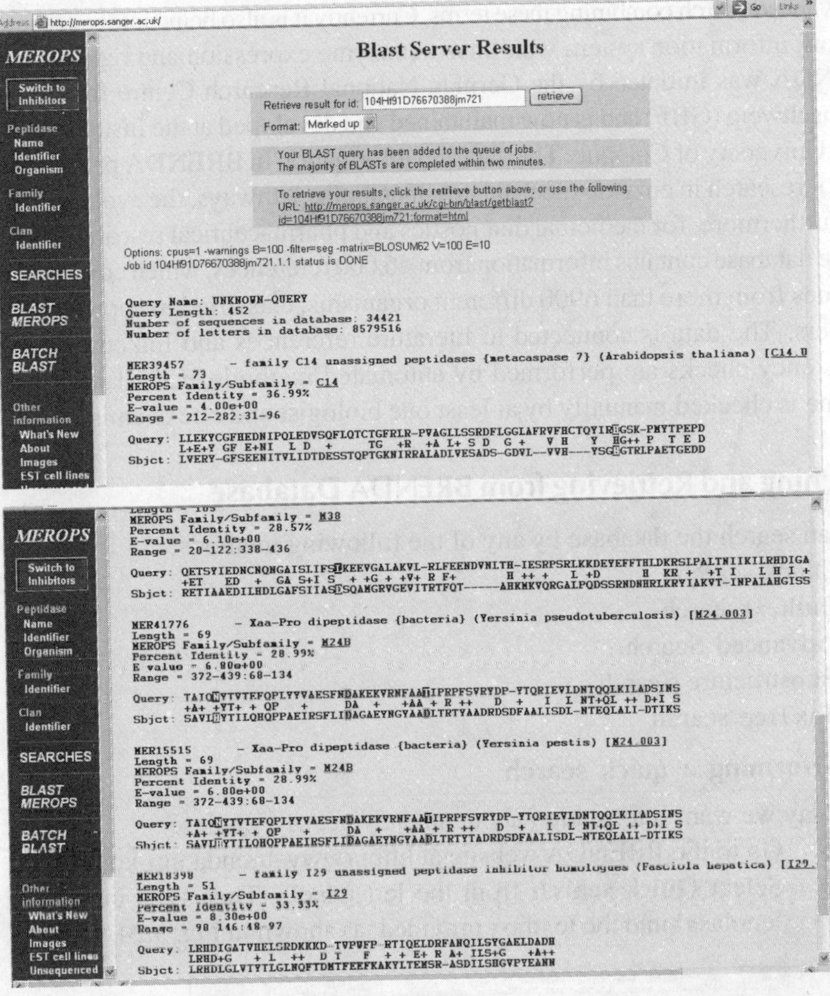

Fig. 3.23

Step 9 Prepare a report on the Xaa-Pro dipeptidase (bacteria) using the following parameters:

(a) metallopeptidase families that are described as being co-catalytic
(b) role of the X-Pro dipeptidase in the cheese industry
(c) distribution of this depeptidase
(d) alignment of the peptidase units with other sequences
(e) substrate cleavage sites

3.3 BRENDA (www.brenda.uni-koeln.de)

(a) History, size, content

BRENDA (BRaunschweig ENzyme DAtabase) is database that contains enzyme functional data. It has facilities to search the database by the EC number, enzyme name, organism, or an advanced search combining these terms. Currently it is also being integrated into a metabolic network information system with links to enzyme expression and regulation information. BRENDA was initiated by the German National Research Centre for Biotechnology in Braunschweig (GBF) and is now maintained and developed at the Institute of Biochemistry at the University of Cologne. The data and information in BRENDA provide a fundamental tool for research in enzyme mechanisms, metabolic pathways, the evolution of metabolism and, furthermore, for medicinal diagnostics and pharmaceutical research.

The database contains information from 46,000 references, which covers 40,000 different enzymes from more than 6900 different organisms, classified in approximately 3900 EC numbers. The data is connected to literature references and linked to PubMed. While consistency checks are performed by automated methods, each dataset on a classified enzyme is checked manually by at least one biologist and one chemist.

TASK Searching and Retrieving from BRENDA Database

You can search the database by any of the following methods:

- Quick search
- Fulltext search
- Advanced Search
- Substructure Search
- TaxTree search

(a) Performing a quick search

Let us say we want information about an enzyme, say, enolase.

Step 1 Go to the BRENDA website at http://www.brenda.uni-koeln.de/

Step 2 Select **Quick search** from the left panel. Then type in the enzyme name 'enolase' into the textbox provided, as shown in Fig. 3.24.

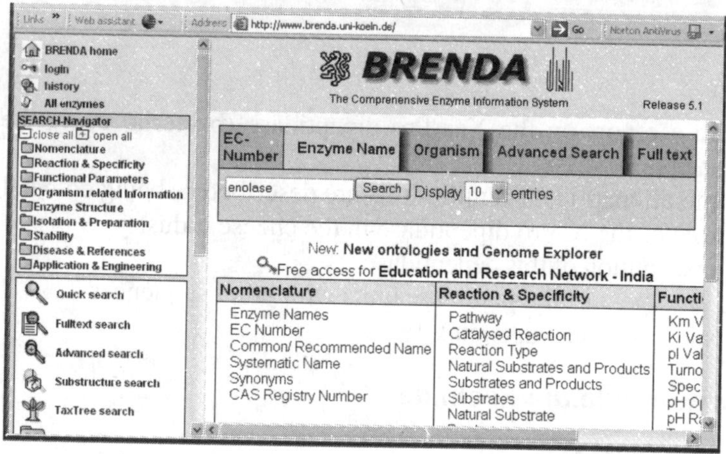

Fig. 3.24

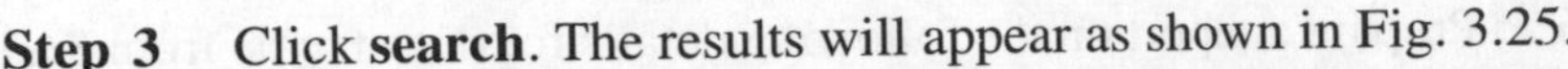

Step 3 Click **search**. The results will appear as shown in Fig. 3.25.

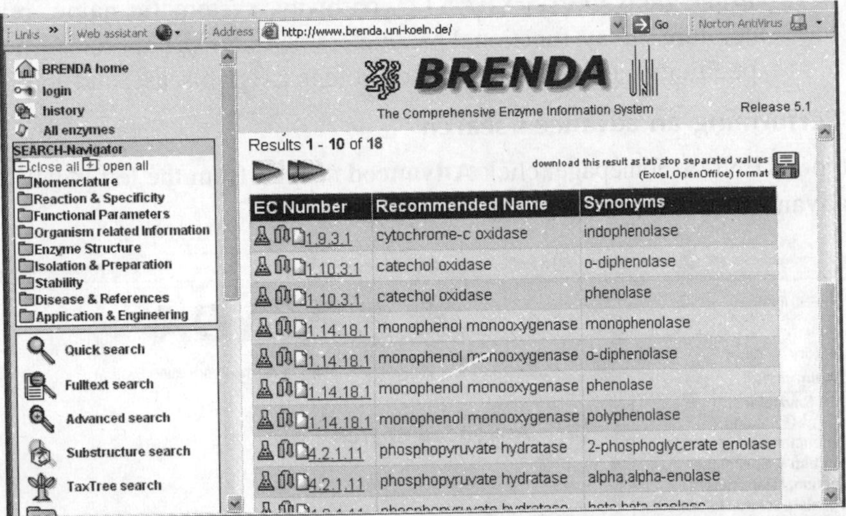

Fig. 3.25

Step 4 Click ***1.9.3.1*** from the EC Number column to look for information pertaining to cytochrome-C oxidase.

Step 5 The results will appear as shown in Fig. 3.26.

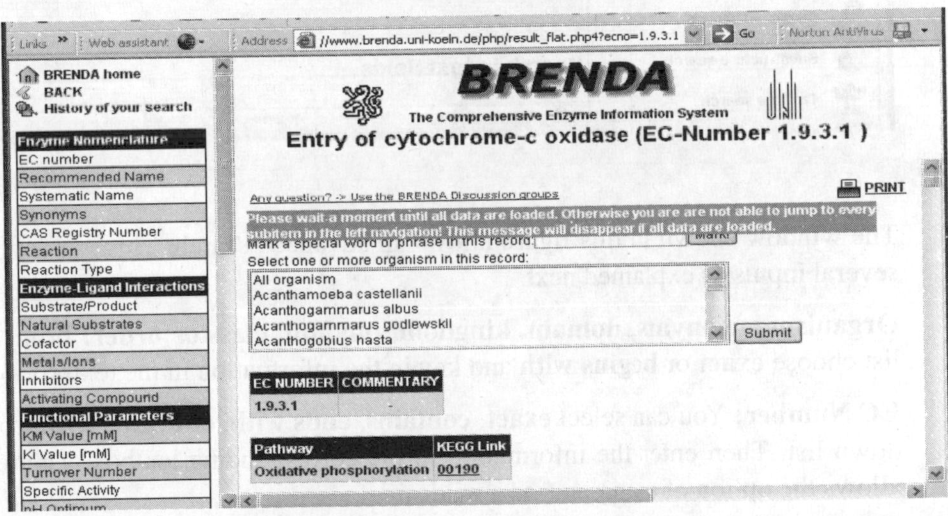

Fig. 3.26

Note

Allow the page to load completely before navigating to collect other data.

Step 6 Prepare a report on the following details for the enzyme from *Saccharomyces cerevisiae* (COX12_YEAST)—pathway, systematic name, reaction type, literature survey, functional parameters (KM, Ki, pH optima, turnover number, specificity), amino acid sequence, role in Leigh disease, and applications.

(b) Performing an advanced search

On the BRENDA homepage, click **Advanced search** from the left panel. This display the advanced search page (see Fig. 3.27).

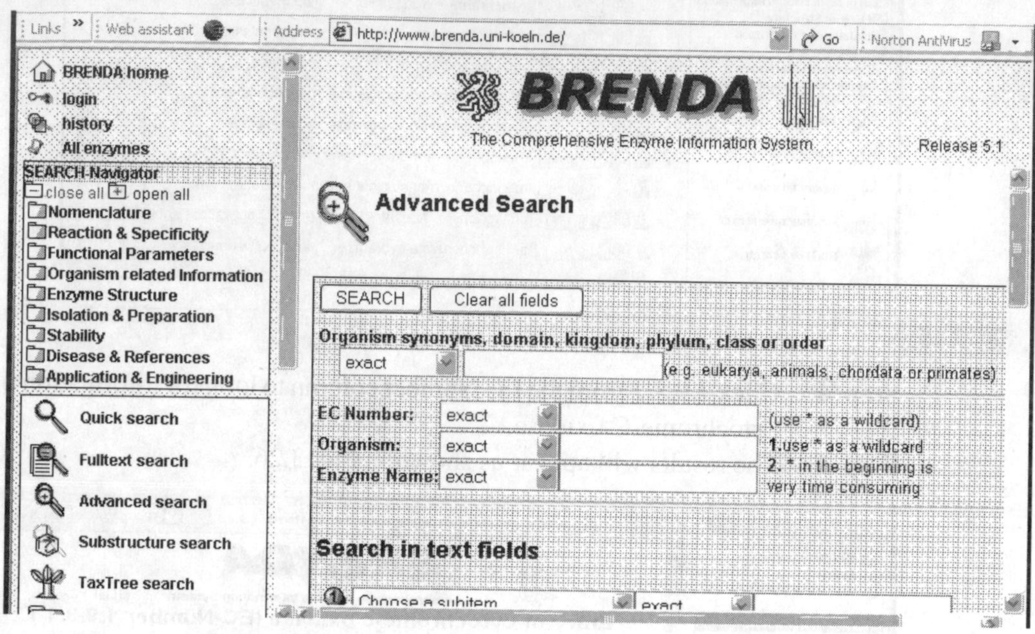

Fig. 3.27

The window shown in this figure allows an advanced search. This window can accept several inputs as explained next.

Organism synonyms, domain, kingdom, phylum, class or order: From drop down list choose **exact** or **begins with** and key in the information in the textbox.

EC Number: You can select **exact**, **contains**, **ends with**, or **begins with** from the drop down list. Then enter the information in the corresponding textbox. The textbox also allows the option of using an * as a wildcard.

> **Note**
>
> Avoid using * at the beginning as it is very time consuming.

Organism: You can select **exact**, **contains**, **ends with**, or **begins with** from the drop down list and enter the information into the textbox provided. The textbox allows the option of using an * as a wildcard.

Select Enzyme: The drop down list allows one to select **exact**, **contains**, **ends with**, or **begins with**. The textbox allows the option of using an * as a wildcard.

Search in text fields: Choose a subitem offers the following options: **Activating Compound**, **General Stability**, **Cofactor**, **Application**, **Inhibitor**, **Ligands**, **Localization**, **Metals/Ions**, **Natural Substrate**, **Natural Product**, **Organic Solvent Stability against**, **Oxidation Stability**, **Prosttranslation Modification**, **Product**, **Reference (Title)**, **Source/Tissue**, **Storage Stability**, **Subunits**, **Substrate**, **Amino acid sequence**. An additional drop down list box allow one to select from **exact**, **contains**, **ends with**, or **begins with option**. You can also add another search field to fine tune your search.

Search in numeric fields: Choose a subitem offers the following options: **KM value**, **Specific Activity**, **Temperature Optimum**, **Temperature Range**, **Temperature Stability**, **Turnover Number**, **pH Optimum**, **pH Range**, **pH Stability**, **Molecular Weight**, **KI weight**. It is possible to set the values as >, <, =, and between.

Search in type as: You can check the chechbox against **Cloned**, **Crystallized**, **Engineering**, **Purified**, **Renatured**, **PDB entry**.

Search in application fields: The **Application** drop down list box offers the following options: agriculture, analysis, biotech, degradation, detergent, diagnostic, diagnostics, environmental protection, industry, medicine, molecular biology, more, nutrition, paper production, pharmaceutical industry, pharmacology, steel industry, synthesis.
In addition to the above mentioned Search options, search can be done using **Cofactor**, **Localization**, **Metals / Ions**, **Organic Solvent Stability against**, **Source Tissue**, and **Posttranslational Modification**.

TASK Searching and retrieving from BRENDA database (Advanced search)

Let us search for all enzymes in *Saccharomyses cervisiae* where the substrate is ADP, cofactor is ATP and inhibitor is L-phenylalanine. The activator in the case of this particular enzyme is D-fructose 1,6-diphosphate. The enzyme is known to have a KM of < 0.3 mM, pH optimum in the range of 6.8–7.3, and is known to have a molecular weight of 100000 kDa.

Step 1 Go to the BRENDA web page (http://www.brenda.uni-koeln.de/)

Step 2 On this page, click the **Advanced search** option to display the advanced search page.

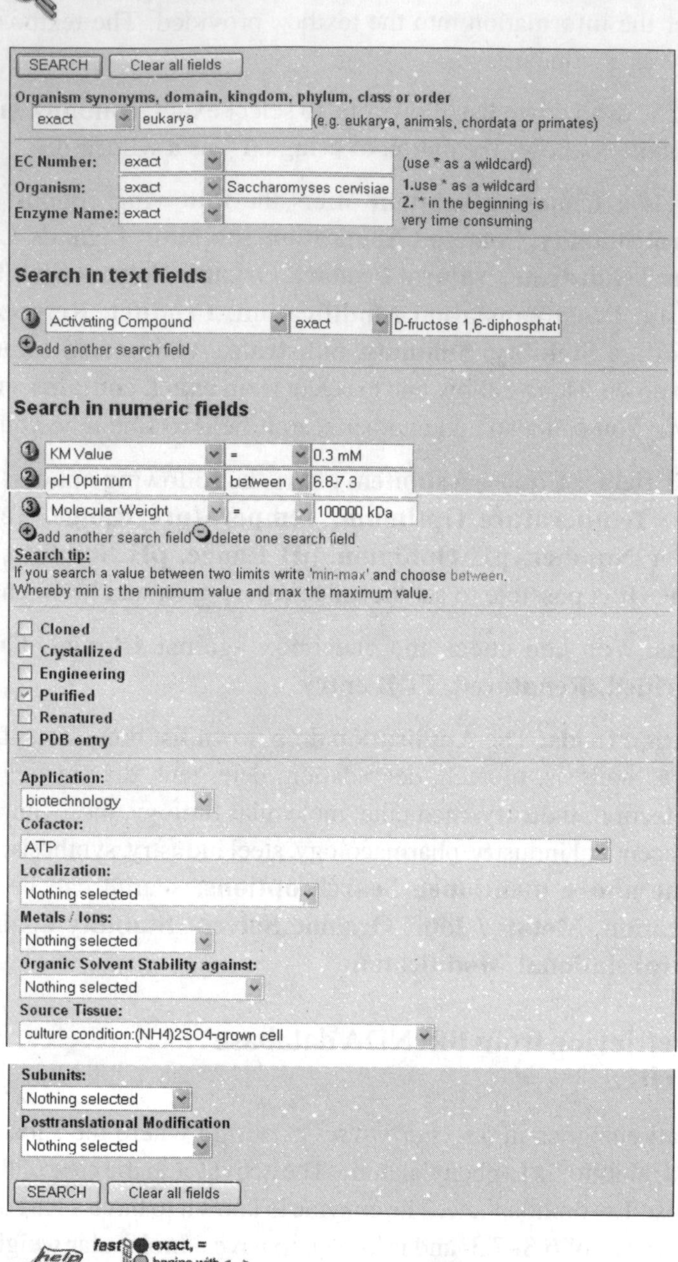

Fig. 3.28

Step 3 Under **Organism synonyms**, **domain**, **kingdom**, **phylum**, **class or order**, choose **exact** and enter 'eukarya' in the adjacent textbox.

Step 4 Enter 'Saccharomyses cervisiae' in the text next to **Organism**.

Step 5 Under **Search in text fields**, choose **substrate** and **exact** and enter 'ADP' in the textbox. Next choose **activator** and exact and enter 'D-fructose 1,6-diphosphate in the textbox. (*Note:* You will have to use the **Add another search field** option.)

Step 6 Under the **Search in numeric fields**, enter 'KM' '<' as '0.3 MM'; enter 'pH optimum' 'range of' as '6.8–7.3'; and enter 'molecular weight' 'exact' as '100000 kDa'. (Note: Here once again you will have to use the **Add another search field** option.)

Step 7 Chick the **Purified** option and click the **SEARCH** button.

Step 8 One or more EC listings should appear. To enter the database, click the EC number of the correct enzyme in the query table. Prepare a report that would include the results of all the parameters that you have searched for. In addition, look for amino acid sequences of all the entries specific to the enzyme 'alcohol dehygrogenase'.

Linking to other resources

The Catalytic Site Atlas (CSA) is a database of enzyme active sites and catalytic residues in enzymes using structural data. Catalytic residues are defined as those residues thought to be directly involved in some aspect of the reaction catalysed by an enzyme. CSA contains two types of entries: original hand-annotated entries, derived from the primary literature, and homologous entries, found by PSI-BLAST alignment to one of the original entries. Access to CSA is via the PDB code, Swiss-Prot entry, or EC number. Each CSA entry lists the catalytic residues found in that entry, using PDB residue numbering, and each active site can be visualized using Rasmol. It has been made available on the internet by the European Bioinformatics Institute (EBI).

The Enzyme Structures Database contains all of the known enzyme structures deposited in the Brookhaven Protein Data Bank. Using the Mar 2003 v.30.0 release of the ENZYME Data Bank, the enzyme structures are classified by their EC number. A link is provided to the PDBsum database where the enzyme structures can also be accessed and which provides summaries and structural analyses of PDB data files. The Enzyme Structures Database was created by Roman Laskowski and Andrew Wallace of the Biomolecular Structure and Modelling Group at the University College London.

ENZYME is part of ExPASy (**Ex**pert **P**rotein **A**nalysis **Sy**stem), the proteomics server of the Swiss Institute of Bioinformatics. It provides information relating to protein nomenclature, based on the recommendations of the Nomenclature Committee of the International Union of Biochemistry and Molecular Biology (IUBMB). ENZYME describes each type of a characterized enzyme for which an EC (Enzyme Commission) number has been provided. An enzyme of interest may be accessed via its EC number, enzyme class, description (official name) or alternative name(s), chemical compound, and cofactor. It has been made available on the internet by the Canadian Bioinformatics Resource (CBR), and mirrored in Switzerland, China, USA, Australia, and Taiwan.

3.4 Pathway Databases: CAZy

Carbohydrate-**A**ctive en**Zy**mes (CAZy) database consists of families of enzymes that act upon, modify, or create glycosidic bonds. These enzymes have structurally related catalytic and carbohydrate-binding functional domains.

(a) History, size, content

It is made available on the internet by *Architecture et Fonction des Macromolécules Biologiques, Centre National de la Recherche Scientifique*. The database has classified and organized enzymes into the following categories:

Glycosisdases and transglycosidases: The glycosisdases (O-glycoside hydrolases EC 3.2.1.-) and transglycosidases family consists of members that hydrolyse the glycosidic bond between two or more carbohydrates or between a carbohydrate and a non-carbohydrate moiety.

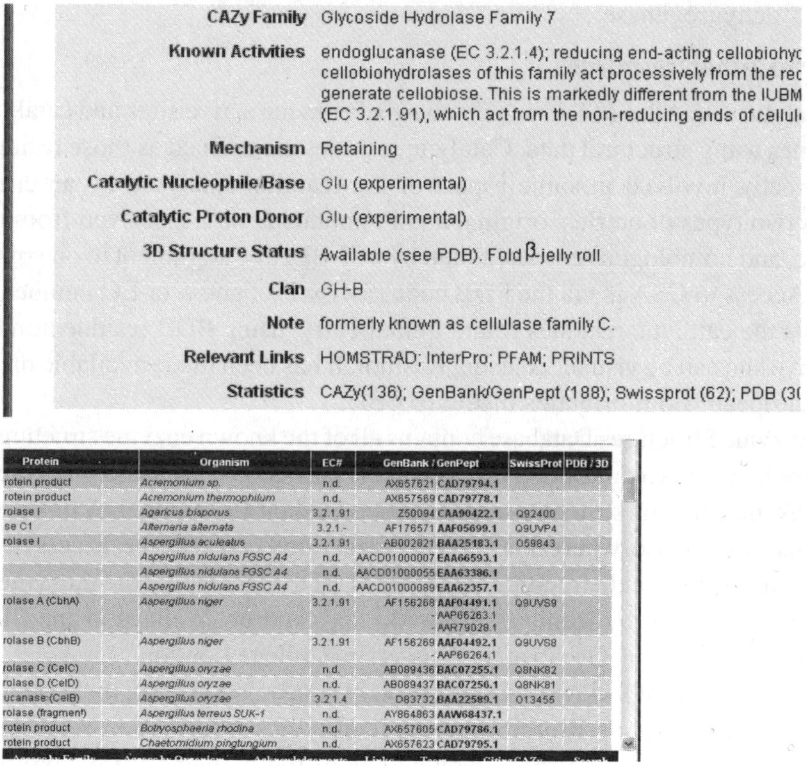

CAZy Family	Glycoside Hydrolase Family 7
Known Activities	endoglucanase (EC 3.2.1.4); reducing end-acting cellobiohyc cellobiohydrolases of this family act processively from the rec generate cellobiose. This is markedly different from the IUBM (EC 3.2.1.91), which act from the non-reducing ends of cellul
Mechanism	Retaining
Catalytic Nucleophile/Base	Glu (experimental)
Catalytic Proton Donor	Glu (experimental)
3D Structure Status	Available (see PDB). Fold β-jelly roll
Clan	GH-B
Note	formerly known as cellulase family C.
Relevant Links	HOMSTRAD; InterPro; PFAM; PRINTS
Statistics	CAZy(136); GenBank/GenPept (188); Swissprot (62); PDB (3(

Protein	Organism	EC#	GenBank / GenPept	SwissProt	PDB / 3D
rotein product	*Acremonium sp.*	n.d.	AX657621 **CAD79794.1**		
rotein product	*Acremonium thermophilum*	n.d.	AX657569 **CAD79778.1**		
rolase I	*Agaricus bisporus*	3.2.1.91	Z50094 **CAA90422.1**	Q92400	
se C1	*Alternaria alternata*	3.2.1.-	AF176571 **AAF05699.1**	Q9UVP4	
rolase I	*Aspergillus aculeatus*	3.2.1.91	AB002821 **BAA25183.1**	O59843	
	Aspergillus nidulans FGSC A4	n.d.	AACD01000007 **EAA66593.1**		
	Aspergillus nidulans FGSC A4	n.d.	AACD01000055 **EAA63386.1**		
	Aspergillus nidulans FGSC A4	n.d.	AACD01000089 **EAA62357.1**		
rolase A (CbhA)	*Aspergillus niger*	3.2.1.91	AF156268 **AAF04491.1** - AAP66263.1 - AAR79028.1	Q9UVS9	
rolase B (CbhB)	*Aspergillus niger*	3.2.1.91	AF156269 **AAF04492.1** - AAP66264.1	Q9UVS8	
rolase C (CelC)	*Aspergillus oryzae*	n.d.	AB089436 **BAC07255.1**	Q8NK82	
rolase D (CelD)	*Aspergillus oryzae*	n.d.	AB089437 **BAC07256.1**	Q8NK81	
ucanase (CelB)	*Aspergillus oryzae*	3.2.1.4	D83732 **BAA22589.1**	O13455	
rolase (fragment)	*Aspergillus terreus SUK-1*	n.d.	AY864863 **AAW68437.1**		
rotein product	*Botryosphaeria rhodina*	n.d.	AX657605 **CAD79786.1**		
rotein product	*Chaetomidium pingtungium*	n.d.	AX657623 **CAD79795.1**		

Fig. 3.29

Unfortunately, the EC classification is based on the substrate specificity and molecular mechanism of action and thus cannot be extrapolated to describe the structural features of these enzymes. Thus enzymes that show conserved folds of proteins are grouped in 'clans' placed in tables. The members of the clans of families exhibit

(a) relationship to more than one family,
(b) increase in sensitivity of sequence comparison methods, or
(c) structural features that show resemblance between members of different families.

These tables are very handy while conducting a search, since a user can directly access the classified sequences by looking for the family number, clan of related families, or the EC number. Since some glycoside hydrolases are multifunctional enzymes containing catalytic domains belonging to different glycoside hydrolases, their GenBank, Swiss-Prot, and/or PDB entries are also included. An example of a typical entry is shown in Fig. 3.29.

Glycosyltransferases: Carbohydrate-binding molecules glycosyltransferases (EC 2.4.*x*.*y*) are enzymes that catalyse the biosynthesis of disaccharides, oligosaccharides, and polysaccharides. These enzymes catalyse the transfer of sugar moieties from activated donor molecules to specific acceptor molecules, forming glycosidic bonds. In CAZy glycosyltransferases are placed into separate sequence-based families using nucleotide diphospho-sugars, nucleotide monophospho-sugars, and sugar phosphates (EC 2.4.1.x) and related proteins. Thus all members of each of the families will have a similar protein fold. Just as for the glycoside hydrolases, several of the families defined on the basis of sequence similarities have similar three-dimensional structures and therefore form clans. Weak local similarities suggest similarities between several families.

Polysaccharide lyases EC (4.2.2.): These are a group of enzymes that cleave polysaccharide chains by beta-elimination mechanism thus establish a double bond at the newly formed non-reducing end. The members of this family bear similarity based on their amino acid sequences.

Carbohydrate esterases: These enzymes catalyse the de-O or de-N-acetylation of substituted saccharides. Recall that an ester = acid + alcohol, therefore the sugar moiety may act as either an acid or an alcohol.

Carbohydrate-binding module family: This class consists of those enzymes that contain an active A carbohydrate-binding module (CBM). A CBM is a contiguous amino acid sequence with a separately identifiable fold that has carbohydrate-binding activity. CBMs exist as separate identities within larger enzyme sets. This feature separates this family from other non-catalytic sugar-binding proteins such as lectins and sugar transport proteins. This classification thus helps in predicting binding specificity, identifying functional residues, predicting polypeptide folds, and establishing evolutionary relationships. Because the fold of proteins is better conserved than their sequences, some of the CBM families are clustered into superfamilies or clans.

TASK

Searching and retrieving from CAZY database

Step 1 Open the CAZy search page at http://afmb.cnrs-mrs.fr/CAZY/

Step 2 Click the **search glycosyltransferase** option. The link page will look like as shown in Fig. 3.30.

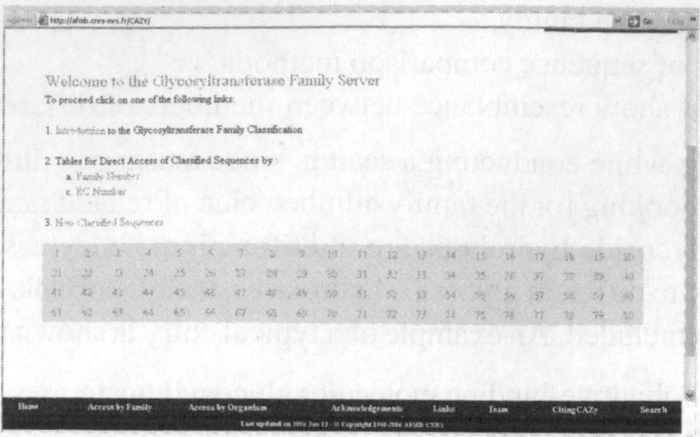

Fig. 3.30

Step 3 Let us search for the glycosyltransferase family 71. Since we do not know its EC number, we click **Tables for Direct Access of Classified Sequences by** and **Family number**.

Step 4 On the next page, go to **Direct Link to Families** and click **number 71**.

Step 5 The result page will look like as shown in Fig. 3.31.

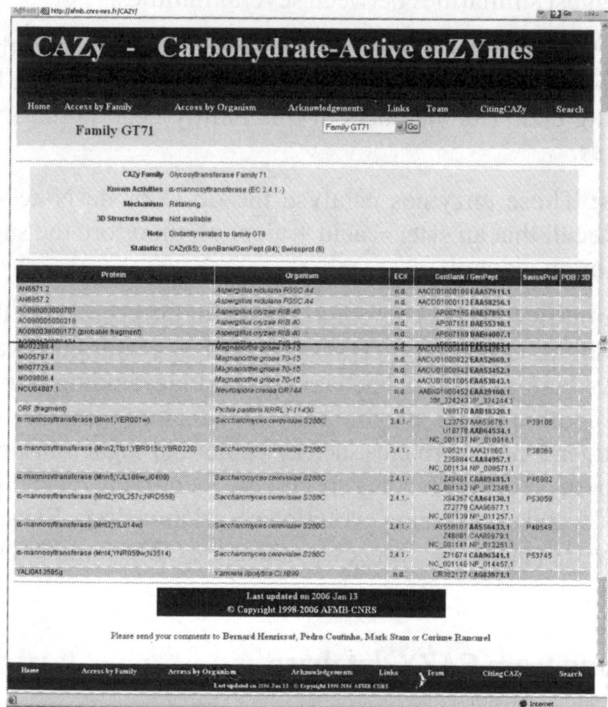

Fig. 3.31

Step 6 Prepare a report on the following for *Saccharomyces cerevisiae*:

(a) known activities for the enzyme
(b) mechanism of action
(c) 3D structure
(d) other enzymes encoded by genes in the family
(e) details about the MNN1 gene from the organism
(f) accession and GI numbers for the enzyme in NCBI .
(g) pathway
(h) sub-cellular localization
(i) length of the protein sequence and its molecular weight
(j) level of protein expression

3.5 Disease Databases

There are several disease databases that are commonly looked at while dealing with diseases, genes, or proteins. In this book we will have a closer look at the most common database OMIM that deal with inherited diseases. Other databases that we will take up here are GeneCards and Medline Plus. Your exposure to these three databases will prepare you to browse and squeeze data from many database you come across in this book.

3.5.1 OMIM (http://www.ncbi.nlm.nih.gov/entrez/query.fcgi?db=OMIM)

Online Mendelian Inheritance in Man (OMIM) is a daily updated online database of human genes and genetic disorders that are inherited or heritable. The database is based upon the textbook *Mendelian Inheritance in Man*. The book and the database are authored, edited, and developed by Dr Victor A. McKusick and his colleagues at Johns Hopkins and hosted by the National Center for Biotechnology Information (NCBI). The database contains textual information and reference links to Medline and related resources of NCBI and elsewhere. It does not contain entries related to genetic mutations or conditions that are caused by a chromosomal aberration, such as monosomy or trisomy. OMIM is not a truly relational database and is maintained primarily as formatted text in ASN.1 format. Some aspects of record tracking are managed in a relational database such as MIM number, create date, update dates, etc. OMIM is intended primarily for physicians and professionals concerned with genetic disorders, researchers, and advanced students in science and medicine. Though it is open to the public, it is desirable to consult and query a physician for diagnosis, information, suggestions, and guidance about personal medical or genetic condition. The updated statistics of these entries can be checked at http://www.ncbi.nlm.nih.gov/Omim/mimstats.html. Let us know about the information that can be obtained from an OMIM entry.

(a) OMIM entry

Data mining from OMIM can be done by typing in single or multiple text searches into the textbox provided in the Entrez interface you have learnt before. Advanced search can be done using the **Limits**, **Preview/Index**, **History**, **Clipboard**, or **Details** options of the

Entrez interface. You can use Boolean algebra while combining searches or implementing some logic. The result lists the responsible genes and its related information like the disorder—direct or related, gene map locus, morbid and gene map. It also provides links to the sequence, protein, journals, and many other databases of either NCBI or otherwise. Each OMIM entry is indexed with a unique six-digit number called MIM number. The first digit indicates the mode of inheritance of the gene involved. Refer to the note that lists the first digit of the six-digit number and what it represents other than the gene being phenotype.

Note

1: Autosomal dominant loci or phenotypes (entries created before May 15, 1994)
2: Autosomal recessive loci or phenotypes (entries created before May 15, 1994)
3: X-linked loci or phenotypes
4: Y-linked loci or phenotypes
5: Mitochondrial loci or phenotypes
6: Autosomal loci or phenotypes (entries created after May 15, 1994)

Each entry or entries can be displayed in different format such as Title, Detail, ASN.1 or XML. One can also view the allelic variants and clinical synopsis, if present, for an entry. Links to resources external to Entrez can be seen through **LinkOut** option. Options for related Entrez records can be accessed through links such as **Related Entries, PubMed, Protein, Nucleotide, Structure, Genome**.

Note

Allelic variants are given a 10-digit number, the 6-digit MIM number of the parent locus, followed by a decimal point and a unique 4-digit variant number. For most genes, only selected mutations are included as allelic variants. There are various criteria for inclusion of allelic variants such as discovering the first mutation, high population frequency, distinctive phenotype, historic significance, unusual mechanism of mutation, unusual pathogenetic mechanism, and distinctive inheritance such as dominant with some mutations and recessive with other mutations in the same gene. Most allelic variants represent disease-producing mutations, and a few show polymorphisms of positive statistical correlation with particular common disorders.

Every MIM number of the entry is preceded by a symbol *, #, +, %, or ^ that broadly categorizes the gene.

Note

Asterisk (*): A gene of known sequence.
Hash (#): A descriptive entry of a phenotype that does not represent a unique locus but has a known molecular basis.

Plus (+): A gene of known sequence and a phenotype.
Percent (%): An entry with a confirmed mendelian phenotype or phenotypic locus for which the underlying molecular basis is not known.
No symbol: A phenotype entry with mendelian basis suspected but not clearly established or that the distinction of this phenotype from that in another entry is unclear.
Caret symbol (^): The entry no longer exists because it was removed from the database or moved to another entry as indicated.

Genes in OMIM entries are represented in Gene Map when cytogenetic locations of those genes are published in the cited references. Gene map disorders and diseases are available from OMIM Gene Map and OMIM Morbid Map, respectively. Let us look at each of them.

OMIM gene map The OMIM gene map is a single file, presented in a tabular format, listing genes of OMIM entries along with their cytogenetic locations from the p telomere of chromosome 1 through the q telomere of chromosome 22, followed by p telomere of chromosome X through the q telomere of chromosome Y. In some entries a gene is not mapped to a specific cytogenic band and instead it is localized to a chromosome in general. Such an entry is shown at the end of the genes on that chromosome. The mapping of the loci to chromosomes or the linkage between two loci is assigned status codes as per the certainty of the assignments such as confirmed (C), provisional (P), inconsistent (I) and limbo (L), i.e., tentative. The methods that are used for gene mapping are some biological experiments and observations. These methods are represented in the gene map using symbols listed in the OMIM repository ftp://ftp.ncbi.nih.gov/repository/OMIM/genemap.key. The gene map file itself can be browsed through on the web or can be downloaded via FTP at ftp://ftp.ncbi.nih.gov/repository/OMIM/genemap. The web version of this gene map can be searched by gene symbol, chromosomal location, or by disorder keyword that return up to 20 entries at a time such that the first entry of the result matches with the search characters. Search by chromosomal location can be done using a number, say 6, or by terms like 1pter, Xq, etc. The **Find Next** button is used to find subsequent instances of the term. The symbols used in the disorder column of the map are given in the note box. This data is also included in the Genes_Cytogenetic map of the Entrez Map Viewer, presenting a graphical display of various cytogenetic, genetic linkages, sequence, radiation hybrid, and other maps.

OMIM morbid map The OMIM morbid map is a single file, presented in a tabular format, listing diseases alphabetically described in OMIM and their corresponding cytogenetic locations. This file can be browsed through on the internet or can be downloaded using FTP at ftp://ftp.ncbi.nih.gov/repository/OMIM/morbidmap.

Here also, the web version of this gene map can be searched using a gene symbol, chromosomal location, disorder keyword, or by a number. The search result displays 20 entries at a time such that the first entry of the result matches with the search characters. Search by chromosomal location can be done using terms such as 1pter, Xq, etc. and not by numbers unlike in the case of OMIM gene map. Thus, search by number will not search the file for a chromosome, instead it will consider the number as part of the text in a column. The **Find Next** button is used to find the subsequent instances of the search term. The symbols used in the disorder column of the map are explained in the note box below. The graphical view of a morbid map is available from Entrez Map Viewer that shows the disease genes in positional order.

- **Genetic variations in brackets [ ]:** These indicate certain 'nondiseases', mainly genetic variations that lead to apparently abnormal laboratory test values (e.g., histidinemia, dysalbuminemic euthyroidal hyperthyroxinemia).
- **Diseases in braces, "{ }":** These indicate examples of mutations that lead to universal susceptibility to a specific infection (diphtheria, polio), to frequent resistance to a specific infection (vivax malaria), protection from nicotine addiction, as well as some other susceptibilities.
- **Question mark:** It is used before the disease name and is the equivalent of L (in limbo) for mapping status.
- **A number in parentheses (-):** It is used after the name of disorder to indicate whether the mutation was positioned by mapping with one of the following,
  1. The wildtype gene such as neuropathy, paraneoplastic sensory
  2. The disease phenotype itself such as breast cancer, parkinson
  3. Both (1) and (2). The wildtype gene combined with demonstration of a mutation in that gene in association with the disorder like diabetes insipidus, hepatitis, hyperthyroidism

(b) Sample OMIM format

A typical OMIM entry is as shown below:

```
Mim-entry ::= {
  mimNumber "600850" ,
  mimType pound ,
  title "SCHIZOPHRENIA 4; SCZD4" ,
  copyright "Copyright (c) 1966-2005 Johns Hopkins University" ,
  symbol "PRODH, SCZD4" ,
  locus "22q11.2, 22q11-q13" ,
  aliases {"SCHIZOPHRENIA SUSCEPTIBILITY LOCUS, CHROMOSOME 22-
RELATED"},
  seeAlso {{ number 15 , author "Schizophrenia Collaborative Linkage Group
(Chromosome 22)", others "" , year 1996 } } ,
```

```
 text {{ label "TEXT" , text "A number sign (#) is used with this entry because
schizophrenia-4 can be caused by mutation in the proline dehydrogenase
gene (PRODH; {606810})." },},
 hasSummary FALSE ,
 hasSynopsis TRUE ,
 clinicalSynopsis {{
    key "Neuro" , terms {"Schizophrenia susceptibility" } } ,
   {key "Lab" , terms {"Interstitial deletion of 22q11" } } ,
   {key "Inheritance" , terms {"Autosomal dominant (22q11)" } } } ,
 editHistory {{author "mgross" , modDate
                {year 2005 , month 4 , day 4 } } , } ,
 creationDate {author "Victor A. McKusick" ,
   modDate {year 1995 , month 10 , day 18 } } ,
 references { {number 1 , origNumber 1 , type citation ,
   authors { {name "Coon, H." , index 1 } ,
   ..............................................................
      {name "Byerley, W." , index 1 } } ,
    primaryAuthor "Coon" , otherAuthors "et al." ,
    citationTitle "Genomic scan for genes predisposing to schizophrenia." ,
    citationType 0 ,
    volume "54" ,
    journal "Am. J. Med. Genet." ,
    pubDate {year 1994 ,month 0 , day 0 } ,
    pages { { from "59" , to "71" } } ,
    pubmedUID 7909992 ,
    ambiguous FALSE ,
    noLink FALSE } ,
},
 numGeneMaps 2 ,
 medlineLinks {num 15 , uids
"7909992,12217952,7485264,7726207,7644464,11930015,8546160,9259369,76
44462,7909990,8178837,8040660,8950408,8678112,9514586",
  numRelevant 0 } ,
 proteinLinks {num 1 , uids "19924111" , numRelevant 0 } ,
 nucleotideLinks {num 1 , uids "19924110" ,numRelevant 0 } }
```

(c) Abbreviation key

To understand the format of an entry, it is quite useful to understand each and every field and the subentries. The computer system stores fieldnames in an abbreviated form.

MIM number: It stores the unique number for each OMIM entry.

MIM type: It stores the type of OMIM entry in the form of symbols like #, %, +, *. It also represents a type when no symbol is stored.

Title: It lists the title of the gene.

Symbol: It stores the symbols of the genes, the parent gene and the mutated gene.

Locus: It refers to the position of the chromosome where the gene is mapped.

Aliases: It lists the gene description with its corresponding chromosome number.

Chromosome: The chromosome onto which a gene or disorder has been mapped, as reported in the OMIM Gene or Morbid Map.

EC/RN No: It indicates the number assigned by the Enzyme Commission or Chemical Abstract Service (CAS) to designate a particular enzyme or chemical, respectively.

Copyright: Stores copyright information such duration and the copyright owner.

Text: It gives textual information of the gene.

See also: It provides Link to one of the reference that is directly related to the entry.

Allelic variant: It describes a subset of disease-producing mutations. Not all entries will have allelic variants as its property.

References: It gives references to the articles cited in the OMIM entry.

Contributors: It gives names of persons who contributed to this entry.

Creation date: It lists the date on which an OMIM record was created (in the format YYYY/MM/DD).

Edit history: This entry lists of the names of the editors and years when this entry was edited in the form of editors, edit or modification history, last modification date.

Summary: It stores the summary of the entry.

Clinical synopsis: It gives clinical features of a disorder and the mode of inheritance (e.g., autosomal dominant, autosomal recessive, X-linked), if known. Not all entries will have clinical synopsis as its property.

Number of gene map: It gives the number of gene map locus for the gene.

Filter: It is used to retrieve subsets of records that contain calculated links and curated links.

Note

- Calculated links: Cross links to other Entrez databases such as Related Entries, PubMed, Protein, Nucleotide, Structure, Genome.
- Curated links: These are of two types:
  - Entrez databases such HUGO, RefSeq, GenBank, Protein, UniGene.
  - LinkOuts to external (non-Entrez) resources such as medical or molecular biology database.

TASK

Let us search OMIM for a disease called 'schizophrenia'.

Step 1 Go to the search page of OMIM (shown in Fig. 3.32) at http://www.ncbi.nlm.nih.gov/omim/

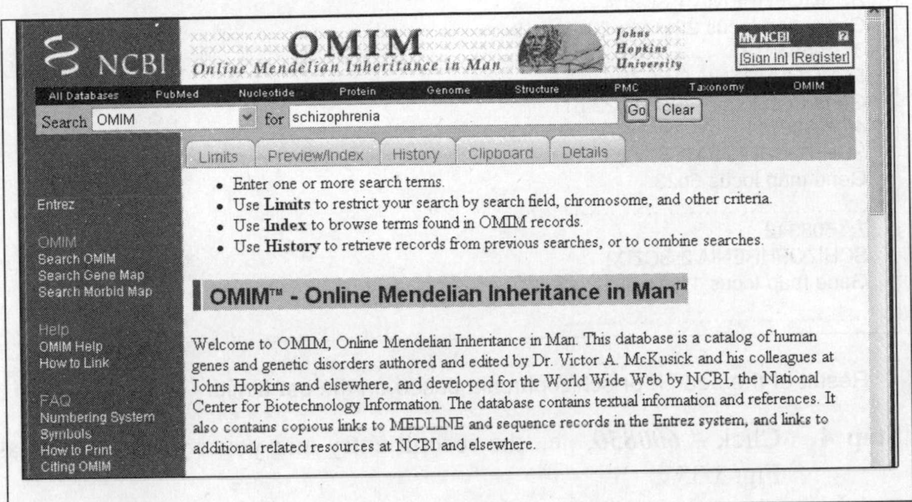

Fig. 3.32 OMIM page in NCBI

Step 2 Enter 'Schizophrenia' into the textbox and click on the **Limits** tab to select fields, chromosome type, creation date and modification date (if any). The next page looks like as shown in Fig. 3.33.

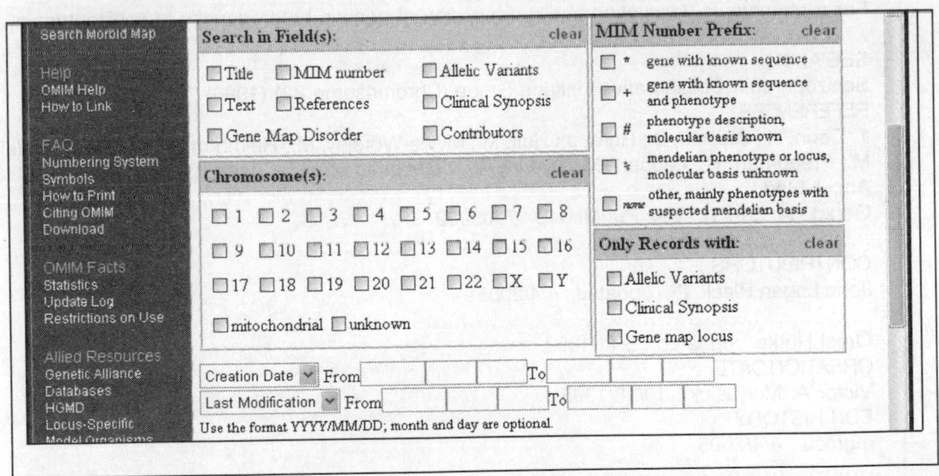

Fig. 3.33 Limits page in OMIM

Step 3 Click **Go** to get the result as shown in Fig. 3.34.

1: #181500
SCHIZOPHRENIA; SCZD
Gene map locus 22q12.3, 22q12.3, 22q12.3, 22q11.2, 22q11-q13, 22q11, 1q42.1, 1q42.1, 18p, 15q15, 14q32.3, 13q34, 13q32, 13q14-q21, 12q24, 11q14-q21, 1q21-q22, 10q22.3, 8p21, 6q13-q26, 6p22.3, 6p23, 3p25
2: #600850
SCHIZOPHRENIA 4; SCZD4
Gene map locus 22q11.2, 22q11-q13
3: %603013
SCHIZOPHRENIA 6; SCZD6
Gene map locus 8p21, 8p22-p11
4: %600511
SCHIZOPHRENIA 3; SCZD3
Gene map locus 6p23
....
7: 603342
SCHIZOPHRENIA 2; SCZD2
Gene map locus 11q14-q21
....

Fig. 3.34 Result of the search or schizophrenia based on the set Limits

Step 4 Click ***# 600850***, i.e., the second entry and you will get results as shown in Fig. 3.35.

#600850 SCHIZOPHRENIA 4; SCZD4
Alternative titles; symbols
SCHIZOPHRENIA SUSCEPTIBILITY LOCUS, CHROMOSOME 22-RELATED Gene map locus 22q11.2, 22q11-q13
TEXT
A number sign (#) is used with this entry because schizophrenia-4 can be caused by mutation in the proline dehydrogenase gene (PRODH; 606810).
For a phenotypic description and a discussion of genetic heterogeneity of schizophrenia, see 181500.
....
SEE ALSO
Schizophrenia Collaborative Linkage Group (Chromosome 22) (1996)
REFERENCES
1. Coon, H.; Jensen, S.; Holik, J.; Hoff, M.; Myles-Worsley, M.; Reimherr, F.; Wender, P.; Waldo, M.; Freedman, R.; Leppert, M.; Byerley, W. : Genomic scan for genes predisposing to schizophrenia. Am. J. Med.
Genet. 54: 59-71, 1994. PubMed ID : 7909992
....
CONTRIBUTORS
John Logan Black, III - updated : 4/4/2005
....
Orest Hurko - updated : 5/11/1998
CREATION DATE
Victor A. McKusick : 10/18/1995
EDIT HISTORY
mgross : 4/4/2005
....
mark : 10/18/1995
Copyright © 1966-2005 Johns Hopkins University

Fig. 3.35 Detail entry of SCZD4

See the Note below to get an idea of each term in the search result.

> **Note**
>
> **Alternative titles; symbols:** It lists MIM number, alternative title of the gene, gene symbol, and gene description with its corresponding chromosome number.
>
> **Gene map locus:** Position of the chromosome where the gene is mapped.
>
> **TEXT:** Textual information of the gene.
>
> **SEE ALSO:** Link to one of the reference that is directly related to the entry.
>
> **REFERENCES:** References to the PubMed literature.
>
> **CONTRIBUTORS:** People who contributed to this entry.
>
> **CREATION DATE:** Date this record was created.
>
> **EDIT HISTORY:** List of names and year when this entry was edited.
>
> **Copyright:** Copyright duration and copyright owner.

Step 5 Go back to Step 3 result page (Fig. 3.34). Click ***22q11.2***, i.e., the gene map locus of the second entry corresponding to SCZD4 gene. You will get a result page as shown in Fig. 3.36.

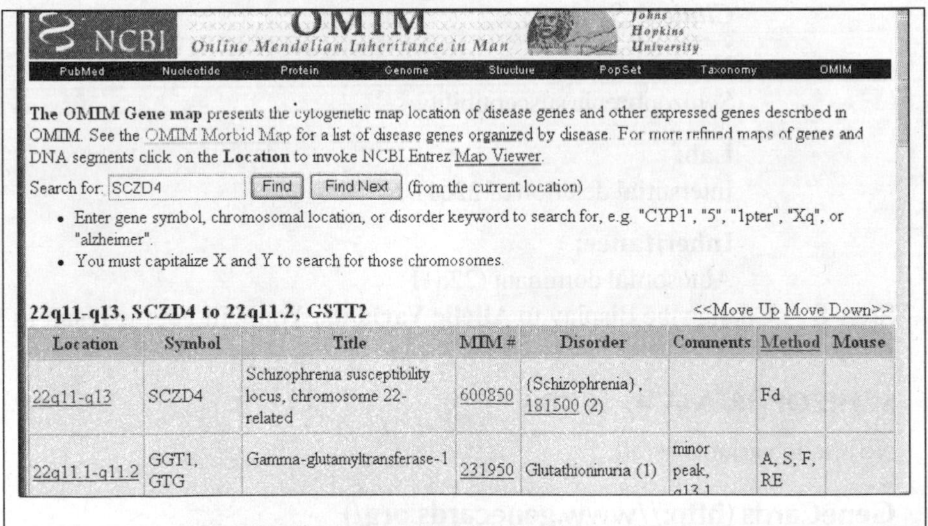

Fig. 3.36 OMIM gene map and result for SCZD4

Step 6 Click ***OMIM Morbid Map*** and search for 600850 MIM number. You will get a result page as shown in Fig. 3.37.

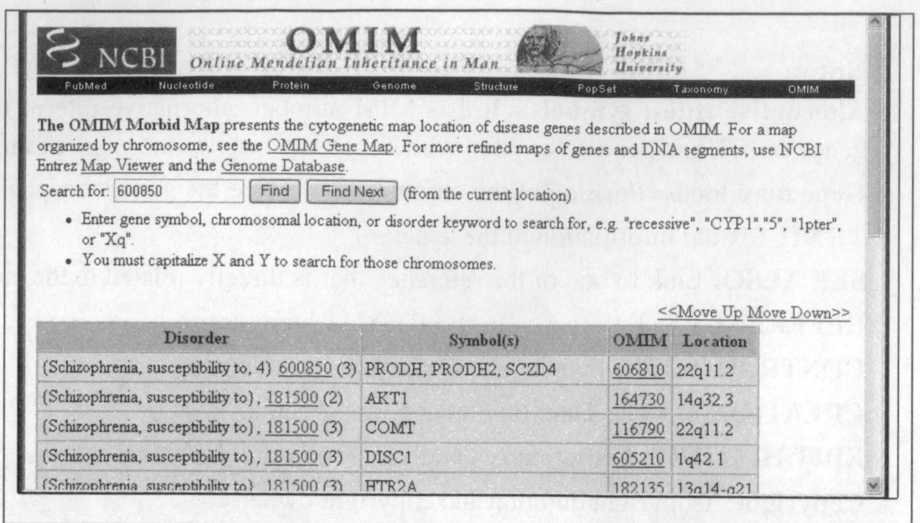
NCBI OMIM Online Mendelian Inheritance in Man Johns Hopkins University

PubMed Nucleotide Protein Genome Structure PopSet Taxonomy OMIM

The OMIM Morbid Map presents the cytogenetic map location of disease genes described in OMIM. For a map organized by chromosome, see the OMIM Gene Map. For more refined maps of genes and DNA segments, use NCBI Entrez Map Viewer and the Genome Database.

Search for: 600850 [Find] [Find Next] (from the current location)

- Enter gene symbol, chromosomal location, or disorder keyword to search for, e.g. "recessive", "CYP1","5", "1pter", or "Xq".
- You must capitalize X and Y to search for those chromosomes.

<<Move Up Move Down>>

Disorder	Symbol(s)	OMIM	Location
{Schizophrenia, susceptibility to, 4} 600850 (3)	PRODH, PRODH2, SCZD4	606810	22q11.2
{Schizophrenia, susceptibility to}, 181500 (2)	AKT1	164730	14q32.3
{Schizophrenia, susceptibility to}, 181500 (3)	COMT	116790	22q11.2
{Schizophrenia, susceptibility to}, 181500 (3)	DISC1	605210	1q42.1
{Schizophrenia, susceptibility to}, 181500 (3)	HTR2A	182135	13q14-q21

Fig. 3.37 OMIM morbid map and result for SCZD4

Step 7 Go back to Step 4 result (Fig. 3.37). Change the display from **Detail** to **Clinical Synopsis**. You will get the result as below:

Clinical Synopsis

Neuro:
Schizophrenia susceptibility

Lab:
Interstitial deletion of 22q11

Inheritance:
Autosomal dominant (22q11)

Step 8 Change the display to **Allelic Variants**. You will get the result as below:

SCHIZOPHRENIA 4; SCZD4

No allelic variants listed.

3.5.2 GeneCards (http://www.genecards.org/)

GeneCards is an integrated database of human genes with genomic, proteomic, and transcriptomic information. It also stores related information such as orthologues, disease relationships, SNPs, gene expression, gene function, and more. The data is broadly categorized into protein coding, pseudogenes, and RNA genes, There are categories for genetic loci, gene cluster, HGNC reserved symbols, uncategorized. The database is developed at the Crown Human Genome Center and Weizmann Institute of Science with the objective to integrate all gene related information into one web-based knowledge base. GeneCards database is a model of an electronic encyclopedia of biological and medical information based on human–computer interaction research and intelligent knowledge navigation technology. The database uses standard nomenclature and approved symbols. Figure 3.38 shows the homepage of genecards.org.

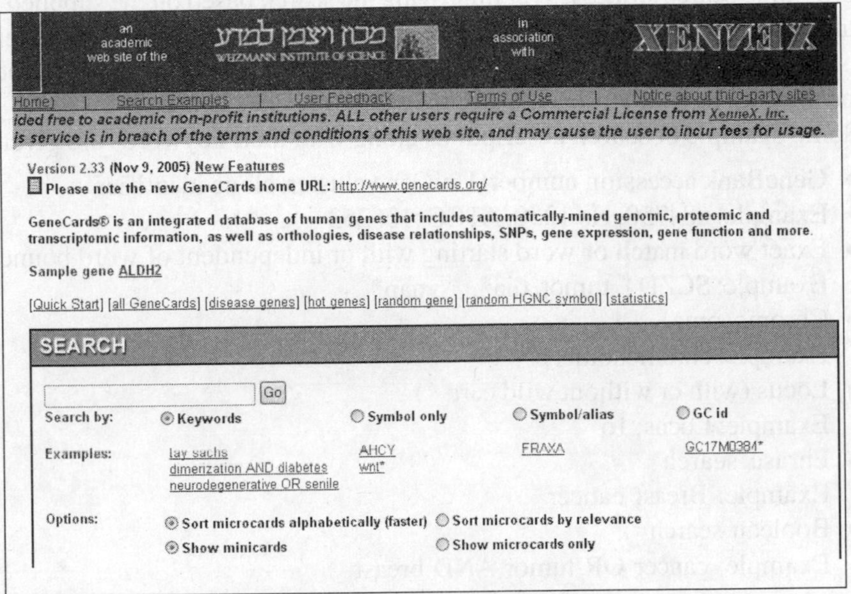

Fig. 3.38 Homepage of GeneCards

(a) GeneCards entry

Entries of GeneCards database are referred as genecards, which integrate the knowledge base of a gene and its related information from medical aspects. The data is stored and indexed in such a manner that it provides user-friendly navigation. The detailed entries of the genecards are either the information or the links as mentioned below:

- Genes, gene IDs, gene synonyms, homologous genes, orthologues
- Cytogenic locus, genomic regions of the genes, gene coordinates
- Gene products such as proteins and their features such as cellular functions, expression patterns, similarities with other proteins, and involvement in diseases
- Gene disorders and mutations
- Titles of related literature with links to the abstract and full citation
- Medical applications such as new therapies and diagnoses

(b) Searching GeneCards

GeneCards provide fast access to information by using the search engine that is accessible from many points in the database. This service is free for academic and non-profit institutions. Other users can access the service against a commercial License from XenneX,

Inc. Search can be initiated by using keywords of gene, protein, disease, cell type, tissue, chromosome, or a related word. The search engine provided in this database returns relevant verbose extracts of data and highlights words matching the query. This helps in fast browsing and helps the user to select the GeneCards of interest or relevance. It also provides tools that guide the user to the information he or she needs in an intelligent way by providing search strategies for improving the search based on the supplied keyword. In addition, direct links to the search engines of other databases are also provided along with the short descriptions of the GeneCards-related information that can be found there. This helps the user to look at the database relevant to his/her query.

Some example of search descriptions along with their keywords are given below:

- GeneBank accession number, UniGene cluster, clone identifier
  Example: x84750, Hs.1230, ATCC :106253
- Exact word match or word starting with or independent of word boundary
  Example: SCZD4, tumor, dia*, *synap*
- Chromosome
  Example: chromosome: 6
- Locus (with or without wild card *)
  Example: Locus: 1q*
- Phrase search
  Example: Breast cancer
- Boolean search
  Example: cancer OR tumor AND breast

TASK

Step 1 Let us search with the keyword 'Schizophrenia' clicking **Keywords** as the **Search by** option. You will get the result grouped into microcards and minicards as shown in Fig. 3.39.

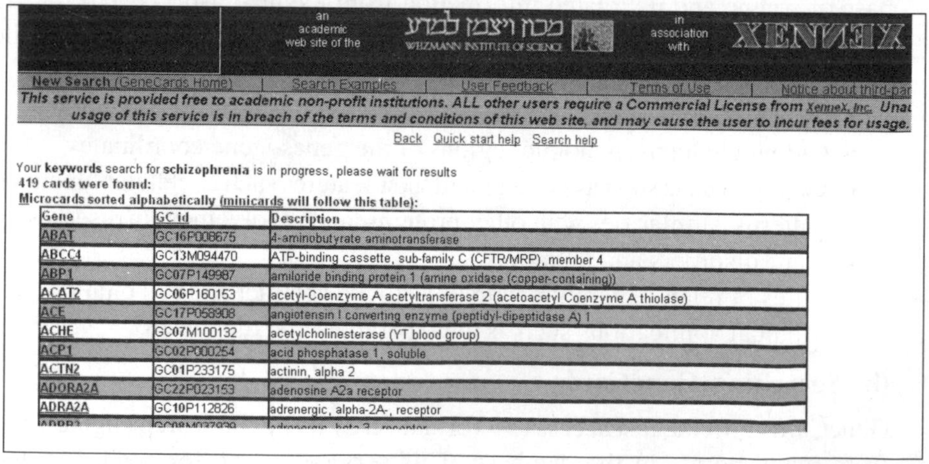

an academic web site of the מכון ויצמן למדע WEIZMANN INSTITUTE OF SCIENCE in association with XENNEX

New Search (GeneCards Home) | Search Examples | User Feedback | Terms of Use | Notice about third-par

This service is provided free to academic non-profit institutions. ALL other users require a Commercial License from XenneX, Inc. Unau usage of this service is in breach of the terms and conditions of this web site, and may cause the user to incur fees for usage.

Back Quick start help Search help

Your **keywords** search for **schizophrenia** is in progress, please wait for results
419 cards were found:
Microcards sorted alphabetically (minicards will follow this table):

Gene	GC id	Description
ABAT	GC16P008675	4-aminobutyrate aminotransferase
ABCC4	GC13M094470	ATP-binding cassette, sub-family C (CFTR/MRP), member 4
ABP1	GC07P149987	amiloride binding protein 1 (amine oxidase (copper-containing))
ACAT2	GC06P160153	acetyl-Coenzyme A acetyltransferase 2 (acetoacetyl Coenzyme A thiolase)
ACE	GC17P058908	angiotensin I converting enzyme (peptidyl-dipeptidase A) 1
ACHE	GC07M100132	acetylcholinesterase (YT blood group)
ACP1	GC02P000254	acid phosphatase 1, soluble
ACTN2	GC01P233175	actinin, alpha 2
ADORA2A	GC22P023153	adenosine A2a receptor
ADRA2A	GC10P112826	adrenergic, alpha-2A-, receptor

Fig. 3.39 Search result for the keyword 'Schizophrenia'

Step 2 Scroll down to minicards and look at the details of GeneCard SCDZ4. You will get the result as shown in Fig. 3.40.

GeneCard: SCZD4 schizophrenia disorder 4 (GC22U990043; Chromosome 22)

The following lines in the GeneCard text contribute to matching your query:

- GENE: SCZD4 | **schizophrenia** disorder 4

- ALIASES: Schizophrenia susceptibility locus, chromosome 22-related (EG)| **schizophrenia** disorder 4 (GDB)

- OMIM: Schizophrenia susceptibility locus, chromosome 22-related; SCZD4 | 600850 | **{Schizophrenia}**, 181500

- LITERATURE: Sequential strategy to identify a susceptibility gene for **schizophrenia:** reportof potential linkage on chromosome 22q12-q13.1: Part 1. . . .

- ENTREZGENE: chr=22 | contig= | cytoLoc=22q11-q13 | dbXrefs=MIM:600850 | egHugo= | egId=6379 | end=0 | geneName=**schizophrenia** disorder 4 | genomicNucAcc= | go= | protAcc= | proteinAliases= | refseq= | rnaNucAcc= | start=0 | status=O | synonyms=**Schizophrenia** susceptibility locus, chromosome 22-related | type=unknown

- HUGO: accs= | aliases= | dateApproved=15/05/1997 | dateModified=17/09/1998 | dateNameChanged= | description=**schizophrenia** disorder 4 | egCuratedId= | egMappedId=6379 | enzymeIds= | gdbId=1387047 | geneFamName= | hgnc=10652 | loc=22q11-q13 | mgdId= | omimId=600850 | prevNames= | prevSymbols= | pubmedIds=8178837 | refseqId= | refseqMappedId= | status=Approved | symbol=SCZD4 | uniprotId=

Fig. 3.40 Minicard view of SCDZ4

Step 3 Click SCDZ4 and you will get the result as shown in Fig. 3.41.

3.5.3 MedlinePlus (http://medlineplus.gov/)

Look at the MedlinePlus home page shown in Fig. 3.42. Now that you have gained expertise in technical disease databases, you can know more about this database that is targeted at general mass. Click **About MedlinePlus** to know about the background of this database. Click FAQs to read the the common questions that occur in any mind while surfing this database. Try exploring all the links provided such as **Health Topics**, **News**, **Drugs & Supplements**. Learn more about the terms that you come across in your medical reports from links like **Medical Encyclopedia** and **Dictionary**. Search for doctors and hospitals from **Directories**. And if you want to understand various test procedures or gain knowledge generally, go through **Interactive Tutorials**.

LeanCard for genetic locus SCZD4
GC22U990043

schizophrenia disorder 4
Symbol approved by the HUGO Gene Nomenclature Committee (HGNC) database

Services DBs Literature Disorders Loc Aliases

Aliases and Descriptions (According to [1]HGNC, [2]Entrez Gene, [3]UniProt/Swiss-Prot, [4]UniProt/TrEMBL, [5]GDB, [6]OMIM, and/or [7]GeneLoc) *About Top*	• **Schizophrenia** susceptibility locus, chromosome 22-related • **schizophrenia** disorder 4
Genomic Location (According to GeneLoc and/or HGNC, and/or Entrez Gene (NCBI build 35), Genomic Views According to UCSC and Ensembl) *About Top*	*Chromosome:* 22 *Entrez Gene cytogenetic band:* 22q11-q13 *Ensembl cytogenetic band:*
Disorders & Mutations (in which this Gene is Involved, According to OMIM, UniProt, Genatlas, GeneTests, HGMD, GAD, GDPInfo, BCGD, and/or TGDB.) *About Top*	**OMIM:** 600850 search databases for these OMIM-named disorders: **(Schizophrenia), 181500**
Research Articles (in PubMed. Associations of this gene to articles via bioalma, HGNC, Entrez Gene, UniProt, and/or GAD) *About Top*	**8 PubMed articles:** ***the following papers are cited by 2 GeneCards sources:*** • Sequential strategy to identify a susceptibility gene for **schizophrenia**: reportof potential linkage on chromosome 22 q13.1: Part 1. *(1994)* ***the following papers are cited by 1 GeneCards source:*** • Linkage of a composite inhibitory phenotype to a chromosome 22q locus in eightUtah families. Myles-Worsley M. .. Byerley W. *(1999)* • A genome-wide search for schizophrenia susceptibility genes. Shaw S.H. ... DeLisi L.E. *(1998)* • Schizophrenia susceptibility loci on chromosomes 13q32 and 8p21. Blouin J.L. ... Pulver A.E. *(1998)* • Schizophrenia and chromosomal deletions within 22q11.2 Lindsay E.A. ... Pulver A.E. *(1995)* • Schizophrenia susceptibility associated with interstitial deletions of chromosome22q11. *(1995)* • Genomic scan for genes predisposing to schizophrenia. *(1994)* • Velo-cardio-facial syndrome associated with chromosome 22 deletions encompassingthe DiGeorge locus. Scamble ... Burn J. *(1992)* **Search PubMed for:** **Gene symbol** **Your search terms** ☑ SCZD4 OR ☐ schizophrenia Search

SCZD4 in Other Genome Wide Resources: (According to GDB, Entrez Gene, HGNC, AceView, euGenes, Ensembl, ECgene, and/or GeneLynx) *About Top*	GDB: 1387047 Entrez Gene: 6379 HGNC: 10652 euGenes: HUgn6379 ECgene: SCZD4
Services (Reagents available from Applied Biosystems, Clones available from RZPD) *About Top*	AB Applied Biosystems **Products for SCZD4:** › Free SNP selection tool

Aliases Loc Disorders Literature DBs Services

Fig. 3.41 Detail entry of SCDZ4

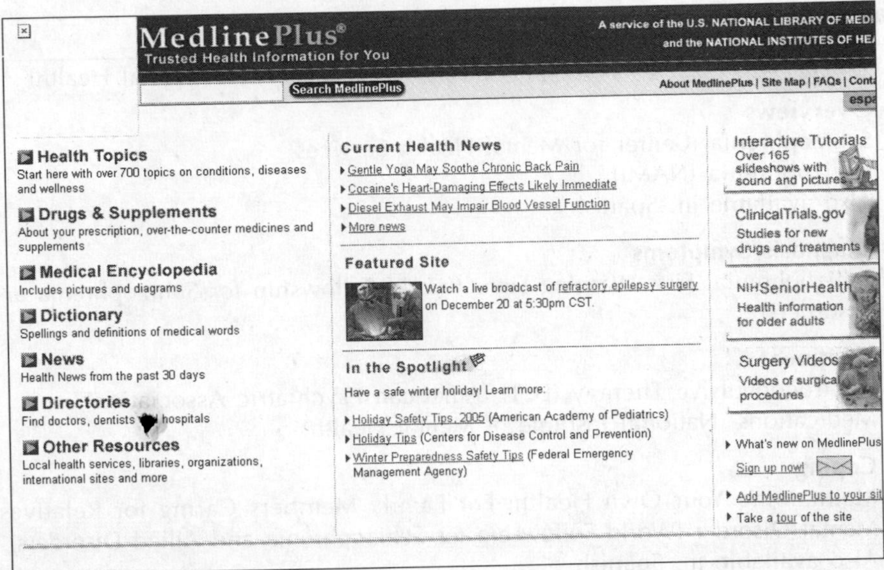

Fig. 3.42 Home Page of Medline Plus

(a) Medline entry

Each Medline entry records the latest news, information from the National Institutes of Health, overviews, diagnosis/symptoms, treatment, coping of the disease. It also provides specific conditions, related issues, clinical trials and research related information. Every record lists the organizations associated with the disease, children specific information, and information from the Medical encyclopedia.

TASK

Step 1 Let us search for keyword 'Schizophrenia' as we did before. The kind of result you get will look like as shown in Table 3.2.

Table 3.2 Result of MedlinePlus for schizophrenia

Latest News
Creativity Tied to Sexual "Success" (12/07/2005, Reuters Health)
Heavy Marijuana Use Damages Adolescent Brains (11/30/2005, Reuters Health)
Ignoring Useless Information Aids Memory (11/23/2005, Reuters Health)
Teens with Deletion Syndrome Confirm Gene's Role in Psychosis (10/23/2005, National Institute of Child Health and Human Development, National Institute of Mental Health)

From the National Institutes of Health
Schizophrenia (National Institute of Mental Health)
Schizophrenia Research at the National Institute of Mental Health (National Institute of Mental Health)

(contd)

Table 3.2 *(contd)*

When Someone Has Schizophrenia (National Institute of Mental Health)

Overviews
Schizophrenia (Center for Mental Health Services)
Schizophrenia (NAMI)
Also available in: Spanish

Diagnosis/Symptoms
Schizophrenia: First Warning Signs (World Fellowship for Schizophrenia and Allied Disorders)

Treatment
Electroconvulsive Therapy (ECT) (American Psychiatric Association)
Medications (National Institute of Mental Health)

Coping
Maintaining Your Own Health: For Family Members Caring for Relatives with a Mental Disorder (World Fellowship for Schizophrenia and Allied Disorders)
Also available in: Spanish
Schizophrenia -- Dealing with a Crisis (World Fellowship for Schizophrenia and Allied Disorders)
Also available in: Spanish
Schizophrenia: How Should One Behave? (World Fellowship for Schizophrenia and Allied Disorders)
Also available in: Spanish

Specific Conditions
Schizoaffective Disorder (National Mental Health Association)
Schizophrenia and Suicide (World Fellowship for Schizophrenia and Allied Disorders)

Related Issues
Choosing a Mental Health Provider: How to Find One Who Suits Your Needs (Mayo Foundation for Medical Education and Research)

Clinical Trials
ClinicalTrials.gov: Schizophrenia (National Institutes of Health)
NIMH Genetic Study of Schizophrenia (National Institute of Mental Health)

Research
Brain Scans Reveal How Gene May Boost Schizophrenia Risk (04/21/2005, National Institute of Mental Health)
NIMH Study to Guide Treatment Choices for Schizophrenia (09/16/2005, National Institute of Mental Health)
Schizophrenia Gene Variant Linked to Risk Traits (08/11/2004, National Institute of Mental Health)

Organizations
NAMI
Also available in: Spanish
National Institute of Mental Health

(contd)

Table 3.2 (*contd*)

National Mental Health Association World Fellowship for Schizophrenia and Allied Disorders Also available in: Spanish
Children Childhood-Onset Schizophrenia: An Update (National Institute of Mental Health) Early Onset Schizophrenia (NAMI)
Information from the Medical Encyclopedia Schizophrenia

Step 2 Click on the links provided to search and explore further.

3.6 Literature Databases

3.6.1 PubMed (http://www.ncbi.nlm.nih.gov/entrez/query.fcgi?DB=pubmed)

PubMed (encompassing Medline) is the database of citations and abstracts for millions of articles from thousands of biomedical and life science journals that are meant for researchers and clinicians. It also provides links to other NCBI resources, PMC articles, and access to a publisher's site for the full text through **LinkOut**. In addition, PubMed provides a single/batch citation matcher, which allows users to match their citations to PubMed citations using bibliographic information such as the journal title, volume, issue, page number, and year. But it does not have citations for PMC material such as book reviews. Publishers participating in PubMed submit their citations electronically to NCBI either prior to or at the time of publication. PubMed records can be accessed through the Entrez retrieval system developed by the National Center for Biotechnology Information (NCBI) at the National Library of Medicine (NLM), and located at the National Institutes of Health (NIH).

PubMed Central (PMC) (www.pubmedcentral.nih.gov/) is a digital archive of life sciences journal literature, primary research papers, review articles, commentaries, editorials, and letters in English at the U.S. National Institutes of Health (NIH) developed by NCBI, the centre of NIH. It is managed by NCBI at the National Library of Medicine (NLM). PMC is not a journal publisher but instead is a library that provides unrestricted access to the electronic literature and ensures the durability and best use of the archive. It is open to publishers who can participate after qualifying the test for the scientific and editorial quality of its content and the technical quality of its digital files. PubMed Central makes the best use of the services of NCBI developed for the worldwide scientific community to archive data in a common XML, SGML, or PDF format that conforms to any established DTD (document type definition). It aims to become a true digital counterpart to NLM's extensive collection of print journals. It is also in the process of digitizing earlier print issues of the journals that are already in PMC. Unfortunately, not all journals offer immediate free access to the full text for some period of time after publication. The value of PubMed Central, in addition to its role as an archive, lies in what can be done when data from diverse sources is stored in a common format in a single repository. It facilitates locating relevant material from the entire body of full-text articles from any source through a common interface. It also integrates the literature with a variety of other information resources such as sequence databases and other factual databases. Look at Fig. 3.43 to know what all is available in the homepage of PubMed Central (http// www.pubmedcentral.nih.gov/).

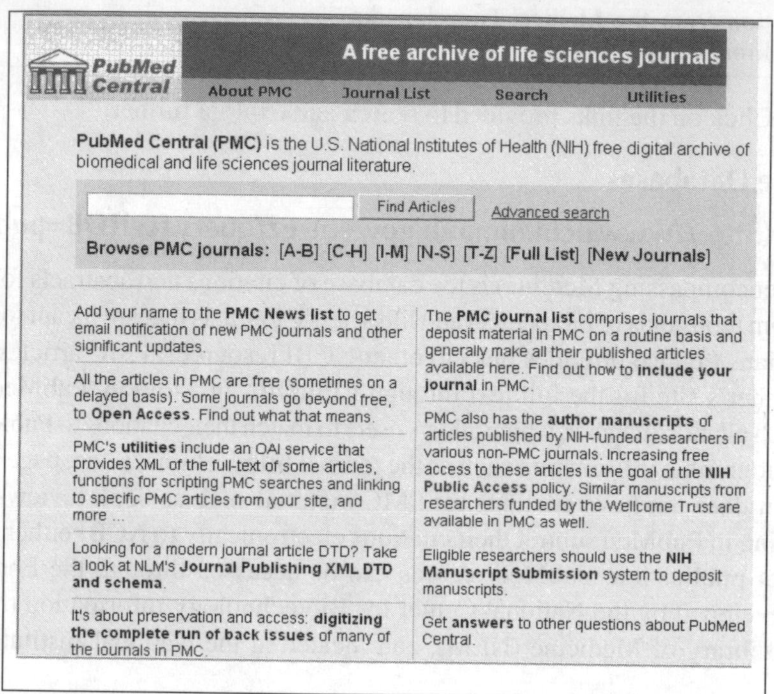

Fig. 3.43 Homepage of PubMed Central

(a) PubMed search

A PubMed entry includes all kinds of bibliographic information such as article title, author(s), journal title, volume, issue, page number, and year. It also keeps track of records with abstract only or with full text options. You can fine time your search by humans or animals studies, age group, gender, languages, publication types, dates, and by other parameters. Let us look at each of them:

1. The **Humans** or **Animals** limit restricts the search to a human or animal study.
2. The **Ages** limit restricts the search to a specific age group for a human study.
3. The **Gender** limit restricts the search to a specific gender for a human study.
4. The **Languages** limit restricts the search to articles written in a particular languages, such as English, French, German, Italian, Japanese, Russian, and Spanish.

5. The **Publication Type** limit will restrict the search based on the type of material the article represents, e.g., Clinical Trial, Editorial, Letter, Meta-Analysis, Practice Guideline, Randomized Controlled Trial, and Review.
6. The **Entrez Date** is the date a citation was added to PubMed. It restricts the search to citations entered into PubMed within a specific time frame, e.g., the last 10 days, the last 5 years. You can select the time frame from the **Entrez Date** pull-down menu or specify your own. PubMed displays search results in the descending Entrez data order, i.e., last in, first out.
7. **Publication Date** will allow you to enter the date in the **From** and **To** boxes in the format yyyy/mm/dd. The month and day are optional, e.g., 1964/06/23, or 1964/06, or 1964. Use the **Publication Date** limit to restrict your search to citations published within a specific time frame (e.g., 2005).
8. The subject subsets restrict retrieval to specific topics such as AIDS, Bioethics, Cancer, Complementary Medicine, History of Medicine, Space Life Sciences, and Toxicology.
9. The citation status indicates the processing stage of an article in the PubMed database, e.g., publisher, in process, medline, pubmednotmedline, and oldmedline

TASK

Step 1 Let us search with keyword 'schizophrenia' and set Limits as shown in Fig. 3.44.

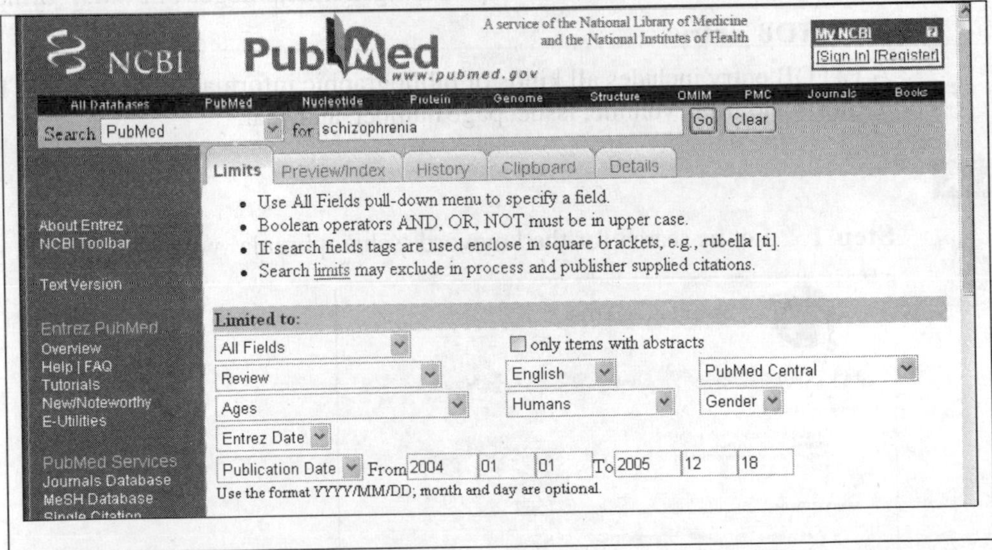

Fig. 3.44 Limit page of PubMed

Step 2 Click **Go** and you will get the results as displayed in Table 3.3.

Table 3.3 Result page of 'Schizophrenia'

1: Kirov G, O'Donovan MC, Owen MJ.
Finding schizophrenia genes.
J Clin Invest. 2005 Jun;115(6):1440-8. Review.
PMID: 15931379 [PubMed - indexed for MEDLINE]

2: Hanson DR, Gottesman II.
Theories of schizophrenia: a genetic-inflammatory-vascular synthesis.
BMC Med Genet. 2005 Feb 11;6:7. Review.
PMID: 15707482 [PubMed - indexed for MEDLINE]

3: Sacco KA, Bannon KL, George TP.
Nicotinic receptor mechanisms and cognition in normal states andneuropsychiatric disorders.
J Psychopharmacol. 2004 Dec;18(4):457-74. Review.
PMID: 15582913 [PubMed - indexed for MEDLINE]

4: McGrath J, Saha S, Welham J, El Saadi O, MacCauley C, Chant D.
A systematic review of the incidence of schizophrenia: the distribution ofrates and the influence of sex, urbanicity, migrant status and methodology.
BMC Med. 2004 Apr 28;2:13. Review.
PMID: 15115547 [PubMed - indexed for MEDLINE]

3.6.2 LITDB (http://www.genome.jp/dbget-bin/www_bfind?litdb)

LITDB is a literature database that can be searched using search engine DBGET. The search page can be accessed at http//www.genome.jp/dbget-bin/www_bfind?litdb

(a) LITDB entry

A LITDB entry includes all kinds of bibliographic information such as LITDB ID, title, author(s), journal, volume, issue, page number, and year.

TASK

Step 1 Let us search for the keyword 'schizophrenia' with **bfind mode** selected.

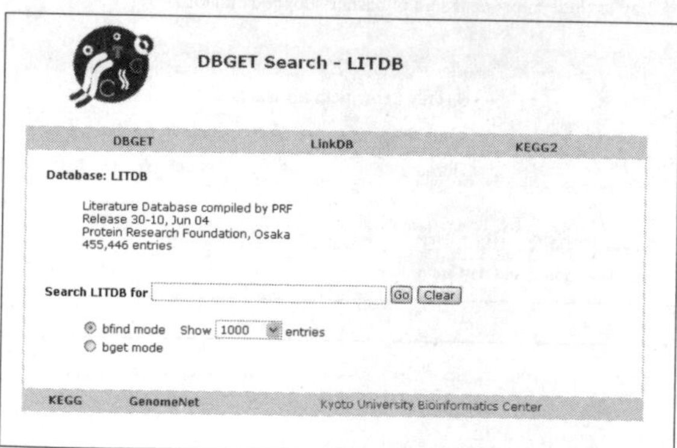

Fig. 3.45 Searchpage of LITDB through DBGET search

Step 2 Click **Go** and you will get result as shown in Fig. 3.46.

```
Database: LITDB
Search term: schizophrenia  (1 - 100 ) Next
lit:0312008 Tripeptide plays role in schizophrenia.
lit:0403010 gamma-Aminobutyric acid (Gaba) and the dopamine pypothesis of  schizophrenia.
lit:0406505 (Des-1-Tyr)-gamma-endorphin in schizophrenia.
lit:0408453 Effects of naloxone on schizophrenia - Reduction in hallucinations  in a subpopulation of subjects.
lit:0412016 Psychopharmacology of morphinomimetic peptides in relation to  schizophrenia.
lit:0412505 Rapid inactivation of enkephalin-like material by C.S.F. in chronic  schizophrenia.
lit:0506541 Sinthetic enkephalin analogue in treatment of schizophrenia.
lit:0506542 C.S.F. beta-endorphin in schizophrenia.
lit:0506543 Schizophrenia and neuroactive peptides from food.
lit:0508019 Naloxone in schizophrenia. Negative result.
lit:0508553 Brain tryptophan metabolism in schizophrenia. A post mortem study of  metabolites on the serotonin and kynurenine pathways
lit:0509006 Some observations on the opiate peptides and schizophrenia.
lit:0509506 C.S.F. beta-Endorphin in schizophrenia.
lit:0510454 Schizophrenia and degradation of endorphins in cerebrospinal  fluid.
lit:0510508 Plasma beta-endorphin immunoreactivity in schizophrenia.
lit:0511348 On beta-5-Leu-endorphin and schizophrenia.
lit:0605517 beta-Endorphin, blood-brain barrier, and schizophrenia.
lit:0607304 Hemodialysis, endorphins, and schizophrenia.
lit:0609292 beta-Endorphin and schizophrenia.
lit:0610262 Cerebrospinal fluid endorphins in schizophrenia.
lit:0612049 Schizophrenia as an inborn error in the degradation of  beta-endorphin. A hypothesis.
lit:0701522 Antipsychotic effect of gamma-type endorphins in schizophrenia.
lit:0701523 Des-1-Tyr-gamma-endorphin in schizophrenia.
lit:0702513 beta-Endorphin and schizophrenia.
```

Fig. 3.46 Result page of LITDB for keyword 'schizophrenia'

3.7 Other Specialized Databases

Though the genome sequencing projects are churning out volumes and catalogues of genes, it is not easy to understand their functional implications. Certain databases, such as PROSITE, do store functional data (in the form of the feature tables of sequence databases and in motif libraries) that relate to sequence information. However, these basically represent sequence-function relationships of single molecules which are isolated, individual components of a biological system. They do not go to the next level of complexity, which is genetic and molecular interactions in a system. Thus the construction of specialized databases such as those storing information about chemicals, expressions, and micro arrays was required.

3.7.1 Chemical Databases

Medicinal chemists and researchers in the field of *in silico* drug discovery and modelling need to play with chemical structures. They often optimize potency, selectivity, and pharmacokinetic properties in parallel, which helps save time and money during lead optimization. This calls for a multidisciplinary approach so that in addition to the above said areas of work a mechanism of exchange and access to information is facilitated between researchers in other fields such as high throughput screening, computational chemistry, combinatorial chemistry, ADME informatics, cheminformatics, toxicology, and metabolic modelling. It is therefore not surprising that several chemical databases have sprung up to fulfil this need.

(a) Kyoto Encyclopedia of Genes and Genomes (KEGG): (www.genome.jp/kegg/)

KEGG was formed in 1995 under the Japanese Human Genome Program. Its primary objective was to computerize the current knowledge of molecular interactions namely, metabolic pathways, regulatory pathways, and molecular assemblies. It also contains gene catalogues for all the organisms that have been sequenced and links each gene product to a component on the pathway. Therefore all chemical compounds in living cells and links of each compound to a pathway component are found in KEGG. Figure 3.47 shows the KEGG interface.

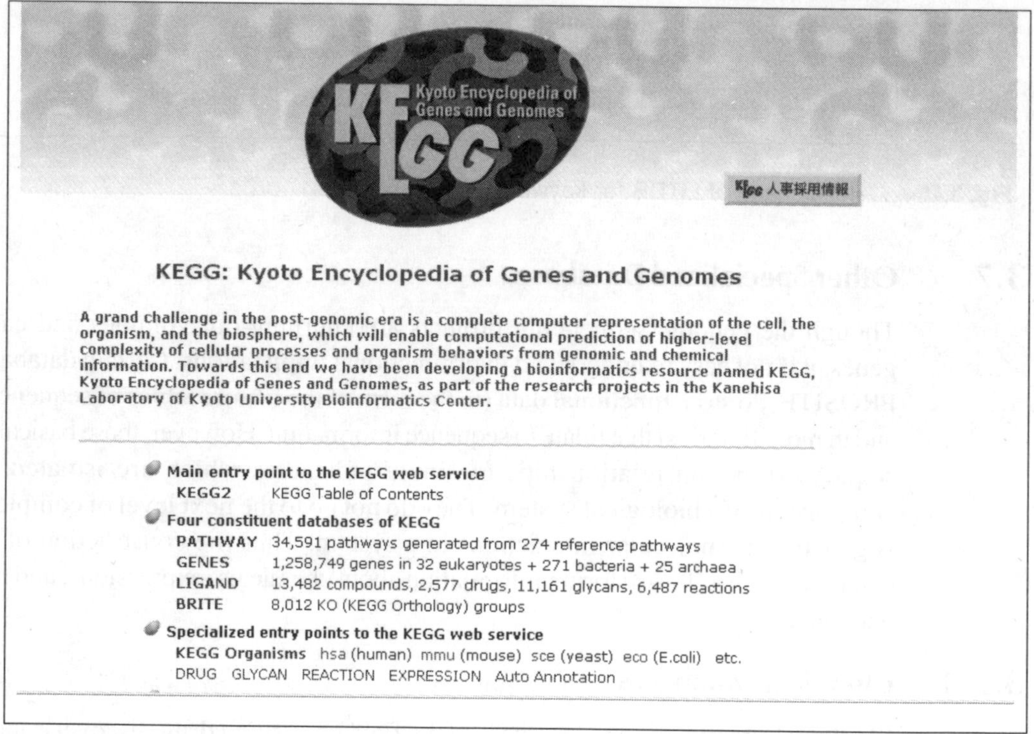

Fig. 3.47 The KEGG interface

KEGG is now a suite of databases and associated software which integrates molecular interaction networks in biological processes (PATHWAY database), information on genes and proteins (GENES/SSDB/KO databases), and the information on chemical compounds and reactions (COMPOUND/DRUG/GLYCAN/REACTION databases). The current statistics of KEGG databases are as shown in Table 3.4.

Table 3.4 Current statistics of KEGG database

Type of Database	Description	Number of Entries
PATHWAY database	Number of pathways	34,591
PATHWAY database	Number of reference pathways	274
PATHWAY database	Number of orthologue tables	87
GENOME database	Number of organisms	374
GENES database	Number of genes	1,258,749
SSDB database	Number of orthologue clusters	38,655
KO database	Number of curated KO groups	8,012
COMPOUND database	Number of chemical compounds	13,482
DRUG database	Number of drugs	2,577
GLYCAN database	Number of glycans	11,161
REACTION database	Number of chemical reactions	6,487
REPAIR database	Number of reactant pairs	5,162

(*Source:* KEGG at http://www.genome.ad.jp/kegg/ligand.html)

The five types of data stored by KEGG are as follows:

Pathway maps represented by graphical diagrams. These contain information pathways of interacting molecules or genes.

Ortholog group tables are alternative representations of the pathway maps. These HTML tables contains information of a conserved, functional unit in the molecular pathway, a list of genes for the functional unit in different organisms, and the positional information of how genes are clustered in a genome. These are a very useful tool for functional prediction.

Molecular catalogues represented by HTML tables or hierarchical texts. These tables contain information on functional aspects of biological macromolecules (including proteins, RNAs), small chemical compounds, and their assemblies.

Genome maps represented by Java applets. These maps contain positional information of genes, such as an operon structure, and its relationship with pathways. These applets are very useful tools and assist in the genes-to-function prediction.

Gene catalogues represented by hierarchical texts. These contain classifications of all the known genes for each organism.

Although housed separately, these five types of data are integrated via an integrated database retrieval system, known as DBGET/LinkDB (Fig. 3.48). Currently it provides access to

- nucleic acid sequences in GenBank, DDBJ, and EMBL;
- protein sequences in SWISS-PROT, PIR, PRF, and PDBSTR;
- 3D structures in PDB;
- sequence motifs in PROSITE, EPD, and TRANSFAC;
- enzyme reactions and chemical compounds in KEGG LIGAND;
- metabolic and regulatory pathways in KEGG PATHWAY;
- gene catalogues for organisms in KEGG GENES;
- amino acid mutations in PMD;
- amino acid indices in AAindex;
- genetic diseases in OMIM;
- literature in Medline (link only) and LITDB; and
- link information in LinkDB.

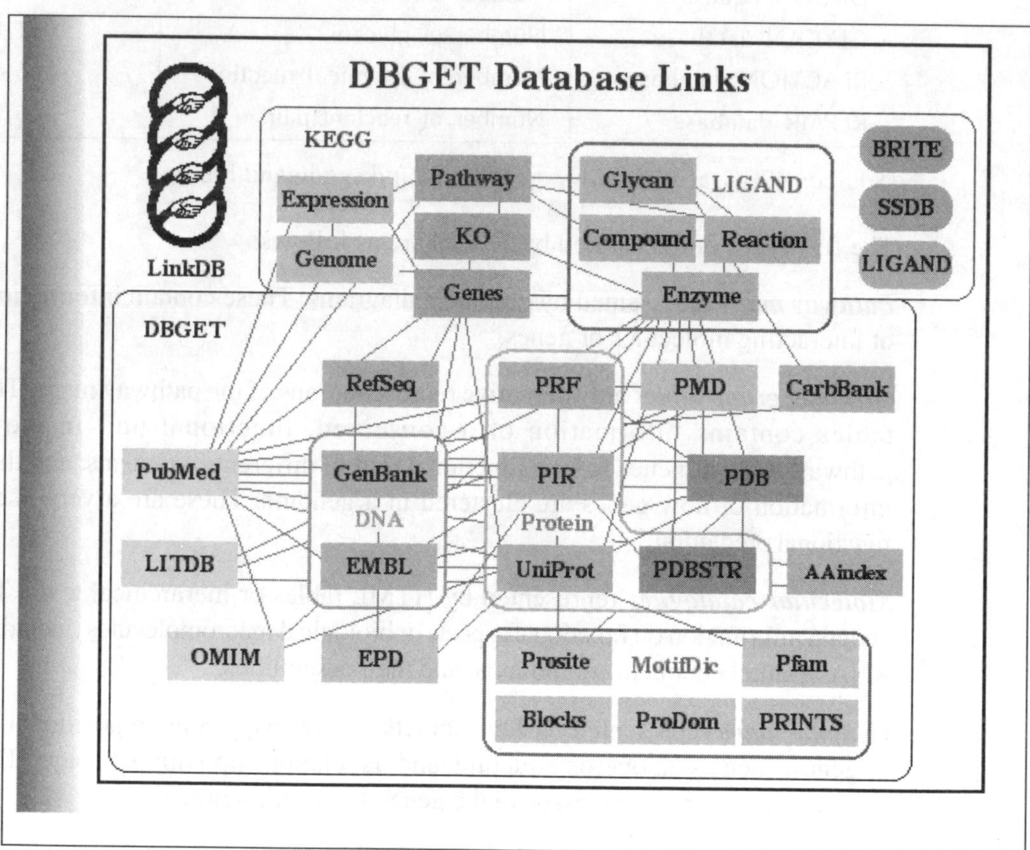

Fig. 3.48 DBGET database links to other databases

In addition to links and searches to above mentioned databases, the best feature of KEGG is its 'inference capabilities', that is, logical reasoning. Given a list of enzymes (EC numbers) of an organism, KEGG automatically generates the organism-specific pathways. This 'map' can then be used to ascertain the correctness of function in the gene catalogue. A missing element implies either the gene catalogues is wrong or there is an unknown reaction pathway.

Search

KEGG can be searched by

- EC numbers for enzymes,
- compound numbers for chemical compounds, and
- by sequence similarity (this is especially useful for identifying orthologues and reconstructing pathways from the gene catalogue),
- gene accessions for specific genes.

If the search is combined with the KEGG grouping or the hierarchical classification, this is as if a relational join operation is being performed.

> **Note**
>
> This feature becomes of special interest as in when searches are conducted for EC numbers from a specific group in the superfamily table (or the SCOP table) and searching them against the pathway diagrams. Such searches can give immediate results where a tendency of similar genes to appear in clusters on the pathway, that is, an indication of gene duplications in the pathway formation, is highlighted.

Let us look at each of these a little more closely in the next subsection.

TASK Searching and retrieving from KEGG

(a) Search for a known pathway

Let us first try to search for a known pathway.

Step 1 Open the Search page by entering the url http://www.genome.jp/kegg/ in your browser window.

Step 2 Link to 'KEGG Gene Universe'; then click on the link for **Pathways database**. The interface will look like as shown in Fig. 3.49.

Step 3 Submit your query in the textbox **Search *pathway* for** as 'lysine biosynthesis'.

Step 4 Select the **bfind** mode.

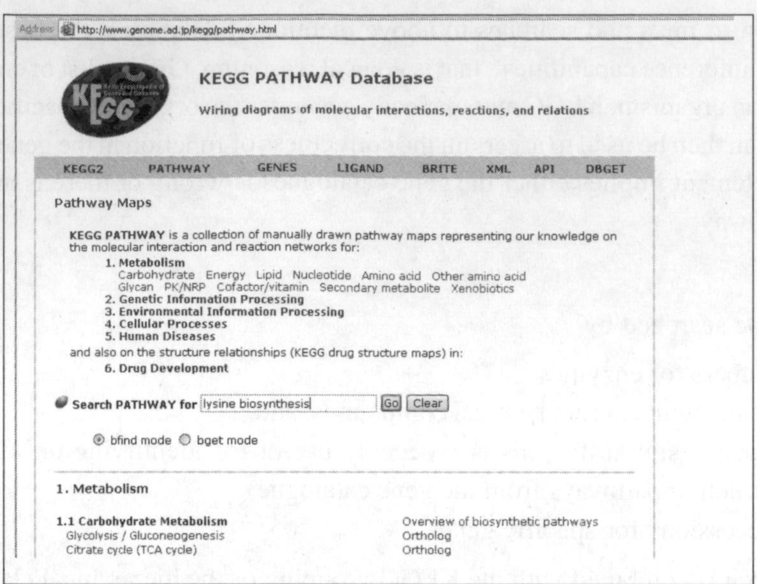

Fig. 3.49

Step 5 The result page, containing 335 hits will look like a table. Further exploration for *Saccharomyces bayanus*, by clicking on the link ***pathdsba:00300***, which is the pathway of our interest, will lead us to a new page that looks like Fig. 3.50.

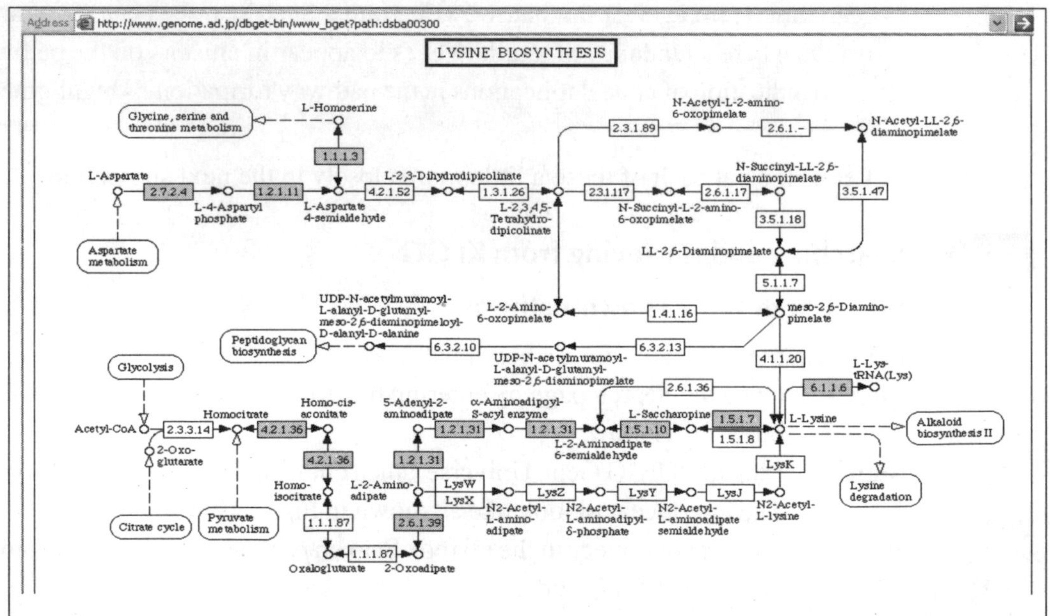

Fig. 3.50

Note

Organism-specific pathways (coloured/shaded pathways) are computationally generated based on the KO assignment in individual genomes.

Step 6 Study the pathway page carefully using the notations shown in Fig. 3.51.

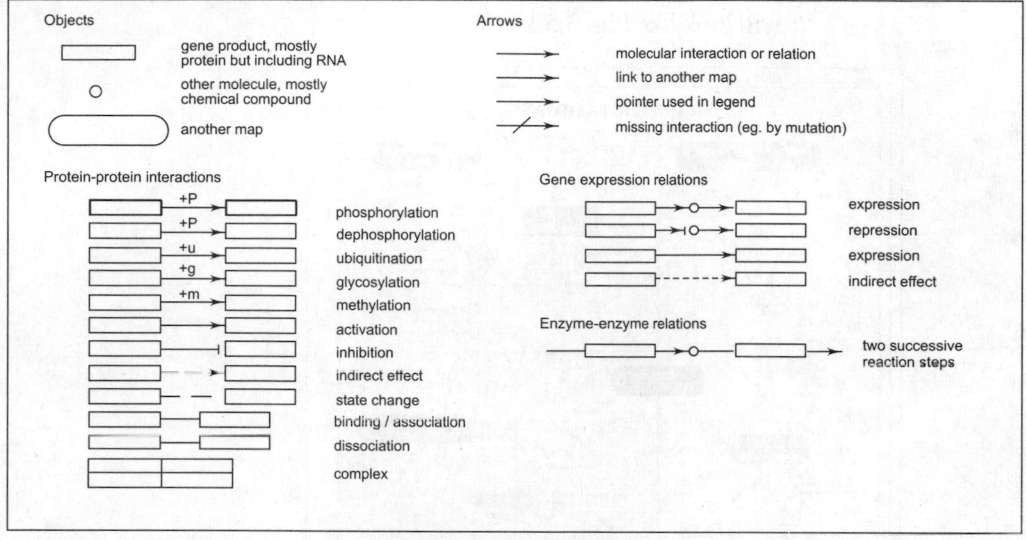

Fig. 3.51 Notations used for constructing pathway maps in KEGG

Step 7 Click on the box labelled **1.1.1.3**. This enzyme EC:1.1.1.3 catalyses the third step in the common pathway for methionine and threonine biosynthesis. Prepare a report on the following subheads:

- Entry (accession number given by the original database)
- Gene name (primary gene name)
- Definition (description of the gene or its product together with the EC numbers for enzyme genes)
- KO (identifier for functional orthologues)
- Pathway (role of the gene product in a pathway context)
- Class (hierarchical classification of gene functions according to the KO hierarchy, in which the third level corresponds to individual KEGG pathway maps)
- Sequence similarity database (SSDB) (result of pairwise genome comparisons for all protein-coding genes in the KEGG GENES database)
- Position (chromosomal position of the gene)
- AA seq (number of amino acids and the sequence in FASTA format)
- NT seq (number of nucleotides and the sequence)

3.7.2 Assigning a Pathway to a Set of Genes and Gene Products

Step 1 Open the search page by entering the url : http://www.genome.ad.jp/ in your browser window.

Step 2 Click **Metabolic Pathways**. The hierarchical text can be expanded by clicking **categories and subcategories**.

Step 3 Click on **metabolism** to switch to the graphical hierarchy. The resultant interface will look like Fig. 3.52.

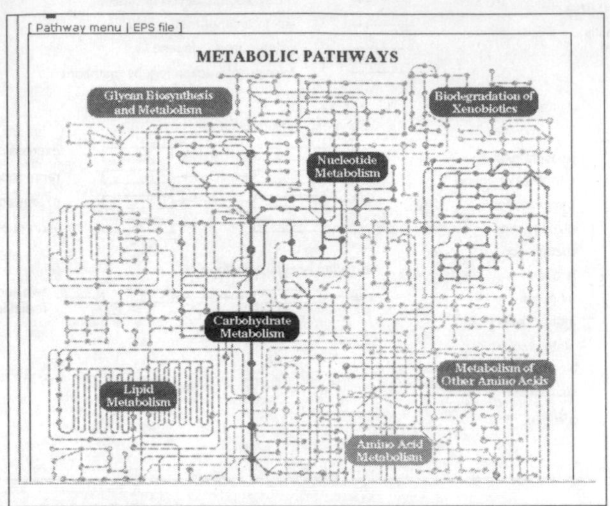

Fig. 3.52

Step 4 Click **Carbohydrate Metabolism** to expand it. The page that is displayed next is shown in Fig. 3.53.

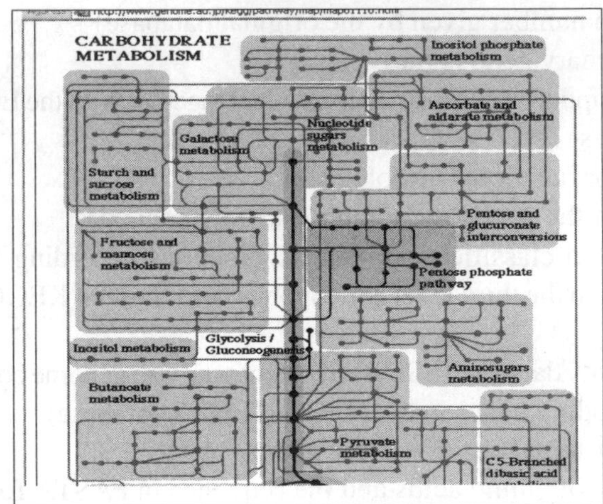

Fig. 3.53

Step 5 Click **Inositol metabolism** to see how it is represented in KEGG. The result page is shown in Fig. 3.54.

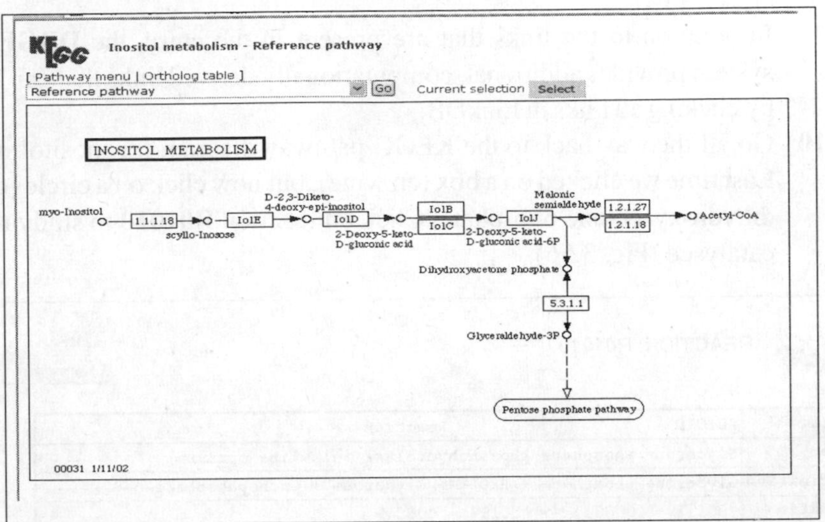

Fig. 3.54

Note that an enzyme with the EC number inside is represented by a box and a metabolic compound is represented by a circle. Both are clickable and can be used to retrieve more detailed molecular information.

Step 6 Click *5.3.1.1*. This takes you to *LIGAND*, which serves as a gateway to several other databases. Examine the substrate information in the COMPOUND section of LIGAND by clicking on the D-glyceraldehyde 3-phosphate [CPD:**C00118**] link. This gives information on the compound, including the chemical structure in GIF format and the CAS registry number as shown in Fig. 3.55.

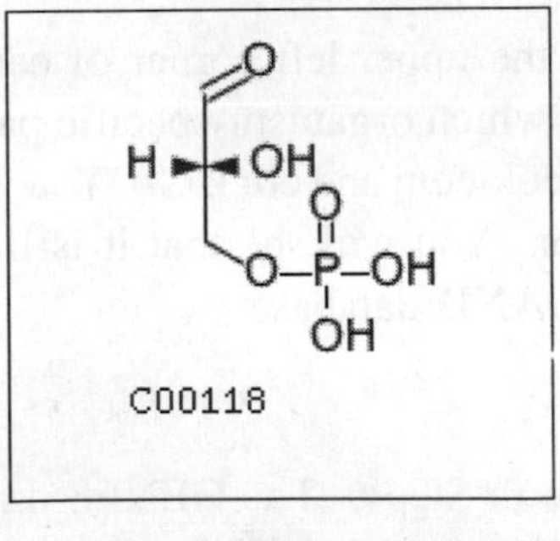

Fig. 3.55

Step 7 Go back to **EC 5.3.1.1**. Examine the *Candida albicans* orf by clicking ***orf19.6745***(TPI1)(listed beside CAL:, under the **ORTHOLOG GENES** heading).

Step 8 Go back to **5.3.1.1**. There are several links to protein data bank (PDB) given. Follow one of the links to examine the three-dimensional structure of this enzyme: click **1YP1**.

Step 9 In addition to the links that are present in the entry, the **DBGET/LinkDB** system provides additional, computationally derived links that can be retrieved by clicking all links in LinkDB.

Step 10 Go all the way back to the KEGG pathway diagram for inositol metabolism. Last time we clicked on a box (enzyme), but now click on a circle (compound), dihydroxyacetone phosphate. Click on reaction **R01010** to study the pathway catalysed (Fig. 3.56).

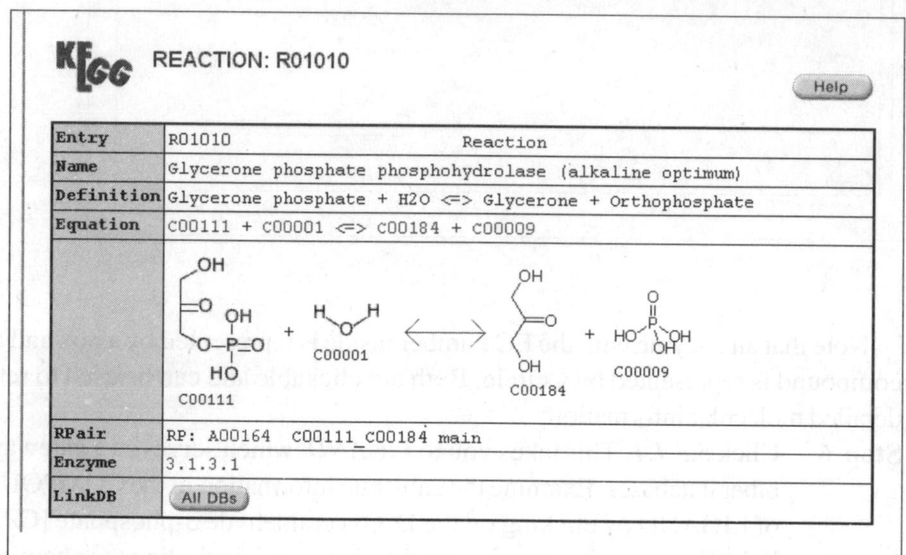

Fig. 3.56

Step 11 Go back to the pathway diagram. In the upper left corner of each pathway map, there is a drop-down menu from which organism-specific pathways can be selected. Choose Dictyostelium discoideum and click **Go**. You will observe two grey boxes. Click the **5.3.1.1**. box. You will see that it is linked to the GENES database rather than the **LIGAND** database.

Note

This is the same GENES entry we saw before in Step 6. The GENES database is linked to the graphical GENOME map in KEGG, which can be invoked by clicking **POSITION**. The genome map contains an overall view window and an enlarged window. The latter shows each gene that is linked to the GENES database.

BioCyc database collection (http://www.biocyc.org/)

The BioCyc database collection is a collection of pathway/genome Databases (PGDBs). It has several metabolic pathway databases that contain information about both predicted and experimentally determined pathways, reactions, compounds, genes, and enzymes. Each of the bacterial PGDB includes predicted operons for the corresponding species. In addition, the database contains specialized software for displaying large-scale data such as microarray gene expression or proteomic data as biochemical pathways. Figure 3.57 shows the various connections available within a cell addressed by BioCyc.

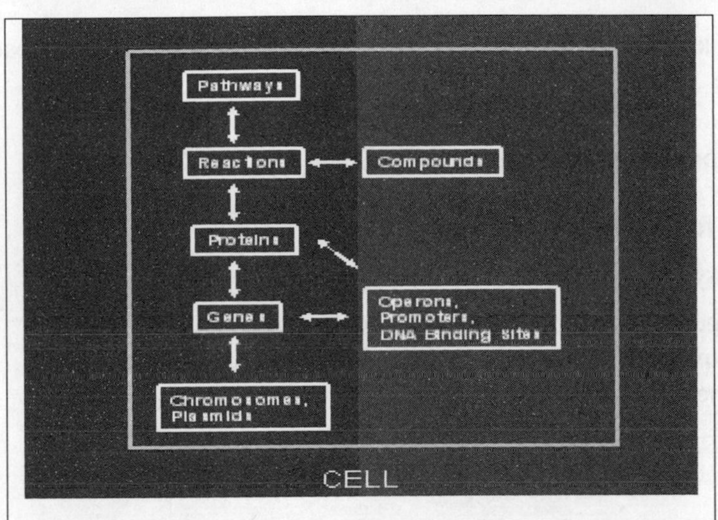

Fig. 3.57 Connections within a cell addressed by BioCyc

The BioCyc collection has several tiers within itself.

Tier 1 databases These databases have been created through intensive manual effort and are updated continuously. These include the following:

EcoCyc, which describes *Escherichia coli* K-12.

MetaCyc, which describes enzymes and metabolic pathways for more than 300 organisms. This database unlike KEGG does not give the complete metabolic network of any one organism, instead it gives a collection of experimentally demonstrated metabolic pathways.

BioCyc open compounds database (BOCD), an open collection of chemical compound data from the BioCyc databases. It includes metabolites, enzyme activators, inhibitors, and cofactors. This database includes compounds that act as either substrates in enzyme-catalysed reactions, enzyme activators, inhibitors, or cofactors. The database stores chemical structures for these and lists of synonyms. In the year 2005 this database had 3558 chemicals, 3274 structures, and 8843 chemical names (including synonyms).

Tier 2 databases These databases have been computationally generated by the PathoLogic program. This tier has as many as 12 databases. These include *Agrobacterium tumefaciens, Bacillus anthracis, Caulobacter crescentus, Francisella tularensis, Escherichia coli O157:H, Helicobacter pylori, Homo sapiens, Mycobacterium tuberculosis CDC155, Mycobacterium tuberculosis H37Rv, Plasmodium falciparum, Shigella flexneri, and Vibrio cholerae.*

Tier 3 databases These databases have been computationally generated by the PathoLogic program and have undergone no review or updating. There are 191 databases in this tier.

Let us look at the BioCyc open compounds database a little more closely in the next subsection.

TASK Searching and retrieving from BOCD

(a) Search for a known pathway

Let us first try to search for a known pathway.

Step 1 Open the search page by entering the url http://biocyc.org/server.html in your browser window.

Step 2 The **BioCyc Query Page** interface is as shown in Fig. 3.58.

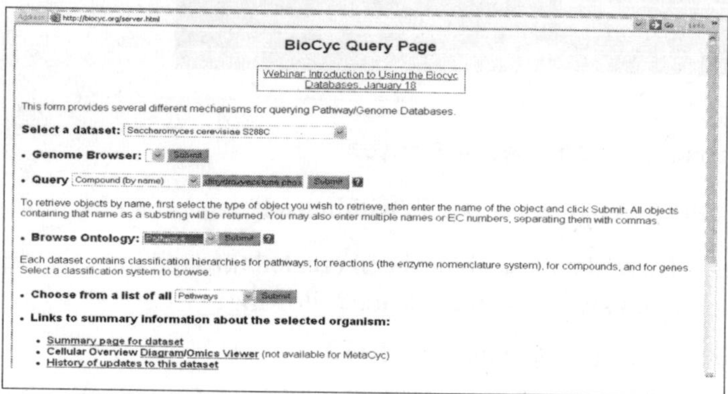

Fig. 3.58

Step 3 Select a dataset from the drop-down list provided. Let us select **Saccharomyces cerevisiae**.

Step 4 The **Query** box has several options available such as **All (by name or EC)**, **Protein (by name or EC)**, **Pathway (by name)**, **reaction (by name or EC)**, **compound (by name), gene (by name)**, and **RNA (by name)**. Select the option **compound (by name)** and write 'Protein tyrosine kinases' in the textbox.

Step 5 The **Browse Ontology** drop-down list has several options available such as **Pathways**, **EC Hierarchy**, **Compounds**, and **Gene**. Select **Compounds**.

Step 6 **Choose from a list of all** There are several options available such as **Pathways**, **Proteins, Compounds, Gene** and **Pseudogene**. Select **Compounds**.

Step 7 Click the result page, **Summary page for dataset**. You will get as shown in Fig. 3.59.

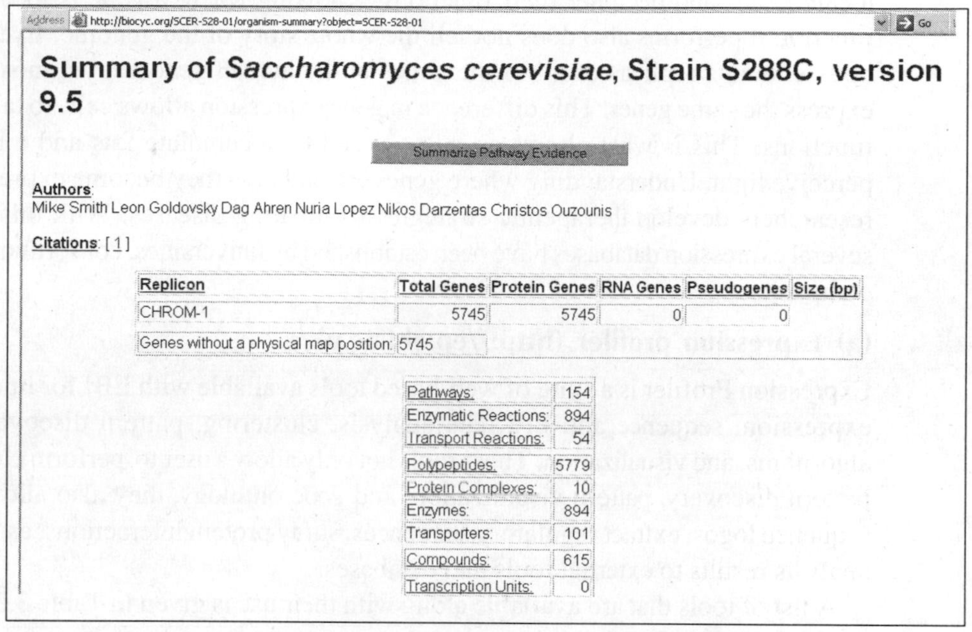

Fig. 3.59

Step 8 Study the pathway summary page carefully and prepare a report on

- Pathways
- Enzymatic reactions
- Transport reactions
- Polypeptides
- Protein complexes
- Enzymes
- Transporters
- Compounds, especially S-adenosyl methionine
- Transcription units
- tRNAs

3.7.3 Expression Databases

February 2001 saw International Human Genome Sequencing Consortium and Celera Genomics announcing the completion of the sequencing of the human genome. Is it safe to conclude that we now know the identity of every single base pair in our genes? Not really. The genome sequence has in the past been compared to a phone directory. You may be familiar with the names in the phone book, but there is not much that you can learn about the city's people. Similarly, you will not learn much about the human genome by reading the bases that constitute it.

To be able to use the genome, researchers have to completely annotate it; they must locate genes and decipher their functions. Knowing where a gene is located and what function it performs also does not tell the whole story of the genome. Every cell in our body contains replicas of all of the genes in the human genome, yet most cells do not express the same genes. This difference in gene expression allows cells to take on varying functions. This is what allows an adipose cell to accumulate fats and a retinal cell to perceive light. Understanding where genes are and how they become expressed can help researchers develop therapeutic strategies for treating diseases. With this background, several expression databases have been established by universities, consortiums, and private companies.

(a) Expression profiler (http://ep.ebi.ac.uk/)

Expression Profiler is a suite of web-based tools available with EBI for microarray gene expression, sequence and PPI data analysis, clustering, pattern discovery, statistics, algorithms, and visualization. These tools not only allow a user to perform cluster analysis, pattern discovery, pattern visualization, and gene ontology, they also allow to generate sequence logos, extract regulatory sequences, study protein interactions, as well as to link analysis results to external tools and databases.

A list of tools that are available along with their use is given in Table 3.5.

Table 3.5 List of tools and their use

S. No	Tool	Use
1	EPCLUST	Expression data matrix handling and visualization
2	URLMAP	Contents mapping between different web-based applications
3	EP:GO	A browser for controlled vocabularies produced by Gene Ontology^{TM} annotation project. It allows the extraction of genes associated to each Gene Ontology category and analysis of gene expression, regulatory sequence, and protein-protein interaction data.
4	GENOMES	Extraction of information about the gene annotations, upstream sequences, etc., from a species-specific database
5	EP:PPI	correlating protein–protein interaction data with expression data
6	PATMATCH	Visualizing sequence patterns (motifs) on the biological sequences

(contd)

Table 3.5 (*contd*)

S. No	Tool	Use
7	SPEXS	A web-based interface to the C++ implementation of the pattern discovery algorithm developed by Jaak Vilo
8	SEQLOGO	Visualization of information content of patterns

Let us first try to use the tool EPCLUST, which is **E**xpression **P**rofile data **CLUST**ering. The process flow for the use of this tool is given in Table 3.6.

Table 3.6 The process flow

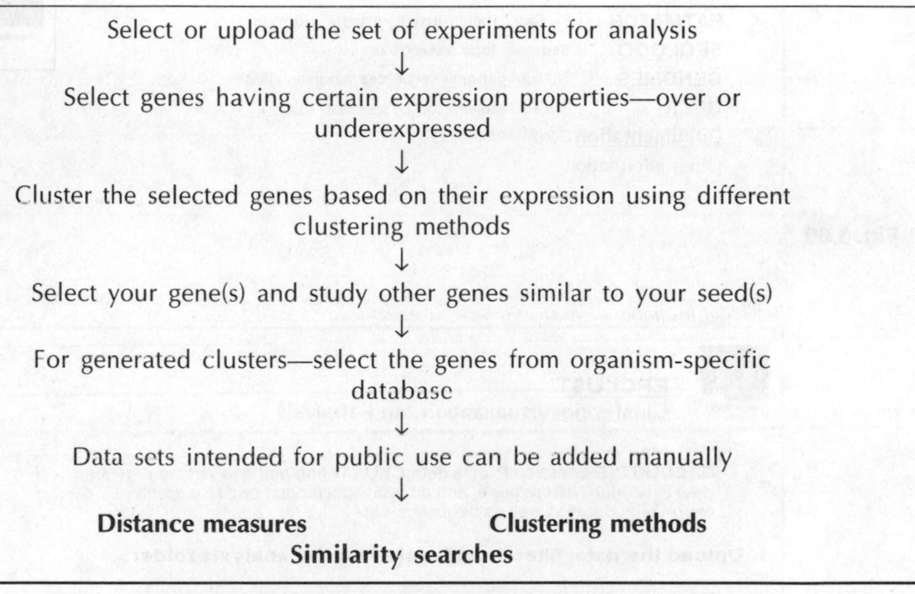

Select or upload the set of experiments for analysis

↓

Select genes having certain expression properties—over or underexpressed

↓

Cluster the selected genes based on their expression using different clustering methods

↓

Select your gene(s) and study other genes similar to your seed(s)

↓

For generated clusters—select the genes from organism-specific database

↓

Data sets intended for public use can be added manually

↓

Distance measures **Clustering methods**

Similarity searches

Step 1 Open the search page by entering the url http://ep.ebi.ac.uk/EP/ in the browser window. The interface will look like as shown in Fig. 3.60.

Step 2 Select a dataset from the list provided. Let us select **EPCLUST**. The next page that is displayed is shown in Fig. 3.61.

Step 3 Let us use the option **Upload the data**, **filter it**, and **store into the analysis folder** using public datasets for *Saccharomyces cerevisiae*. Click **Proceed**. This will lead us to the page shown in Fig. 3.62.

Step 4 Check the box for **Test_and_Demo**. Let us now pick up one gene which was over-expressed and one gene which was under-expressed. Enter MAPX in the textbox at the bottom of the page and click **Select the corresponding experiments**. The next window is shown in Fig. 3.63.

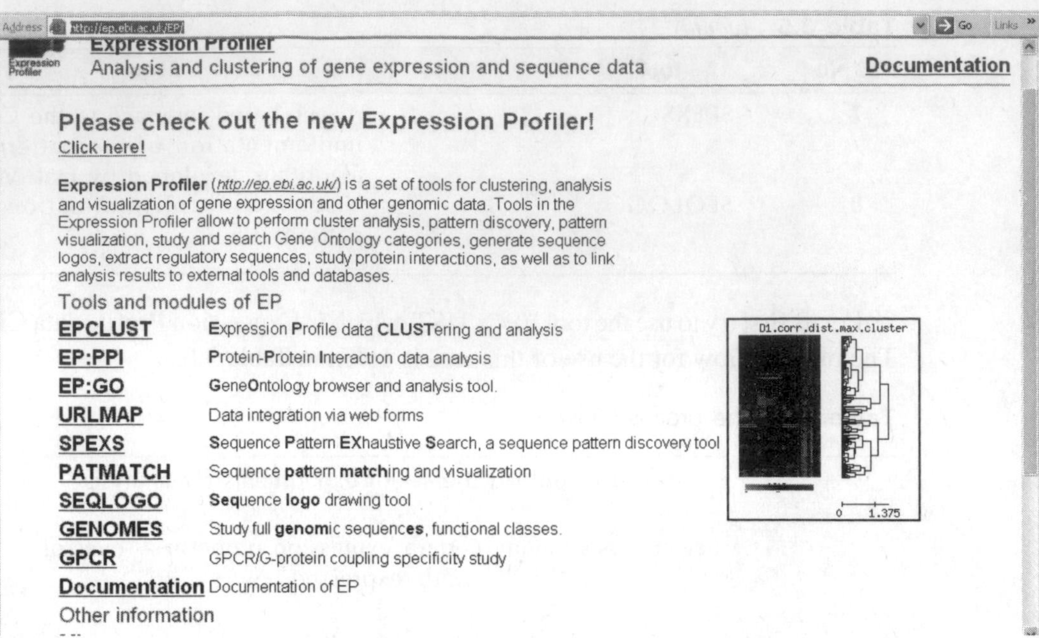

Fig. 3.60

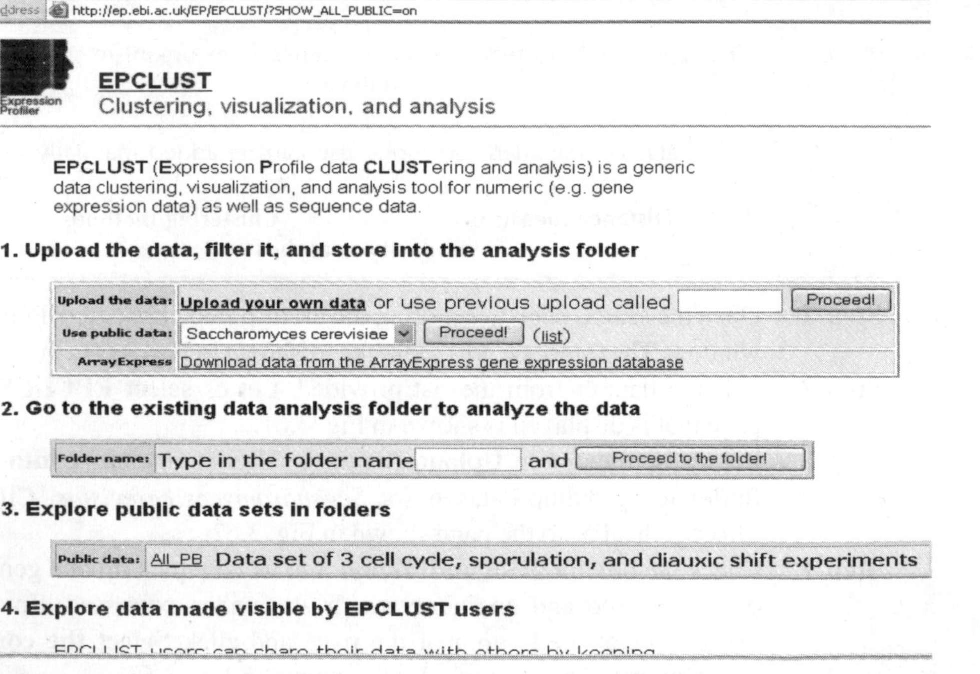

Fig. 3.61

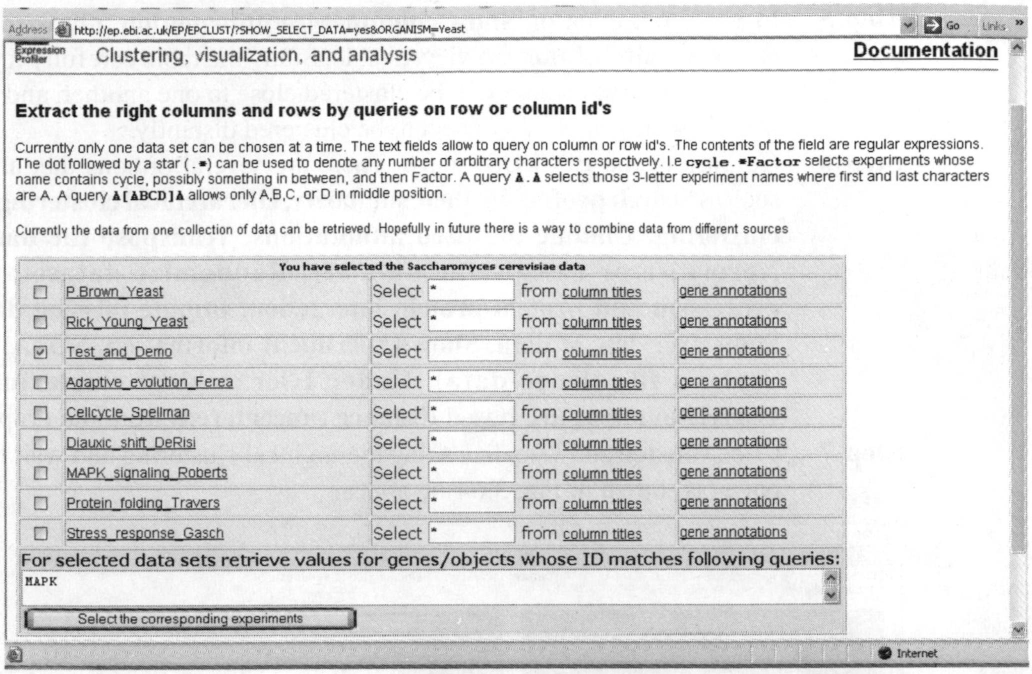

Fig. 3.62

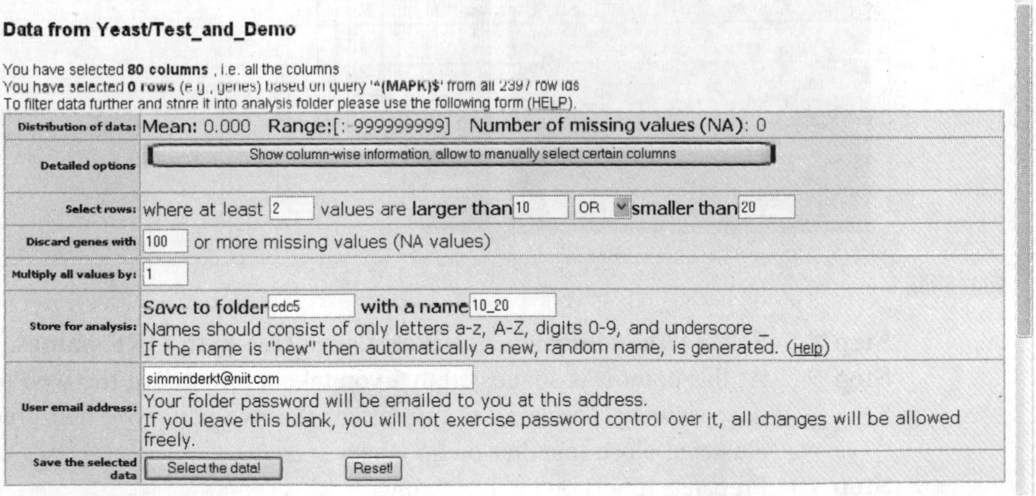

Fig. 3.63

The data is stored in a file **10_20** has in folder **cdc5** now. This can be either saved or emailed to the user.

Step 5 In Fig. 3.63, click on **Show column-wise information, allow to manually select certain columns** to view a table. Study the table carefully try to predict one pair of genes which will be clustered close to one another, and one pair of genes which you would expect to be clustered distantly.

Step 6 Go back to the EPCLUST page. There are several options that are available such as **Search profiles by their similarity, Hierarchical clustering, K-means clustering, Change the used annotations, Transpose the data matrix, Normalise or rescale the data set, Randomize dataset, Compare expression and protein-protein interaction, Impute missing data values, Show overview of data, Show experiment information, SOTA, Delete ols analysis file (leave data). Under Hierarchical clustering, choose Correlation measure based distance (uncentered)** from the drop-down list.

Step 7 Click **Cluster** and visualize using hierarchical clustering and wait for the next page (as shown in Fig. 3.64) to appear.

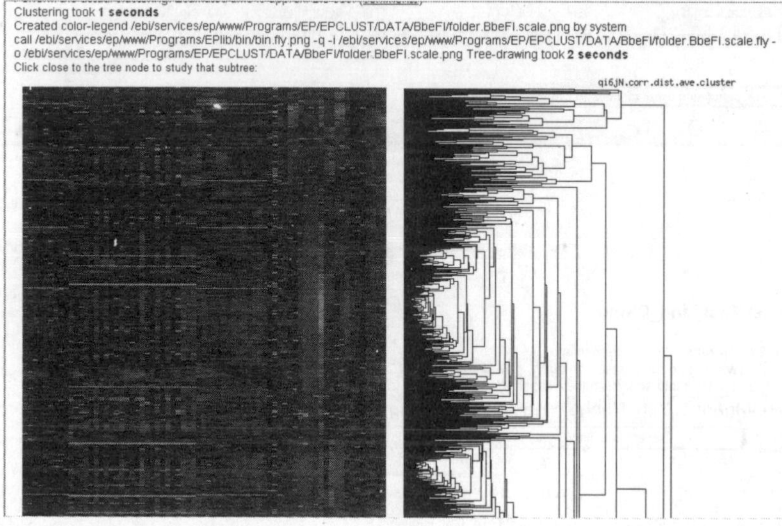
Clustering took **1 seconds**
Created color-legend /ebi/services/ep/www/Programs/EP/EPCLUST/DATA/BbeFI/folder.BbeFI.scale.png by system call /ebi/services/ep/www/Programs/EPlib/bin/bin.fly.png -q -i /ebi/services/ep/www/Programs/EP/EPCLUST/DATA/BbeFI/folder.BbeFI.scale.fly -o /ebi/services/ep/www/Programs/EP/EPCLUST/DATA/BbeFI/folder.BbeFI.scale.png Tree-drawing took **2 seconds**
Click close to the tree node to study that subtree:
qi6jN.corr.dist.ave.cluster

Fig. 3.64

Step 8 Select **Other formats** and click **Large tree with ORF names.**

Step 9 At this point it is suggested that you take a printout of the web page you are now viewing. Circle all the genes for the two pairs you had predicted to be clustered close together or far apart.

Step 10 Prepare a report on the following:

(a) Study the clustering tree and re-look at your prediction in Step 5. How does your prediction compare with the results?

(b) On the printout, draw a line that will slice the tree to split your genes into two similarly-sized groups such that each group contains genes with similar expression profiles.

Step 11 Click **Cluster** and visualize with **K-means**. Take a printout of the web page that appears (see Fig. 3.65). Do the two clusters correlate with the two groups you suggested earlier?

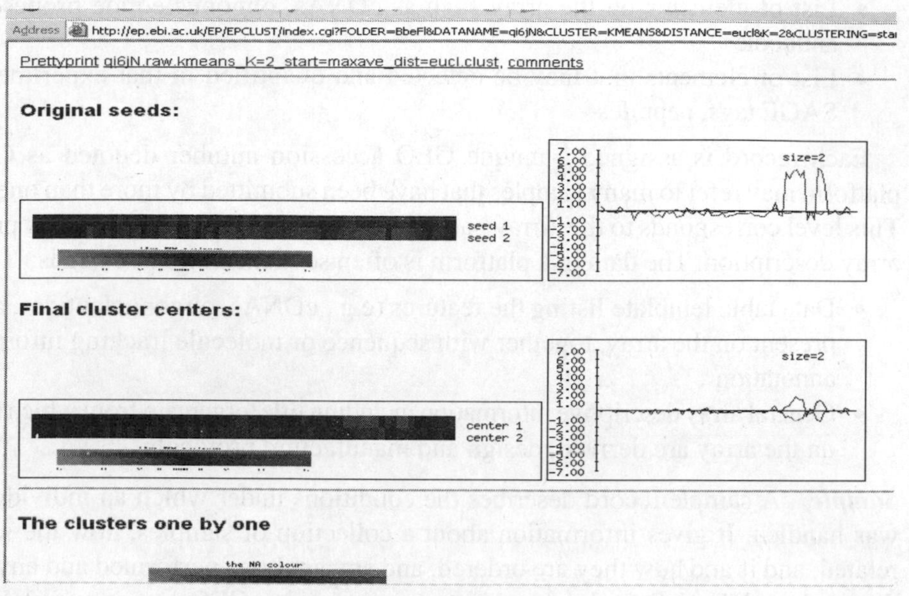

Fig. 3.65

(b) Gene expression omnibus (http://www.ncbi.nlm.nih.gov/geo/)

Gene Expression Omnibus (or GEO) is a database of high-throughput experimental data. The data includes microarray-based experiments measuring mRNA, genomic DNA, and protein quantity. In addition, non-array techniques such as serial analysis of gene expression (SAGE) and mass spectrometry proteomic data are also contained in this database. This database has a flexible and open design that allows submission, storage, and retrieval of many types of datasets.

History, size, content, quality The Gene Expression Omnibus at the National Center for Biotechnology Information (NCBI) collects data on four main levels. There are several features that are provided to examine data from both experiment- and gene-centric perspectives using user-friendly web-based interfaces accessible to those without computational or microarray-related analytical expertise. Its data is stored in a relational database partitioned into three upper-level entity types. These are given below.

Submitter All submissions are MIAME-compliant. Submissions are validated syntactically according to a set of criteria and are subject to basic curation, ensuring that records contain meaningful information and are organized correctly. Submitters have editorial control and are responsible for the content and quality of their records as outlined by the Microarray Gene Expression Data (MGED) Society board.

Platform This level contains information about the physical reagents which are being used to 'query' the sample in a high-throughput manner. A record in platform describes the following elements:

- List of elements on the array such as cDNAs, oligonucleotide probesets, ORFs, antibodies
- List of elements that may be detected and quantified in that experiment such as SAGE tags, peptides

Each record is assigned a unique GEO accession number denoted as GPLxxx. A platform may refer to many samples that have been submitted by more than one submitter. This level corresponds to the Array package in MAGE-OM. In a nutshell, it provides the array description. The data in a platform is often sectioned into two heads;

- Data table template listing the features (e.g., cDNAs, oligonucleotides, antibodies) present on the array, together with sequence or molecule tracking information and annotation.
- General array descriptive information including title, organism from which the features on the array are derived, design and manufacture protocols.

Sample A sample record describes the conditions under which an individual Sample was handled. It gives information about a collection of samples, how the samples are related, and if and how they are ordered, and any analyses performed and any clustering data gathered. Each Sample record is assigned a unique GEO accession number denoted by GSMxxx. It corresponds to the Experiment package in MAGE-OM. A sample entity must reference only one Platform and may be included in multiple series.

Note

Unlike metadata that is stored in designated fields within database tables, platform and sample data tables are not fully granulated, but are stored as text objects. This design allows GEO to remain adaptable and responsive to the developing technology trends, as it permits optimal flexibility in the quantity and type of data stored.

Series A series record contains information about the mRNA sample being queried and the experimental conditions, as well as the gene expression measurement data generated from the experiment. It defines a set of related samples that form a group, their relationship, and ordering. Series records may also contain tables that contain extracted data, summary conclusions, or analyses. Like the platform, each series record is assigned a unique and stable GEO accession number denoted by GSExxx. It corresponds to the BioAssay package in MAGE-OM.

Figure 3.66 shows the architecture of GEO showing the relation between submitter, platform, sample, and series.

A GEO dataset or GDSxxx is a record that represents a group of biologically and statistically comparable GEO samples. Samples within a GDS refer to the same platform and therefore share a common set of probe elements. All measurements for each sample

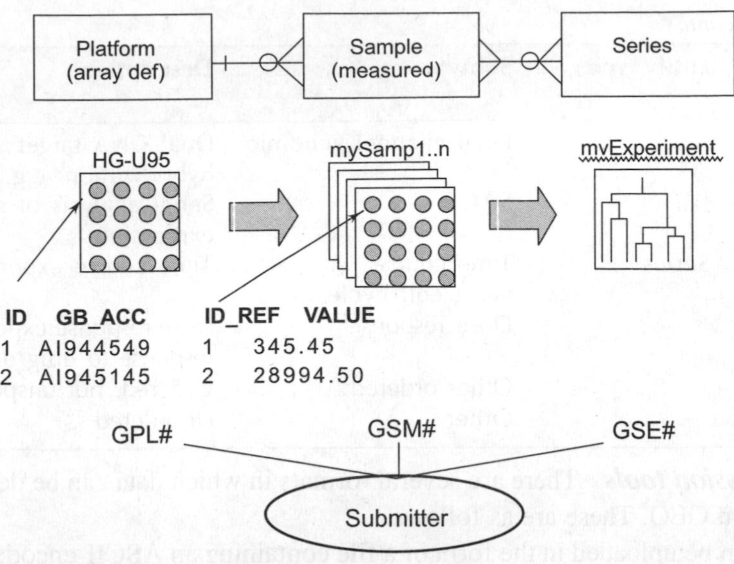

Fig. 3.66

within a GDS are assumed to be calculated in an equivalent and consistent manner. Datasets are curated form the basis of GEO's data display and analysis tools. You can search for experimental design through GDS subsets. Table 3.7 summarizes the entity prefixes, types, and subtypes in the GEO database.

Table 3.7 Entity prefixes, types, and subtypes in the GEO database (*source:* http://www.ncbi.nlm.nih.gov)

Accession prefix	Entity type	Subtype	Description
GPL	Platform	Commercial nucleotide array	Commercially available nucleotide hybridization array
		Commercial tissue array	Commercially available tissue array
		Commercial antibody array	Commercially available antibody array
		Non-commercial nucleotide array	Nucleotide array that is not commercially available
		Non-commercial tissue array	Tissue array that is not commercially available
		Non-commercial antibody array	Antibody array that is not commercially available
GSM	Sample	Dual channel	Dual mRNA target sample hybridization
		Single channel	Single mRNA target sample hybridization

(*contd*)

Table 3.7 *(contd)*

Accession prefix	Entity type	Subtype	Description
		Dual channel genomic	Dual DNA target sample hybridization, e.g., array CGH
		SAGE	Serial analysis of gene expression
GSE	Series	Time course	Time course experiment, e.g., yeast cell cycle
		Dose response	Dose response experiment, e.g., response to drug dosage
		Other ordered	Ordered, but unspecified
		Other	Unordered

Data submission tools There are several formats in which data can be deposited and retrieved from GEO. These are as follows.

(a) Data can be uploaded in the form of a file containing an ASCII-encoded text table and metadata fields can be entered through a series of web forms. The submission process using the web as an interface is outlined on the GEO website. It can be summarized as depicted in Fig. 3.67. This process is most useful for quick and easy submission of records.

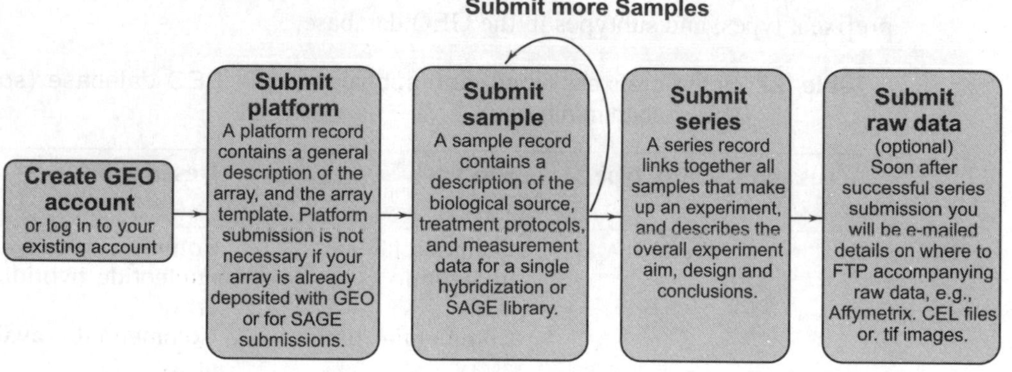

Fig. 3.67 GEO submission process as outlined on (*source:* http://www.ncbi.nlm.nih.gov/projects/geo/info/depguide.html)

(b) Simple Omnibus Format in Text (SOFT) is a format used to upload data and metadata for platforms, samples, or series. In SOFT, metadata appear as label-value pairs and are associated with the tab-delimited text tables of platforms and samples. SOFT has been designed for easy manipulation by readily available line-scanning software and may be quite readily produced from, and imported into, a spreadsheet, database, and analysis software.

> **Note**
>
> Metadata is literally 'data about data'. This form of data gives a description of the content, quality, contact, condition, and other characteristics of data. This description is organized in a standardized format using a common set of terms. An analogy to metadata is library catalogue records that maintain a record of books such as title, author, and publisher in a standard way to facilitate easy retrieval.

Search (keyword search, advanced options) As outlined earlier, GEO accession numbers are used to retrieve data from this database. The accession number may be from a literature reference, such as from a publication citing data deposited to GEO, or through a query interface through NCBI's Entrez ProbeSet interface.

This unique in Entrez ProbeSet is the GEO sample, fused with its affiliated platform and series information. The indexing process iterates through all platforms in the GEO database, extracting metadata and the data table, and searches for any sequence-based identifiers such as GenBank Accession, ORFs, Clone IDs, or SAGE tags. Each sample belonging to that platform is in turn assigned a new UID and indexed with the above platform information plus any related series metadata. GenBank accession numbers, PubMed references, and taxonomy information appear in the Links section of the display. **Neighbors** refers to related intra-Entrez database links which are generated for UIDs sharing the same GEO platform or series.

GEO provides several tools to assist with the visualization and exploration of GEO data. Let us try exploring them. Let us try and retrieve information about how budding *Saccharomyces cerevisiae* cells become committed to sporulation.

Step 1 Let us go to the GEO home page at http://www.ncbi.nlm.nih.gov/geo/. The interface that appears is shown in Fig. 3.68.

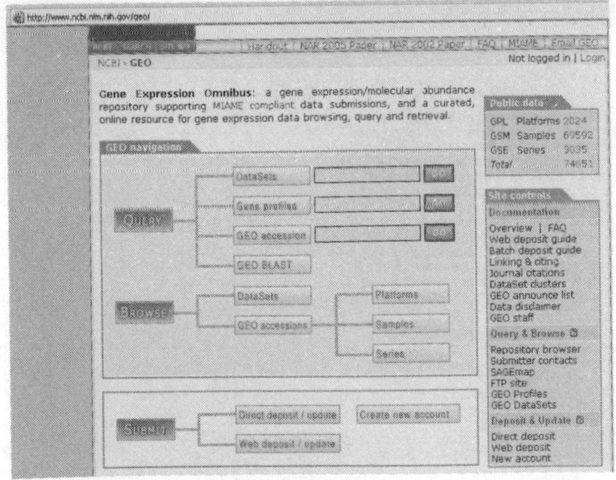

Fig. 3.68

Step 2 Select the link for GSE series from **Public data**. This will open a new page that looks like Fig. 3.69.

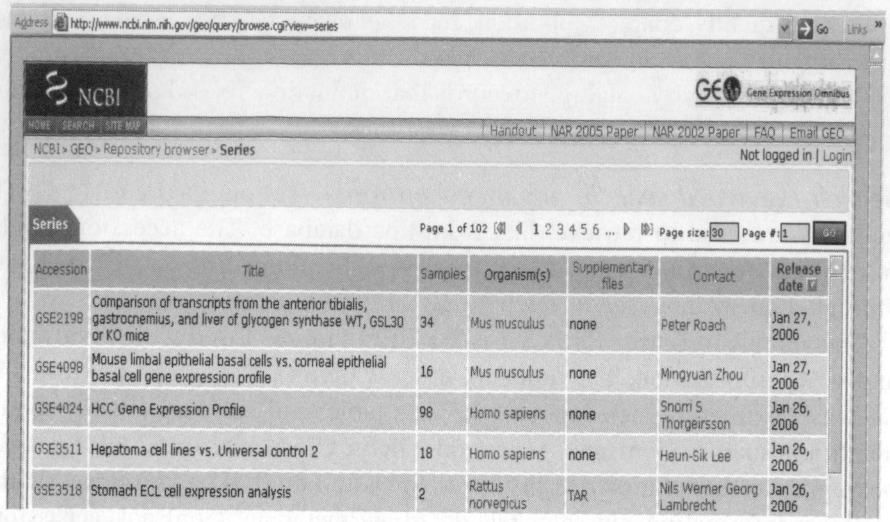

Fig. 3.69

Step 3 Click accession **GSE 3820** with title **Sporulation_transfer_to_YPA2**.

Step 4 The description of GSE 3280 on the Accession Display will result in a summary assessment of the data (as shown in Fig. 3.70). This dataset can be downloaded in SOFT format.

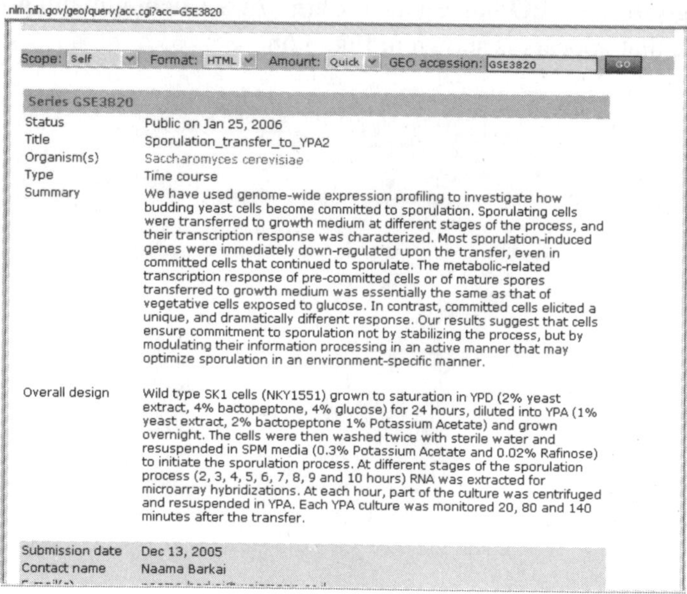

Fig. 3.70

Step 5 To download this data, enter the attributes as shown in Fig. 3.71and click **GO**.

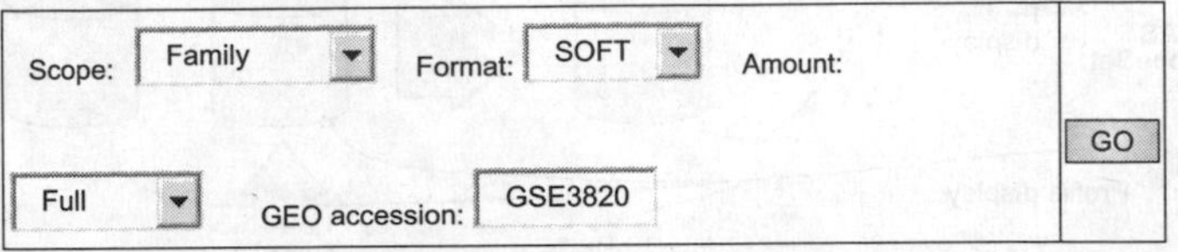

Fig. 3.71

Three types of display options are available:

- **Scope** allows you to display the GEO accession numbers that you want to target for display. You can choose to display the GEO accession, which is typed into the **GEO accession** box itself (**Self**). You can choose any of **Platform**, **Samples**, or **Series** or all (**Family**) of the accessions related to an accession. The Family setting will retrieve all accessions (of different types) related to self including the self.
- **Format** allows you to display the GEO accession in human-readable, linked HTML form or in machine-readable, **SOFT** form.
- **Amount** allows you to control the amount of data that you will see displayed. Brief displays the accession's metadata only. **Quick** displays the accession's metadata and the first 20 rows of its dataset. **Full** displays the accession's metadata and the full dataset. **Data** omits the accession's metadata, showing only the links to other accessions as well as the full dataset.

Step 6 The browser takes only seconds to download the 2.3 Mb file (Fig. 3.72)..

```
GSE3820 - Notepad
File Edit Format View Help
^SERIES = GSE3820
!Series_title = Sporulation_transfer_to_YPA2
!Series_geo_accession = GSE3820
!Series_status = Public on Jan 25 2006
!Series_submission_date = Dec 13 2005
!Series_summary = We have used genome-wide expression profiling to investigate how budding yeast cells become committed to sp
!Series_overall_design = wild type SK1 cells (NKY1551) grown to saturation in YPD (2% yeast extract, 4% bactopeptone, 4% gluc
!Series_type = Time course
!Series_sample_id = GSM87634
!Series_sample_id = GSM87635
!Series_sample_id = GSM87636
!Series_sample_id = GSM87637
!Series_sample_id = GSM87638
!Series_sample_id = GSM87639
!Series_sample_id = GSM87640
!Series_sample_id = GSM87641
!Series_sample_id = GSM87642
!Series_sample_id = GSM87643
!Series_sample_id = GSM87644
!Series_sample_id = GSM87645
!Series_sample_id = GSM87646
!Series_sample_id = GSM87647
!Series_sample_id = GSM87648
!Series_sample_id = GSM87649
!Series_sample_id = GSM87650
!Series_sample_id = GSM87651
!Series_sample_id = GSM87652
!Series_sample_id = GSM87653
!Series_sample_id = GSM87654
!Series_sample_id = GSM87655
!Series_sample_id = GSM87656
!Series_sample_id = GSM87657
!Series_sample_id = GSM87658
!Series_sample_id = GSM87659
!Series_sample_id = GSM87660
!Series_contact_name = Naama,,Barkai
!Series_contact_email = naama.barkai@weizmann.ac.il
!Series_contact_institute = Weizmann Institute
!Series_contact_address = P.O box 26
!Series_contact_city = Rehovot
!Series_contact_zip/postal_code = 76100
!Series_contact_country = Israel
!Series_platform_id = GPL3245
^PLATFORM = GPL3245
!Platform_title = UHN Yeast 6.4kv4
!Platform_geo_accession = GPL3245
!Platform_status = Public on Jan 25 2006
!Platform_submission_date = Dec 11 2005
!Platform_technology = spotted DNA/cDNA
```

Fig. 3.72

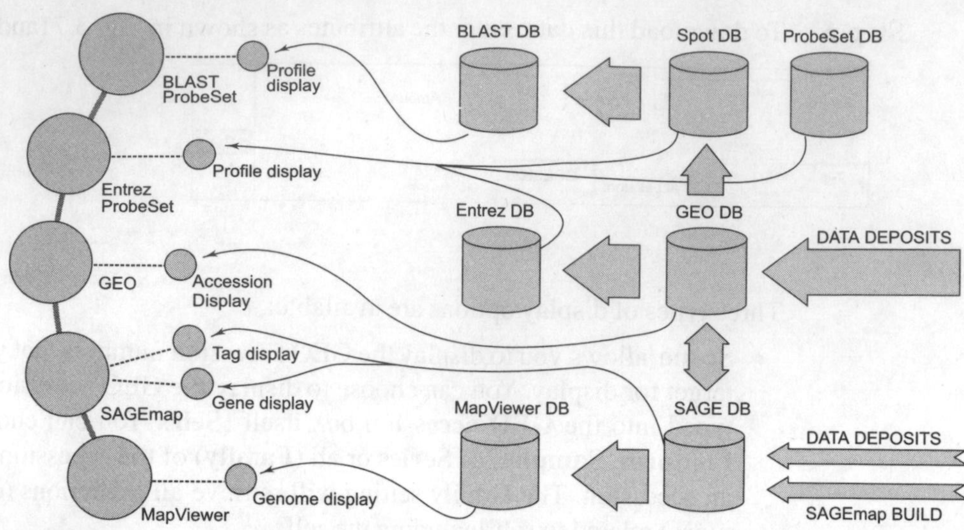

Fig. 3.73 NCBI gene expression resources. Blue spheres indicate websites; orange cylinders indicate primary NCBI databases; green cylinders indicate secondary databases; and yellow cylinders indicate tertiary NCBI interface databases.

The GEO database aims to improve its indexing, linking, searching, and display capabilities to allow vigorous data mining. Figure 3.73 shows its links to other databases in NCBI.

3.7.4 Microarray Database

Microarray (also known as DNA microarray, gene chip, DNA chip, or biochip) is a collection of microscopic DNA spots or probes in the form of an array which are attached to a base solid surface. The base surface can be of a glass, plastic, or silicon chip. These DNA spots can serve the purpose of expression profiling and/or monitoring expression levels for thousands of genes simultaneously in a single experiment. Microarrays are used to measure gene expressions and to study and identify a disease by comparing the gene expression in disease and normal cells.

The microarray informatics group at the EBI handles managing, storing, and analysing microarray data. This group was established in May 2000. ArrayExpress is a public database for microarray gene expression data. It stores the Minimum Information About Microarray Experiment (MIAME)-compliant data in accordance with the Microarray Gene Expression Data Society (MGED) recommendations of gene expression experiments that groups related hybridizations to address a biological question. It is based at the EBI and funded by the EU and EMBL.

The data relating to each microarray project in the ArrayExpress database is subdivided into three main components as follows:

Array gives information about the design and manufacture of the array itself.

Experiment gives information on the experimental factors and the actual data obtained

Protocol describes the procedures used in the production of the array or the execution of the experiment.

Note

Due to the biological complexity of a gene expression, the considerations of experimental design for expression profiling are of critical importance if statistically and biologically valid conclusions are to be drawn from the data.

The lack of standardization in arrays presents an interoperability problem in bioinformatics, which hinders the exchange of array data. The 'Minimum Information About a Microarray Experiment' (MIAME) XML based standard for describing a microarray experiment is followed to standardize microarray results.

TASK

Let us try to explore the database.

Step 1 You can log onto the website www.ebi.ac.uk/arrayexpress/ as a 'guest' by default and can view publicly available datasets. Click **Advanced Query interface**. The query interface will let you retrieve data that will provide you with information based on various criteria for the following:

- Identify an array design
- Provide links to any experiments which used that array
- Provide links to the protocol used to manufacture the array

Note

Arrays, experiments, and protocols have accession numbers in the format A-XXXX-n, E-XXXX-n, P-XXXX-n, respectively (where X is an alphabetic character, n is a number, and the A/E/P prefix denotes the accession number type).

Step 2 The section of the main query page which is used to search the database for experiments is shown in Fig. 3.74.

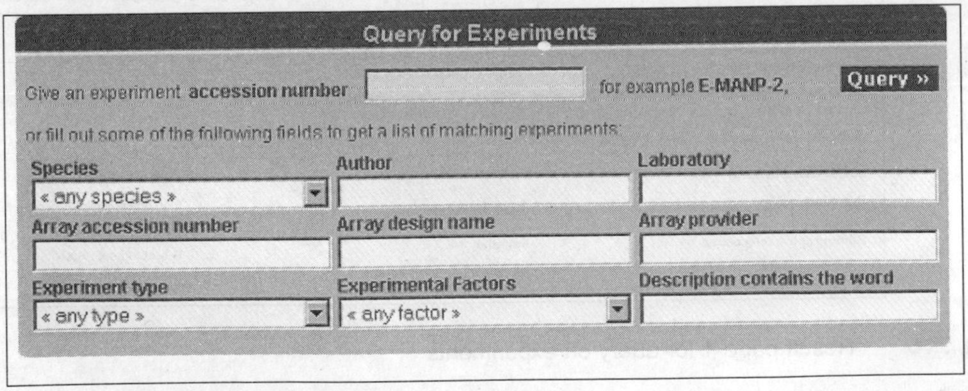

Fig. 3.74 Query section for experiments

The attributes of an experiment which can be queried for are as follows:

- **Experiment accession number** is a unique identifier assigned to each experiment that is associated with multiple hybridizations. The format is E-XXXX-n.
- **Species** is the Latin names for species from the NCBI taxonomy database. You may limit your query results to a particular species using the pull-down menu. If the species is not in the list, then there is no public data available for that species. Let us select **Homo sapiens**.
- **Author** is the name(s) of the researcher(s) associated with the experiment and the original submitter of the experiment.
- **Laboratory** is the institutions(s) where the authors work.
- **Experiment type** is the experiment that is of interest in your search. You may limit your query results to a particular species using the pull-down menu.
- **Experimental factors** are the parameters which are varied by the experimenter during the experiment. These can relate to the sample, the treatment of the sample during the experiment (e.g., the protocols used), or some other methodological factor such as the equipment used. Let us select **Cancer site**.
- **Description** is the experiment description and includes free text information provided by the submitter and an automatically generated summary. You may enter the keyword.

  Click **Query**.

Step 3 You will get a result page as shown in Fig. 3.75.

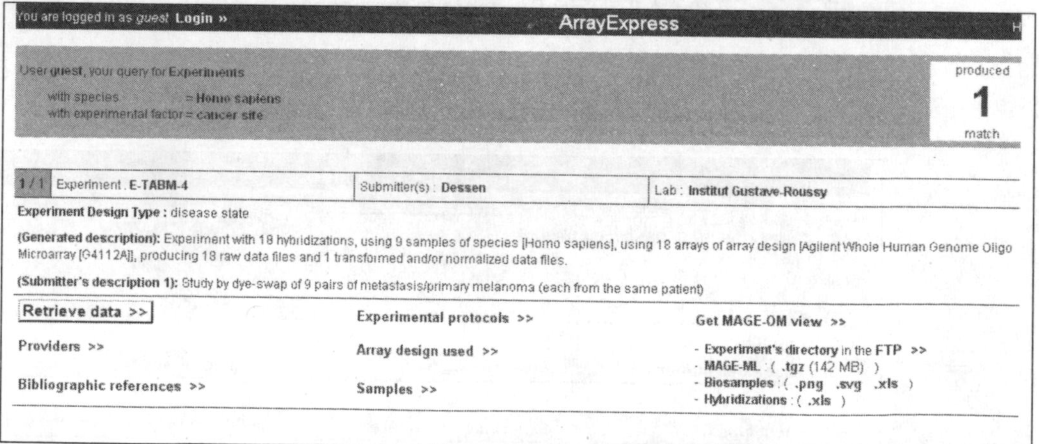

Fig. 3.75 Result page 1 for query on experiments

Step 4 Find out the array design used for this experiment. You will get a result page as shown in Fig. 3.76.

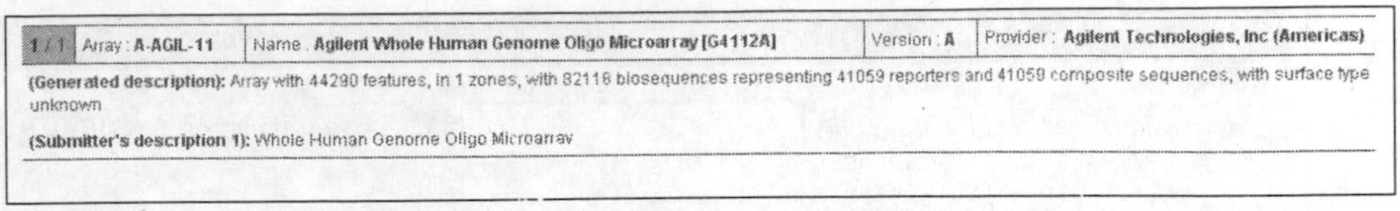

1/1 Array : A-AGIL-11	Name : Agilent Whole Human Genome Oligo Microarray [G4112A]	Version : A	Provider : Agilent Technologies, Inc (Americas)

(Generated description): Array with 44290 features, in 1 zones, with 82118 biosequences representing 41059 reporters and 41059 composite sequences, with surface type unknown

(Submitter's description 1): Whole Human Genome Oligo Microarray

Fig. 3.76 Result page 2 for query on experiments

Step 5 Retrieve and export data in a Microsoft Excel comma separated values file.

Step 6 Let us understand the section on the main query page which is used to search the database for arrays as shown in Fig. 3.77.

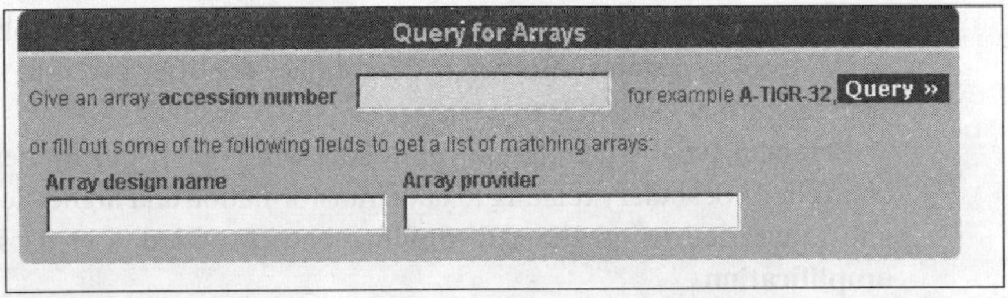

Query for Arrays

Give an array **accession number** [ ] for example **A-TIGR-32**, **Query »**

or fill out some of the following fields to get a list of matching arrays:

Array design name [ ] **Array provider** [ ]

Fig. 3.77 Query section for arrays

The attributes of an array which can be queried for are as follows:

- **Array accession number** is a unique identifier given to the array design. It has the format A-XXXX-n, where XXXX is a code representing the person or organization which submitted the array design, but not necessarily the organization originally responsible for the production of the array.
- **Array design name** is the name provided for the array design by the manufacturer. This could be the name of a commercial design or a name given to an in-house design supplied by a specific institute or lab.
- **Array provider** is the name of the person or institution who supplied the array(s) used in the experiment. This is usually the manufacturer of the arrays, either a commercial manufacturer or an institute or lab. Let us type 'Biosciences'. Click **Query**.

Step 7 You will get a result page with the name of the providers as 'Biosciences'. Prepare a report grouping the provider's name, mentioning the array design name with the number of features, zones, biosequences, reporters, composite sequences, and the surface type. What are the array spreadsheets available in the result page? In what formats are these spreadsheets available? (***Hint:*** Look for ADFs that are used to define layout of reporters on the array.)

Step 8 The panel on the main query page which is used to search the database for protocols is shown in Fig. 3.78.

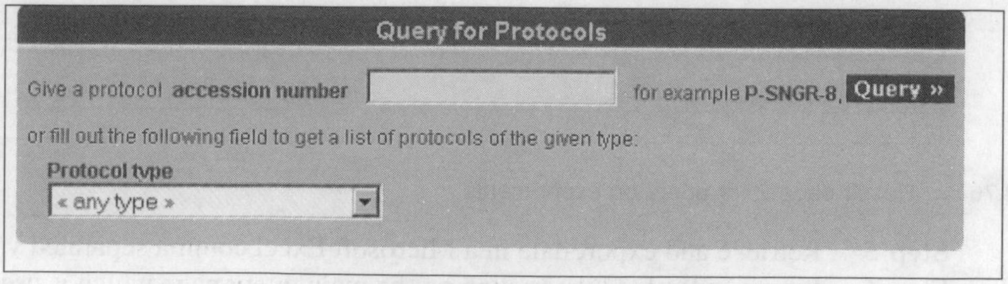

Query for Protocols

Give a protocol **accession number** [ ] for example **P-SNGR-8**, **Query »**

or fill out the following field to get a list of protocols of the given type:

Protocol type

« any type »

Fig. 3.78 Query section for protocols

The attributes of a protocol which can be queried for are as follows:

Protocol accession number is the unique identifier given to the protocol with the format P-XXXX-n.

Protocol type is the class of protocols which is of our interest. It uses a controlled vocabulary relating to data transformation and higher-level analysis and is queried using the pull-down menu provided. Let us select **PCR amplification.**

Click **Query**.

Step 9 Protocols are supplied for the treatments of the samples and experimental procedures including sample treatments, labelling, hybridization, image acquisition, and data analysis. You will get a result page as shown in Fig. 3.79.

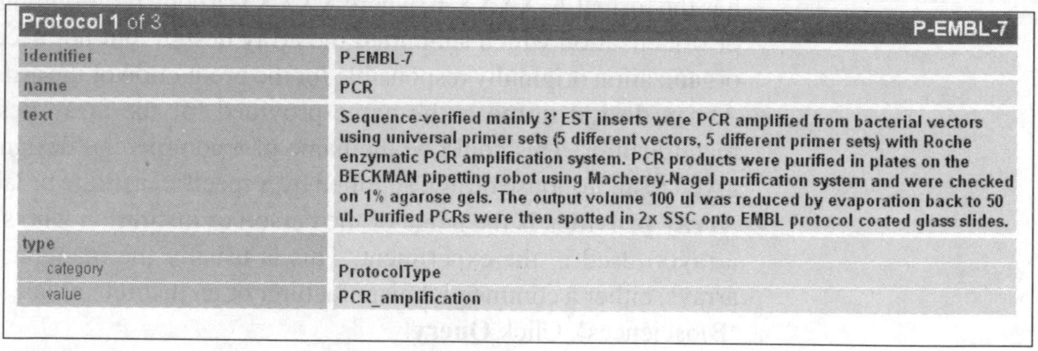

Protocol 1 of 3	**P-EMBL-7**
identifier	P-EMBL-7
name	PCR
text	Sequence-verified mainly 3' EST inserts were PCR amplified from bacterial vectors using universal primer sets (5 different vectors, 5 different primer sets) with Roche enzymatic PCR amplification system. PCR products were purified in plates on the BECKMAN pipetting robot using Macherey-Nagel purification system and were checked on 1% agarose gels. The output volume 100 ul was reduced by evaporation back to 50 ul. Purified PCRs were then spotted in 2x SSC onto EMBL protocol coated glass slides.
type	
category	ProtocolType
value	PCR_amplification

Fig. 3.79 Result page of protocol query

3.7.5 Genome Database and Genome Browser

Ensemble is a genome browser project developed jointly by EMBL-EBI and the Sanger Institute. It provides free access to all the data produced by the project as well as the tools used to analyse and present the data. It has a software system that produces and maintains automatic annotation on selected eukaryotic genomes. Some data and software is subject to third-party constraints. This project was primarily funded by the Wellcome Trust.

The browser area is broadly grouped into **Mammalian genomes** and **Other species**. you can click on a species name to browse the data. Additionally, you can also use Ensembl to do the following:

- Run a BLAST search
- Search ensemble
- Perform datamining
- Export data
- Download data

TASK

Let us try to explore the *Homo sapiens* genome. We need to go to http://www.ensembl.org/index.html

Step 1 Click on ***Homo sapiens*** under ***Mammalian genome***. It is a ***Karyotype*** genome with 22 chromosomes and X, Y, MT chromosomes as shown in Fig. 3.80.

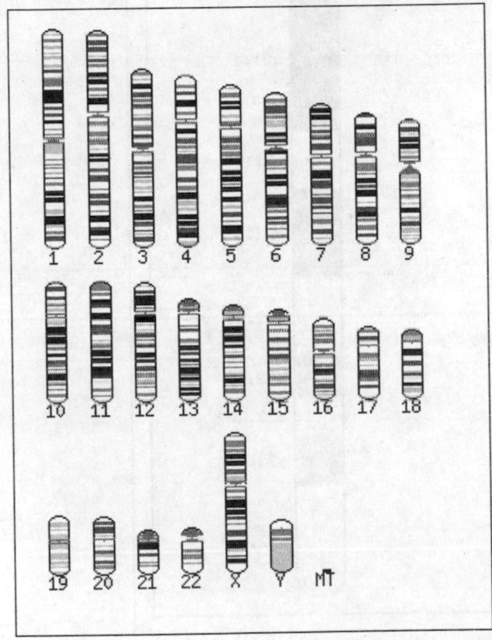

Fig. 3.80 Chromosomal view of *Homo sapiens*

Note
You may also click **Vega**, which will provide data from the manual annotation. As the manual annotation is ongoing, you will get data on more limited regions of the concerned species. Data may be either from the entire chromosomes or from small sections of chromosomes (one or more clones) or individual loci or regions that have not been annotated at all.

Step 2 You can click on any of the chromosome for a closer view. Let us click on chromosome 21 and get a consolidated view of **Known Genes**, **% GC, SNPs**, and **Chromosome 21** with p and q in general as shown in Fig. 3.81.

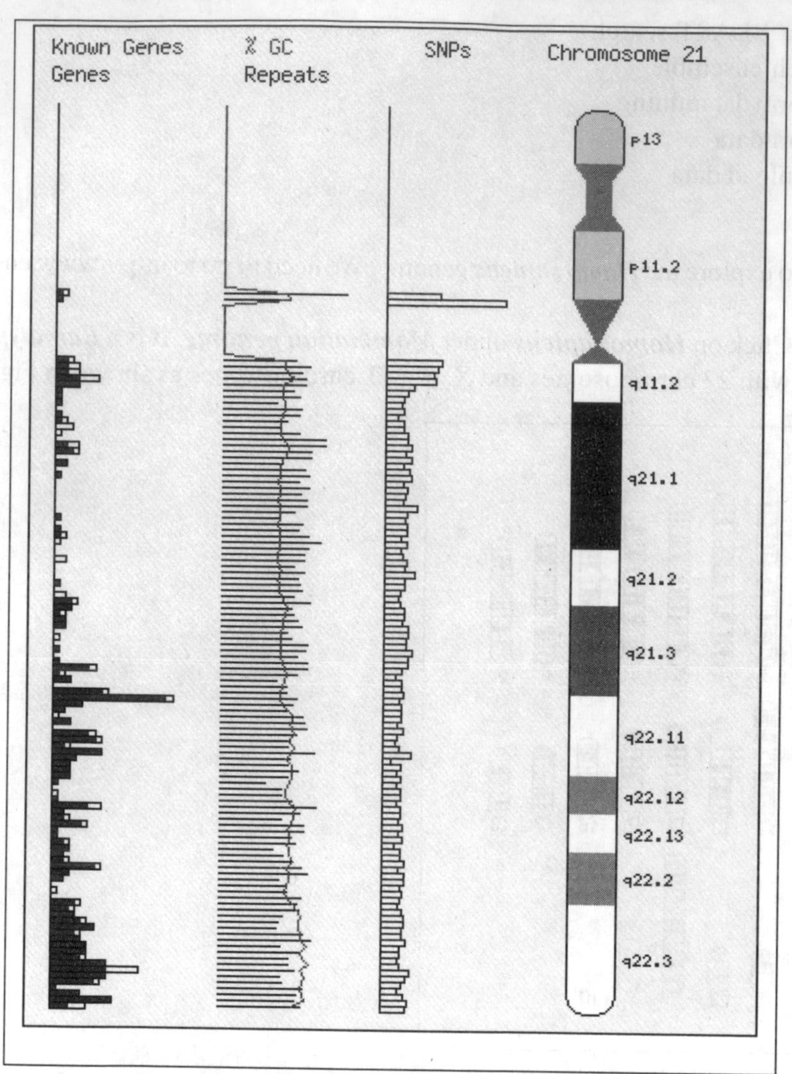

Fig. 3.81 Map view of chromosome 21

Step 3 Click on the **q21.3** region of the chromosome. You will get the contig map of Chromosome 21 between 28,626,632 and 28,726,633 base pairs as shown in Fig. 3.82.

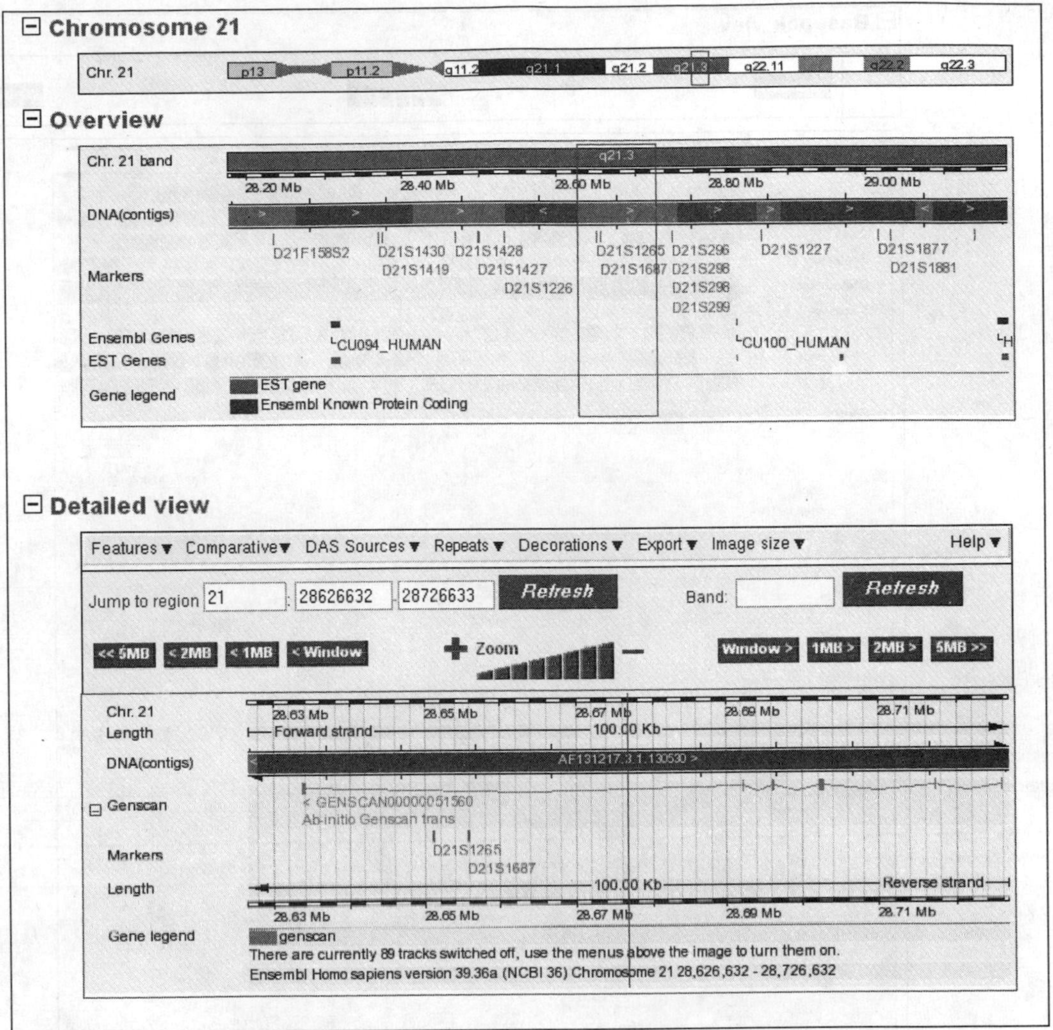

Fig. 3.82 Overview and detailed view of the contig map

Pointing your mouse pointer on the Overview region, you will find the tool tip **Click for menu** on certain regions. If you click on the region where the tool tip does not appear, you will get the current zoom and the options of changing the zoom of the image. On clicking any of the Ensemble or EST gene, you will get information like gene name, gene ID, position in bp, length, and gene type. If you click on any of the DNA contig band, you will get various options of contig, clone, and supercontig. Detailed view has several drop-down options, which you can check on your own, other than the **click for menu** and zooming options as in the **Overview** region.

The **Basepair** view is as shown in Fig. 3.83.

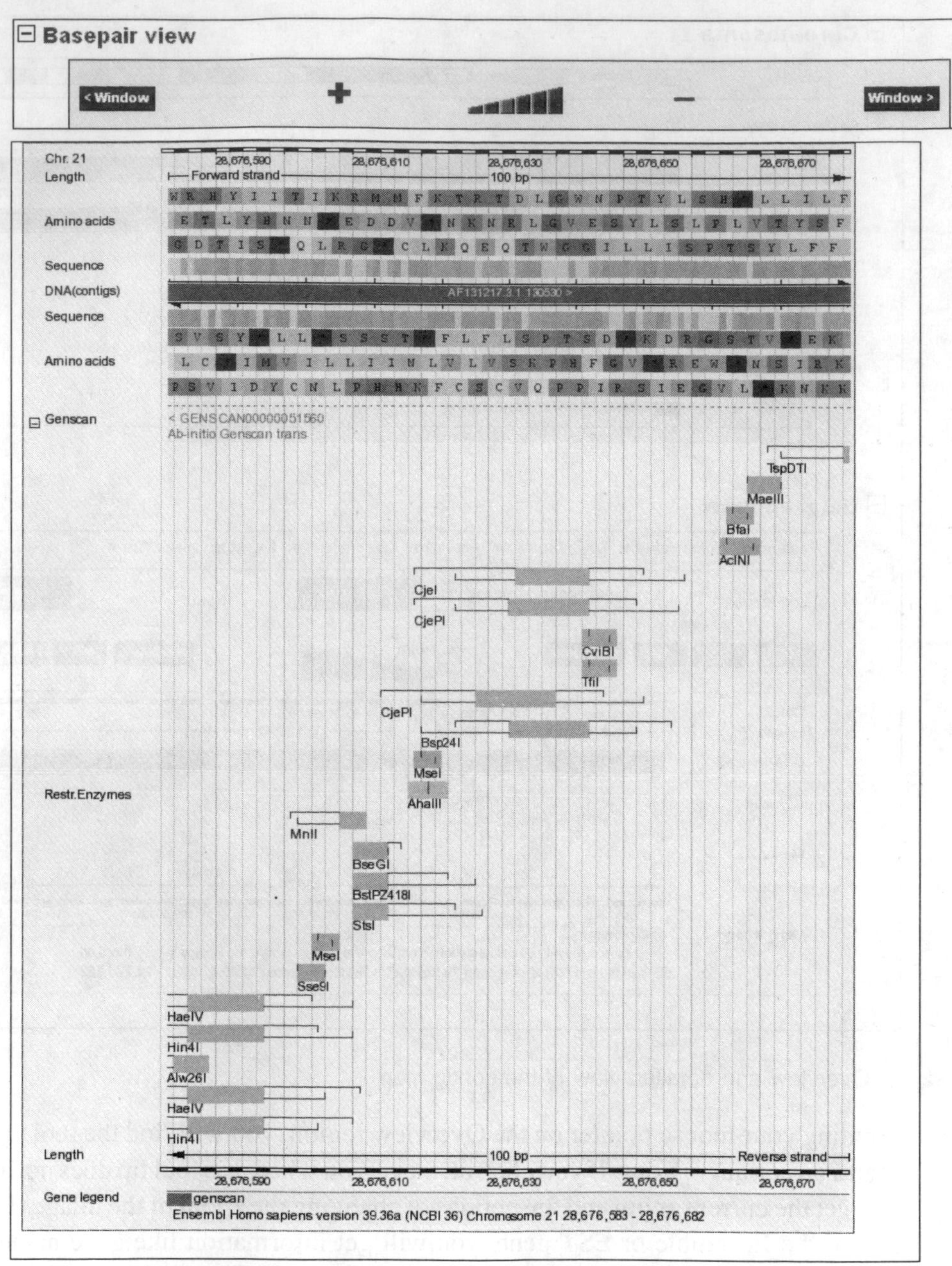

Fig. 3.83 The Basepair view of the contig map

Step 4 Click on an Ensemble known protein coding gene CU100_human and click its respective ID. You will get a result page as shown in Fig. 3.84.

Ensembl Gene Report for ENSG0000178446

Gene	CU100_HUMAN (UniProtKB/Swiss-Prot ID) . To view all Ensembl genes linked to the name click here. This gene is a member of the human CCDS set: CCDS13581
Ensembl Gene ID	**ENSG0000178446**
Genomic Location	This gene can be found on Chromosome 21 at location H28,833,511-28,834,547. The start of this gene is located in Contig AF165147.1.1.146735.
Description	No description
Prediction Method	Genes were annotated by the Ensembl automatic analysis pipeline using either a GeneWise/Exonerate model from a database protein or a set of aligned cDNAs followed by an ORF prediction. GeneWise/Exonerate models are further combined with available aligned cDNAs to annotate UTRs (For more information see V.Curwen et al. , Genome Res. 2004 14:942-50.)

Transcripts

ENST0000032 4961	ENSP0000032 6733	CU100_HU MAN	[Transcript i nfo]	[Exon in fo]	[Peptide i nfo]

Features ▼

☐	SNPs
☐	Regulatory regions
☐	Vega genes
☐	Ensembl genes
☐	ncRNA genes
☐	EST genes
☐	Genscan

CLOSE MENU ▲

Orthologue Prediction

The following gene(s) have been identified as putative orthologues:

Species	Type	Gene identifier
Pan troglodytes	1 to	ENSPTRG00000013816 (C21orf100)

Fig. 3.84

(contd)

(contd)

		1	[MultiContigView] [Align] No description
	Macaca mulatta	1 to 1	ENSMMUG00000006520 (CU100_HUMAN) [MultiContigView] [Align] No description
	View alignments of homologies. **This gene can be viewed in genomic alignment with other species** view genomic alignment with **5 eutherian mammals MLAGAN** view genomic alignment with **7 amniota vertebrates MLAGAN** view genomic alignment with **3 primates MLAGAN** view genomic alignment with **Rattus norvegicus** view genomic alignment with **Macaca mulatta** view genomic alignment with **Canis familiaris** view genomic alignment with **Pan troglodytes** view genomic alignment with **Mus musculus** view genomic alignment with **Bos taurus** view genomic alignment with **Gallus gallus**		

⊟ Gene DAS Report

⊟ DAS	

Sources	☐ AltSplice (Alternative splice database) ☐ AltTrans (Alternative Transcript Diversity Database) ☐ ArrayExpress (Gene Expression Database) ☐ GAD (Genetic Association Database) ☐ HGNC (HUGO Gene Nomenclature Committee) ☐ HUGO_text (PubMed text-mining via HUGO symbol) ☐ Phenotypes (Associated directly or via orthologues or protein families) ☐ Protonet (Global classification of proteins into hierarchical clusters) ☐ RZPD verif. cDNA (RZPD sequence verified non-redundant cDNA clone sets) ☐ RZPD esiRNA (RZPD gene silencing (RNAi) resources) ☐ RZPD Prot Exp (RZPD clones ready for protein expression) ☐ Reactome (Knowledgebase of biological processes) ☐ UniProt (Protein knowledgebase)
	Update

Fig. 3.84 *(contd)*

(contd)

⊟ Transcript ENST00000324961

Transcript	**CU100_HUMAN** (UniProtKB/Swiss-Prot ID) . To view all Ensembl genes linked to the name click here. This transcript is a member of the human CCDS set: CCDS13581
Transcript information	**Exons:** 2 **Transcript length:** 869 bps **Protein length:** 55 residues [Further Transcript info] [Exon information] [Protein information]
Similarity Matches	**This Ensembl entry corresponds to the following database identifiers:**

CCDS:	CCDS13581.1
UniProtKB/Swiss-Prot:	CU100_HUMAN [Target %id: 100; Query %id: 100] [align]
RefSeq peptide:	NP_659470.1 [Target %id: 100; Query %id: 100] [align]
RefSeq DNA:	NM_145033.2 [align]
UniProtKB/TrEMBL:	Q494S0_HUMAN [Target %id: 98; Query %id: 98] [align] Q494S4_HUMAN [Target %id: 100; Query %id: 100] [align]
EntrezGene:	118421
Agilent CGH:	A_14_P128592 [Target %id: 6; Query %id: 100]
Agilent Probe:	A_23_P377839 [Target %id: 6; Query %id: 98]
EMBL:	AY063458 [align] AY063459 [align] BC101414 [align] BC101416 [align] BC101419 [align] BC101420 [align]
IPI:	IPI00168278.1 [Target %id: 100; Query %id: 100]
Protein ID:	AAI01415.1 [align] AAI01417.1 [align] AAI01420.1 [align] AAI01421.1 [align] AAL60598.1 [align] AAL60599.1 [align]

Fig. 3.84 *(contd)*

(contd)

	UniGene:	Hs.438549 [Target %id: 99; Query %id: 98]
	Affymx Microarray	1552466_x_at
	U133:	1554405_a_at
		Hs2.208166:1.S1_3p_s
	Affymx Microarray	U95:89348_at
Protein Family	ENSF00000016456 : UNKNOWN This cluster contains 1 Ensembl gene member(s) in this species.	
Transcript structure	Forward strand 1.04 Kb	
Protein features	Peptide Sig.Pep cleavage Scale (aa) 0 6 12 18 24 30 36 42 48	

Fig. 3.84

Step 5 By now you have enough links to click and explore. Find out the Splice information by clicking on the peptide band under **Protein Features.**

Step 6 Find out genes that have been identified as putative orthologueus?

Step 7 Click on the link to explore **Ensemble Exon Report**, or go to the **ExonView page.**

Step 8 Click **ENSF00000016456** to explore **Ensemble Protein Report**, or go to the **ProtView page.**

Step 9 Click **ENST00000324961** to explore **Ensemble Transcript Report**, or go to the TransView.

Conclusion

After going through this chapter you will be in a position to appreciate the comprehensive bibliographic, abstracting, and indexing databases that exist in bioinformatics fields of enzymes, pathways, and literature. All of these comprise different facets of bioinformatics needed to not only to search, store, and manage data but also to analyse and be able to apply it meaningfully.

EXERCISES

Exercise 3.1 *Trichophyton* are the most common agents of mycoses of the stratum corneum, nails, and hair. Secreted proteolytic activity is important for their virulence. Proteases secreted by them are dermatophytes, which are similar to those of other fungi such as *Aspergillus spp.* and are members of large protein families. Two gene families encoding endoproteases of the S8 (subtilisins) and M36 (fungalysins) families) were found in *Trichophyton rubrum* and *Trichophyton mentagrophytes*. Two fungalysins, FL3 and FL4, were detected in culture supernatants of *T. rubrum*. Access MEROPS (http://merops.sanger.ac.uk/) to check out the following:

- The alignment of the family type.
- The catalytic residues are invariant in the alignment.
- The genomes section contains a list of completed genomes. What do the blue and black lines denote?
- How many completed genomes contain peptidase homologues and what is its distribution in the fungi?
- This takes you to a 2D representation of the diagram. Click on the PDB accession, e.g., **1DF0**. This displays the structure and the 2D representation.

Exercise 3.2 Argininosuccinate lyase (ASL) or delta-crystallin catalyses the reversible hydrolysis of argininosuccinate to arginine and fumarate, a reaction important for the detoxification of ammonia via the urea cycle and for arginine biosynthesis. ASL belongs to a superfamily of structurally related enzymes, all of which function as tetramers and catalyse similar reactions in which fumarate is one of the products. Genetic defects in the ASL gene result in the autosomal recessive disorder argininosuccinic aciduria. This disorder has considerable clinical and genetic heterogeneity and also exhibits extensive intragenic complementation. Go to the Genome net server in Japan and enter KEGG at http://www.genome.jp/kegg/. Go to the enzyme number **EC 4.3.2.1** (ASL). Obtain the following information:

- What are the ranges of specific activity that have been measured for this enzyme?
- Are there different specific activities reported for the enzyme from the same organism? If so, why might the specific activity vary?
- Why do you expect the specific activity of the same enzyme isolated from different organisms to be different?
- Go to the section on Km. Pick a substrate that has been used in several studies. What are the ranges of Km that have been measured for that substrate? Explain what leads to variation in the kinetic parameters measured between these different studies.

- What is the *p*H optimum of your enzyme? Relate the *p*H optimum to some aspect of the enzyme mechanism.
- Identify two competitive inhibitors for the enzyme from the table of inhibitors. (***Hint:*** Competitive inhibitors generally look like the natural substrate.)

Now go to ExPASy, then to ENZYME, and compare the information on Biochemical Pathways (map numbers F8 ; G8), PROSITE PDOC00147,

PUMA2, and Kyoto University LIGAND chemical database for enzyme **4.3.2.1**.

Exercise 3.3 Cell walls constrain the final sizes and shapes of plant cells, and therefore the stature and form of plants. Plants invest over 10% of their genomes in building their cell walls. As part of your study on biosynthesis of plant cell walls (at the Plant Biochemistry Laboratory) you are interested in glycosyltransferases. These enzymes are involved in biosynthesis of various carbohydrate polymers such as pectin, xyloglucan, and arabinoxylan. You have discussed this problem with a senior scientist at your laboratory and you have been advised to perform a heterologous expression of the gene of interest. *Arabidopsis* was taken for the study. Go to CaZy (http://afmb.cnrs-mrs.fr/CAZY/) and try to retrieve information about glycosyltransferase family 47.

- What is known about the enzymes encoded by the genes in this family?
- Which families comprise arabidopsis genes?
- Get the sequence of the gene **At2g35100** from NCBI.
- Use a restriction enzyme tool such as Webcutter on the sequence map. Analyse its restriction map.

Exercise 3.4 You are a researcher working on proteins from malaria genome annotation. You have found that on the basis of this annotation, a set of five proteins with specific EC numbers have been found. The set comprises the following:

- EC 2.7.1.1
- EC 5.4.2.8
- EC 2.7.7.13
- EC 4.2.1.47
- EC 2.7.1.90

Use KEGG (http://www.genome.jp/kegg/) to make a pathway and predict the missing links.

Exercise 3.5

(a) You will submit a project in genomic and proteomic analysis of *Mycobacterium tuberculosis*. Prepare a reference list from PubMed and LitDB.

(b) Go to Ensemble and browse Saccharomyces Cerevisiae under 'Other Species'.

   i. What type of genome is this?

ii. Give the recent statistics.
iii. Give the information for length in bps, known protein-coding genes, ncRNA genes, rRNA genes, tRNA genes of chromosome mito.
iv. Prepare a gene report of SGD gene ID tP(UGG)Q.

Exercise 3.6 You are a doctor with special interest in inherited diseases. Go to the NCBI website and do the following:

(a) Search OMIM for schizophrenia in **Gene Map Disorder** field for a gene with known sequence with allelic variants.
(b) Prepare a report on the MIM number, gene name and gene map locus.
(c) Search OMIM for schizophrenia in **Gene Map Disorder** field for a gene with known sequence and phenotype with allelic variants.
(d) Write the following for the result gene that you got after step (c):
    i. MIM number
    ii. Chromosome number
    iii. Loci
    iv. Gene ID
    v. Gene name
    vi. Species name
(e) Search OMIM gene map for the same gene.
    i. What is the disorder?
    ii. What are the methods?
    iii. In which chromosome is this gene present in mouse genome.
(f) Search for OMIM morbid map.
(g) What disorder is closely associated with the disorder you found earlier that is Mendelian phenotype or locus whose molecular basis is unknown?
(h) What is its gene ID?

Exercise 3.7 You are interested in genes responsible for tuberculosis. Search GeneCard and note the following:

(a) Annexin genes.
(b) Genes related to chemokine (C-C motif) ligands and receptors.
(c) What is the GC ID of tyrosine kinase-2?
(d) What is the genomic location of MTBS1?
(e) What is the OMIM ID for this disorder?
(f) Mention two gene IDs from MiniCards of Chromosome 1.

Exercise 3.8

(a) You are a teacher in a higher secondary school interested to prepare a booklet on 'Health and Wellness' for your students to make them health conscious. Go to http://www.medlineplus.gov/ and search for relevant data for your booklet.

(b) You have been invited to give a talk on hair care. You need to prepare your talk regarding the genetic basis of the following topics:

- Hair loss
- Excess hair
- Graying of hair

PART C
Tools

Chapter 4: Data Submission Tools

Chapter 5: Data Analysis Tools

Chapter 6: Prediction Tools

Chapter 7: Modelling Tools

C Tools

Introduction

While working through the first part of this book, you would have realized that the recent years have seen a dramatic surge in the amount of information stored in databases. Bioinformatics, the discipline that generates computational tools, databases, and methods to support genomic research, has grown up. Bioinformatics acts as an interface among biology, medicine, mathematics, and computer science. It faces a singular major challenge, which is to elucidate the relationship between a sequence, its structure, and its function in genomes. Having paid so much of attention to the accumulation of data and its management, the next logical step is to use this data in a most efficient possible manner. The data produced by functional genomics in the form of gene expression, disease profiling, secondary structure prediction, and cell modelling would obviously require unique, and sometimes new, methods for analysis. As a result, several tools have been developed to extract hidden information from the available data, such as patterns, etc. Software for data mining is often called 'siftware' because it helps to sift through vast datasets.

C.1 Need for Tools

The objective of data mining is not only to look at patterns that are visible in a given dataset but also to identify the structure or the basis of the dataset that has given rise to that particular pattern. Thus today's data-mining tools involve aspects of engineering, statistics, biology, and computer science. There is an enormous amount of data that is available in the form of sequence databases, micro-array expression data, proteomic structure data, protein interaction data, and literature data (see Fig. C.1). Development of tools that can be used to extract knowledge from this data is rather a tedious process.

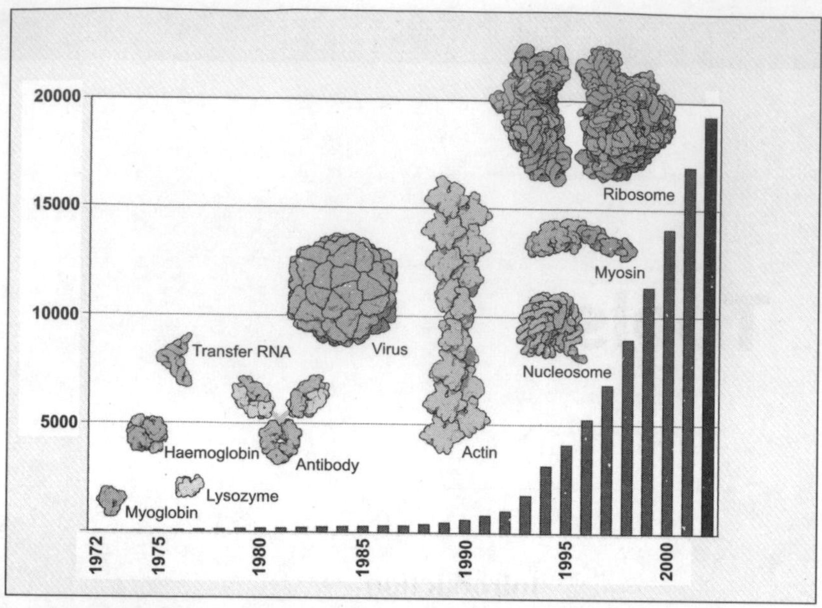

Fig. C.1 Data growth over the last few decades

All human cells have 23 pairs of chromosomes, each holding about 3.5 billion pairs of nucleotides (A, C, G, and T). However, only less than 3% of the genome code for actual genes that make proteins; the rest is genetic noise. Gene hunting is made more challenging by the fact that their protein-coding elements are scattered, as are the genetic signals.

The Oak Ridge National Laboratory came forward with the first gene-finding program that used artificial neural networks [a type of artificial intelligence (AI) program], called GRAIL, which can pick out genes. Since then many other gene-finding programs have been developed and are available online to researchers. Today pattern-recognition programs are used not only for discovering genes, but also to give clues about the function of these genes.

C.2 Knowledge Discovery

Data mining and knowledge discovery are terms that have been coined to describe a variety of techniques to identify nuggets of information or decision-making knowledge from large bodies of data. Once identified, the data is extracted and put to use in areas such as decision support, prediction, forecasting, and estimation. Both bioinformatics and computational biology deal with the biological problems where knowledge can be gained from the information that we can extract from biology.

Extraction of knowledge from data through the process of discovering solutions to problems, unfortunately, is not a simple process. It entails knowledge of several other domains, and within bioinformatics it spans domains such as databases, pattern recognition, machine learning, data visualization, optimization, and high-performance computing. To do so, data-mining techniques use a variety of data analysis tools to discover patterns and relationships in data which may be used to make valid predictions.

Figure C.2 illustrates the knowledge-discovery process which takes in raw data and leads to knowledge. In this process, data mining is one of the principal steps.

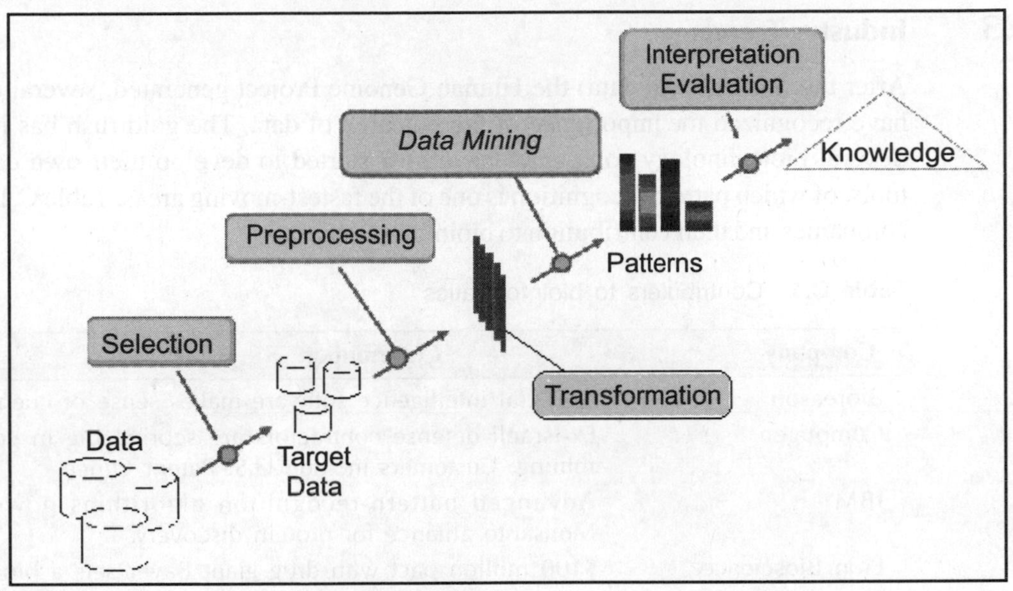

Fig. C.2 Process of knowledge discovery

The knowledge-discovery process is sequential and follows the path outlined below.

Selection The data is selected or segmented according to some criteria.

Preprocessing This is the data-cleansing stage where certain information, which is thought to be unnecessary and which may slow down query processing, is removed.

Transformation The data is not merely transferred across but transformed so that overlays may be added.

Data mining This activity is concerned with extraction of patterns from data.

Interpretation and evaluation The patterns identified by the system are interpreted into knowledge, which can then be used to support human decision-making.

In the chapters in this part, you will learn bioinformatics tools that are used extensively to work out each of the steps elucidated above. For example, database submission tools are used to select and preprocess data before it is deposited into shared databases, and annotation tools perform the function of transformation. There are various other tools that carry out data mining and analysis. Here, analysis includes interpretation and evaluation and covers tools for analysis, prediction, and modelling data. Mostly we use the term 'data mining' for not just mining or searching data but to analyse, i.e., interpret and evaluate it as well. So much so that data mining is also used synonymously with knowledge discovery. Analysis comes from evaluation and interpretation of the mined data. Interpretation and evaluation are the last steps (Fig. C.2) in the process of knowledge discovery.

C.3 Industry Trends

After the interest (and data) the Human Genome Project generated, several companies have recognized the importance of the potential of data. The gold rush has just begun! Several biotechnology companies have now started to develop their own data-mining tools, of which pattern recognition is one of the fastest-moving areas. Table C.1 lists some companies and their contribution to bioinformatics.

Table C.1 Contributers to bioinformatics

Company	Contribution
Bioreason	Artificial intelligence software makes sense of chemistry data.
Compugen	Ex-Israeli defense contractors are scoring big in genetic data mining. Customers include U.S. Patent Office.
IBM	Advanced pattern-recognition algorithms power a 1997 Monsanto alliance for protein discovery.
Lion Bioscience	$100 million pact with drug giant Bayer sets a bioinformatics record.
Molecular Mining	Raised $2 million in start-up funds from venture capitalists in March 1999.
Neomorphic	Hidden Markov models are among this 1996 start-up's advanced gene-finding tools.
Partek	Neural networks specialists moved into biology market in 1998.
Silicon Genetics	Stanford spin-off mines gene data for profit.
Silicon Graphics	Mine Set visual data-mining tool is popular in financial, telecom, and drug industries.

The fastest growing market of bioinformatics is expected to increase from $1.02 billion in 2002 to $3.0 billion in 2010, at an average annual growth rate (AAGR) of 15.8%. (*source: Bioinformatics: Technical Status and Market Prospects, BCC Market Research Report*). This boom in bioinformatics is driven by bioinformatics technologies that help reduce the amount of time and money required for drug discovery and development. Bioinformatics market can be segmented based on application areas. The main market segments by application areas are genomics (50%), proteomics (25%), pharmacogenomics (10%), and cheminformatics (15%). While genomics currently represents the largest market segment for bioinformatics, pharmacogenomics and cheminformatics have significant growth opportunities over the coming five years. By 2006 it is forecasted that the bioinformatics market will be equally divided among the four segments (*source: Avendus Estimates*). Thus it is only logical to see interest in this sector not only by companies engaged in biotechnology and agro-biotechnology, but also by traditional technology and IT companies. Traditional technology and IT companies have scaled up to offer products and services for the bioinformatics market. For the most part, these companies have taken the complementary approach of providing infrastructure that supports various solutions offered by specialized bioinformatics agencies. These companies have developed tools to mine and analyse data belonging to every domain of bioinformatics (see Fig. C.3). Unfortunately, most of them are proprietary in nature.

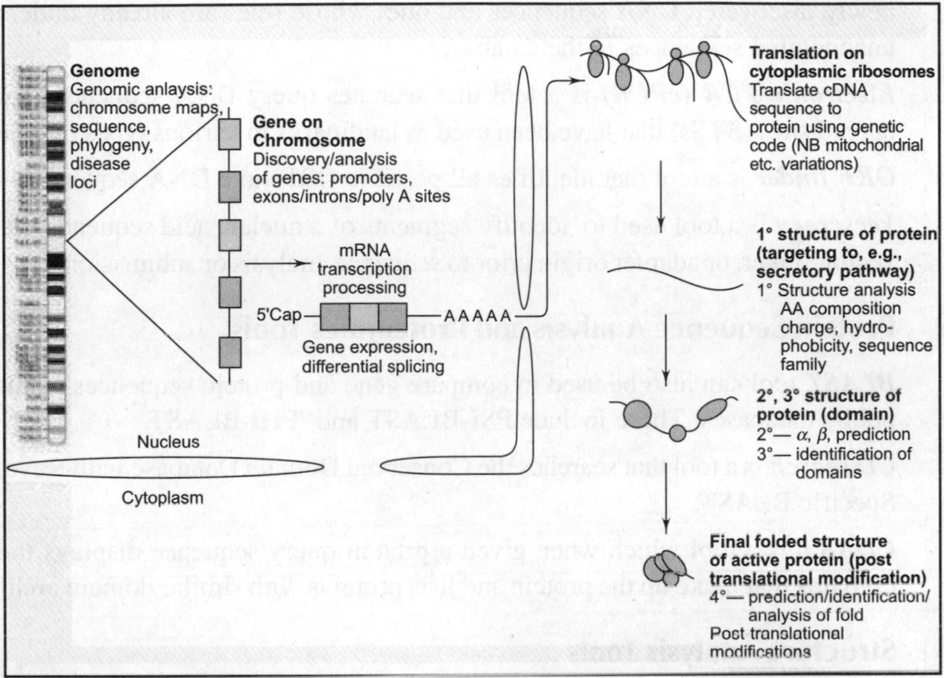

Fig. C.3 A few common applications of the use of data-mining tools in bioinformatics

C.4 Data-mining Tools

Data-mining and knowledge-discovery tools are used not only for fundamental biological research but also for development. The objective of the chapters in this part is to familiarize you with a set of tools most widely used and accepted by the bioinformatics community world wide. The focus will be on the web-based tools freely available online. The availability of these online tools provides everybody with the opportunity to derive considerable amounts of useful information. The focus of these chapters will also be on tools that pertain to bioinformatics data mining. For convenience, these have been grouped as tools for data submission, analysis, prediction, and modelling. Needless to say that these tools help in 'finding needles in haystacks'. (Haystacks being the databases for DNA or protein sequences, protein structures, genomes, proteomes, pathways, etc., and the needles represent the information of importance and are worth more than all the hay put together.) We will discuss these tools next in relation to their functions.

C.4.1 Sequence Analysis Tools

Basic local alignment search tool (BLAST) is the most widely used program at NCBI. It receives 50,000 hits every day from researchers searching for similarities between newly discovered DNA sequences and ones whose roles are already understood or the unannotated sequences in the databases.

Electronic PCR (ePCR) is a tool that searches query DNA sequences for sequence-tagged sites (STSs) that have been used as landmarks in various types of genomic maps.

ORF finder is a tool that identifies all possible ORFs in a DNA sequence.

Vecscreen is a tool used to identify segments of a nucleic acid sequence that may be of vector, linker, or adapter origin prior to sequence analysis or submission.

C.4.2 Protein Sequence Analysis and Proteomics Tools

BLAST tool can also be used to compare gene and protein sequences against others in public databases. These include PSI-BLAST and PHI-BLAST.

CD search is a tool that searches the Conserved Domain Database with Reverse Position Specific BLAST.

CDART is a tool which when given a protein query sequence displays the functional domains that make up the protein and lists proteins with similar domain architectures.

C.4.3 Structure Analysis Tools

Cn3D is a web browser application that allows one to view 3D structures from NCBI's Entrez retrieval service.

***CD* search** is a tool that searches the Conserved Domain Database with Reverse Position Specific BLAST.

C.4.4 Genome Analysis and Gene Expression Tools

Clusters of orthologous groups (COGs) is a tool that uses a natural system of gene families from complete genomes. It is used to derive the phylogenetic classification of proteins encoded in genomes.

Map viewer is a tool used to visualize integrated views of chromosome maps for many organisms and is particularly useful for the identification and localization of genes and other biological features.

Gene expression omnibus (GEO) is a suite of tools that assist in the visualization and exploration of datasets. This high-throughput gene expression/molecular abundance data repository also serves as a curated, online resource for gene expression data browsing, query, and retrieval.

C.4.5 Data Submission Tools

Bankit is a submission tool available to feed data into GenBank. It involves utilization of the internet and is meant for those submissions that are limited to one or a few sequences.

Sequin is a stand-alone tool developed by the NCBI for submitting data to three databases, namely, GenBank, EMBL, and DDBJ. This tool can also be used to update these sequence databases. The tool allows submission in the form of single, short mRNA sequences and complex, long sequences, as well as multiple annotations, segmented sets of DNA, or phylogenetic and population studies.

Webin is another internet-based tool used for submission of nucleotide sequences to the EMBL database. Its special feature is the inclusion of a security feature that protects the privacy of the data submitted. It also has a linked tool that screens data for vector contamination prior to its submission.

Sakura is a nucleotide sequence data submission system that operates through the internet and feeds into the DDBJ database. The system can be used to submit nucleotide and translated amino acid sequences and their functions, features, and references.

Spin is a web-based tool for submitting directly sequenced protein sequences and their biological annotations to the UniProtKB/Swiss-Prot database. A unique accession number is assigned to each sequence, which is submitted only after approval by curators.

AutoDep is a web-based tool used to submit data to the PDB repository. It includes capture of information such as coordinates in the PDB format, unit-cell data, collection details, sequence database reference for the biological molecules, description of heterogens, and other relevant data.

C.4.6 Prediction Tools

Phylip is a commonly used free package of programs for deducing phylogenies. It uses a number of methods to arrive at the phylogeny such as parsimony, distance matrix, and likelihood methods, including bootstrapping and consensus trees. Data can be fed in the form of molecular sequences, gene frequencies, restriction sites and fragments, distance matrices, and discrete characters.

Alibee utilizes the old GeneBee algorithm of multiple alignment. The phylogenetic construction proceeds in four steps: construction of pairwise motifs, construction of multiple motifs, formation of supermotifs from multiple motifs, and construction of multiple alignments from previously obtained motifs and supermotifs, which is consequently followed by the selection of the best alignment.

GenScan is one of the best gene-finding algorithms. It was developed by Burge at MIT. It is one of the most popular gene-finding programs available today. It uses an ab-initio approach to finding genes and other gene-related features within a DNA sequence.

Genefinder was developed by Colin Wilson, LaDeana Hilyer, and Phil Green. It is a species-specific gene-prediction tool. It uses maximum likelihood estimation to predict gene structure and can be configured for a variety of species. This software tool is designed to predict putative, internal protein coding exons in genomic DNA sequences.

Prosite is a database of protein families and domains which houses biologically significant sites, patterns, and profiles that help to reliably identify to which known protein family (if any) a new sequence belongs. This tool requires a protein sequence as input; however, DNA/RNA may be translated into a protein sequence using Transeq and then queried.

3D PSSM is a protein-fold-recognition server that predicts protein 3D structure and its probable function. It uses 1D (sequence profiles built from relatively close homologues) and 3D (more general profiles containing more remote homologues) sequence profiles to predict structure. In addition, it uses secondary structure and solvation potential information also.

ChouFasman algorithm is used for the prediction of protein secondary structure. It is one of the most widely used predictive schemes. It works by assigning a set of prediction values to a residue and then applying a simple algorithm to the conformational parameters and positional frequencies.

GOR (Garnier-Osguthorpe-Robson) method uses both information theory and Bayesian statistics for predicting the secondary structure of proteins. The algorithm is very similar to ChouFasman. Its methods of prediction are also based on the analysis of amino acids implied in known secondary structures.

Predict protein (PP) is an automatic service for protein database searches and prediction of aspects of protein structure. A submitted amino acid sequence can be used to derive a variety of information such as a multiple sequence alignment, ProSite sequence motifs, ProDom domain assignments. These can then be used to predict the secondary structure, solvent accessibility, globular regions, and transmembrane helices in coiled-coil regions.

DomPred is a server designed to predict protein domains for a given protein sequence. Since polypeptide chains often fold into one or more distinct regions of structure and such domains are considered to be the basic units of folding, function, and evolution, it is important to 'know' them. DomPred identifies similarities to Pfam-A domain sequences and uses them to predict domains.

Pfam is a collection of protein families and domains. It contains multiple protein alignments and profile HMMs of these families and is used to predict protein structure and domains.

HMMER is a software package that uses hidden Markov models (HMMs). HMMs are statistical models of the primary structure consensus of a sequence family. HMMs are similar to profiles or flexible patterns and are used to predict secondary protein structures.

Modeler is a tool that is extensively used for comparative modelling of protein three-dimensional structures. It takes in an alignment of a sequence to be modelled with known related structures and develops a model containing all non-hydrogen atoms. In addition, it can be used to derive *de novo* modelling of loops in protein structures, optimization of various models of protein structure with respect to a flexibly defined objective function, multiple alignment of protein sequences and/or structures, clustering, searching of sequence databases, comparison of protein structures, etc.

Swiss model is a fully automated protein structure homology-modelling server developed at the Biozentrum Basel within the Swiss Institute of Bioinformatics. It consists of annotated three-dimensional comparative protein structure models. These models are obtained by the homology-modelling pipeline Swiss-model. It has only theoretically calculated models, which may contain errors. It houses tools such as *DeepView* (Swiss PDB viewer), ANOLEA, and SWISS model workspace.

Pred TMR is a tool used to predict transmembrane domains in proteins using information contained in the sequence itself. The algorithm uses hydrophobicity analysis with the detection of potential termini of transmembrane regions.

Tmpred is used to predict membrane-spanning regions and their orientation. The algorithm working bchind this tool uscs statistical analysis of TMbasc. TMbasc is a databasc of naturally occurring transmembrane proteins. Prediction is made using a combination of several weight matrices for scoring.

C.4.7 Modelling Tools

Rasmol is molecular graphics program intended for the visualization of proteins, nucleic acids, and small molecules. It is available for Windows, Macintosh, and UNIX platforms and is used extensively as an educational and research tool. It is easy to use. Rasmol reads in molecular coordinate files in a number of formats and interactively displays the molecule on the screen in a variety of colour schemes and representations. Currently supported input file formats include PDB, Mol2, MDL, etc.

SPDV stands for Swiss PDB viewer. It helps in viewing, manipulating, and modelling of multiple molecules in layers. It helps to manipulate interactive Ramachandran-plot for protein modelling. It has an interface to the SwissModel database for homology modelling from known sequences and the output to Quickdraw3D and POVRay for publication-grade imaging. It also has interfaces to molecular dynamics (MD) software packages such as AMBER, CHARMM, for energy minimization.

Hyperchem is a molecular modelling software used for 3D visualization and animation with quantum chemical calculations, molecular mechanics, and dynamics. It is known for its quality, flexibility, and user friendliness. The new version of Hyperchem offers modules such as Raytrace, Rms Fit, Sequence Editor, Crystal Builder, Sugar Builder, Conformational Search, Qsar Properties, and Script Editor.

Conclusion

Knowledge discovery and data mining involve various steps (such as data selection, data cleaning, data transformation and reduction, visualization, interpretation, and evaluation) with the objective being to incorporate the mined 'knowledge' into the larger decision-making processes. Methods from databases, statistics, algorithmic complexity, and optimization are used to build efficient, scalable systems, which are seamlessly integrated with the relational database structures. This enables database developers to easily access data and successfully apply data-mining technology.

4 Data Submission Tools

Learning Objectives

- To gain insight into nucleotide sequence submission
- To gain insight into protein sequence submission
- To gain insight into protein structure submission

Introduction

Information about bioinformatics is being generated at an exponentially increasing rate since the completion of the Human Genome Project, which sequenced and mapped the complete human genome. Though the project itself is now complete, other derivative work is far from over. The structure, function, and molecular mechanisms of all the genetic elements comprising the human genome are still being discovered. Bioinformatics, which intensively uses computing tools, has been put to use in analysing the collected data.

A plethora of bioinformatics tools has been developed today and it exists on the web as well. The internet has proved to be a very good resource for specialized bioinformatics tools. Added to this are worldwide research activities producing publicly available data, and new technologies, e.g., high-throughput devices such as microarrays. Thus, data mining and analysis require comprehensive integration of heterogeneous data, which is typically distributed across many sources on the web and often structured only to a limited extent.

Even with the use of new inter-operability technologies such as XML and web services, data integration is still difficult and mostly manual. This is especially so due to semantic heterogeneity and varying data quality. In addition, there are specific application requirements that are enforced. One of the easiest and best ways to avoid this is to store data in a database in a particular format. Over a period of time, each database has developed certain tools that allow easy and complete data submission. These tools allow both individual and project-based data to flow to the database. Today, using these tools, one can submit large microarray-based experiments in the form of a spreadsheet description with associated data files. The basic aim of this chapter is to familiarize you with some of the frequently used tools for data submission.

4.1 Nucleotide Sequence Submission Tools

Sequences today are either submitted to databases prior to or after their publication. Prior submission to databases affords to build authenticity and helps clear away repetition, if any. Upon submission of a sequence, a unique identifier is assigned to it by the database. It is now a standard practice to have a reference to this number built into the publication. In fact, the protocol for authoring requires references such as accession numbers.

A large number of tools use the web as the medium of submission of sequences to a database. This brings in an element of insecurity. However, most of these tools have in-built security features to protect data and allay fears of insecurity.

4.1.1 BankIt for GenBank (http://ncbi.nih.gov/BankIt/)

NCBI has developed a World Wide Web (www) form called BankIt to facilitate quick submission of sequence data. This form allows entering sequence information and editing and adding biological annotations (e.g., coding regions, mRNA features). The submitted data is then transformed into GenBank format. Once reviewed and completed by the submitter, it becomes a part of GenBank.

The form is compatible with Netscape clients for Unix computers, Macintosh, and PCs.

(a) Submission to BankIt

A BankIt submission is a multi-step process (Benson, D.A. et al 1999). It requires the submitter to confirm that the concerned sequence is not an update or a duplicate of an already existing entry in the GenBank database. The interface will look like Fig. 4.1.

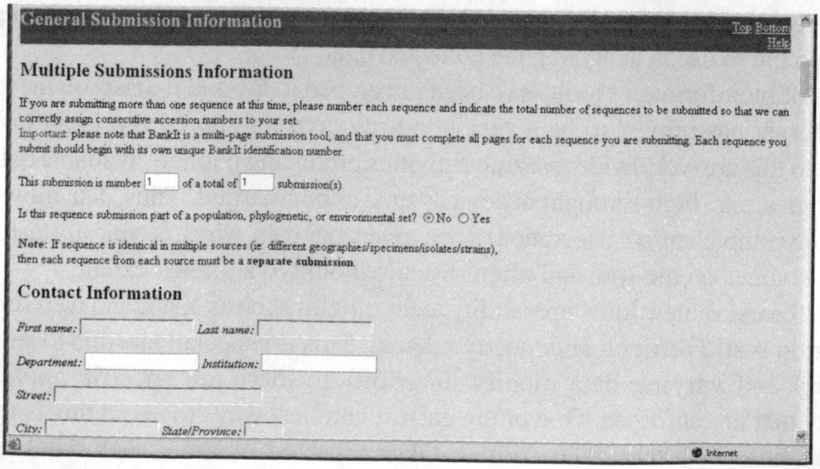

General Submission Information

Top Bottom
Help

Multiple Submissions Information

If you are submitting more than one sequence at this time, please number each sequence and indicate the total number of sequences to be submitted so that we can correctly assign consecutive accession numbers to your set.
Important: please note that BankIt is a multi-page submission tool, and that you must complete all pages for each sequence you are submitting. Each sequence you submit should begin with its own unique BankIt identification number.

This submission is number 1 of a total of 1 submission(s).

Is this sequence submission part of a population, phylogenetic, or environmental set? ⊙No ○Yes

Note: If sequence is identical in multiple sources (ie: different geographies/specimens/isolates/strains), then each sequence from each source must be a **separate submission**.

Contact Information

First name: *Last name:*

Department: *Institution:*

Street:

City: *State/Province:*

Internet

Fig. 4.1 General submission information interface

Once the contact information has been filled in, release information is asked for. Details about when EMBL can release the sequence record can be provided here. However, all sequences must be released when the accession number or any portion of the sequence is published. The next step involves declaring whether the data submitted is primary or not. In case the data is not primary sequence data, the submitter is required to provide the accession number(s) only of the DBJ/EMBL/GenBank primary sequence(s) that were used. Citation of the published work is also asked for. The submitter is then prompted to fill in details about the source organism, as shown in Fig. 4.2.

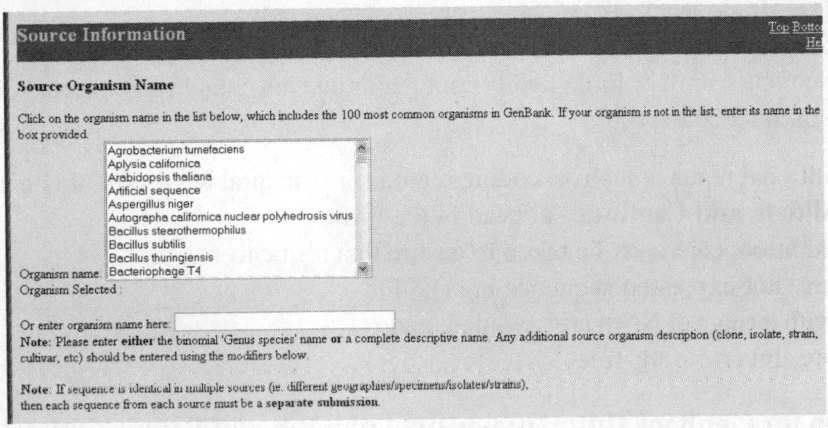

Fig. 4.2 Details about the source organism

In case the sequence was derived through PCR, the details of the primer need to be submitted. In case the source is a lineage, organelle, chromosome, or map, the details about the same are to be deposited too, e.g., the details about the location of the sequence in the genome, chromosome number, and map location.

The next step requires details about the DNA sequence to be submitted, as shown in Fig. 4.3.

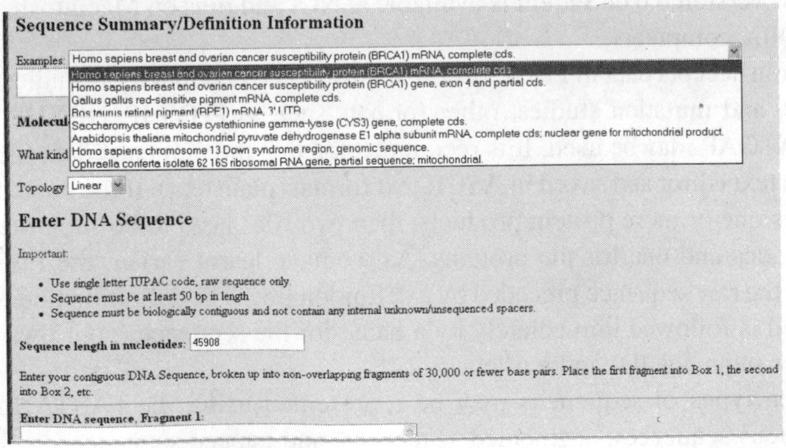

Fig. 4.3 Details of the DNA sequence submission

The details about the molecule type, such as genomic DNA, genomic RNA, cDNA to mRNA, pre mRNA, mRNA, tRNA, and snRNA, are to be submitted too. Along with this, the topology, circular or linear, is also asked for.

The actual submission of the DNA sequence requires the submission of the contiguous DNA sequence, broken up into non-overlapping fragments of 30,000 or fewer base pairs.

Note

All submissions require the following to be kept in mind:

- Use single letter IUPAC code, raw sequence only.
- Sequence must be at least 50 bp in length.
- Sequence must be biologically contiguous and not contain any internal unknown/ unsequenced spacers.

Additional features such as coding regions or structural RNAs should be added under the **Validate and Continue** subhead in the form.

In addition, care must be taken to ensure that sequences

- are not expressed sequence tag (EST),
- with strings of NN's are avoided, and
- are linker-/vector-free.

4.1.2 Sequin for GenBank (http://www.ncbi.nlm.nih.gov/Sequin/index.html)

Sequin is a suite of forms for allowing submitting authors to enter sequences and information about the organism, gene, and protein names, viewing the complete submission, and editing and annotating the submission.

This tool has been developed by NCBI for submitting and updating entries to GenBank, EMBL, or DDBJ sequence databases. It is capable of handling simple submissions that contain a single short mRNA sequence, and complex submissions containing long sequences, multiple annotations, segmented sets of DNA, or phylogenetic and population studies. Version 6.0 of Sequin is available at NCI and runs on Macintosh, PC/Windows, and UNIX computers.

Sequin accepts data in FASTA format. However, for population studies, phylogenetic studies, and mutation studies, other formats such as PHYLIP, NEXUS, MACAW, or FASTA+GAP may be used. It is recommended that the sequence data files be prepared using a text editor and saved in ASCII text format (plain text). If the nucleotide sequence encodes one or more protein products, then two files need to be submitted: one for the nucleotides and one for the proteins. As we have learnt earlier, the FASTA format is simply the raw sequence preceded by a definition line. The definition line begins with a > sign and is followed immediately by a name for the sequence (your own identification code, or sequence ID) and a title.

Three types of sequences may be represented using the FASTA format: single, contiguous sequences; segmented sequences; and gapped sequences.

(b) Single sequence

A single sequence contains the definition line followed by the sequence data. A sample single sequence file is shown here:

```
>ABC-1 [organism=Saccharomyces cerevisiae][strain=ABC][clone=1]
ATTGCGTTATGGAAATTCGAAACTGCCAAATACTATGTCACCATCATTGA
TGCACCTGGACACAGAGATTTCATCAAGAACATGATCACTGGTACTT
```

(c) Segmented nucleotide sequences

Sometimes the submission need to be in the form of a set of segmented sequences, e.g., genomic DNA. The genomic DNA segmented set could include encoding exons along with fragments of their flanking introns. To import nucleotides in a segmented set, each individual sequence must be in FASTA format with an appropriate definition line, and all sequences should be in the same file. An example follows:

```
[
>s_cere_seg1 [organism= Saccharomyces cerevisiae] Saccharomyces cerevisiae
NADH
dehydrogenase ...
ATGGAGCATACATATCAATATTCATGGATCATACCGTTTGTGCCACTTCCAATTCCTATTTT
TTGGACTCCTACTTTTTCCGACGGCAACAAAAAATCTTCGTCGTATGTGGGCTCTTCCCA
GTTAAGTATAGTTATGATTTTTTCGGTCGATCTGTCCATTCAGCAAATAAATAAAAGTTCTATC
TATGTATGGTCTTGGACCATCAATAATGATTTTTCTTTCGAGTTTGGCTACTTTATTGATTCGCTT
AGTTCGAATTTGATACAAATTTATATTTTTTGGGAATTAGTTGGAATGTGTTCTTATCTATTAATAG
TTTGGTTCACACGACCCGCTGCGGCAAACGCCTGTCAAAAAGCATTTGTAACTAATCG
TGGTTTATTATTAGGAATCTTAGGTTTTTATTGGATAACGGGAAGTTTCGAATTTCAAGATTTGTT
ATATTTAATAACTTGATTTATAATAATGAGGTTCAGTTTTTATTTGTTACTTTATGTGCCTCTTTATTA
> s_cere _seg2
GGTATAATAACAGTATTATTAGGGGCTACTTTAGCTCTTGC
TCAAAAAGATATTAAGAGGGGTTTAGCCTATTCTACAATGTCCCAACTGGGTTATATGATG
GGTATGGGGTCTTATCGAGCCGCTTTATTTCATTTGATTACTCATGCTTATTCGAAGGCATTG
TAGGATCCGGATCCGTTATTCATTCCATGGAAGCTATTGTTGGATATTCTCCAGATAAAAG
GGTTTTTATGGGCGGTTTAAGAAAGCATGTGCCAATTACACAAATTGCTTTTTTAGTGGGTA
CTTTGTGGTATTCCACCCCTTGCTTGTTTTTGGTCCAAAGATGAAATTCTTAGTGACAGCTG
> s_cere _seg3
TCAATAAAACTATGGGGTAAAGAAGAACAAAAAATAATTAACAGAAATTTTCGTTTATCTC
]
```

In this case there is a square open bracket on a line by itself before the first segment and a square close bracket on a line by itself after the last segment. These square brackets are required if multiple segmented sequences are to be submitted.

Some modifier names predefine formats. For example, the organism name should be the unabbreviated scientific name such as [organism=Saccharomyces cerevisiae]. Molecule should be either DNA or RNA such as [molecule=DNA]. The Moltype should use one of the following values such as [moltype=Genomic DNA or Genomic RNA or Precursor RNA or mRNA [cDNA] or Ribosomal RNA or Transfer RNA or Small nuclear RNA or Small cytoplasmic RNA or Other-Genetic or cRNA or Small nucleolar RNA]. The next modifier is location. It should use one of the following values: [location=mitochondrion or genomic or chloroplast or kinetoplast or plastid or macronuclear or extrachromosomal or plasmid or proviral or virion etc.]. The collection date should be in the form of DD-MM-YYYY such as [collection-date=2006].

Some modifiers use only TRUE or FALSE such as [transgenic=TRUE]. Other modifiers that use the True/False mode are environmental-sample, germline, rearranged and transgenic.

In case the sequence being submitted is a protein sequence, modifier names that are to be included in the definition lines for protein files are the following: gene, protein, prot_desc. Sometimes a single sequence upon alternate splicing may result in distinct protein products. The following is the suggested format for a hypothetical submission.

```
Nucleotide Sequence:
>eIF4E [organism=Saccharomyces cerevisiae] [strain=ATCC 2043] Saccharomyces
...
CGGTTGCTTGGGTTTTATAACATCAGTCAGTGACAGGCATTTCCAGAGTTGCCCTGTTCA ...
Protein Sequences:
>4E-I [gene=eIF4E] [protein=eukaryotic transciption factor 4E-I]
MQSDFHRMKNFANPKSMFKTSAPSTEQGRPEPPTSAAAPAEAKDVKPKEDPQETGEPAGN ...
>4E-II [gene=eIF4E] [protein=eukaryotic transciption factor 4E-II]
MVVLETEKTSAPSTEQGRPEPPTSAAAPAEAKDVKPKEDPQETGEPAGNTATTTAPAGDD ...
```

In case a gapped sequence needs to be submitted, a different method to represent it is followed. A gap is represented by a line that starts with '>?' and is immediately followed by either a length (for gaps of known length) or "unk100" for gaps of unknown length. For example, ">?100". The next sequence segment continues on the next line, with no separate definition line or identifier.

Difference between a gapped sequence and a segmented sequence

Gapped sequences use a single identifier and can specify known length gaps. This is not true of segmented sequences, which is why gapped sequences are preferred to segmented sequences.

A sample of the gapped sequence follows.

Nucleotide Sequence:

```
>?100
GGTTTTTATGGGCGGTTTAAGAAAGCATGTGCCAATTACACAAATTGCTTTTTTAGTGGA
CTTTGTGGTATTCCACCCCTTGCTTGTTTTTGGTCCAAAGATGAAATTCTTAGTGACAGC
>?unk100
TCAATAAAACTATGGGGTAAAGAAGAACAAAAAATAATTAACAGAAATTTTCGTTTAT
TATTAACGATGAATAATAATGAGAAGCCATATAGAATTGGTGATAATGTAAAAAAAG
```

The Sequin interface looks like as shown in Fig. 4.4.

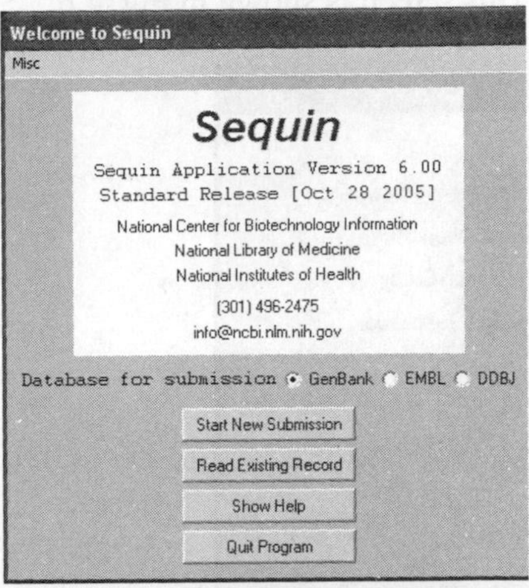

Fig. 4.4 Sequin interface

Select the database that you would want to submit data to from among **GenBank**, **EMBL**, and **DDBJ**. Click on **Start New Submission**. The next screen that appears will look like Fig. 4.5.

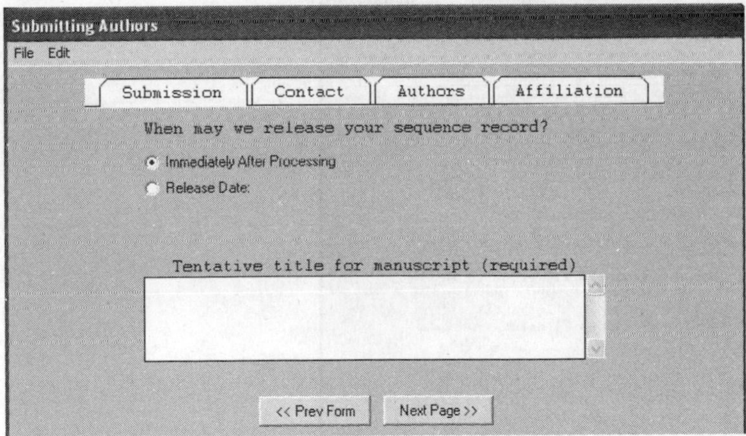

Fig. 4.5 Screen that is displayed by selecting the start new submission option

This form asks for details pertaining to submission, contact, authors, and affiliation areas. The release date details are also captured in this form.

After filling in all the information, click **Next Page>>**. The next form that appears captures details about the sequence such as the type of submission (single, gapped, segmented, population, phylogenetic, mutation, environmental samples, and batch submission). The data format, as described earlier, has to be FASTA. The submission category may be either original or TPA. This screen is shown in Fig. 4.6.

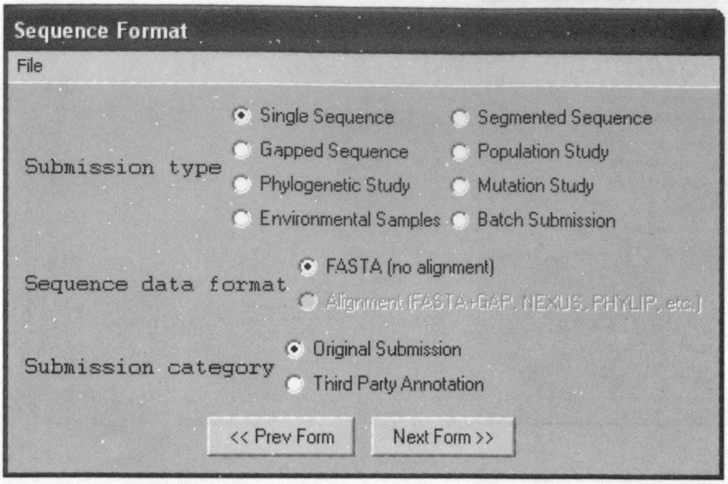

Sequence Format

File

Submission type
- Single Sequence
- Segmented Sequence
- Gapped Sequence
- Population Study
- Phylogenetic Study
- Mutation Study
- Environmental Samples
- Batch Submission

Sequence data format
- FASTA (no alignment)
- Alignment (FASTA+GAP, NEXUS, PHYLIP, etc.)

Submission category
- Original Submission
- Third Party Annotation

<< Prev Form | Next Form >>

Fig. 4.6 Sequence format form

The next form, shown in Fig. 4.7, is designed to capture details on the nucleotide, organism, proteins, and annotation.

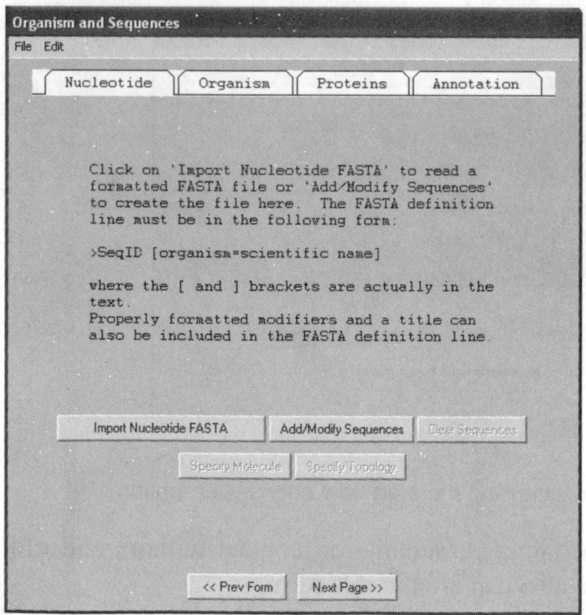

Fig. 4.7 The organism and sequences form

You can choose to either import data that is locally stored on your machine or add/modify existing sequences. The interface for these actions is shown in Fig. 4.8.

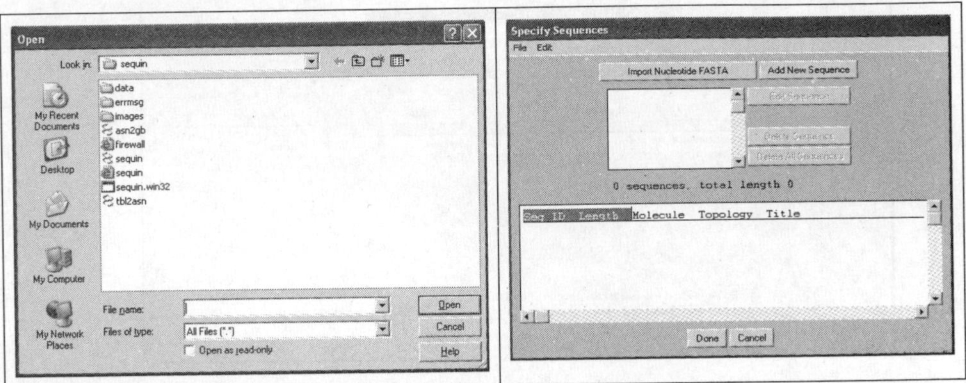

Fig. 4.8 The open dialog box

Once submitted, the interface will look like Fig. 4.9.

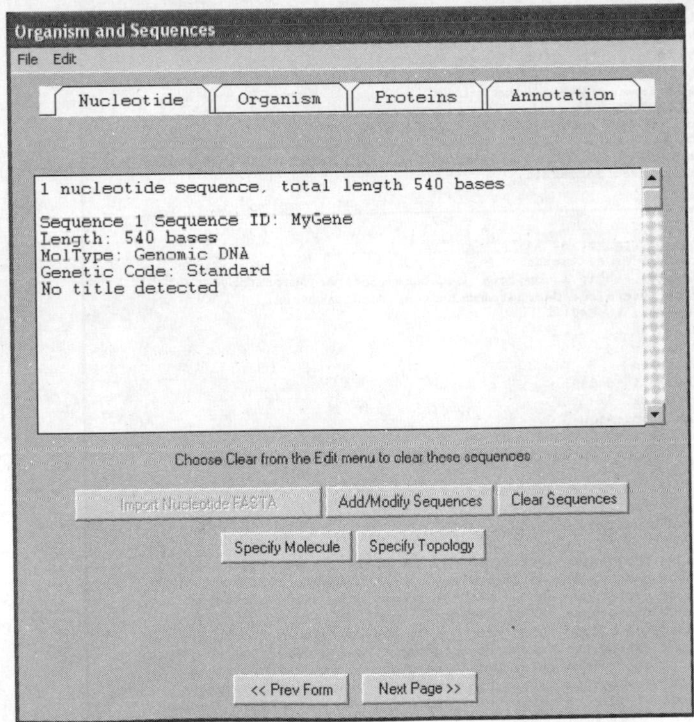

Fig. 4.9 The interface after submission

The details of the organism can be selected from this interface as shown in Fig. 4.10; click on the **Accept** button if all entries are correct.

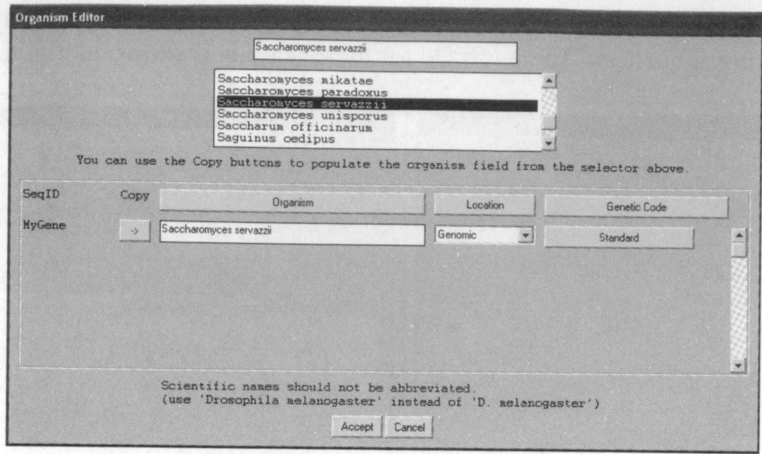

Fig. 4.10 Organism editor showing the details of the organism

Once all details have been filled in, the interface will look like Fig. 4.11.

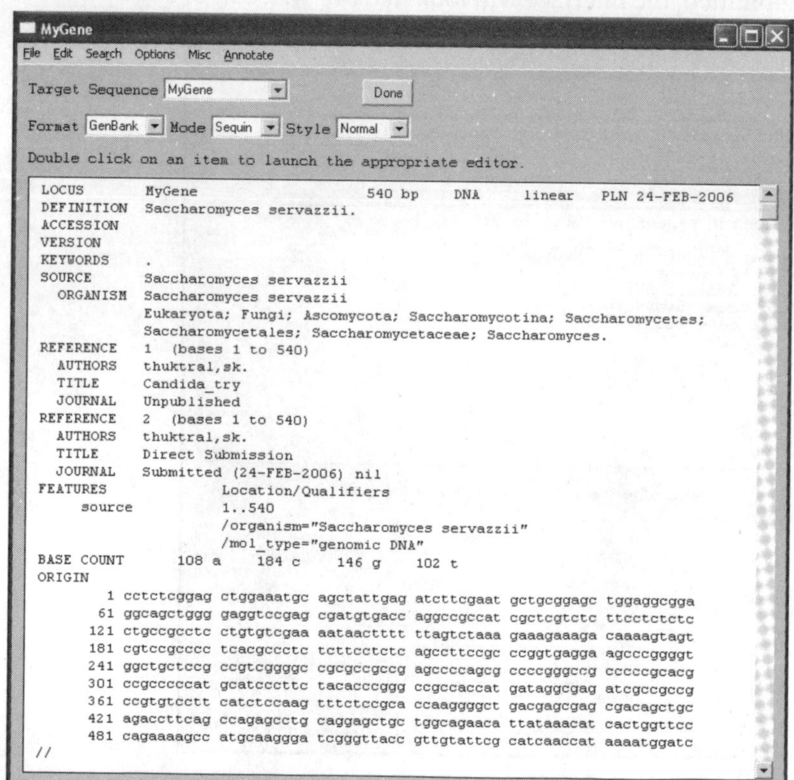

Fig. 4.11 The view page interface

On the basis of the information provided in your DNA sequence files, any coding regions will be automatically identified and annotated.

4.1.3 Webin for EMBL (http://www.ebi.ac.uk/embl/Submission/webin.html)

Webin is a service offered by the European Bioinformatics Institute (EBI) of the European Molecular Biology Laboratory (EMBL). It was originally developed by Kruszewska and launched in June 1997. Webin is the preferred mode of data submission to the EMBL database that uses the internet. It recommends the usage of version 4 or above browsers (Netscape 4.5 or Internet Explorer 4). The data inflow comes from two major sources, namely, project groups and individuals. Project groups are sources where the data flow not only happens in bulk but may also be submitted by these groups into collaborations or consortiums.

Once a sequence is submitted to a database, the author may decide to either release it immediately or hold it back till a date that is decided by him. However, upon publication in a journal, the data is not withheld. EMBL also gives the submitter permission to update records at any time.

(a) Submissions to Webin

Submission of sequences to Webin can be made in three ways. These are discussed next.

General submissions Upon submission of a new sequence to Webin, a unique identification number is allotted. This number is used for tracking the submission. This Webin ID has to be used for one submission only, and each time you make a new submission, a new Webin ID is allocated. The Webin ID is especially helpful if your session crashes and you need to recover the submission. The interface of the Webin submission is shown in Fig. 4.12.

> **Note**
>
> The Webin ID is not an accession number. An accession number is sent only after a submission has been processed and reviewed by a curator. Once you have received an accession number, you should cite it in every further communication with EBI.

The red dots in the Webin interface indicate mandatory fields. The following information is needed:

- Submitter contact information
- Release date
- Citations
- Features
- Third party annotation details (where applicable)

Once information is filled in, a summary form can be accessed that will allow the submitter to add features and be able to view the EMBL flat file. From this flat file summary you can make additions, deletions, and corrections by using the hyperlinks within the flat file summary.

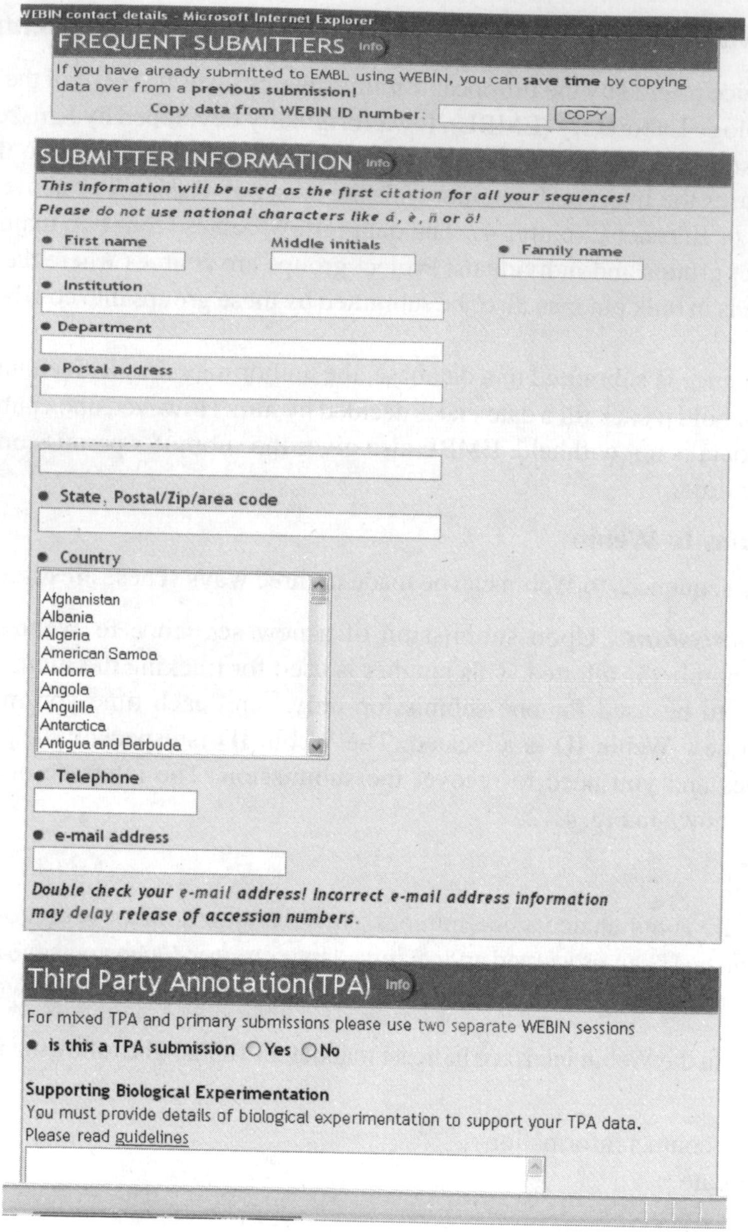

WEBIN contact details - Microsoft Internet Explorer

FREQUENT SUBMITTERS Info

If you have already submitted to EMBL using WEBIN, you can **save time** by copying data over from a **previous submission!**

Copy data from WEBIN ID number: COPY

SUBMITTER INFORMATION Info

This information will be used as the first citation for all your sequences!

Please do not use national characters like á, è, ñ or ö!

- First name
- Middle initials
- Family name
- Institution
- Department
- Postal address
- State, Postal/Zip/area code
- Country

Afghanistan
Albania
Algeria
American Samoa
Andorra
Angola
Anguilla
Antarctica
Antigua and Barbuda

- Telephone
- e-mail address

Double check your e-mail address! Incorrect e-mail address information may delay release of accession numbers.

Third Party Annotation(TPA) Info

For mixed TPA and primary submissions please use two separate WEBIN sessions

- is this a TPA submission ○ Yes ○ No

Supporting Biological Experimentation

You must provide details of biological experimentation to support your TPA data.
Please read guidelines

Fig. 4.12 Interface of the Webin submission

Bulk submissions This mode of submission is to be used only when an individual needs to submit 25 or more related sequences (for instance, the same gene sequenced in a large number of different organisms). In case the submission requires you to use more than three templates for any one dataset, it is advised to contact EBI directly at datasubs@ebi.ac.uk. This mode of submission affords greater efficiency and saves on time. The bulk method of submission meets the efficiency and time-saving criteria because after the initial submission it asks the submitter to only define the differences between each of the sequences to be submitted in bulk. Upon the completion of such a submission, a curator reviews the data profile and advises on the rest of the submission.

Third party annotation submissions Not all the data submitted to a database is unique or it may not be being submitted for the first time. Third party annotation (TPA) submissions refer to those sequences that may not be completely sequenced by a submitter. These submissions may either complete sequences already present in the database or correct/enhance annotation of existing entries. Such submissions are recorded and linked back to the documents that contain data (usually in a peer reviewed journal).

TPA submissions that are accepted by EMBL may pertain to re-annotation of sequences present in the primary databases of DDBJ/EMBL/GenBank. EMBL also accepts TPA submissions that help in filling the gaps in HTG sequences (DNA sequences that are generated by the high-throughput sequencing and are 'unfinished'). These sequences are available to the scientific community for homology searches. EBI also accepts TPA submissions based on trace sequences from Ensembl/NCBI trace archive available at http://trace.ensembl.org/

While EBI accepts data from the consensus of sequences already present in the database, it does not accept consensus sequences from multiple organisms.

> **Note**
>
> Unlike the general and bulk submissions, the release date for TPA cannot be added. TPA submissions are not released until publication.

(b) Webin features and flat file summary

Webin provides a prototype that contains all the features which can be used to describe a sequence. The options that are provided are the following: continuous, segmented protein coding genes (CDSs), structural RNA, and annotation by adding special features such as signal features and/or additional miscellaneous features. It is possible to add as many different features (e.g., a promoter, a CDS, and a poly-A signal) or as many of the same feature (e.g. several independent CDS features) as required. An example of annotation is shown in Fig. 4.13. It is necessary to provide source and destination locations for all features.

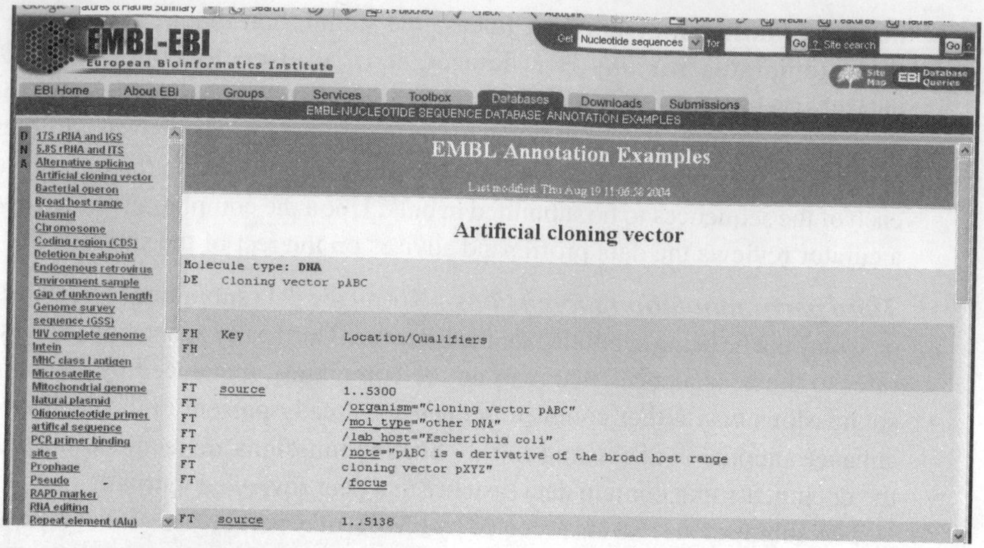

Fig. 4.13 EMBL annotation example

A complete description of all the features of a sequence is mandatory. For example, if the sequence submitted is incomplete then even details such as incomplete at which end (5′ or the 3′ end) must be captured. In case the sequence submitted is incomplete at the 5` end, the submitter also needs to submit the reading frame.

The flat file allows the submitter to review and correct submissions. A submitter can use the hypertext links provided within the flat file to

- correct submitter information,
- correct existing citations,
- add additional citations,
- correct source information,
- correct sequence information,
- correct existing features,
- add additional features, and
- delete existing features.

4.1.4 Sakura for DDBJ (http://sakura.ddbj.nig.ac.jp/)

As already explained in the previous section, DDBJ, EMBL, and GenBank have built a consortium of international nucleotide database that collects the nucleotide sequences experimentally determined and constructs the database. The database also includes the data from Japan Patent Office (JPO), European Patent Office (EPO), and the United States Patent and Trademark Office (USPTO). Sakura works on Internet Explorer V5.5, V6.0 (Windows), Netscape V4.7 (Macintosh, Linux), V6.2, V7.X (Windows, Macintosh, Linux), and Mozilla V1.7.2 (Linux).

DDBJ originally developed two internet interface-oriented systems, Sakura and Yamato II; Sakura is used for data submission, whereas Yamato II is utilized in data annotation and management. The mass scale submission system consists of four separate parts: (i) the web data submission system, (ii) large-scale data submission system/off-line (installed at the submitter side), (iii) data submission management system, and (iv) data storing system. An overview of data submission to and access at DDBJ is given in Fig. 4.14.

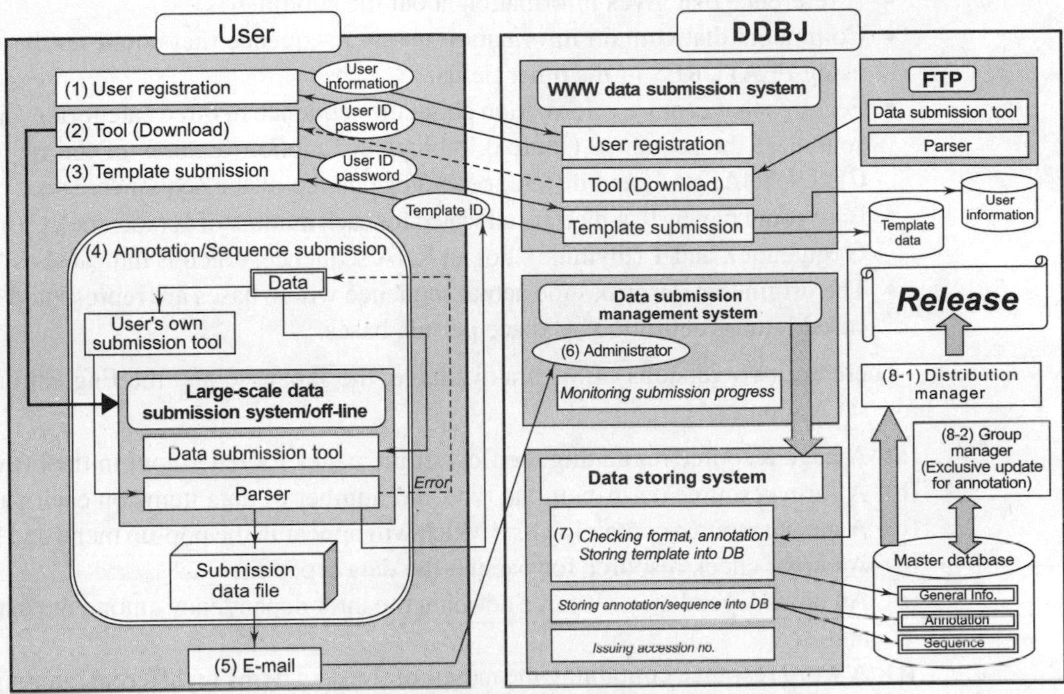

Fig. 4.14 An overview of data submission and access at DDBJ

Two types of files are used for submission: one is used for annotation and the other for recording sequences. The file for annotation is in tabular format that can be handled by word processors, spreadsheets, and database management systems. Nucleotide sequences are submitted in the FASTA file format. Each independent sequence submitted (entry submitted) to DDBJ is processed and made public in the DDBJ format for distribution. This flat file includes the sequence and the information on submitters, references, source organisms, and feature information, etc.

To recapitulate, each DDBJ flat file contains information about the following:

- A locus name, which is a unique ID for the entry in the database. In DDBJ, the locus name assigned is the same as the accession number.
- A definition that briefly describes the information about the sequence and contains some information about the organism name, its scientific name, gene name, product name, and the complete CDS.

- A unique accession number issued to the data submitter by each of the three databases.
- The keywords that have information about the category of the data (EST, HTG, etc), gene names, product names, information about experimental method, etc.
- The organism that gives the scientific name of the source organism and an organelle type of the sequence, that is, an organellar.
- A reference that gives information about the submitter(s).
- Comments that contain information about a sequence that could not be captured using FEATURES or the other fields.
- Features that capture information about the sequence in three categories: biological source of the sequence (source), biological function features of the region (e.g., CDS, rRNA, etc.), and difference and/or change of the sequence data.
- Base count having 9, which are allocated for each number of A (adenine), C (cytosine), G (guanine), and T (thymine). For an RNA sequence, uracil is indicated as 'T'.
- The origin that describes the actual sequence where bases are represented as lower cased letters, delimited by space per ten bases.

There are two versions of Sakura available, the Japanese and the English one. Both have six resources:

(a) A page resource regulating the flow of the pages for data input in the browser;
(b) A form resource describing the type and number of data items on each page;
(c) A menu resource containing lists which will appear in the pop-up menu and list-box;
(d) An error check resource for parsing the data typed in;
(e) An item-dependency resource defining the inter dependency among the data items; and
(f) A word resource containing the names of the data items in different languages. The interface of Sakura is shown in Fig. 4.15.

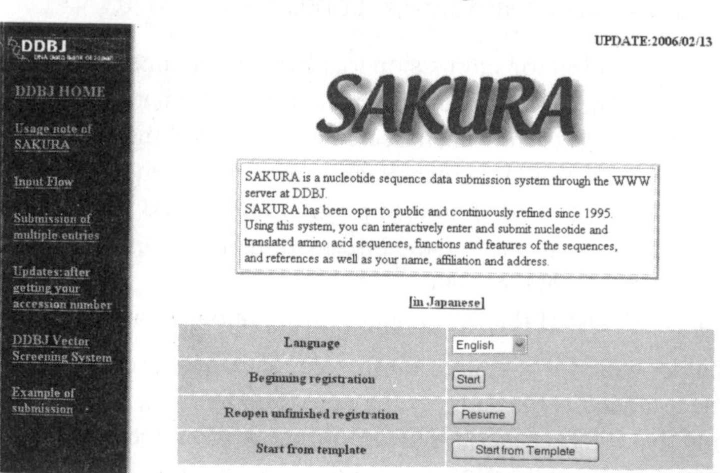

Fig. 4.15 Sakura interface

The maximum sequence length that can be handled by Sakura is 500,000 bp. If a sequence longer than 500,000 bp needs to be submitted, the mass submission system needs to be used. The mass submission system/off-line is publicly available by downloading the program from an FTP server at DDBJ.

4.2 Protein Submission Tools

Protein submission follows the same pattern and process as the nucleotide sequence submission.

4.2.1 Spin (Sequence Submission) for Swiss-Prot

Spin is a new web-based submission tool for the Swiss-Prot protein database. It is designed to allow scientists to submit directly sequenced proteins with functional annotations. Once the UniProt knowledgebase curators approve a sequence, a unique accession number is assigned to it, which is to be quoted and included in any published article.

The use of a browser version 4 or above (e.g., Netscape 4.5 or Internet Explorer 4) is strongly recommended better results. In case the browser crashes during submission to Spin, it is recommended that the browser's javascript be turned off.

Note

For Spin to operate optimally, there are certain parameters that need to be set on the browser. These are

Javascript—set to 'enabled' (it is assumed that you will be using Internet Explorer or Netscape versions 4.0 and above). If javascript is not enabled, Spin will work at a slower pace.

Cache—set to 'check every time'. This is strongly recommended.

Cookies—set to 'accept cookies'. This is strongly recommended and will help you track your submission.

Style sheets—set to 'enable style sheets'.

A new user can register by clicking on the Register button and fill in the submitter form. Once registered, the interface that he gets is shown in Fig. 4.16. Please note that there is a green tick that appears next to **Submitter**.

Sequence submission can be done in various modes as follows:

- Several proteins from one organism; however, a separate entry form for each protein/peptide is to be used.
- Same protein from several organisms; again a separate entry form for each organism is to be used.
- Several fragments from one protein/peptide. Here only one entry form is to be used for all fragments.

- Several fragments from different proteins. For such a submission separate entry forms are to be used for each fragment.

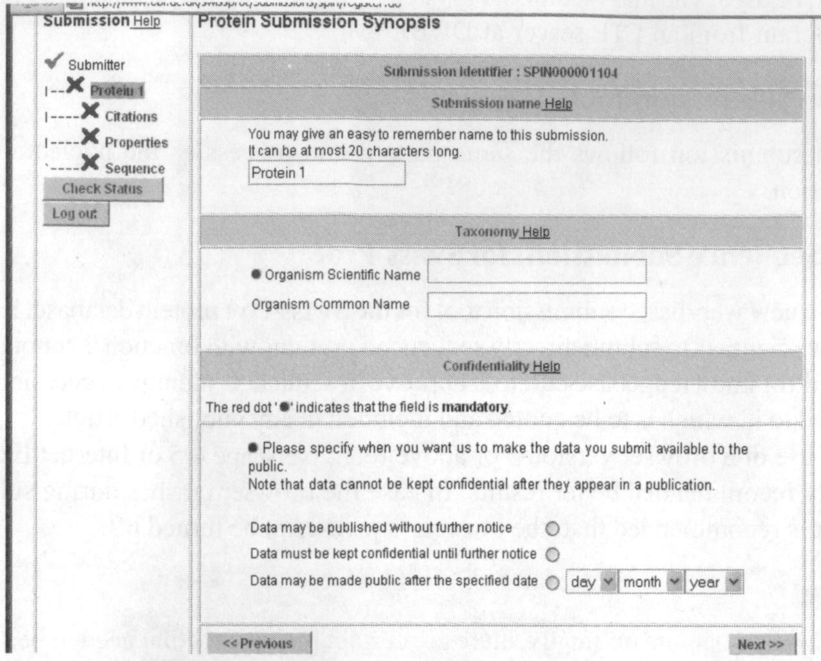

Fig. 4.16 The interface after registration

> **Note**
>
> Sometimes a security virtual firewall is used in certain environments. Therefore, you may not be able to access your session, since another computer may be accessing the pages on the internet on your behalf. To circumvent this, you need to deactivate the proxy server and change settings to **Direct connection to the internet** in your browser configuration.

(a) Submission synopsis

The synopsis for submission to Spin has three parameters: submission name, organism, and confidentiality.

Submission name It is the name given to the submission to make it easy to identify and recall; it is limited to 20 characters.

Organism The scientific name of the organism from which the sequence was obtained has to be specified along with any common name, e.g., *Saccharomyces cerevisiae* (baker's yeast).

Confidentiality The sequence data can be either released immediately (after being checked by a curator) or kept confidential initially. However, once the data is published in a journal, it cannot be kept confidential. Spin, however, gives you three options for choosing the release date, namely, to be published without further notice, to be kept confidential until publication, and to be made public after a specified date. Sequences have to be entered and made part of the database only within a week of them being assigned an accession number.

(b) Citations

For practical reasons it is always better to submit data to Spin before sending it for publication or at least before you receive galley proofs. The previous interface will lead you to the **SPIN Citation Form** (see Fig. 4.17).

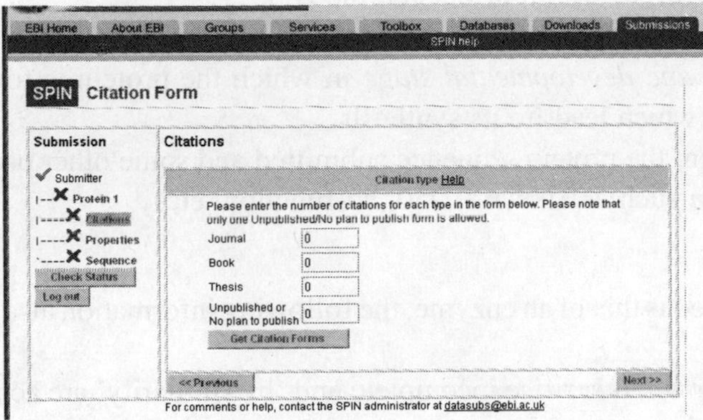

EBI Home | About EBI | Groups | Services | Toolbox | Databases | Downloads | Submissions

SPIN help

SPIN Citation Form

Submission
- Submitter
- Protein 1
- Citations
- Properties
- Sequence

Check Status
Log out

Citations

Citation type Help

Please enter the number of citations for each type in the form below. Please note that only one Unpublished/No plan to publish form is allowed.

Journal 0
Book 0
Thesis 0
Unpublished or No plan to publish 0

Get Citation Forms

<< Previous   Next >>

For comments or help, contact the SPIN administrator at datasubs@ebi.ac.uk

Fig. 4.17 Spin citation form

Clicking on **Get Citation Forms** will open another page that looks like Fig. 4.18.

Journal

The red dot '●' indicates that the field is **mandatory**.

● Citation Details — Year, Volume, Pages ... to ...

● Title

● Author — Initial(s), Surname

Add 0 more names

● Journal Name - SELECT ONE -

Other Journal Name

Source

Strain/ Cultivar / Variety

Fig. 4.18 Page obtained upon clicking on get citation forms

The submitter is required to enter the number of citation forms and provide the publication details on this page. In case you do not wish/plan to publish your data, a list of author names and a title for the submission need to be provided. The title selected should be similar to what would have been used if you published the sequence in a journal.

(c) Properties of the protein

For the protein sequence to be submitted, certain details need to be submitted. These include the following:

- *Name of the protein* If it is unknown, then experimental details need to be provided.
- *Functions* that the protein performs.
- *Mass spectrometry details* about the method used, accuracy in daltons, and quaternary structure.
- *Subcellular localization* within or outside the cell.
- *Posttranslational modifications* which may have occurred.
- *Tissue specificity and developmental stage* in which the protein is found.
- *Induction*, if any, which leads to its synthesis.
- *Similarity* between the protein sequence submitted and some other proteins.
- *Other information* such as known LD50, co-function, etc.

(d) Enzyme data

If the sequence submitted is that of an enzyme, the following information also needs to be submitted:

- *EC number*, if known; otherwise incomplete and 'by similarity' are acceptable.
- *Catalytic activity* in terms of the reactions catalysed by the enzyme.
- *Cofactor and nature of prosthetic groups* involved.
- *Enzyme regulation* or a list of activators and inhibitors that affect the enzyme activity.
- *Pathway* that the enzyme may function within.

(e) Sequence and features

Spin allows the submission of three kinds of sequences as listed below.

- *Submission of a complete sequence* Choose the option **This sequence is complete** if you have to submit a complete sequence..
- *Submission of a fragment* When submitting a fragment, choose the option **This sequence is not complete**. The fragment's location needs to be specified, for example, if it is from the N-terminal, internal, or C-terminal.
- *Submission of several fragments* When submitting several fragment, choose the **This sequence is not complete** option. Specify the number of fragments that have been sequenced by using the **ADD** button and the provided box. The fragments' locations need to be specified, i.e., whether a fragment is from the N-terminal, internal, or C-terminal. In addition, the order of the fragments—known, unknown, or uncertain—has to be specified. Specify the C-terminal ends of each fragment (in numbers) in the box provided.

In all the three cases the length of the sequence too needs to be specified. The submitter will have to either manually type in the sequence in the box provided, using single-letter amino acid codes, or load the sequence from a file. The submitter is encouraged to provide any additional detail as listed below: *Description from the options provided:* some of the descriptions (such as signal) apply only to sequences for which the nucleotide sequence translation is also available.

- *Comments* so that more specific details about the feature are included.
- *Span*, which identifies the start point of a feature and its 'end point'.
- *Residues* which are involved in imparting a particular feature.
- *Evidence*, which defines the basis of the feature; either experimental or by similarity.
- *Other*, which captures the details of the sources of the data in the free text description.

Additionally, below given features can also be captured by Spin:

- Posttranslational modification (MOD_RES)
- Disulfide bond (DISULFID)
- Glycosylation site (CARBOHYD)
- Involved in enzymatic activity (ACT_SITE)
- Uncertainty in the sequence (UNSURE)
- Propeptide (PROPEP)
- Metal ion binding site (METAL)
- Chemical group binding site (BINDING)
- Released active peptide (PEPTIDE)
- Lipid moiety (LIPID)
- Selenocysteine (SE_CYS)
- Signal sequence (SIGNAL)
- Transit peptide (TRANSIT)
- DNA-binding region (DNA_BIND)
- Nucleotide phosphate binding (NP_BIND)
- Zinc finger (ZN_FING)
- Transmembrane region (TRANSMEM)
- Internal sequence repetition (REPEAT)
- Calcium-binding region (CA_BIND)
- Sequence variant (VARIANT)

4.2.2 AutoDep Structure Submission Tool for PDB (www.ebi.ac.uk/msd-srv/autodep4/index2.jsp)

PDB files are generally perceived as no more than a long and rather boring list of xyz co-ordinates and some other experimental data. However, this data is presented in a consistent format and can be easily handled by computer programs.

AutoDep is the software system used for online automatic submission of protein and macromolecular structural data into the Brookhaven Protein Data Bank. It's programs are written in various programming languages, such as PERL, JAVAScript, FORTRAN, and C++. These make use of standard web tools such as HTML and CGI. Like other submission tools, the purpose of AutoDep is to simplify the data submission procedure by taking advantage of internet tools. First it will ask you to enter a password of your choice and take you to a page to fill in the details of author as shown in Fig. 4.19.

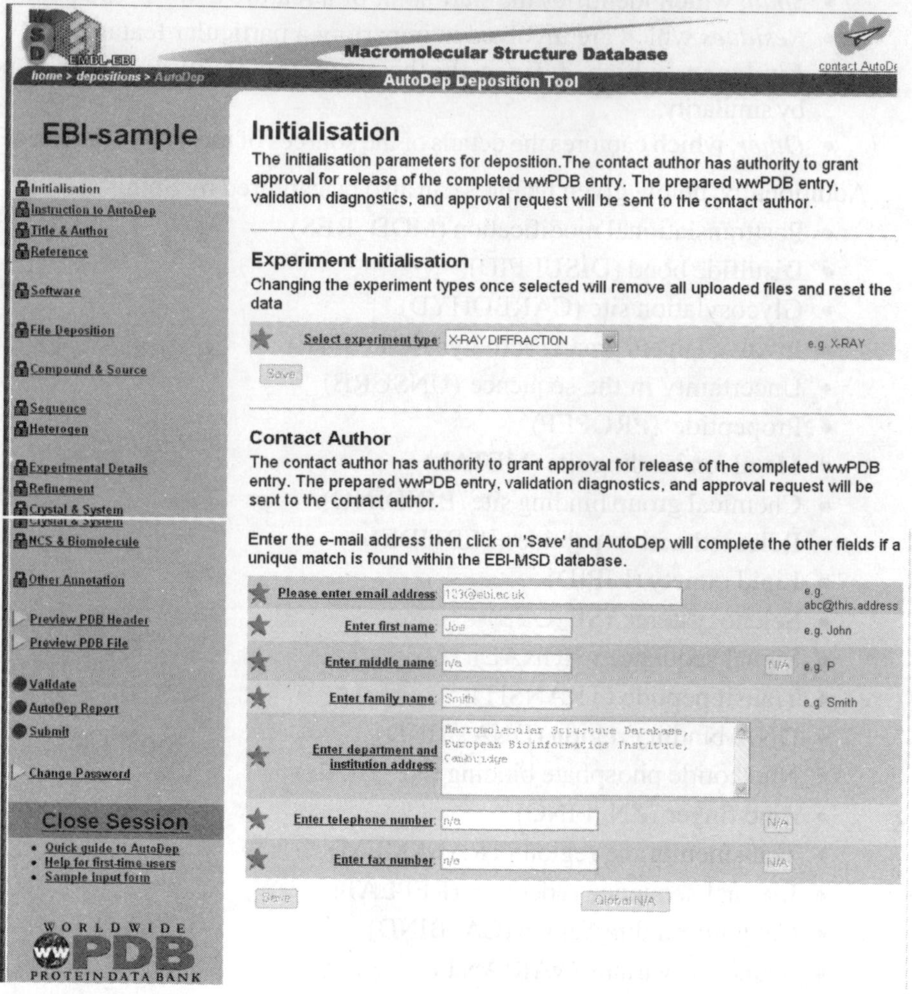

Fig. 4.19

Submitting crystallographic data (Fig. 4.20) to PDB is a rather time consuming series of steps, wherein red stars turn into green stars, turn into red arrows, and finally into green dots. PDB processing programs are run on each new entry. Once the process of verification is complete, an output (in the form of WHAT_CHECK output) is returned to the submitter, and summarized in the associated report file. Verification by itself is a meticulous task and includes checks of stereochemistry, bonded/non-bonded interactions, crystallographic information, noncrystallographic transformations, primary sequence data, secondary structure, and heterogen groups.

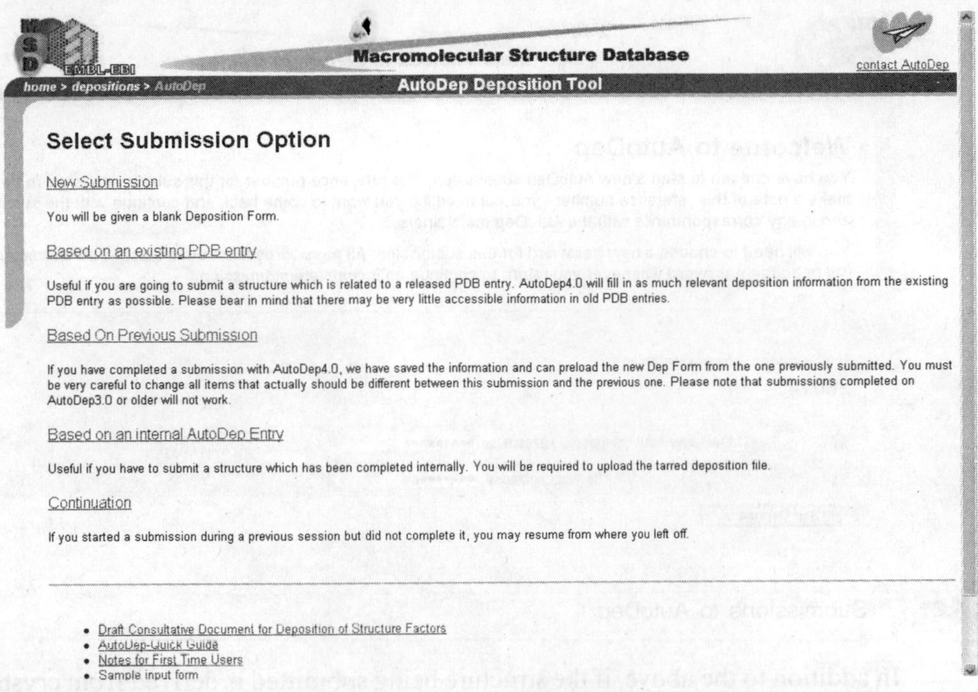

Fig. 4.20 Crystallographic data submission form

The Layered Release Protocol for deposition of structural data to the PDB began in 1998 using AutoDep (version 2.1). Layered Release was initiated in order to speed up the availability of new entries in the PDB. Computer programs associated with AutoDep 2.1 now give feedback to the depositor ensure enforcement of standard nomenclature and representation, and automatically return a set of diagnostics about the structure. The depositor in turn re-examines the data and its structure and approves it. Once approved, the entry is given an ID code. The main advantage of the layered release is that information can be released as soon as it is deposited. Following this LAYER-1 release, the PDB staff process the entry. The entries are checked and normalized . This entry, after author approval, is equivalent to the traditional PDB entry and is referred to as LAYER-2 (see Fig. 4.21).

All submitters to AutoDep must come prepared with the following:

- HARVEST files from CCP4 or CNS.

- A file containing the atomic coordinates in PDB format.
- Sequence database reference for the biological molecule(s).
- A complete description of the heterogens.
- A copy of their manuscript, which describes this structure.

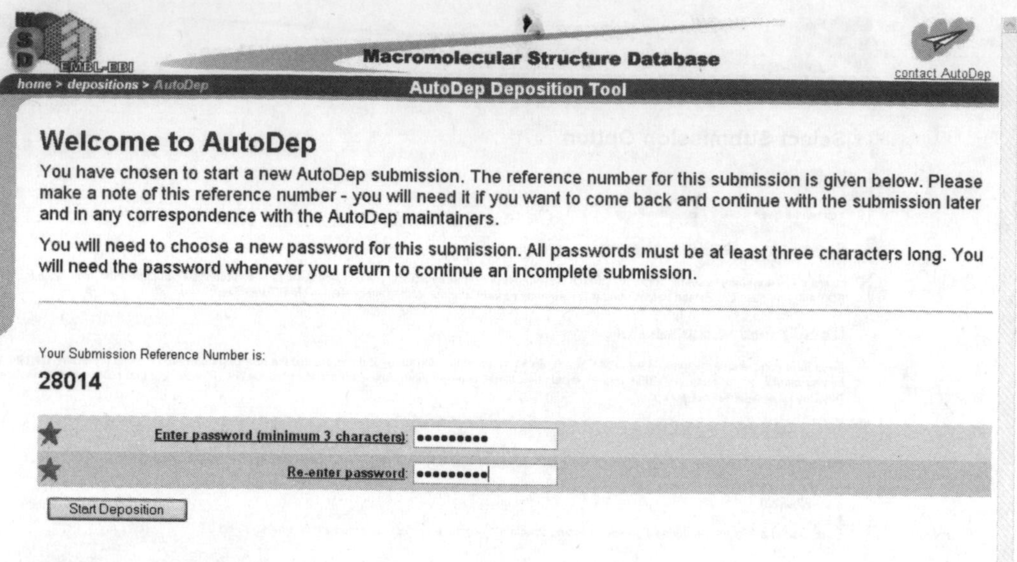

Fig. 4.21 Submissions to AutoDep

In addition to the above, if the structure being submitted is derived from crystallographic studies, most of this data is picked up automatically when the HARVEST files are uploaded. The submission also should contain details about the following:

- Unit cell data
- Data collection details
- Structure refinement statistics output from refinement program
- Crystallographic and biologically significant non-crystallographic symmetry matrices

In case NMR spectroscopic data is being submitted, the deposition file should contain the constraints used to carry out the final refinement of the structure. In addition, such data should carry the following in details:

- The file containing statistics for the structures and constraints.
- The file containing all chemical shift assignments (preferably in NMR-STAR format). The creation of an NMR-STAR chemical shift assignment file, using the BMRB protein or nucleic acid atom table builder is depicted in Fig. 4.22.

http://www.bmrb.wisc.edu - BMRB NMR-STAR Atom Table Generator for Amino Acid Chemical Shift Ass - Microsoft Interne...

BMRB

BMRB NMR-STAR Atom Table Generator for Amino Acid Chemical Shift Assignments

This form is used to generate tables of atoms for a given protein. The tables are produced in NMR-STAR format and are suitable for including with submissions to BMRB. The NMR-STAR format contains only ASCII characters and can be easily edited. The output will also include the sequence as entered and the sequence in NMR-STAR format.

Presently, the Generator can produce tables for only the common 20 amino acids. If another monomer occurs in the protein, please place an "X" at that point in the sequence.

Please select the NMR-STAR format you wish to use, by default 3.0 will be used:

⊙ NMR-STAR 3.0 format ○ NMR-STAR 2.1 format

By default, the Generator produces tables containing every atom in every residue in the protein it's given. Tables containing subsets of the possible atoms can be produced by making choices that limit the output to specific atoms or specific residues. For more information on producing these subsets, see Help.

Enter the residue sequence here, using one-letter codes.

1 5 10 15 20 25 30 35 40 45 50 55 60

Select which atoms to include:

⊙ Atoms Observed in Routine NMR Studies **(Default Condition)**
○ All Atoms (incl. individual methyl protons)

List only these nuclei....

H	☐ All Protons	☐ Alpha Protons	☐ Backbone Amide Protons	☐ Side Chain Protons
C	☐ All Carbons	☐ Alpha Carbons	☐ Carbonyl Carbons	☐ Side Chain Carbons
N	☐ All Nitrogens	☐ Backbone Amide Nitrogens		☐ Side Chain Nitrogens

☐ Backbone Atoms (CA, C, N, HA, H)
☐ Heavy Atoms (carbon and nitrogen)
☐ Side Chain Atoms (any atom not listed in "Backbone Atoms")

List only these residues...

☐ Alanine	☐ Arginine	☐ Asparagine	☐ Aspartic Acid
☐ Cysteine	☐ Glutamine	☐ Glutamic Acid	☐ Glycine
☐ Histidine	☐ Isoleucine	☐ Leucine	☐ Lysine
☐ Methionine	☐ Phenylalanine	☐ Proline	☐ Serine
☐ Threonine	☐ Tryptophan	☐ Tyrosine	☐ Valine

☐ All residues *except* those selected in the table
☑ Include description of ambiguity codes

Create Atom Table | Clear Form | NMR-STAR 3.0 example | NMR-STAR 2.1 example

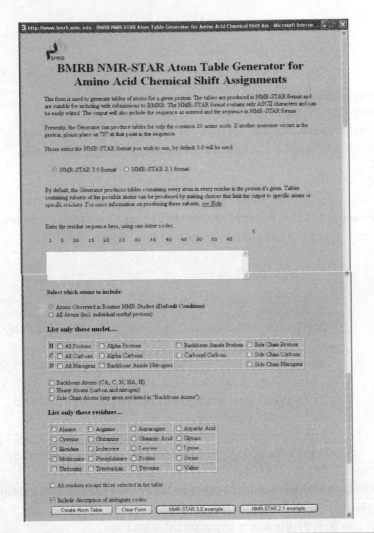

Fig. 4.22 Creation of an NMR-STAR chemical shift assignment file

An example of an NMR-Star 3.0 output file as follows.

Example of output from the Table Generator
NMR-STAR 3.0 format
The initial sequence was: AVR

```
#
# The original sequence entered was:
# AVR
#
# Expressed in NMR-STAR 3.0, this sequence is:
_Entity.Polymer_seq_one_letter_code
```

```
;
AVR
;
loop_
    _Entity_comp_index.ID
    _Entity_comp_index.Auth_seq_ID
    _Entity_comp_index.Comp_ID
1   @     ALA
2   @     VAL
3   @     ARG
  stop_
  ######################################################################
  #     Chemical Shift Ambiguity Code Definitions                       #
  #                                                                     #
  #   Codes          Definition                                         #
  #                                                                     #
  #     1              Unique                                           #
  #    2           Ambiguity of geminal atoms or geminal methyl #
  #                       proton groups                                 #
  #    3           Aromatic atoms on opposite sides of the ring #
  #                    (e.g. Tyr HE1 and HE2 protons)                   #
  #    4           Intraresidue ambiguities (e.g. Lys HG and    #
  #                      HD protons)                                    #
  #    5           Interresidue ambiguities (Lys 12 vs. Lys 27) #
  #    9           Ambiguous, specific ambiguity not defined    #
  #                                                                     #
  ######################################################################
  # INSTRUCTIONS
  # 1) Replace the @ - signs with appropriate values.
  # 2) Text comments concerning the assignments can be
     supplied in the full deposition.
  # 3) Feel free to add or delete rows to the table as needed.
  #   The row numbers (_Atom_chem_shift.ID values)
     will be re-assigned to sequential values by BMRB
  #
  # The atom table chosen for this sequence is:
  loop_
     _Atom_chem_shift.ID
     _Atom_chem_shift.Auth_seq_ID
     _Atom_chem_shift.Auth_comp_ID
     _Atom_chem_shift.Auth_atom_ID
```

```
      _Atom_chem_shift.Atom_type
      _Atom_chem_shift.Chem_shift_val
      _Atom_chem_shift.Chem_shift_val_err
      _Atom_chem_shift.Chem_shift_ambiguity_code
      _Atom_chem_shift.Chem_shift_occupancy
   #
   # Auth  Auth  Auth       Chem   C.S.   C.S.   Chem
   # seq   comp  atom  Atom shift  val    ambig  shift
   #ID     ID    ID    ID   type   val    err    code   occup
   #--------------------------------------------------
    #
    1 1    ALA   H     H    @      @      @      @
    2 1    ALA   HA    H    @      @      @      @
    3 1    ALA   HB    H    @      @      @      @
    4 1    ALA   C     C    @      @      @      @
    5 1    ALA   CA    C    @      @      @      @
    39      3    ARG   NE   N      @      @      @      @
    40      3    ARG   NH1  N      @      @      @      @
    41      3    ARG   NH2  N      @      @      @      @
stop_
# The following loop is used to define sets of Atom-shift assignment IDs
that
 represent related ambiguous assignments taken from the above list of
 assigned chemical shifts.  Each element in the set should be separated
by a
 comma, as shown in the example below, and is the assignment ID for a
chemical
 shift assignment that has been given as ambiguity code of 4 or 5.
Each set
 indicates that the observed chemical shifts are related to the defined
 atoms, but have not been assigned uniquely to a specific atom in the
set.
loop_
  _Atom_chem_shift.Atom_shift_assign_ID_ambiguity
#
#   Sets of Atom-shift Assignment Ambiguities
                   #
#   ------------------------------------------
# Example:   5,4,7
#
                @
stop_
```

4.3 tbl2asn (Command Line Tool for GenBank) http://www.ncbi.nlm.nih.gov/Genbank/tbl2asn2.html)

GenBank offers special batch procedures for large-scale sequencing groups to facilitate data submission. It is called tbl2asn. tbl2asn is a command-line program for submission to GenBank at NCBI. tbl2asn generates **.sqn** files for submission to GenBank. Its advantage is that additional manual editing is not required before submission.

tbl2asn is available by anonymous FTP (ftp://ftp.ncbi.nih.gov/toolbox/ncbi_tools/converters/by_program/tbl2asn/), as shown in Fig. 4.23.

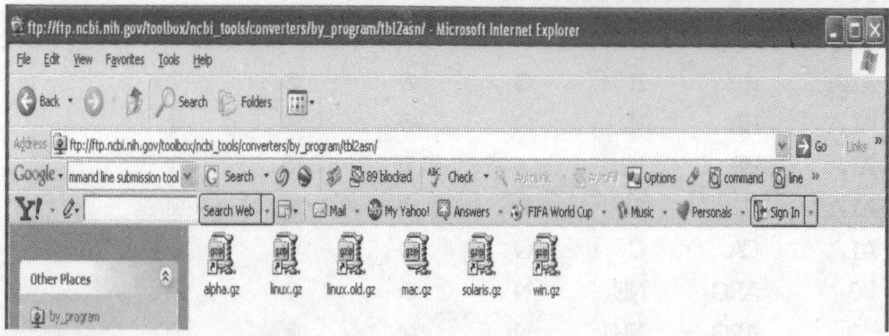

Fig. 4.23 tbl2asn, available by anonymous FTP

These files are available as per the platform that you are operating on (Linux, Solaris, Windows). You can download these files on an uncompress thema, and rename them as tbl2asn. Its interface looks like any other command line program (see Fig. 4.24).

Fig. 4.24 Interface for tbl2asn

The input file can be submitted using six different files: a template file containing text ASN.1 (extension **.sbt**), nucleotide sequence data in FASTA format (extension **.fsa**), feature table (extension **.tbl**), protein sequence (extension**.pep**), source table (extension **.src**), and quality scores (extension **.qvl**).

Table 4.1 shows the command line arguments that can be used while submitting files to tbl2asn.

Table 4.1 Command line arguments for submitting files to tbl2asn

	Command line arguments
-p	Path to the directory. If the files are in the current directory -p. should be used.
-r	Path for the resulting **.sqn** submission file (if the -r argument is not used, the **.sqn** files will be saved in the source directory).
-t	Specifies the template file (**.sbt**). If the **.sbt** file is in a different directory, the full path must be specified.
-s	Instructs tbl2asn to read multiple FASTA components in one file as a set of unrelated sequences. This command compiles multiple submissions into a single file. (1000 sequences per file is the maximum.)
-c	Instructs tbl2asn to annotate the longest open reading frame (ORF) if a **.tbl** file is not provided.
-m	Allows alternative start codons to be used in ORF searches.
-v	Validates the data records. The output is saved to files with a **.val** extension.
-b	Generates GenBank flat files with a **.gbf** extension.
-I	Creates single submission from indicated **.fsa** file in a directory of multiple **.fsa** files.
-j	Allows the addition of source qualifiers, which will be the same for each submission. Example: -j "[organism=Saccharomyces cerevisiae] [strain=S288C]".
-o	Creates a single submission from multiple fasta files.
-l	Reads one or more FASTA+GAP alignments to create one or more phylogenetic sets.
-y	Adds a COMMENT to each submission. Example: -y "Contigs larger than 2 kb have been annotated, representing approx. 87% of the total genome."
-Y	Like -y, but adds a COMMENT to each submission from a file.

Conclusion

Science is a cooperative venture, and therefore the fruits of its labour must also be treated the same way. Sometimes the only rationale for decoding a protein from its nucleotide sequence, all the way to its structure, is to enable others to use it and further the cause of science. There are cases in which sequences are known but structures are not and vice versa. The easiest way to collaborate is to submit/deposit the data to/in a database/bank so that it may be accessed by others. Bioinformaticians have set standards for such submissions.

EXERCISES

Exercise 4.1

1. Which are the two tools available for data submission in GenBank?
2. What is the basis for selecting a tool for submission? What condition will you check for while selecting any of these tools?
3. Which types of submissions are not accepted by GenBank?
4. Prepare a dummy submission report for an already existing DNA sequence as per the format given below.
   (a) Contact information: name, address, phone number, fax number, and e-mail address of the submitter
   (b) Release date information
   (c) Reference information:
      (i) Sequence authors: list of authors credited with the sequence
      (ii) Citation(s) associated with the sequence
   (d) Source description
      (i) Scientific name (Genus species) of the source organism. If the genus and species names are not known, provide a general description of the organism (e.g., uncultured bacterium). For synthetic sequences, provide a specific name (e.g., cloning vector pRB223).
      (ii) Unique source modifiers (e.g., clone, strain, isolate, cultivar, specimen voucher name). These are especially important if the scientific name is not known.
      (iii) Identification of the organelle from which any non-nuclear nucleotide sequence originates (e.g., chloroplast, mitochondrion).
   (e) Input DNA sequence
      (i) A contiguous nucleotide sequence of at least 50 base pairs, sequenced by the submitter(s)
      (ii) Type of molecule sequenced (e.g., genomic DNA, genomic RNA, mRNA)
      (iii) Description of the sequence
5. What are the important points one should remember while entering a DNA sequence?

Exercise 4.2

(a) What is Sequin?
(b) Which is its latest version?
(c) What is new in its latest version?
(d) Is it a downloadable or web-based tool?
(e) What are the advantages of Sequin?
(f) What are the two modes in Sequin?
(g) What is SequinMacroSend?

Exercise 4.3 Prepare dummy reports on the annotation of already existing genomes.

(a) Prokaryotic genome
- Gene features
- locus_tag
- CDS (coding region) features
- protein_id

(b) Eukaryotic genome
- locus_tag and protein_id
- Gene features
- CDS (coding region) features
- Partial coding regions in incomplete genomes
- Gene fragments
- Split genes on two contigs
- Ribosomal RNA, tRNA and other RNA features
- mRNA features
- Alternatively spliced genes

Exercise 4.4

(a) What are the different kinds of submissions supported by Webin?
(b) What is the URL for Webin submission?
(c) How is the sequence for vector contamination checked?
(d) What is the maximum length of a sequence that can be submitted through Webin?
(e) What should not be submitted through Webin?
(f) What is Spin?
(g) List one similarity and one difference between Webin and Spin.

Exercise 4.5

(a) In which server is the Sakura submission tool available?
(b) In which cases can Sakura accept a modification request before assigning an accession number?
(c) What does the data fields specify when
   (i) marked with M
   (ii) marked with R
   (iii) not marked
(d) What is the Mass Submission System?
(e) What are the two tools that the DDBJ Vector Screening System uses?
(f) What are the differences between Feature Key and Qualifier Key?

Exercise 4.6

(a) What is the AutoDep tool used for?
(b) What is Submission Reference Number?
(c) What are the five mandatory sections that need to be filled up before uploading a file?

(d) What are insertion codes?
(e) COMPLETE a dummy AUTODEP DEPOSITION FORM of an already existing submission in the following format.
  (i) Atom names must be given in two parts: the atom name and its alternate location indicator. A question mark (?) must be used as a placeholder if the alternate location indicator is null.
  (ii) Residues names must be given in four parts: the residue name, chain identifier, sequence number, and insert code, using '?' as a placeholder.

Exercise 4.7

(a) What is tbl2asn?
(b) What are the six types of input data files?
(c) What is the extention of a template file?
(d) What is the extention of fasta defline?
(e) How can you create a **.sqn** file?
(f) What are the minimum requirements of fasta defline?
(g) What are feature tables?
(h) What are the source table and quality scores table formats?

5 Data Analysis Tools

Learning Objectives

- To learn and locate various analysis tools for nucleotide sequences
- To learn and locate various analysis tools for amino acid sequences
- To set the attributes for each tool and process sequence data
- To analyse the result obtained from each tool

Introduction

Waterman describes the current situation with databases in the following way: 'It is probably important to realize from the very beginning that the databases will never completely satisfy a very large percentage of the user community. The range of interest within biology itself suggests the difficulty of constructing a database that will satisfy all the potential demands on it. There is virtually no end to the depth and breadth of desirable information of interest and use to the biological community.' The aim of this chapter is to give a practical introduction to the use of the available tools that use data from various databases. The chapter also presents the bioinformatics resources that are available (locally or through the Internet) and link together theoretical notions with the behaviour of programs and the results they produce.

We will discuss the data analysis tools available for nucleotide and amino acid sequences. For ease of learning, each tool is introduced in a pre-defined format under several sub-headings. It is better to actually log in to the Internet and work on these tools while reading this chapter.

5.1 Tools for Nucleotide Sequence Analysis

Various tools are available for nucleotide sequence analysis. Most of them are available freely on the Web and many are downloadable also. To make this study convenient as well as to maintain consistency, we have mostly picked up chromosome I data from an organism called *Saccharomyces cerevisiae*. Only in tools where the length of the sequence is a constraint, we have picked up genes of smaller sizes from the same organism. When all the tools from this section have been studied along with the guided tasks, the reader will end up doing a mini project on 'data analysis of nucleotide sequences'.

5.1.1 Transeq

Tool for Transeq translates nucleic acid sequences into the corresponding peptide sequences. It can translate in any of the three forward or three reverse sense frames, in all three forward or reverse frames, or in all six frames.

Link http://www.ebi.ac.uk/emboss/transeq/

Institute EMBL-EBI: EMBOSS.

Type of tool Web-based.

OS required Windows.

Algorithm used Transeq translates in all or one of the six frames using the standard or 'alternative genetic codes'.

Tool screenshot Figure 5.1 shows a screenshot of Transeq.

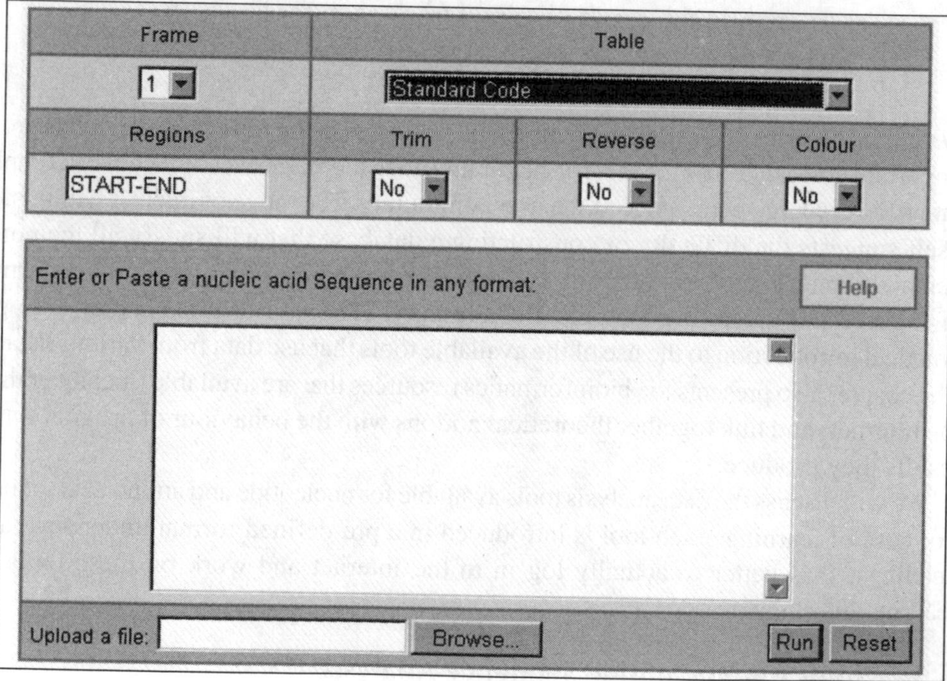

Fig. 5.1 Screenshot of Transeq

Input data The data that are fed into the tool is a nucleotide sequence in any format.

(a) Attributes

This tool has a small number of attributes, described below.

Frame 1, 2, and 3 are the first, second, and third forward reading frames. F represents all forward reading frames. -1, -2, and 3 are the first, second, and third reverse reading frames. R represents all reverse reading frames. 6 represents all six forward and reverse reading frames.

Table There are several codon tables available as mentioned below: the **Standard Code** is the default one.

- The Standard Code
- Vertebrate Mitochondrial
- Yeast Mitochondrial
- Mitochondrial and Mycoplasma/Spiroplasma
- Invertebrate Mitochondrial
- Ciliate, Dasycladacean and Hexamita Nuclear
- Echinoderm Mitochondrial
- Euplotid Nuclear
- Bacterial and Plant Plastid
- Alternative Yeast Nuclear
- Ascidian Mitochondrial
- Flatworm Mitochondrial
- Blepharisma Nuclear
- Chlorophycean Mitochondrial
- Trematode Mitochondrial

Regions This option allows one to specify the regions of the nucleotide sequence to be translated.

Trim When **Trim** is set to **Yes**, it removes the '*' and 'X' (stop and ambiguity) symbols from the end of the translation.

Reverse When **Reverse** is set to **Yes**, one can reverse the sequence.

Colour When set to **Yes**, one can colour the translation.

Upload a file If the input sequence file is large and stored in one's computer, one can upload a file from the computer in formats such GCG, FASTA, EMBL, GenBank, PIR, NBRF, and Phylip. It is preferable to store the files in text mode, as Word stores hidden characters that may give erroneous results.

Result Upon clicking on the **Run** button, a result page with the peptide sequence is obtained.

TASK

Input data The data that are fed in Transeq is as shown below:

```
gi|50593113|ref|NC_001133.5| Saccharomyces cerevisiae chromosome I, complete
chromosome sequence
CCACACCACACCCACACACCCACACACCACACCACACACCACACCACACCCACACACACACATCCTAACACTACC
CTAACACAGCCCTAATCTAACCCTGGCCAACCTGTCTCTCAACTTACCCTCCATTACCCTGCCTCCACTCGTTACCCTGTCC
CATTCAACCATACCACTCCGAACCACCATCCATCCCTCTACTTACTACCACTC
```

Attribute settings The following attributes are selected.

- **Frame** 1
- **Table:** The default code, i.e., **Standard Code** is selected
- **Regions: Start-Stop**
- **Trim: Yes**
- **Reverse: No**
- **Colour: Yes**
- **Upload a file: No**

Note

In case of short sequences, the sequence is selected and copied from the sequence file and then pasted in the sequence box. Some input sequence boxes allow to enter the identifier or accession number. For both of these cases, upload a file is 'No'. Alternatively it is 'Yes' and you can select a sequence file already stored in your machine as the input file.

Result Upon giving these input data and setting these attributes as the results shown in Figs 5.2 and 5.3 are obtained.

Transeq Results	
Frame	1
Translation table	Standard (0)
Regions	START-END
Trim	no
Reverse	no
Color	yes
Transeq output	transeq-20050725-12481912612582.output
SUBMIT ANOTHER	

Fig. 5.2 Result page 1 of Transeq

```
>NC_001133.1_1 Saccharomyces cerevisiae chromosome I, complete
chromosome sequence
PHHTHTPTHHTTHHTTPTHTHPNTTLTQP*SNPGQPVSQLTLHYPASTRYPVPFNHTTPN
HHPSLYLLPLTHRYPPITHIQPTATYPTITLPSTMTYSPYCSSTHHIETLTNDRK*HTRA
YPTTLYHHHMPYSPSLVY*FYVRTRMLQYIPSQTYPTLRFHFTPWPISH*ISTKCTHIIM
HGTCLSGLYPVPFTHNAHHYPHFDIYISFGGPKYCITALNTYVIPLLHHILTTPFIYTYV
NITEKSPQKSPKHKNILLFNNNT*TYWLVVATLSWYH*RKSSSILQFA*TDAISEYFVLT
QAIH*NNMSHHCRNTLYSPSNNTVVAQTHAGAMIQLYLISIPIC*PQYPKSITDASLILY
VTLLIRRDYI*SRRYCDRYVI**DL*RNVK*FYGNITYQRRILKRTLRYCLTSSYHPLSY
C**NTNPSALFLVTVTQKTMPTQKS*YFTCQKMRVSK*EFGTMTCNSHCPDLQSCS*K*R
IFYTARRDAPKNEKRSSDSFLFKDKGCEAAHFQFHCCLLDIHC*LYYRPRFFYNSVEVSF
LCSSYS*NASRTPSLIK*VYNINIHLYNLRYLYHQKKVVFLFYFVR*FSISMETRS*NWR
LSLVCDSVDTVLG*STGDGWL*SAGVPWNTGDHSGHLVWSNTGQHGGEVTVVENGFSNFD
WVGFSWVGGLEHVVLG*VSSDIRDVDTQFHQVDSFVRLS*SGGCRSSSSDGSDTSGD*S*
FDHCICFVC*C*YKLNRKGKNKDIFSKAYS*SSSIYTHSLMGCCYLNDR*LAPVPHQIFS
ISHLSHNLIISMEMLLFLNES*IFHRFRMWSTVLWRLCVFVCAECGNANYRGAEVPYKTL
FCACDISFFGQKEYPNFRFGPSYRSLLSKPEFSLL*TASAEEIFPSLEFVQH*TCVGSRI
LLGSVNL*TLGKCLGAIT*F*PLRSGW**DKLISNRYIFKRGYR*FSRAVLL*FDMYG*L
------------------------------------------------------------
EYRWSERWPWCRVLWRVS*MMYCLRYICLNWICLH*ICLHFNISMEGM*HYGK*HVVDGD
CRWMVG*VVVRVG*DILGRG*MVVGVW*WIVSG**VDGWWSGGMRQGMGW*GKCRGL**W
RGRVVDMELELGQC*C*C*Y*GVVCGCGVGVGVGVGVGVVWCVGVVWVWCVWX
```

Fig. 5.3 Result page 2 of Transeq

5.1.2 CpGPlot/CpGReport

Link http://www.ebi.ac.uk/emboss/cpgplot/

Institute This tool belongs to the European Molecular Biology Open Software Suite (EMBOSS). It is a free Open Source software analysis package specially developed for the needs of the molecular biology (e.g., EMBnet) user community.

Type of tool Web-based.

OS required Windows.

> **Note**
>
> CpG islands are unmethylated regions of the genome that are associated with the 5' ends of most housekeeping genes and many regulated genes. Identification of potential CpG islands during sequence analysis helps to define the extreme 5' ends of genes, something that is difficult with cDNA-based approaches. As they are associated with genes, CpG islands tend to be unique sequences and are therefore very useful in genome mapping projects.

There are two tools involving CpG islands, each serving a specific purpose.

CpGPlot

Tool for CpG plot is a tool used to plot the CpG-rich areas of the nucleotide sequence.

Algorithm used The algorithm used has been described by Gardiner-Garden and Frommer (1987). The method is based on the calculation of a running average in a window that moves along the sequence at a specified interval. The observed/expected CpG ratio and the percentage of C's and G's are calculated within this window to produce the two numerical arrays plotted by the program.

Tool screenshot A screenshot of CpGPlot is shown in Fig. 5.4.

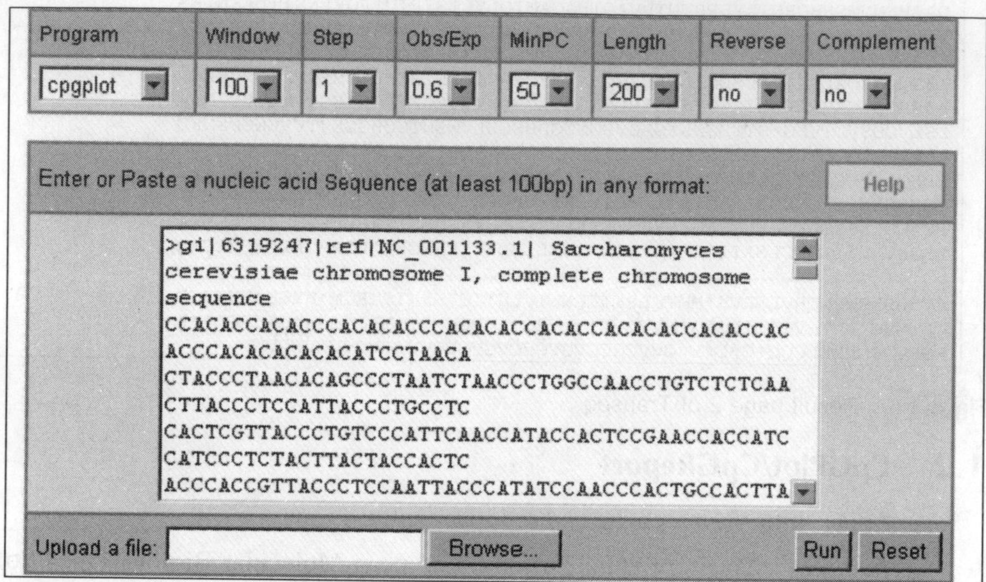

Fig. 5.4 Screenshot of CpGPlot

Input data The nucleotide sequence in raw format can be cut and pasted into the text-box called **Sequence Input Window**. The sequence can also be pasted in GCG, FASTA, EMBL, GenBank, PIR, NBRF, or Phylip format. However, partially formatted sequences are not accepted also. One should also avoid copying and pasting directly from word processors, as this gives unpredictable results because of the hidden/control characters present in these files. It is a good practice to add a return at the end of the sequence to ensure no application ignores the last line. The minimum sequence length required for the algorithm to carry out the correct calculation is 100, which of course does not guarantee that islands will get detected. A file containing a valid nucleotide sequence can be uploaded from a computer in any of the above-mentioned formats. Uploading should be done using Netscape or Internet Explorer.

Attributes

The attributes of CpGplot are few and are described below.

Window The percentage CG content and the observed frequency of CG are calculated within a window whose size is set by this parameter. The window is moved down the sequence and these statistics are calculated for each position that the window is moved to.

Step This parameter determines the number of bases that the window is moved by each time after the values of the percentage CG content and the observed frequency of CG are calculated within the window.

OBS/EXP This option sets the minimum average observed to expected ratio of C plus G to CpG in a set of 10 windows that are required before a CpG island is reported.

MINPC This parameter sets the minimum average percentage of G plus C in a set of 10 windows that are required before a CpG island is reported.

LENGTH This parameter sets the minimum length that a CpG island has to have before it is reported.

REVERSE This option allows the sequence being analysed to be reversed.

COMPLEMENT This option allows the complement of the sequence pasted in the **Sequence Input Window** to be analysed.

Result

The results of using the CpGPlot tool are the following.

1. It will produce a parameter report of unusual CG compositions in which observed/expected ratio > 0.60, per cent C + per cent G > 50.00, and length > 200. It also tells the position of the CpG Island.
2. It will produce graphs depicting the plots for observed/expected ratio of CG composition, % CG, and putative CpG Islands.

TASK

Input data The data that are fed to CpGPlot is shown below:

>gi|6319247|ref|NC_001133.1| Saccharomyces cerevisiae chromosome I, complete chromosome sequence
CCACACCACACCCACACACCCACACACCACACCACACACCACACCACACCCACACACACACATCCTAACACTACC
CTAACACAGCCCTAATCTAACCCTGGCCAACCTGTCTCTCAACTTACCCTCCATTACCCTGCCTCCACTCGTTACCCTGTCC
CATTCAACCATACCACTCCGAACCACCATCCAT

GCCGTGGATTGTGATGATGGAGAGGGAGGGTAGTTGACATGGAGTTAGAATTGGGTCAGTGTTAGTGTTAGTGTTAGTATTAGG
GTGTGGTGTGTGGGTGTGGTGTGGGTGTGGGTGTGGGTGTGGGTGTGGGTGTGGGTGTGGTGTGGTGTGTGGGTGTGGTGTG
GGTGTGGTGTGTGTGGG

Attribute settings The following attributes are selected.

Window	100
Step	1
Obs/Exp	0.6
MinPC	50
Length	200
Reverse	No
Complement	No

Result Upon giving these input data and setting these attributes, following the result is obtained.

Parameter report

```
CPGPLOT islands of unusual CG composition
NC_001133.1 from 1 to 230203

Observed/Expected ratio > 0.60
Percent C + Percent G > 50.00
Length > 200

Length 249 (32118..32366)
Length 262 (32415..32676)
Length 418 (33731..34148)
Length 281 (46227..46507)
Length 250 (61303..61552)
Length 399 (68691..69089)
Length 270 (72721..72990)
Length 263 (108330..108592)
Length 391 (108938..109328)
Length 361 (113782..114142)
Length 357 (114156..114512)
Length 633 (129422..130054)
Length 447 (130060..130506)
Length 277 (131250..131526)
Length 243 (132245..132487)
Length 1858 (189964..191821)
Length 324 (191895..192218)
Length 561 (192862..193422)
Length 229 (221068..221296)
Length 359 (222034..222392)
```

CpG Plots CpG plots are shown in Fig. 5.5.

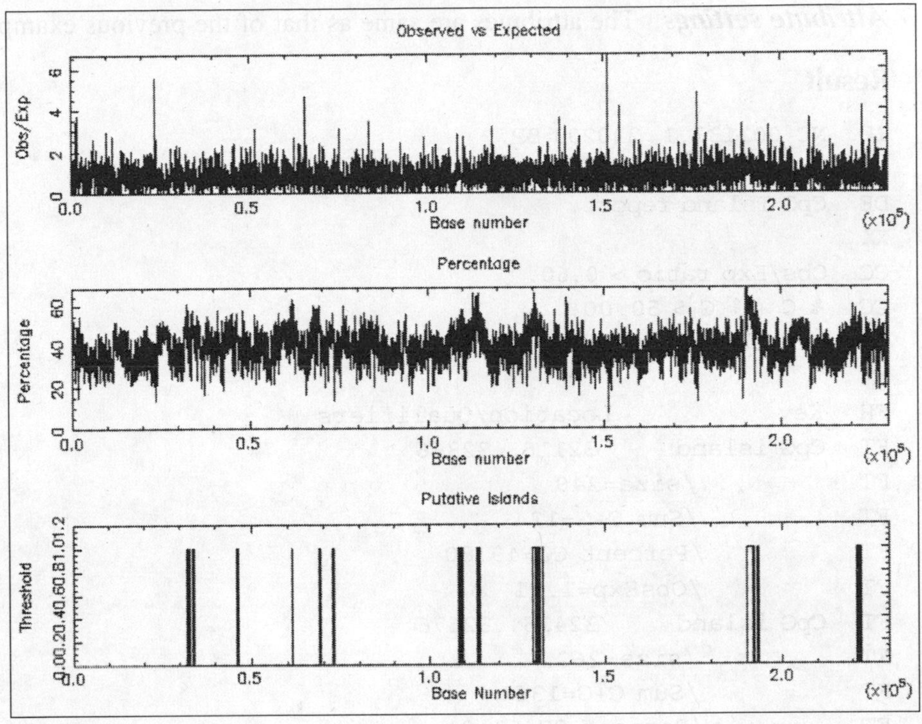

Fig. 5.5 Result page of CpGPlot

CpGReport

Tool for CpGReport produces an EMBL formatted report with a feature table that contains a key for each island found, along with the location/qualifiers that depict the position of the island, its size, the total sum of C+G's in the island, the % CG, and the maximum observed/expected value for it. At the bottom of each report the total number of islands found is printed or **No islands found**.

Tool screenshot The CpGReport screenshot is the same as that of CpGPlot (Fig. 5.4). The only difference is **cpgreport** is selected from the **Program** drop down option.

Input data The input data are the same as those for CpGPPlot.

Attributes The attributes are the same as those of CpGPlot.

Result CpGReport will produce a report in EMBL format with feature table details having CpG islands.

TASK

Input data The input data of >gi|6319247|ref|NC_001133.1| Saccharomyces cerevisiae chromosome I, complete chromosome sequence, is fed in FASTA format.

Attribute settings The attributes are same as that of the previous example.

Result

```
ID   NC_001133.1 230203 BP.
XX
DE   CpG Island report.
XX
CC   Obs/Exp ratio > 0.60.
CC   % C + % G > 50.00.
CC   Length > 200.
XX
FH   Key              Location/Qualifiers
FT   CpG island        32118..32366
FT               /size=249
FT               /Sum C+G=124
FT               /Percent CG=49.80
FT               /ObsExp=1.01
FT   CpG island        32415..32676
FT               /size=262
FT               /Sum C+G=136
FT               /Percent CG=51.91
FT               /ObsExp=0.85
FT   CpG island        33731..34148
FT               /size=418
FT               /Sum C+G=215
FT               /Percent CG=51.44
FT               /ObsExp=0.82
FT   CpG island        46227..46507
FT               /size=281
FT               /Sum C+G=142
FT               /Percent CG=50.53
FT               /ObsExp=0.78
FT   CpG island        61303..61552
FT               /size=250
FT               /Sum C+G=129
FT               /Percent CG=51.60
FT               /ObsExp=1.03
FT   CpG island        68691..69089
FT               /size=399
FT               /Sum C+G=212
FT               /Percent CG=53.13
FT               /ObsExp=1.07
```

```
FT   CpG island       72721..72990
FT               /size=270
FT               /Sum C+G=134
FT               /Percent CG=49.63
FT               /ObsExp=0.91
FT   CpG island       108330..108592
FT               /size=263
FT               /Sum C+G=129
FT               /Percent CG=49.05
FT               /ObsExp=1.14
FT   CpG island       108938..109328
FT               /size=391
FT               /Sum C+G=194
FT               /Percent CG=49.62
FT               /ObsExp=1.00
FT   CpG island       113782..114142
FT               /size=361
FT               /Sum C+G=197
FT               /Percent CG=54.57
FT               /ObsExp=0.86
FT   CpG island       114156..114512
FT               /size=357
FT               /Sum C+G=200
FT               /Percent CG=56.02
FT               /ObsExp=0.86
FT   CpG island       129422..130054
FT               /size=633
FT               /Sum C+G=324
FT               /Percent CG=51.18
FT               /ObsExp=0.99
FT   CpG island       130060..130506
FT               /size=447
FT               /Sum C+G=248
FT               /Percent CG=55.48
FT               /ObsExp=0.97
FT   CpG island       131250..131526
FT               /size=277
FT               /Sum C+G=136
FT               /Percent CG=49.10
FT               /ObsExp=0.93
FT   CpG island       132245..132487
FT               /size=243
FT               /Sum C+G=117
FT               /Percent CG=48.15
```

```
FT              /ObsExp=1.08
FT   CpG island       189964..191821
FT              /size=1858
FT              /Sum C+G=1076
FT              /Percent CG=57.91
FT              /ObsExp=1.19
FT   CpG island       191895..192218
FT              /size=324
FT              /Sum C+G=185
FT              /Percent CG=57.10
FT              /ObsExp=0.89
FT   CpG island       192862..193422
FT              /size=561
FT              /Sum C+G=295
FT              /Percent CG=52.58
FT              /ObsExp=1.01
FT   CpG island       221068..221296
FT              /size=229
FT              /Sum C+G=120
FT              /Percent CG=52.40
FT              /ObsExp=1.06
FT   CpG island       222034..222392
FT              /size=359
FT              /Sum C+G=177
FT              /Percent CG=49.30
FT              /ObsExp=0.97
FT   numislands      20
//
```

Note

The absence of methylation slows CpG decay and therefore CpG islands can be detected in DNA sequences as regions in which CpG pairs occur at close to the expected frequency. The detection process indicates that the corresponding germline DNA has been substantially hypomethylated for an extended period of time.

5.1.3 GCUA

Tool for Graphical Codon Usage Analyser (GCUA) is a tool that displays the codon bias in a graphical manner.

Link http://gcua.schoedl.de/sequential.html

Institute GENEART, Germany; GENEART North America, Canada.

Type of tool Web-based.

OS required Windows.

Algorithm GCUA computes the codon usage frequency against the codon usage table and originating organism.

Tool screenshot The screenshot of GCUA is shown in Fig. 5.6.

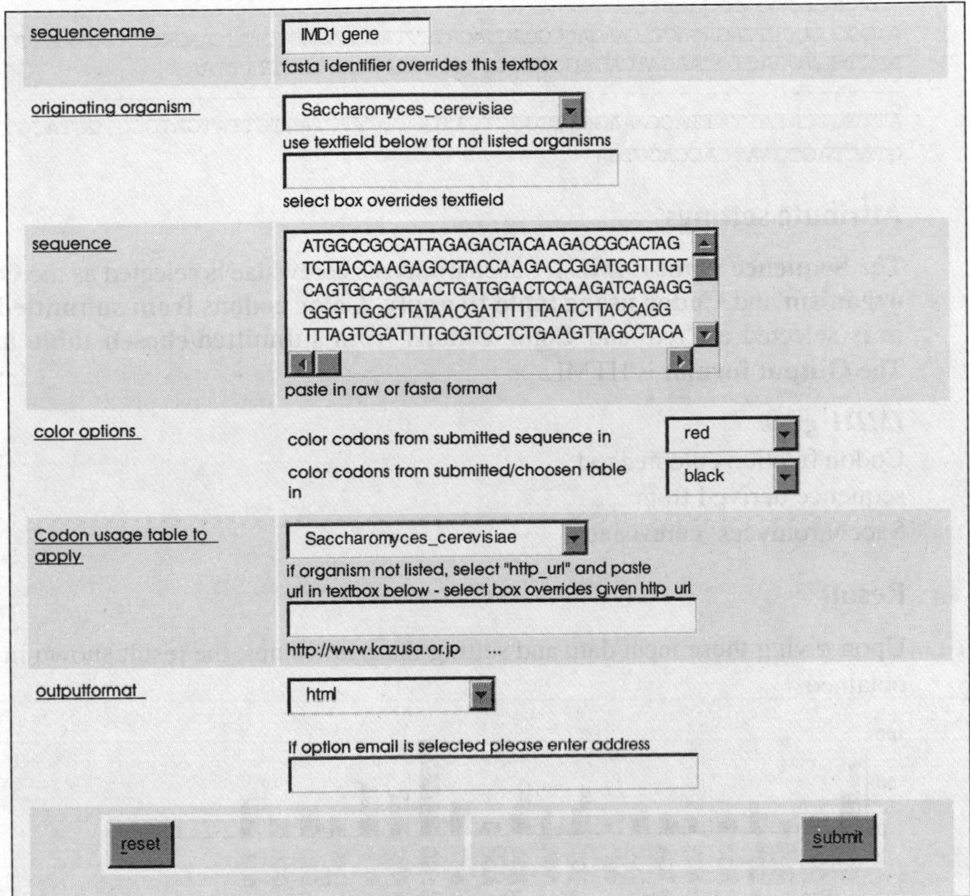

Fig. 5.6 Screenshot of GCUA

Input data Raw sequence cut and pasted from FASTA format is required as input data.

Attributes There are various attributes that need to be selected. The organism name can be selected from **Originating organism** and **Codon usage table to apply**. If the name is not there, the textbox can be used. **Color options** can be selected as per choice. It is better to give the sequence name rather than the description of the sequence.

Result The result is a graph showing the sequence parts of cumulative low codon usage. The triplet-fraction is represented as a column according to the codon usage fraction of the table of the chosen organism. The result is in HTML/PDF/text format.

TASK

Input

>gi|50593113:227733-228944 Saccharomyces cerevisiae chromosome I, complete chromosome sequence
ATGGCCGCCATTAGAGACTACAAGACCGCACTAGATCTTACCAAGAGCCTACCAAGACCGGATGGTTTGTCAGTGCAGG
AACTGATGGACTCCAAGATCAGAGGTGGGTTGGCTTATAACGATTTTTTAAT

ATTGGTCATATTATTACCAAAGCTTTGGCTCTTGGTTCTTCTACTGTTATGATGGGTGGTATGTTGGCCG
GTACTACCGAATCACCAGGTGA

Attribute settings

The **Sequence name** is IMD1. Saccharomyces cerevisiae is selected as the **Originating organism** and **Codon usage table to apply**. **Color codons from submitted sequence** in is selected as 'red' and **Color codons from submitted/chosen table in** is Black. The **Output format** is HTML.

IMD1 gene

Codon fraction coloured red
sequence derived from
Saccharomyces_cerevisiae

Result

Upon giving these input data and setting these attributes, the result shown in Fig. 5.7 is obtained.

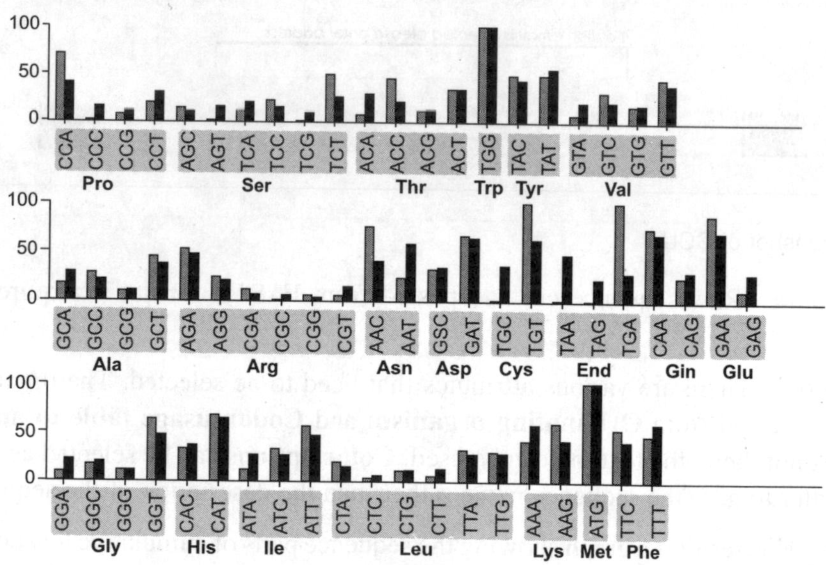

Fig. 5.7 Result page of GCUA

> **Note**
>
> Differences in codon usage among organisms obtained from the codon usage tables can lead to heterologous gene expression. The submitted sequence will be split into codons and the fraction of usage of each codon in the selected organism will be represented as a red column in the graph. Black columns show expected codon usage of the organism as per the selected codon table. One can also submit their own codon usage table other than the codon usage tables already listed in the tool.

5.1.4 BLAST

Tool for Basic Local Alignment Tool (BLAST) is a tool for local alignment of sequences.

Link http://www.ncbi.nlm.nih.gov/BLAST/

Institute NCBI. There are several different distributions of BLAST. For the sake of simplicity here, we will take up the BLAST from NCBI, where it was originally developed. WU-BLAST 2.0 is a distinctly different software package in spite of a common lineage for some portions of their code. The two packages work differently and, consequently, provide different results and different features.

Kinds of BLAST available BLAST is available via a web browser on several sites on the Internet. Its distribution also includes the IMPALA (Integrating Matrix Profiles And Local Alignments) package. The NCBI-BLAST programs include the following.

- **BLASTall**—performs all five kinds of blast comparison:
  BLASTp—protein against protein
  BLASTn—nucleotide against nucleotide
  BLASTx—translated nucleotide against protein
  tBLASTn—protein against translated nucleotide
  tBLASTx—translated nucleotide against translated nucleotide
- **BLASTpgp**—takes a protein query and performs a PSI-BLAST search to create a position-specific matrix using a protein database
- **BLASTclust**—automatically and systematically clusters protein sequences based on pairwise matches found using the BLAST algorithm in the case of proteins or the Mega BLAST algorithm in the case of DNA
- **MEGA BLAST**—uses the greedy algorithm for nucleotide sequence alignment search and concatenates many queries to minimize the time spent in scanning the database
- **RPS-BLAST** (reverse PSI-BLAST)—searches a query sequence against a database of profiles
- **bl2seq**—performs a comparison between two sequences using either the blastn or blastp algorithm
- **fastacmd**—retrieves FASTA formatted sequences from a BLAST database
- **formatdb**—formats FASTA sequence databases for BLAST

Type of tool Web-based downloadable

OS required The NCBI site provides binary files for the following operating systems and platforms:

- Apple MacOS X (ppc32)
- FreeBSD 4.5 (ia32)
- HP HPUX 11 (ia64)
- HP Tru64 5.1 (alpha)
- IBM AIX 5.1 (ppc64)
- Linux (kernel 2.4, glibc 2.2.5) (ia32, ia64, amd64)
- Microsoft Windows 2000 (ia32)
- SGI IRIX 6.5 (mips64)
- Sun Solaris 9 (ia32)
- Sun Solaris 8 (sparc64)

(a) BLASTn

Tool for Local alignment of nucleotide sequences.

Algorithm used The BLAST search algorithm is used.

Tool screenshot The screenshot of the BLASTn tool is shown in Fig. 5.8.

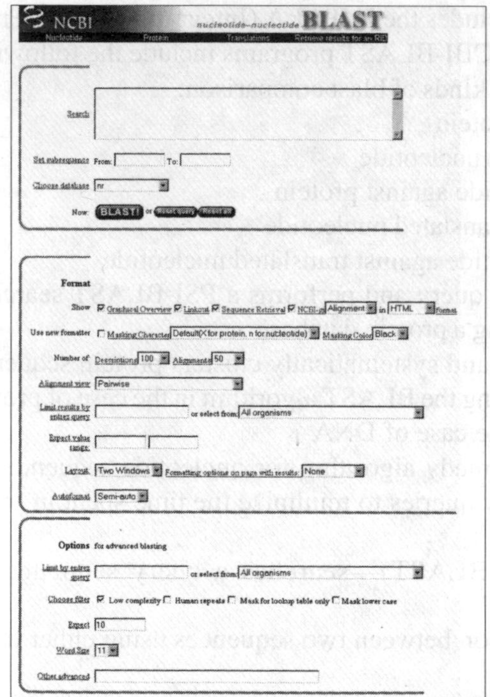

Fig. 5.8 Screenshot of BLASTn

Input data Sequences should be represented using the standard IUB/IUPAC and nucleic acid codes. The input data can be either in FASTA format or in the form of a raw sequence. Lower case letters are accepted and are converted into upper case letters. Any identifier can also be used. The sequences should not have any numerical characters. Missing nucleotides should be represented by the letter code 'N' or 'X' for unknown amino acid residue. A single hyphen or dash can be used to represent a gap of indeterminate length. '*' can be used to denote translation stop. The input is fed into the **Search** textbox.

Attributes There are various attributes that need to be selected.

Set subsequence To limit the region of the match, the required range of the nucleotides or protein residues can be specified in the **From** and **To** boxes. In case one of the limits is out of range, the search will be conducted on the whole query sequence.

Choose database Searches are conducted against several nucleotide databases, such as nr, Refseq_ma, refseq_genomic, est, est_human, est_mouse, est_others, gss, htgs, pat, pdb, month, alu_repeats, dbsts, chromosome, wgs, and env_nt.

Options for advanced BLAST

Limit by Entrez query BLAST search can be limited by using the Boolean logic used in Entrez search.

Select from organisms drop down list The search can also be limited by selecting the organism from the drop down list.

Choose filter Filtering or masking is the process of hiding regions of the (nucleic acid or amino acid) sequence that give spurious high scores. The default filtering option in BLAST 2.0 automatically converts low-complexity sequences into N's. There is a variety of filters to choose from, as follows.

Low complexity Filtering can eliminate segments with low compositional complexity such as hits against common acidic-, basic-, or proline-rich regions that are biologically uninteresting. These segments can be parts of the query sequence or its translation products. The default filtering for BLASTn is DUST. Filtering may mask the whole sequence or may not mask at all, so one should not always expect a result.

Human repeats This option masks human repeats (LINEs and SINEs) and is especially useful for human sequences that may contain these repeats. Filtering for repeats can increase the speed of a search, especially with very long sequences (>100 kb) and against databases which contain large number of repeats (htgs).

Mask for lookup table only This option masks entries of the lookup table used by BLAST, eliminating low-complexity sequences. BLAST, searches consist of two phases. Hits are first based upon a lookup table and second extended to low-complexity sequences. The second phase is still experimental and may change in the near future.

Mask lower case This option lets the filter work on sequences in lower case when upper case characters in FASTA format are pasted in the textbox.

Expect This is a statistical value denoted by the letter 'E' used to report matches against database sequences. Its default value is 10, i.e., at least 10 matches are expected. If the statistical value for the match is greater than the E-value, it will not show any match. The smaller the E-value, the closer the match, leading to few reported number of matches. Fractional values are acceptable.

Word size BLAST works by matching words between the query and database sequences. The regions of word matches are used to initiate extensions, which might lead to alignments. For BLASTn, an exact match of the entire word is required before an extension is initiated. Based on the requirement, the sensitivity of the search can be decreased by increasing the word size. This also increases the speed of the search. The available word sizes in BLASTn are 7, 11, and 15.

Other advanced There are various advanced options available. A list with default values is given in Table 5.1.

Table 5.1 List of advanced options available in BLASTn

Options	Purpose	Default value
-G	Cost to open gap [Integer]	5 for nucleotides, 11 for proteins
-E	Cost to extend gap [Integer]	2 for nucleotides, 1 for proteins
-q	Penalty for nucleotide mismatch [Integer]	-3
-r	Reward for nucleotide match [Integer]	1
-e	Expect value [Real]	10
-W	Word size [Integer]	11 for nucleotides, 3 for proteins
-y	Dropoff (X) for blast extensions in bits (default if zero)	20 for BLASTn, 7 for other programs
-X	X dropoff value for gapped alignment (in bits)	15 for all programs except for BLASTn. Not applicable for BLASTn.
-Z	Final X dropoff value for gapped alignment (in bits)	50 for BLASTn, 25 for other programs
-G	Cost to open gap [Integer]	5 for nucleotides, 11 for proteins

Limited values for gap existence and extension are supported for BLAST programs. Some supported and suggested values are as follows:

Existence	Extension
10	1
10	2
11	1
8	2
9	2

Various format options available under BLAST

Graphical overview In this format the query sequence aligned with database sequences is shown. The range of score is divided into five groups represented by five different colours. Each alignment is indicated by the colour corresponding to its score. In case of multiple alignments of the sequences of the same database, a vertical striped line connects the sequences. When the mouse points to a hit sequence, its definition and score are displayed. Upon clicking on a hit sequence, the associated alignments are displayed.

Linkouts This option provides cross-reference links from BLAST results to other specialized databases of NCBI. If the submitted query matches the sequences of the database specified in **Choose Database** and also finds matches in the LocusLink or UniGene databases, then this option provide links like [L] and [U] in the BLAST result.

NCBI-GI Along with the accession number and locus name, this option causes NCBI GI identifiers to be shown in the output.

Format The BLAST result can be displayed in any format, such as HTML, Plain Text, ASN.1, and XML.

Use new formatter Under this head options for masking character and colour are provided. Characters can be masked either with small letters or with the default character 'X' for proteins and 'n' for nucleotides. The masking colour can be black, grey, or red.

Descriptions This restricts the number of short descriptions of matching sequences reported in the BLAST output. The default value is 100 descriptions.

Alignments The number assigned to this option is the limit on the database sequences with high-scoring segment pairs (HSPs). The default value is 50. If more sequences than the number set satisfy the statistical significance threshold then only the matches with greatest statistical significance are reported.

Alignments view Various alignment views are below.

Pairwise It displays standard BLAST alignment in pairs of query sequence and database match.

Pairwise with identities It displays Pairwise alignment views along with dashes for gaps and single letter nucleotide abbreviation for mismatches.

Query-anchored with identities The databases alignments are anchored (shown in relation to) to the query sequence.

Query-anchored without identities The databases alignments are anchored (shown in relation to) to the query sequence with identities.

Flat query-anchored with identities Here inserts are shown as deletions on the query along with identities.

Flat query-anchored without identities Here inserts are shown as deletions on the query without identities.

Hit table It is possible to speed up search by specifying maximum number of hits to be computed.

Layout This option enables one to display the result either in **two windows** or in **one window**. We select **two windows**.

Formatting options on page with results If no specific formatting option is desired **None** can be selected. Formatting options can be displayed either **At the top** or **At the bottom**.

Autoformat This option allows the selection of one of the following three—Off disables the option, **Semi-auto** and **Full auto** enable the option partly and fully, respectively. When the **Autoformat** option is enabled (checked), clicking on the **Format** button will show the status and time stamps. It will also automatically format BLAST results when they are ready. When the **AutoFormat** option is disabled (unchecked), **Status** shows **Ready** and the background colour is changed to blue. This indicates that the search is complete. However, actual formatting will not be performed till the **Format** button is pressed.

Result

All members of the BLAST family of programs produce a similar output. The output consists of the following:

- program introduction,
- a schematic distribution of the alignments of the query sequence to those in the databases,
- a series of one-line descriptions of the database sequences that have significantly aligned to the query sequence,
- the actual sequence alignments, and
- a list of statistics specific to the BLAST search method and the version number are displayed at the top of the output.

Schematic distribution of alignments Only significant alignments of the query sequence to the database sequences are displayed schematically against the query sequence. The coloured bar coding is used to depict the region of alignment onto the query sequence. A colour legend that represents alignment scores is used, the higher the score the greater its significance. A bar can be selected, which allows a description of that specific sequence to be displayed in the window and causes the browser to jump to that particular alignment.

Line descriptions Similar to the database entries, the line descriptions are useful for identifying biologically interesting database matches and correlating these with the calculated statistical significance of the alignment. The sequences are listed in increasing order of the E (**Expect**) value. The alignments are listed in decreasing order of significance. Identifiers for the database sequences appear in the first column and are hyperlinked to the associated Genbank sequence records. The **Score** is another value attributed to the alignment, the higher this value, the better the match.

Note

The E-value is the probability that the query match is due to randomness. The lower the E-value, the more specific/significant the match. The **Score** (in bits) is a value attributed to the alignment but is independent of the scoring matrix used.

TASK

Input

`Gi|50593113|`

Attribute settings

Choose database nr (non-redundant) nucleotide database is selected.

Set subsequence From 227733 To 228944

Choose filter Low complexity is checked and the remaining left unchecked.

Expect The threshold of statistical significance for reporting matches against database sequences is set to 1.

Word size To increase the sensitivity of the search the word size is kept to 7.

The **Format** options **Graphical Overview, Linkout**, **Sequence Retrieval, NCBI-gi** are checked. The format option of **Alignment** is HTML. **Masking Character** and **Masking Color** are 'n' and Black, respectively. The default values of 100 and 50 are retained for **Descriptions** and **Alignments**, respectively. **Alignment views** is Pairwise with identities.

Layout This enables one to display the result either in two windows or in one window. We select **two windows**.

Formatting options on page with results None.

Autoformat Full auto.

Result

Upon giving these input data and setting these attributes, the result shown below and in Figs 5.9 and 5.10 is obtained.

Query= gi|50593113:227733-228944 Saccharomyces cerevisiae chromosome I, complete chromosome sequence(1212 letters)
All GenBank+EMBL+DDBJ+PDB sequences (but no EST, STS, GSS, environmental samples or phase 0, 1 or 2 HTGS sequences) 2,966,887 sequences; 13,504,363,242 total letters
Distribution of 92 Blast Hits on the Query Sequence

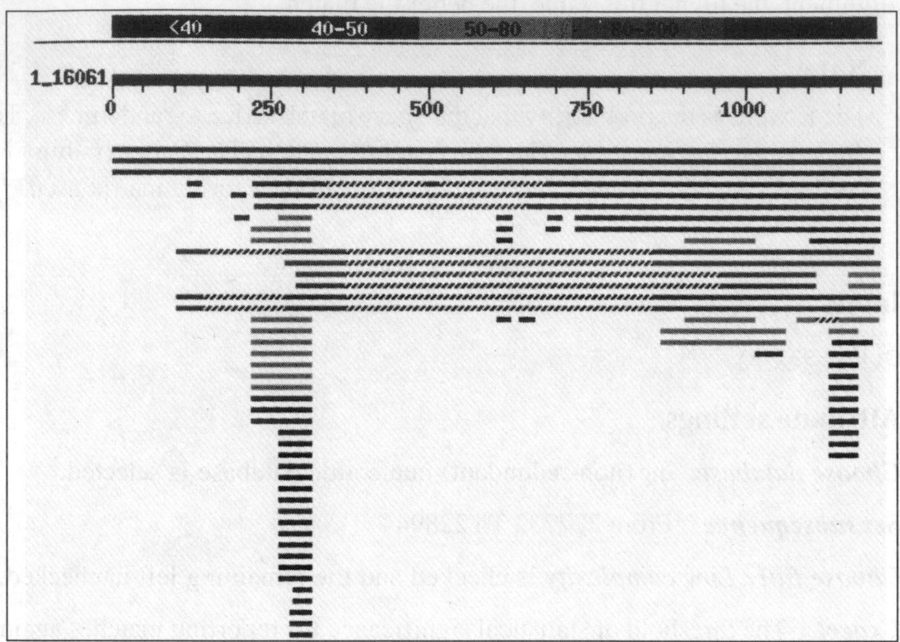

Fig. 5.9 Result page 1 of BLASTn

Sequences producing significant alignments:	Score (bits)	E Value
gi|37584899|gb|L28920.2|YSCCHR1RAA Saccharomyces cerevisiae...	2403	0.0
gi|551322|gb|U00029.1|YSCH9177 Saccharomyces cerevisiae chr...	2074	0.0
gi|665967|gb|U21094.1|YSCL9753 Saccharomyces cerevisiae chr...	1193	0.0
gi|577134|emb|Z46729.1|SC9958 S.cerevisiae chromosome XIII ...	504	e-139

Fig. 5.10 Result page 2 of BLASTn

Statistical details of the search

1. Database: Non-redundant GenBank CDS translations+PDB+Swiss-Prot+SPupdate+PIR
2. Posted date: Feb 26, 2000 10:08 PM
3. Number of letters in database: 142,135,17
   Number of sequences in database: 461,162

4. Lambda K H
   0.313 0.135 0.349
   Gapped Lambda K H
   0.270 0.0470 0.230
5. Matrix: BLOSUM62
6. Gap Penalties: Existence: 11, Extension: 1
7. Number of Hits to DB: 39862250
   Number of Sequences: 461162
   Number of extensions: 1595704
   Number of successful extensions: 8417
   Number of sequences better than 1.0: 86
   Number of HSP's better than 1.0 without gapping: 57
   Number of HSP's successfully gapped in prelim test: 29
   Number of HSP's that attempted gapping in prelim test: 8293
   Number of HSP's gapped (non-prelim): 121
8. length of query: 162
   length of database: 142,135,178
   effective HSP length: 60
   effective length of query: 102
   effective length of database: 114,465,458
   effective search space: 11675476716
   effective search space used: 11675476716
9. T: 11
   A: 40
   X1: 16 (7.4 bits)
   X2: 38 (14.8 bits)
   X3: 64 (24.9 bits)
   S1: 42 (21.9 bits)
   S2: 75 (33.6 bits)

Analysis

(a) From the Entrez query report for gi|50593113:227733-228944 in the nucleotide database, NC_001133 corresponds to IMD1 gene on chromosome I of Saccharomyces cerevisiae.

(b) From the graphical overview shown above, it is apparent that there are several high-scoring sequences that are highly related to the query sequence.
- The top three bars (red) are matches to IMD1 itself.
- The remaining partial-length bars reflect similarity between a portion of the query and a database entry that may be due to a shared domain or motif.

Only 78 descriptions are displayed though the value was set to 100. This shows that there are only 78 sequences that the query sequence matches with. To view any alignment of interest, one can click on the E-value of the first 50 bars in the graphic that are linked to the alignment. To view the details of the matched sequence one can click on the description line. This takes the viewer to the sequence view through NCBI.

(c) The description/definition lines listed above are all those hits whose E-values are less than the threshold, i.e., 1. This does not imply biological significance.

- The description lines reveal that the sequence in the database with greatest similarity to the queried IMD1 gene is IMD1 itself. Saccharomyces cerevisiae chromosome I right arm sequence (**Length** = 54812) with **Score** = 2403 bits (1212), **Expect** = 0.0, **Identities** = 1212/1212 (100%), and **Strand** = Plus/Plus.
- The second hit is to the database entry associated with the Saccharomyces cerevisiae chromosome VIII cosmid 9177 (Length-56097) with a lower score value and the same E-value. Upon clicking on the score value, one can further evaluate any hits of potential interest and examine the corresponding alignments: **Score** = 2074 bits (1046), **Identities** = 1172/1212 (96%), **Gaps** = 4/1212 (0%), and **Strand** = Plus/Plus.
- The third hit is Saccharomyces cerevisiae chromosome XII cosmid 9753 (**Length** = 24761) with **Score** = 1193 bits (602), **Expect** = 0.0, **Identities** = 1061/1212 (87%), **Gaps** = 4/1212 (0%), and **Strand** = Plus/Plus.
- The fourth hit with **Score** 504 is S. cerevisiae chromosome XIII cosmid 9958 with **Length** = 20951. It also has the following values: **Score** = 504 bits (254), **Expect** = e-139, **Identities** = 461/528 (87%), **Gaps** = 4/528 (0%), and **Strand** = Plus/Minus.

(d) The description list is truncated at E = 3.3, though the E-value was set to 1.

(b) BLASTx

Tool for BLASTx is a tool for local alignment of sequences. It compares a nucleotide query sequence translated in all reading frames against a protein sequence database.

Link http://www.ncbi.nlm.nih.gov/BLAST/

Institute NCBI.

Type of tool Web-based/downloadable.

OS required Windows.

Tool screenshot The screenshot of the BLASTx tool is shown in Fig. 5.11.

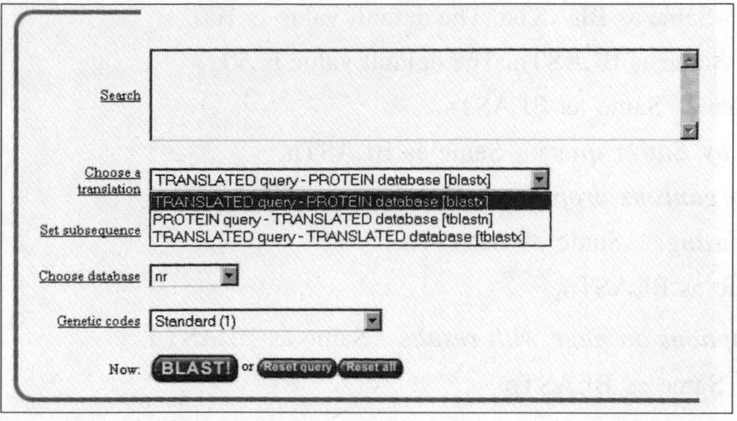

Fig. 5.11 Screenshot of BLASTx

Input data Amino acid sequence.

Attributes Various attributes need to be selected as follows.

Choose a translation Among three different options, **TRANSLATED query-PROTEIN database [blastx]** is selected.

Set subsequence Same as BLASTn.

Choose database The searches are conducted against several protein databases such as nr, Refseq, Swiss-Prot, pat, pdb, env_nr, and month.

Genetic codes There are several genetic codes available as mentioned earlier while discussing the ORF finder.

Other options for advanced BLAST are the following.

Limit by entrez query Same as BLASTn.

Select from organisms drop down list Same as BLASTn.

Choose filter Same as BLASTp.

Expect Same as BLASTn.

Word size Same as BLASTn. The available word sizes in BLASTp are 2 and 3.

Matrix The various options are PAM30/PAM70/BLOSUM80/BLOSUM62/ BLOSUM45. Same as BLASTp.

Gap costs Same as BLASTp.

Other Advanced same as BLASTn.

The following format options are available.

Show Same as BLASTn.

Use new formatter Same as BLASTn.

Descriptions Same as BLASTn. The default value is 100.

Alignments Same as BLASTn. The default value is 50.

Alignment views Same as BLASTn.

Limit results by Entrez query Same as BLASTn.

Select from organisms drop down list Same as BLASTn.

Expect value range Same as BLASTn.

Layout Same as BLASTn.

Formatting options on page with results Same as BLASTn.

Autoformat Same as BLASTn.

Result The results of BLASTx are graphical report of sequence similarity, report of sequence similarity with score in bits and E-value, pairwise alignment.

Input

```
>gi|50593113:227733-228944 Saccharomyces cerevisiae chromosome I, complete
chromosome sequence
ATGGCCGCCATTAGAGACTACAAGACCGCACTAGATCTTACCAAGAGCCTACCAAGACCGGATGGTTTGTCAGT
GCAGGAACTGATGGACTCCAAGATCAGAGGTGGGTTGGCTTATAACGATTTTTTAATCTTACCAGGTTTAGTCGATTTT
GCGTCCTCTGAAGTTAGCCTACAGACCAAGCTAACCAGGAATATTACTTTAAACATTCCATTAGTATCCTCTCCAATG
GACACTGTGACGGAATCTGAAATGGCCACTTTTATGGCTCTGTTGGATGGTATCGGTTTCATTCACCATAACTGTACTC
CAGAGGACCAAGCTGACATGGTCAGAAGAGTCAAGAACTATGAAAATGGGTTTATTAACAACCCTATAGTGATTTC
TCCAACTACGACCGTTGGTGAAGCTAAGAGCATGAAGGAAAAGTATGGATTTGCAGGCTTCCCTGTCACGGCAG
ATGGAAAGAGAAATGCAAAGTTGGTGGGTGCCATCACCTCTCGTGATATACAATTCGTTGAGGACAACTCTTTACTC
GTTCAGGATGTCATGACCAAAAACCCTGTTACCGGCGCACAAGGTATCACATTATCAGAAGGTAACGAAATTCTA
AAGAAAATCAAAAAGGGTAGGCTACTGGTTGTTGATGAAAAGGGTAACTTAGTTTCTATGCTTTCCCGAACTGATTTA
ATGAAAAATCAGAAGTACCCATTAGCGTCCAAATCTGCCAACACCAAGCAACTGTTATGGGGTGCTTCTATTGGGA
CTATGGACGCTGATAAAGAAAGACTAAGATTATTGGTAAAAGCTGGCTTGGATGTCGTCATATTGGATTCCTCTCAAG
GTAACTCTATTTTCCAATTGAACATGATCAAATGGATTAAAGAAACTTTCCCAGATTTGGAAATCATTGCTGGTAACGTT
GTCACCAAGGAACAAGCTGCCAATTTGATTGCTGCCGGTGCGGACGGTTTGAGAATTGGTATGGGAACTGGCTCT
ATTTGTATTACCCAAAAAGTTATGGCTTGTGGTAGGCCACAAGGTACAGCCGTCTACAACGTGTGTGAATTTGCTAAC
CAATTCGGTGTTCCATGTATGGCTGATGGTGGTGTTCAAAAACATTGGTCATATTATTACCAAAGCTTTGGCTCTTGGTT
CTTCTACTGTTATGATGGGTGGTATGTTGGCCGGTACTACCGAATCACCAGGTGA
```

Attribute settings

Choose database nr (non-redundant) nucleotide database is selected.

Genetic codes Standard(1)

Choose filter **Low complexity** is checked and the remaining left unchecked.

Expect The threshold of statistical significance for reporting matches against database sequences is set to 10.

Word size The word size is kept to 3.

Matrix BLOSOM62

Gap costs **Existence** : 11 Extention: 1

The format options **Graphical Overview**, **Linkouts**, **Sequence retrieval**, and **NCBI-gi** are checked. The Format option of **Alignment** is **HTML**. The **Masking character** and **Masking color** are 'n' and **Black**, respectively. The default values of 100 and 50 are kept for **Descriptions** and **Alignments**. **Alignment view** is Pairwise.

Layout **Two windows.**

Formatting option **None.**

Autoformat **Full auto.**

Result

Upon giving these input data and setting these attributes, the result shown below and in Figs 5.12–5.14 are obtained.

Query= gil50593113:227733-228944 Saccharomyces cerevisiae
chromosome I, complete chromosome sequence
(1212 letters)

Database: All non-redundant GenBank CDS
translations+PDB+Swiss-Prot+PIR+PRF excluding environmental samples
2,410,074 sequences; 815,621,784 total letters

(i) Graphical result of hit sequences with colour code:

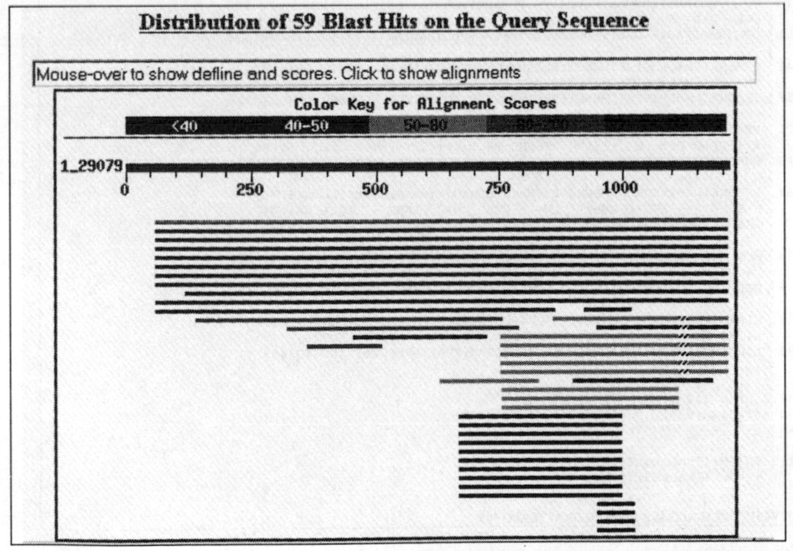

Fig. 5.12 Result page of BLASTx

(ii) Result of hit sequences with score in bits and E-value

```
                                                                    Score     E
Sequences producing significant alignments:                        (bits)  Value

gi|51095067|gb|EAL24310.1|  IMP (inosine monophosphate) dehy...      411   e-125 G
gi|57999523|emb|CAI45968.1|  hypothetical protein [Homo sapi...      411   e-125 G
gi|51095068|gb|EAL24311.1|  IMP (inosine monophosphate) dehy...      411   e-125 G
gi|25014074|sp|P20839|IMD1_HUMAN  Inosine-5'-monophosphate d...      411   e-125 G
gi|106722|pir||A35566  IMP dehydrogenase (EC 1.1.1.205) I - ...      411   e-125 S
gi|13543973|gb|AAH06124.1|  IMP (inosine monophosphate) dehy...      407   e-124 G S
gi|4504689|ref|NP_000875.1|  IMP (inosine monophosphate) deh...      406   e-123 G
gi|307067|gb|AAA36114.1|  IMP dehydrogenase type 1 (EC 1.1.1...      407   e-123 G
gi|47077068|dbj|BAD18464.1|  unnamed protein product [Homo s...      397   e-120 G
gi|16549223|dbj|BAB70780.1|  unnamed protein product [Homo s...      357   e-108 G
gi|51467033|ref|XP_496992.1|  PREDICTED: similar to inosine ...      246   8e-65 G
gi|51467471|ref|XP_497019.1|  PREDICTED: similar to Inosine-...      197   3e-50 G
gi|44979607|gb|AAS50155.1|  IMP dehydrogenase 2 [Homo sapiens]       147   3e-35 G
gi|45708411|gb|AAH03053.1|  GMPR2 protein [Homo sapiens]              76   3e-20 G
gi|50541956|ref|NP_057660.2|  guanosine monophosphate reduct...       76   3e-20 G
gi|50541948|ref|NP_001002002.1|  guanosine monophosphate red...       76   3e-20 G
```

Fig. 5.13 Result page of BLASTx

(iii) Pairwise alignment (with IMP dehydrogenase 1)

```
□
   >gi|51095067|gb|EAL24310.1| G IMP (inosine monophosphate) dehydrogenase 1 [Homo sapiens]
 gi|34328930|ref|NP_000874.2| G inosine monophosphate dehydrogenase 1 isoform a [Homo sapiens]
          Length = 599

 Score =  411 bits (1057), Expect(2) = e-125
 Identities = 207/356 (58%), Positives = 271/356 (76%), Gaps = 4/356 (1%)
 Frame = +1

Query: 61    DGLSVQELMDSKIRGGLAYNDFLILPGLVDFASSEVSLQTKLTRNITLNIPLVSSPMDTV 240
              DGL+ Q+L  S    GL YNDFLILPG +DF + EV L + LTR ITL  PL+SSPMDTV
Sbjct: 101   DGLTAQQLFASA--DGLTYNDFLILPGFIDFIADEVDLTSALTRKITLKTPLISSPMDTV 158

Query: 241   TESEMATFMALLDGIGFIHHNCTPEDQADMVRRVKNYENGFINNPIVISPTTTVGEAKSM 420
              TE++MA  MAL+ GIGFIHHNCTPE QA+ VR+VK +E GFI +P+V+SP+ TVG+
Sbjct: 159   TEADMAIAMALMGGIGFIHHNCTPEFQANEVRKVKKFEQGFITDPVVLSPSHTVGDVLEA 218

Query: 421   KEKYGFAGFPVTADGKRNAKLVGAITSRDIQFV--EDNSLLVQDVMTK--NPVTGAQGIT 588
              K ++GF+G P+T  G   +KLVG +TSRDI F+  +D++ L+ +VMT      V    G+T
Sbjct: 219   KMRHGFSGIPITETGTMGSKLVGIVTSRDIDFLAEKDHTTLLSEVMTPRIELVVAPAGVT 278

Query: 589   LSEGNEILKKIKKGRLLVVDEKGNLVSMLSRTDLMKNQKYPLASKSANTKQLLWGASIGT 768
              L E NEIL++ KKG+L +V++   LV++++RTDL KN+ YPLASK +  KQLL GA++GT
Sbjct: 279   LKEANEILQRSKKGKLPIVNDCDELVAIIARTDLKKNRDYPLASKDSQ-KQLLCGAAVGT 337

Query: 769   MDADKERLRLLVKAGLDVVILDSSQGNSIFQLNMIKWIKETFPDLEIIAGNVVTKEQAAN 948
               + DK RL LL +AG+DV++LDSSQGNS++Q+ M+ +IK+ +P L++I GNVVT  QA N
Sbjct: 338   REDDKYRLDLLTQAGVDVIVLDSSQGNSVYQIAMVHYIKQKYPHLQVIGGNVVTAAQAKN 397

Query: 949   LIAAGADGLRIGMGTGSICITQKVMACGRPQGTAVYNVCEFANQFGVPCMADGGVQ 1116
              LI AG DGLR+GMG GSICITQ+VMACGRPQGTAVY V E+A +FGVP +ADGG+Q
Sbjct: 398   LIDAGVDGLRVGMGCGSICITQEVMACGRPQGTAVYKVAEYARRFGVPIIADGGIQ 453


Score = 46.2 bits (108), Expect(2) = e-125
Identities = 21/32 (65%), Positives = 28/32 (87%)
Frame = +2

Query: 1115 KNIGHIITKALALGSSTVMMGGMLAGTTESPG 1210
              + +GH++ KALALG+STVMMG +LA TTE+PG

Sbjct: 453  QTVGHVV-KALALGASTVMMGSLLAATTEAPG 483
```

Fig. 5.14 Result page of pairwise alignment

(iv) Statistical report

Database: All non-redundant GenBank CDS
translations+PDB+Swiss-Prot+PIR+PRF excluding environmental samples
Posted date: Mar 27, 2005 1:57 AM
Number of letters in database: 815,228,232
Number of sequences in database: 2,408,643

Lambda K H
0.318 0.134 0.401

Gapped
Lambda K H
0.267 0.0410 0.140

Matrix: BLOSUM62
Gap Penalties: Existence: 11, Extension: 1
Number of Hits to DB: 114,885,826
Number of Sequences: 2408643
Number of extensions: 2725523
Number of successful extensions: 7082
Number of sequences better than 10.0: 42
Number of HSP's better than 10.0 without gapping: 6438
Number of HSP's successfully gapped in prelim test: 0
Number of HSP's that attempted gapping in prelim test: 0
Number of HSP's gapped (non-prelim): 7047
length of database: 45,908,306
effective HSP length: 110
effective length of database: 31,608,856
effective search space used: 9261394808
frameshift window, decay const: 40, 0.1
T: 12
A: 40
X1: 16 (7.3 bits)
X2: 38 (14.6 bits)
X3: 64 (24.7 bits)
S1: 41 (21.7 bits)

Forward ePCR

Tool for Forward Electronic PCR (e-PCR) searches the STS database against the query sequence. The STS database of NCBI is UniSTS, which is searched to identify sequence-tagged sites (STSs) in the query DNA sequence(s). STSs that closely match the query primers with correct order, orientation, and spacing represent potential PCR primers. These can be used to generate known STSs.

Link http://www.ncbi.nlm.nih.gov/sutils/e-pcr/

Institute NCBI

Type of tool Web-based/downloadable

OS required Windows

Algorithm used e-PCR implements fuzzy logic behind the search engine.

Tool screenshot The screenshot of Forward e-PCR is shown in Fig. 5.15.

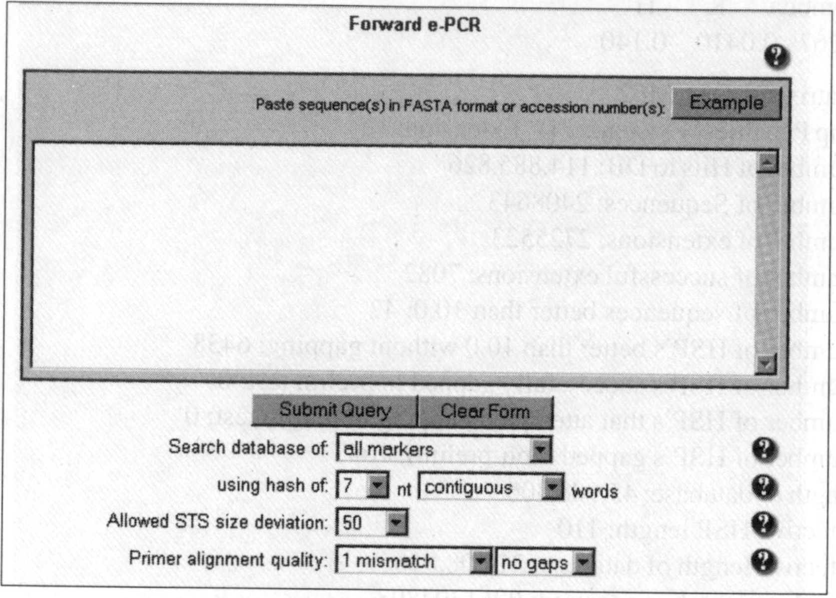

Fig. 5.15 Screenshot of forward e-PCR

Input data A raw sequence cut and pasted from FASTA format or the accession number(s) is required as input data.

Attributes Various attributes need to be selected as follows.

Search database of The options are **All markers/Non-repetitive markers/Mapped markers**. The search is performed in the UniSTS database. If the non-repetitive option is selected, markers that satisfy one of the following criteria are excluded from the search list:

- marker maps to > 1 chromosome
- hits > 1 LocusLink
- hits > 2 places on the contig(s)
- hits > 1 draft genomic sequences

Using hash of The values offered range from 3 to 15. The default value is 7. The database is searched with a nucleotide word size of **Using hash of** value from the 3′ end of each primer. Greater the hash of value, faster the search. Lower hash values increase the sensitivity of the search.

Words The options are **Contiguous/Discontiguous**. Contiguous words do not allow mismatches within the word boundary, thereby improving the sensitivity. Discontiguos words increase the search time by allowing mismatches within the word boundary.

Note

To avoid the chance of missing true STSs due to mismatches, multiple discontigous words can be selected instead of a single exact word. These words will have a group of significant positions separated by 'wildcard' characters representing no match. One can also allow gaps in the primer alignments.

Allowed STS size deviation The deviation of word size set using hash of value can be set between numbers 0/10/50/100/500/1000/5000. The default value is 50. Any value other than 0 allows the detection of hits with a size that is out of the previously reported (expected) PCR product size range.

Primer alignment quality

(a) No mismatches/1 mismatch/2 mismatches/3 mismatches

(b) No gaps/1 gap/2 gaps

The quality of primer alignment depends on the number of mismatches and the number of gaps allowed in each primer sequence. For higher numbers there is the possibility of getting both potential STSs as well as false hits. Reasonable values are **1 mismatch and 1 gap**. Mismatches and gaps are not allowed within the nucleotide's word size from the 3′ end of the primer. Mismatches are only allowed in the case of discontiguous words.

Result For every query sequence, a list of markers, their location, and the originating organisms and size are generated.

TASK

Input

NC_001137

NC_001148

Attribute settings

Search database of Non-repetitive markers.

Using hash of 7.

Words Discontiguous.

Allowed STS size deviation 50.

Primer alignment quality:

(a) 1 mismatch
(b) 1 gap

Result

Upon giving these input data and setting these attributes, the following result is obtained.

Query sequence: NC_001137, 576869 bases long

Saccharomyces cerevisiae chromosome V, complete chromosome sequence

Site (bases)	Marker	Chr.	Organism	Size	L m/g	R m/g
94,915..95,042	AF025868	-	Aegilops caudata	128 (<)	1/1	1/1
116,142..116,285	D17S1248	-	Homo sapiens	144	0/0	0/0
175,893..176,223	AF025876	-	Aegilops caudata	331 (>)	1/1	1/0
186,587..186,852	AF025876	-	Aegilops caudata	266 (<)	1/0	1/1
317,945..318,281	AF025865	-	Aegilops caudata	337 (>)	1/1	1/1
377,179..377,735	AF025874	-	Aegilops caudata	557 (<)	1/1	1/1
377,179..377,735	AF025874	-	Aegilops caudata	557 (<)	1/1	1/1
503,268..503,613	AF025869	-	Aegilops caudata	346	1/0	1/1

Query sequence: NC_001148, 948062 bases long

Saccharomyces cerevisiae chromosome XVI, complete chromosome sequence

Site (bases)	Marker	Chr.	Organism	Size	L m/g	R m/g
5,506..5,905	AF025856	-	Aegilops caudata	400 (<)	1/0	1/1
284,016..284,345	AF025869	-	Aegilops caudata	330	0/1	1/1
341,680..342,036	AF025869	-	Aegilops caudata	357 (>)	1/1	1/1
387,665..387,986	AF025876	-	Aegilops caudata	322 (>)	1/1	1/1
406,764..406,910	AF025868	-	Aegilops caudata	147 (<)	0/1	1/1
428,878..429,353	AF025859	-	Aegilops caudata	476 (<)	1/0	0/1
452,617..452,675	AF023863	-	Parus major	59 (<)	1/0	1/1
452,617..452,675	AF023863	-	Parus major	59 (<)	1/0	1/1
541,164..541,732	AF025874	-	Aegilops caudata	569 (<)	1/1	1/1
728,824..729,145	px-2f6	-	Mus musculus	322 (>)	0/0	1/0
775,757..775,926	px-12f9	-	Mus musculus	170 (>)	0/0	0/0

Analysis

(>) or (<) in the **Size** column means that the product size of the hit is greater or smaller, respectively, than expected for this STSs.

Legend		**Call parameters**
L m/g	Left primer mismatches/gaps	(M) margin: 50
R m/g	Right primer mismatches/gaps	(N) mism: 1
Times		(G) gaps: 1
Started Wed Mar 30 09:24:31 2005		(W) wordsize: 7
Finished Wed Mar 30 09:25:54 2005		(F) wordtype: 1 Contiguous
Elapsed 83		STS set non_warn

5.1.6 ORF Finder

Tool for The ORF (open reading frame) Finder is a graphical analysis tool that finds all open reading frames of the query sequence. It is also useful in preparing a complete and accurate sequence during sequence submissions. We will learn about it when we discuss the sequence submission tool.

Link http://www.ncbi.nlm.nih.gov/gorf/gorf.html

Institute NCBI.

Type of tool Web-based/downloadable.

OS required Windows.

Algorithm used This tool identifies all open reading frames using the standard or 'alternative genetic codes'.

Tool Snapshot The tool snapshot of ORF Finder is shown in Fig 5.16.

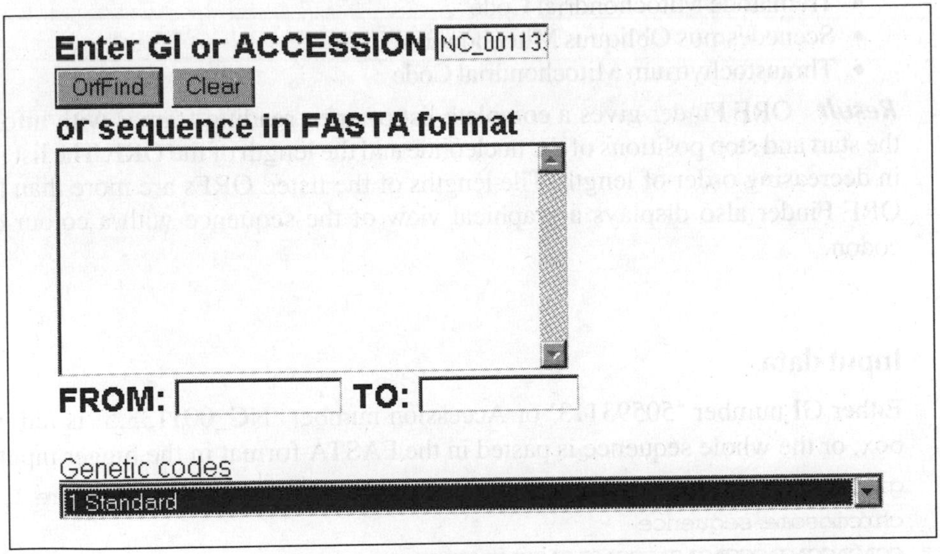

Enter GI or ACCESSION NC_00113

OrfFind Clear

or sequence in FASTA format

FROM: **TO:**

Genetic codes

1 Standard

Fig. 5.16 ORF finder

Input data The data that are fed to the tool is either an identifier of a nucleotide sequence or the sequence itself. The identifiers can be the GI number or the accession number. The sequence needs to be necessarily in FASTA format.

Attributes The attributes are less in number and simple to handle, described as follows.

From, To These attributes limit the region of match of the input query sequence. It can be left blank if the whole sequence is to be considered. Using the **From** and **To** boxes, the position of the nucleotide sequence to be considered is specified.

Genetic Codes There are several genetic codes available as mentioned below, the **Standard Code** being the default one:

- Standard Code
- Vertebrate Mitochondrial Code
- Yeast Mitochondrial Code
- Mold, Protozoan, and Coelenterate Mitochondrial Code and the Mycoplasma/ Spiroplasma Code
- Invertebrate Mitochondrial Code
- Ciliate, Dasycladacean and Hexamita Nuclear Code
- Echinoderm and Flatworm Mitochondrial Code
- Euplotid Nuclear Code
- Bacterial and Plant Plastid Code
- Alternative Yeast Nuclear Code
- Ascidian Mitochondrial Code
- Alternative Flatworm Mitochondrial Code
- Blepharisma Nuclear Code
- Chlorophycean Mitochondrial Code
- Trematode Mitochondrial Code
- Scenedesmus Obliquus Mitochondrial Code
- Thraustochytrium Mitochondrial Code

Result ORF Finder gives a complete list of open reading frames with information on the start and stop positions of the nucleotide and the length of the ORF. The list is displayed in decreasing order of length. The lengths of the listed ORFs are more than length 100. ORF Finder also displays a graphical view of the sequence with a colour-coded start codon.

TASK

Input data

Either GI number '50593113' or Accession number 'NC_001133.5' is fed in the input box, or the whole sequence is pasted in the FASTA format in the bigger input box.

```
gi|50593113|ref|NC_001133.5| Saccharomyces cerevisiae chromosome I, complete
chromosome sequence
CCACACCACACCCACACACCCACACACCACACCACACACCACACCACACCCACACACACACATCCTAACACTACC
CTAACACAGCCCTAATCTAACCCTGGCCAACCTGTCTCTCAACTTACCCTCCATTACCCTGCCTCCACTCGTTACCCTGTCC
CATTCAACCATACCACTCCGAACCACCATCCAT
-----------------------------------------------------------------------------
```

Attribute settings

The attributes for this example are follows:

From, To These attributes are left blank.

Genetic Codes The default code, i.e., the **Standard Code** is set.

Result

Upon giving these input data and setting these attributes, click on the **OrfFind** button to get the result shown in Fig. 5.17.

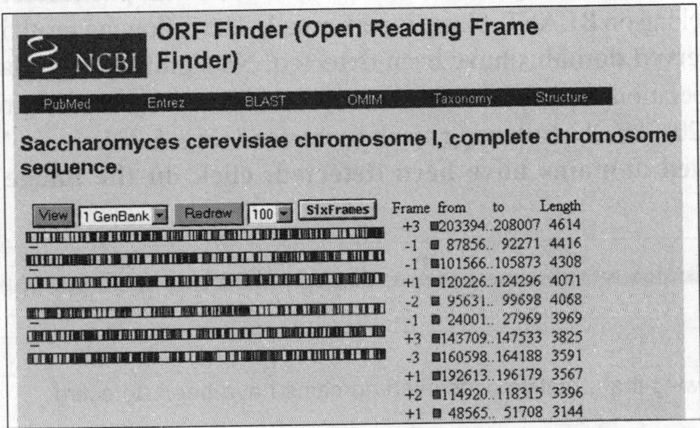

Fig. 5.17 Result page 1 of ORF Finder

On the left, the whole sequence is shown graphically with teal coloured bands representing the start codons. On the right, a list of ORFs is shown mentioning the frame they belong to, the start and stop codon positions and the lengths of the ORFs.

Upon clicking on the teal coloured box of the smallest ORF of length 102, from frame +2 from 872 to 973, it turns pink. The corresponding bands in the graphical bars also turn pink. Information is also displayed as shown in Fig. 5.18. Upon clicking on the **Accept** button, the pink coloured box turns into a green coloured circle. The corresponding bands in the graphical bars also turn green. Then the green circle is clicked; it becomes a pink box. Try clicking on the **Alternative Initiation Codons** button and see the result.

Length: 33 aa

Accept | Alternative Initiation Codons

```
872 atgctatttcagaatatttcgtacttacacaggccatacattaga
     M  L  F  Q  N  I  S  Y  L  H  R  P  Y  I  R
917 ataatatgtcacatcactgtcgtaacactctttattcaccgagca
     I  I  C  H  I  T  V  V  T  L  F  I  H  R  A
962 ataatacggtag 973
     I  I  R  *
```

Fig. 5.18 Result page 2 of ORF finder

ORF Finder also offers BLASTp or tBLASTn for the selected amino acid with the interface as shown in Fig. 5.19.

Fig. 5.19 Interface offering BLASTp and tBLASTn through ORF finder

First, the **Accept** button is clicked; the pink box turns into a green circle. The attributes selected are **Program BLASTp**, **Database** nr, and the **with parameters** option is checked. Upon clicking on **BLAST**, blastp is performed. The following result is displayed: **No putative conserved domains have been detected. No significant similarity found.** When a similar operation is performed on the largest ORF of length 4614, from frame + 3 from 203394 to 208007, the following result is obtained (Fig. 5.20).

Putative conserved domains have been detected, click on the image below for detailed results.

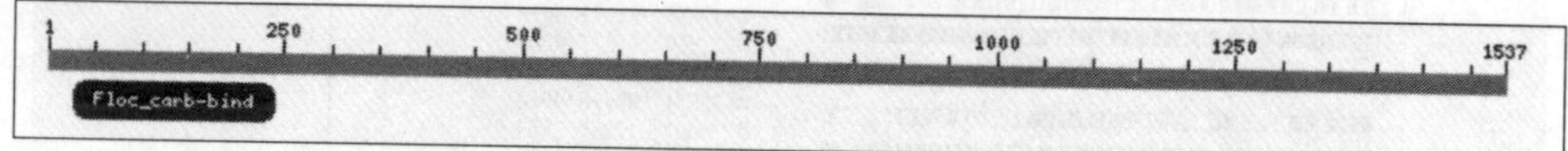

Fig. 5.20 Result page showing that putative conserved domains have been detected

Upon clicking on the image shown in Fig. 5.20 the result shown in Fig. 5.21 is displayed.

Descriptions

Name	Title	PssmId	Accession	E-value
[–]Floc_carb-bind	[Single-domain CD] pfam06660, Floc_carb-bind, Flocculin carbohydrate-bi...	46547	pfam06660	2e-89

[Single-domain CD] pfam06660, Floc_carb-bind, Flocculin carbohydrate-binding. This family represents the carbohydrate-binding N-terminus of yeast flocculin. Flocculin is a cell-surface lectin that causes adhesion of cells in clumps resulting in their sedimentation..

CD Length: 200, **Pct. Aligned:** 100, **Bit Score:** 325.912739, **E-value:** 2e-89

```
                      10        20        30        40        50        60        70        80
                ....*....|....*....|....*....|....*....|....*....|....*....|....*....|....*....|
query        29 CLPAGQRKSGMNINFYQYSLKDSSTYSNAAYMAYGYASKTKLGSVGGQTDISIDYNIPCVSSSGTFPCPQEdsygnwgck 108
consensus     1 CLPAGTRKNGMNINFYQYSLKDSSTYSDPNYMAYGYASHRKLGSVSGQTNISIYYHIPCTPQSGTLPCSYN--------- 71

                      90       100       110       120       130       140       150       160
                ....*....|....*....|....*....|....*....|....*....|....*....|....*....|....*....|
query       109 gmgACSNSQGIAYWSTDLFGFYTTPTNVTLEMTGYFLPPQTGSYTFKFATVDDSAILSVGGATAFNCCAQQQPPITSTNF 188
consensus    72 ---ACSNSQGIAYWSSDLFGFYTTPTNVTVEMTGYFLPPQTGSYTFKFATVDDSAILSVGGGNAFECCAQEQPPITSTNF 148

                     170       180       190       200       210
                ....*....|....*....|....*....|....*....|....*....|..
query       189 TIDGIKPWGGSLPPNIEGTVYMYAGYYYPMKVVYSNAVSWGTLPISVTLPDG 240
consensus   149 TINGIKPWGAKAPTDIEGTVYMYAGYYYPIKIVYSNAVSWGTLPVSVVLPDG 200
```

Search for similar domain architectures

Fig. 5.21 Result page with full result of putative conserved domain

Upon clicking on **Search for similar domain architectures**, a result page with domain relatives is displayed with the help of CDART, as shown in Fig. 5.22.

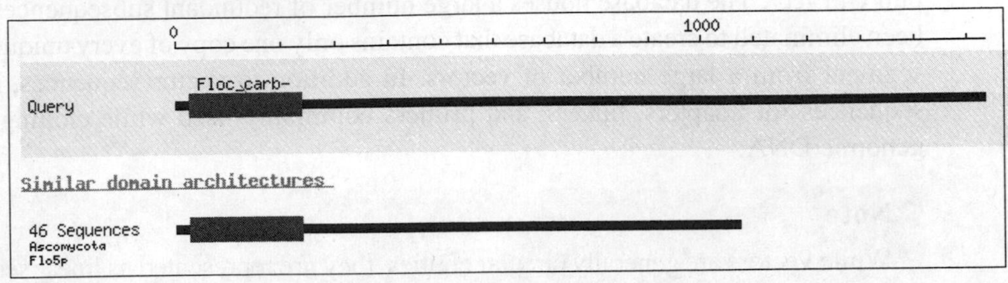

Fig. 5.22 Result page with domain relatives

Upon clicking on the **Format** button of BLASTp, similar sequences, as shown in Fig. 5.23, are displayed.

Sequences producing significant alignments:	Score (bits)	E Value
gb|AAC09499.1| Flo1p (Saccharomyces cerevisiae) >gi|6319341...	3028	0.0
emb|CAA55024.1| flocculation FL01 (Saccharomyces cerevisiae...	3011	0.0
gb|AAX47297.1| flocculin (Saccharomyces cerevisiae)	2964	0.0
pir||S51959 hypothetical protein YAL063c - yeast (Saccharom...	1855	0.0
ref|NP_009338.1| Lectin-like protein with similarity to Flo...	1753	0.0
ref|NP_012081.1| Lectin-like protein involved in flocculati...	1412	0.0
dbj|BAA19915.1| flocculln (Saccharomyces cerevisiae)	868	0.0
gb|AAS53279.1| AFL095Wp (Ashbya gossypii ATCC 10895) >gi|45...	736	0.0
gb|AAX47295.1| flocculln (Saccharomyces cerevisiae)	619	e-175
ref|NP_013028.1| Lectin-like protein with similarity to Flo...	512	e-143
emb|CAG60504.1| unnamed protein product (Candida glabrata C...	507	e-141

Fig. 5.23 Result page with domain relatives

5.1.7 Vecscreen

Tool for Vecscreen is a tool developed by NCBI. It was developed with the intention to clean up public sequence databases and look for accidental inclusion of vectors in them. Some commonly used vectors are plasmids, phages, cosmids, BAC, PAC, and YAC. The tool is presently used by GenBank Annotation Staff on sequences submitted to the database.

Link http://www.ncbi.nlm.nih.gov/VecScreen/VecScreen.html

Institute NCBI.

Type of tool Web-based.

OS required Windows.

Algorithm used Vecscreen searches for vectors in a specially created database called UniVec. UniVec can be obtained from the NCBI FTP directory: ftp://ftp.ncbi.nih.gov/pub/UniVec/. The database houses a large number of redundant subsequences that have been eliminated to create a database that contains only one copy of every unique sequence segment from a large number of vectors. In addition to vector sequences, it also has sequences for adapters, linkers, and primers commonly used while cloning cDNA or genomic DNA.

Note

While vectors are generally circular entities, they are represented as linear sequences for most bioinformatics purposes. BLAST has the limitation of not being able to make an extended match across the ends when presented with this format. Thus, it is relatively easy to miss out a fragment of vector that is placed at the circular junction. To avoid this, a copy of the first 49 bases of the sequence for a circular vector is appended to the end of the sequence before it is processed for addition to UniVec.

Tool snapshot The screenshot of VecScreen is shown in Fig. 5.24.

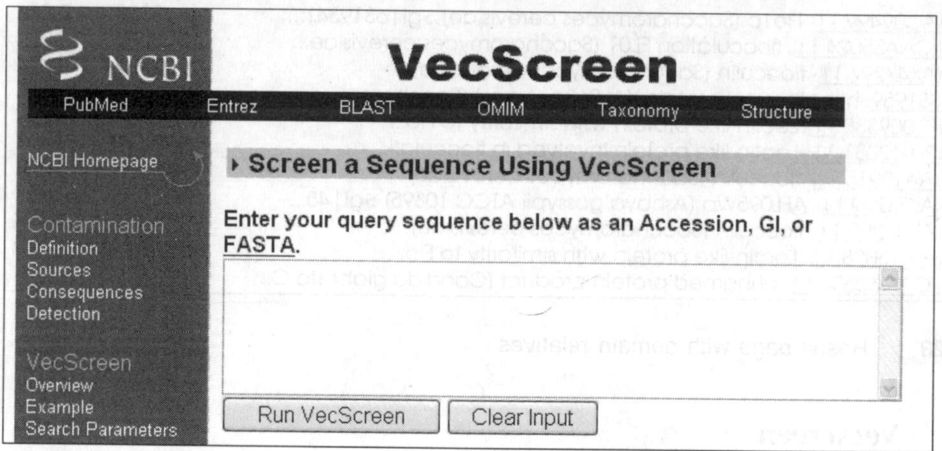

Fig. 5.24 Screen of Vecscreen

Input data Sequence in FASTA format, GI numbers, accession numbers.

Attributes The attributes of Vecscreen use BLASTn with preset parameters. However, VecScreen uses an increased penalty for mismatches, gap penalties that are more tolerant of single-base insertions or deletions (thus sequencing errors are tolerated). It also has low complex filtering caused by unusual composition that can give rise to artifactual alignments.

Result VecScreen reports contamination in terms of matches. Matches, if any, are categorized as strong, moderate, weak, or suspect.

Note

Strong match to vector
(Expect 1 random match in 1,000,000 queries of length 350 kb.)
Terminal match with Score = 24.
Internal match with Score = 30.

Moderate match to vector
(Expect 1 random match in 1,000 queries of length 350 kb.)
Terminal match with Score 19 to 23.
Internal match with Score 25 to 29.

Weak match to vector
(Expect 1 random match in 40 queries of length 350 kb.)
Terminal match with Score 16 to 18.
Internal match with Score 23 to 24.

Segment of suspect origin
Any segment of fewer than 50 bases between two vector matches or between a match and an end.

TASK

Step 1 Paste the FASTA format sequence as mentioned below in the input box and click on **Run Vecscreen**.

GCCGCCACCGCGGTGGAGCTCCCAGCTTTTGTTCCCTTTAGTGAGGGTTGGGccccccTCGAGGTCGACCcccTCGAGGT
CGACCC

Step 2 A new page is displayed, which provides a request ID number as shown in Fig. 5.25. Click the **Format** button.

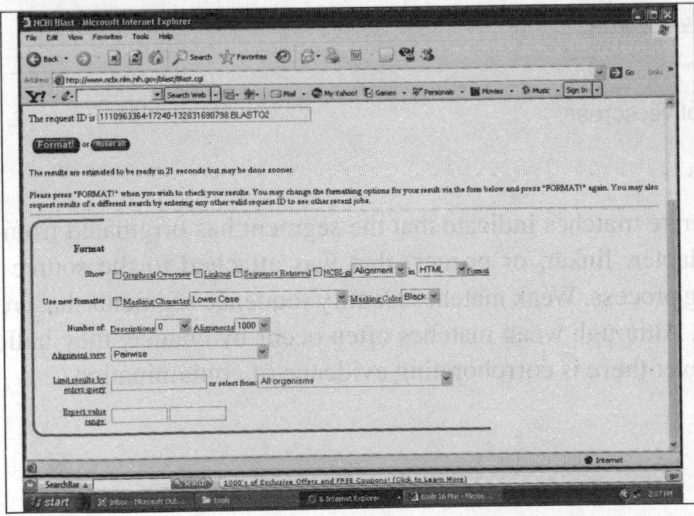

Fig. 5.25 Intermediate result page with request ID number

Step 3 The result shown in Figs 5.26 and 5.27 is thrown up in 21 seconds or lesser.

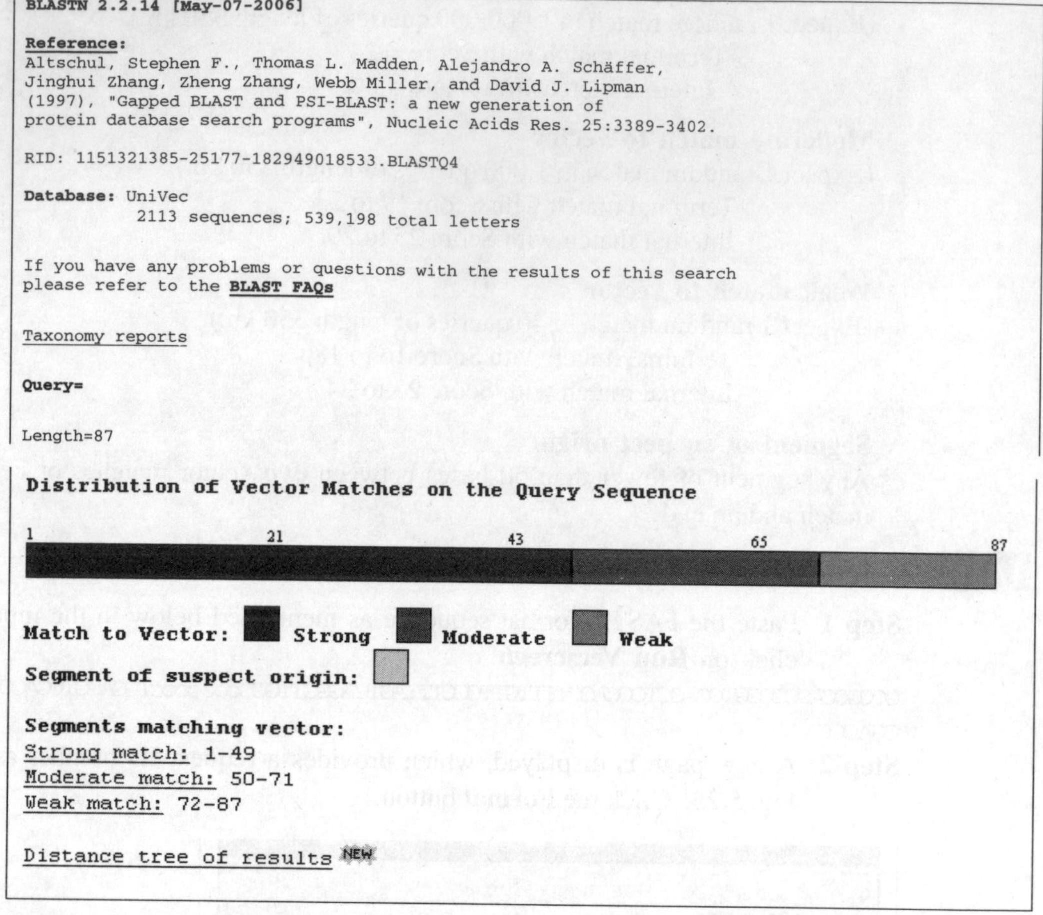

BLASTN 2.2.14 [May-07-2006]
Reference:
Altschul, Stephen F., Thomas L. Madden, Alejandro A. Schäffer,
Jinghui Zhang, Zheng Zhang, Webb Miller, and David J. Lipman
(1997), "Gapped BLAST and PSI-BLAST: a new generation of
protein database search programs", Nucleic Acids Res. 25:3389-3402.
RID: 1151321385-25177-182949018533.BLASTQ4
Database: UniVec
2113 sequences; 539,198 total letters
If you have any problems or questions with the results of this search
please refer to the BLAST FAQs
Taxonomy reports
Query=
Length=87
Distribution of Vector Matches on the Query Sequence
1
21
43
65
87
Match to Vector:
Strong
Moderate
Weak
Segment of suspect origin:
Segments matching vector:
Strong match: 1-49
Moderate match: 50-71
Weak match: 72-87
Distance tree of results NEW

Fig. 5.26 Result page 1 of vecscreen

Step 4 Analysis

Strong and moderate matches indicate that the segment has originated from the foreign DNA (vector, adapter, linker, or primer) that was attached to the source DNA/RNA during the cloning process. Weak matches identify sequence segments that are potentially of foreign origin. Although weak matches often occur by chance, they indicate foreign sequences whenever there is corroborating evidence of contamination.

```
                              Alignments

> gnl|uv|X52328.1:714-808 pBluescript II SK(+) vector DNA, phagemid excised from lambda ZAPII
Length=95

 Score = 84.6 bits (42), Expect = 6e-14, Identities = 48/49 (97%), Gaps = 1/49 (2%)
 Strand=Plus/Plus

Query  1   GCCGCCACCGCGGTGGAGCTCCCAGCTTTTGTTCCCTTTAGTGAGGGTT  49
           ||||||||||||||||||||| |||||||||||||||||||||||||||
Sbjct  27  GCCGCCACCGCGGTGGAGCT-CCAGCTTTTGTTCCCTTTAGTGAGGGTT  74

> gnl|uv|U02448.1:3446-3536 Cloning vector pMAMneo-LUC Length=91

 Score = 44.5 bits (22),  Expect = 0.068
 Identities = 22/22 (100%), Gaps = 0/22 (0%)
 Strand=Plus/Plus

Query  50  GGGcccccccTCGAGGTCGACC  71
           ||||||||||||||||||||||
Sbjct  29  GGGCCCCCCCTCGAGGTCGACC  50

 Score = 34.5 bits (17), Expect = 71, Identities = 17/17 (100%), Gaps = 0/17 (0%)
 Strand=Plus/Plus

Query  70  CCCCCTCGAGGTCGACC  86
           |||||||||||||||||
Sbjct  34  CCCCCTCGAGGTCGACC  50

> gnl|uv|U13860.1:6124-7626-49 pMSG cloning vector Length=1552

 Score = 32.5 bits (16), Expect = 287, Identities = 16/16 (100%), Gaps = 0/16 (0%)
 Strand=Plus/Minus

Query  72    CCCTCGAGGTCGACCC  87
             ||||||||||||||||
Sbjct  1528  CCCTCGAGGTCGACCC  1513

Database:
Posted date:  Dec 12, 2002  2:17 PM
Number of letters in database: 539,198
Number of sequences in database:  2,113
```

```
Lambda     K        H
1.39     0.747     1.38
Gapped
Lambda     K        H
1.39     0.747     1.38
Matrix: blastn matrix:1 -5
Gap Penalties: Existence: 3, Extension: 3
Number of Sequences: 2113
Number of Hits to DB: 943
Number of extensions: 208
Number of successful extensions: 208
Number of sequences better than 700: 82
Number of HSP's better than 700 without gapping: 0
Number of HSP's gapped: 200
Number of HSP's successfully gapped: 114
Length of query: 87
Length of database: 539198
Length adjustment: 12
Effective length of query: 75
Effective length of database: 513842
Effective search space: 38538150
Effective search space used: 1750000000000
A: 0
X1: 10 (20.0 bits)
X2: 14 (28.1 bits)
X3: 24 (48.1 bits)
S1: 10 (20.5 bits)
S2: 16 (32.5 bits)
```

Fig. 5.27

5.2 Tools for Protein Sequence Analysis

Like we have tools for nucleotide sequence analysis, there are various tools available for protein sequence analysis too. Again, almost all of them are available freely on the Web and many are downloadable too. In this section also, we have mostly considered proteins of *Saccharomyces cerevisiae*. The objective again is similar, i.e., when all the tools from this section have been studied along with the guided tasks, the reader will end up doing a mini project on 'data analysis of amino acid sequences'.

5.2.1 BLASTp

Tool for BLASTp is a tool for comparing an amino acid query sequence with a protein sequence database. It detects regions with similar sequences even in functionally unrelated proteins. From these regions, the functions of uncharacterized proteins can be determined.

Algorithm used BLAST uses a heuristic algorithm that seeks local as opposed to global alignments and is therefore able to detect relationships among sequences that share only isolated regions of similarity. The standard protein–protein BLASTp attributes enable users to optimize their BLAST search. BLASTp utilizes sequence filtering and gapped alignments as default settings. The filter option allows the tool to mask regions of a query sequence such that regions of low complexity are excluded (e.g., repetitive elements). Gaps (regions of insertions and deletions) are introduced into alignments as the default setting.

Substitution matrices are used to search for similarities during alignment. The two main types of these matrices that are frequently used by programs such as the BLAST family are the PAM (percent accepted mutation) matrix and BLOSUM (Blocks substitution matrix). PAM matrices are most sensitive to alignments of sequences with evolutionary-related homologues. The greater the number in the matrix name, the greater the expected evolutionary (mutational) distance. BLOSUMs are most sensitive to local alignment of related sequences and are therefore ideal when one is trying to identify an unknown nucleotide sequence. BLOSUM62 is used by default for protein BLAST searches.

Note

Both the BLASTn and BLASTp search tools offer fully gapped alignments while BLASTx and TBLASTn have 'in-frame' gapped alignments and the TBLASTx search tool provides only ungapped alignments.

Tool screenshot The screenshot of the BLASTp tool is shown in Fig. 5.28.

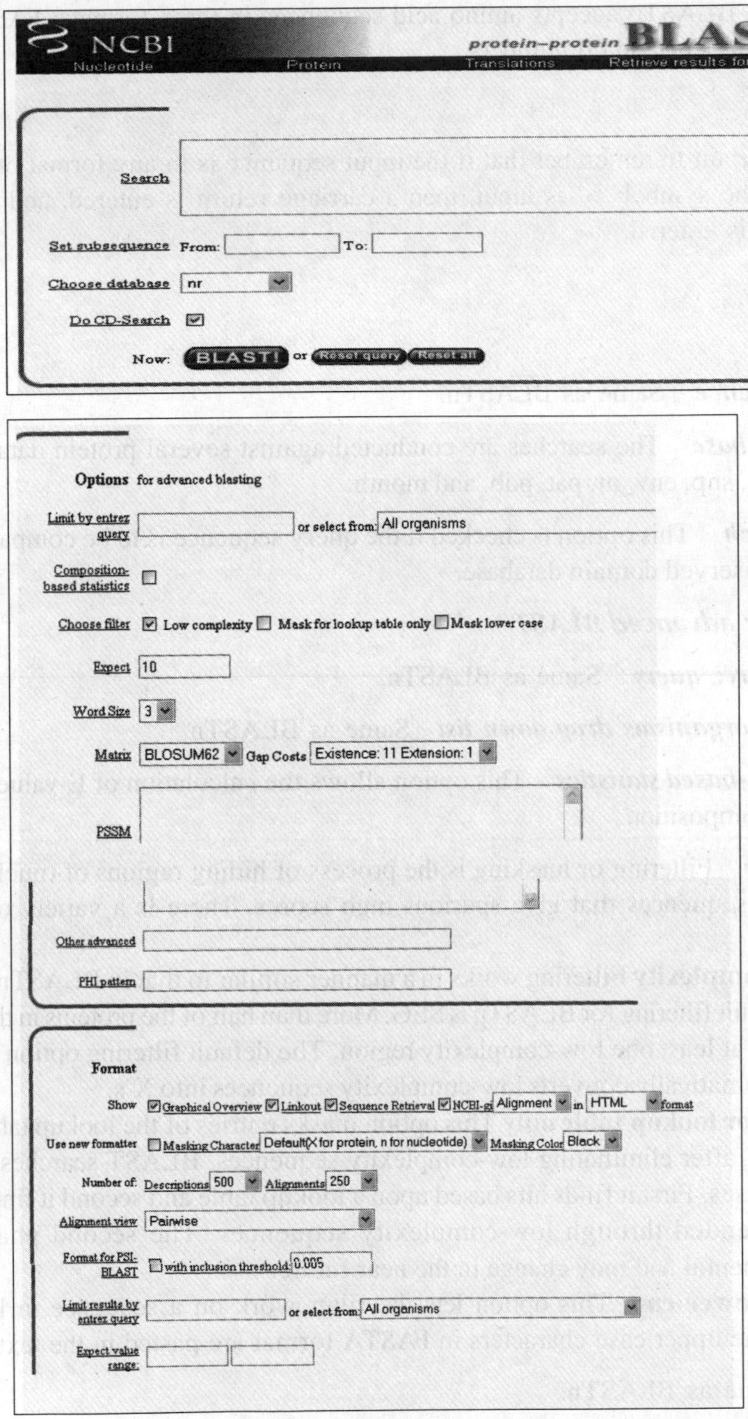

NCBI protein–protein BLAST

Nucleotide | Protein | Translations | Retrieve results for an RID

Search

Set subsequence From: To:

Choose database nr

Do CD-Search

Now: BLAST! or Reset query Reset all

Options for advanced blasting

Limit by entrez query or select from: All organisms

Composition-based statistics

Choose filter Low complexity Mask for lookup table only Mask lower case

Expect 10

Word Size 3

Matrix BLOSUM62 Gap Costs Existence: 11 Extension: 1

PSSM

Other advanced

PHI pattern

Format

Show Graphical Overview Linkout Sequence Retrieval NCBI-gi Alignment in HTML format

Use new formatter Masking Character Default(X for protein, n for nucleotide) Masking Color Black

Number of: Descriptions 500 Alignments 250

Alignment view Pairwise

Format for PSI-BLAST with inclusion threshold 0.005

Limit results by entrez query or select from: All organisms

Expect value range:

Fig. 5.28 Screenshot of the BLASTp tool

Input data BLASTp accepts amino acid sequences in three formats: FASTA, NCBI accession numbers, and GI numbers.

> **Note**
>
> It is important to remember that if the input sequence is in any format other than FASTA, the symbol '>' is input, then a carriage return is entered, and then the sequence is entered.

Attributes

General

Set subsequence Same as BLASTn.

Choose database The searches are conducted against several protein databases such as nr, Refseq, snp, env_nr, pat, pdb, and month.

Do CD Search This option is checked if the query sequence is to be compared against a protein conserved domain database.

Options for advanced BLAST

Limit by entrez query Same as BLASTn.

Select from organisms drop down list Same as BLASTn.

Composition-based statistics This option allows the calculation of E-values based on amino acid composition.

Choose filter Filtering or masking is the process of hiding regions of (nucleic acid or amino acid) sequences that give spurious high scores. There is a variety of filters to choose from.

- **Low-complexity** Filtering works in a manner similar to that in BLASTn. However the default filtering for BLASTp is SEG. More than half of the proteins in the database contain at least one low-complexity region. The default filtering option in BLAST 2.0 automatically converts low-complexity sequences into X's.
- **Mask for lookup table only** This option masks entries of the lookup table used by BLAST after eliminating low-complexity sequences. BLAST searches consist of two phases. First it finds hits based upon a lookup table and second it finds hits that are extended through low-complexity sequences. The second phase is still experimental and may change in the near future.
- **Mask lower case** This option lets the filter work on a sequence in lower case when the upper case characters in FASTA format are pasted in the textbox.

Expect Same as BLASTn.

Word size The available word sizes in BLASTp are 2 and 3.

Matrix This refers to the substitution matrix, which assigns a score for aligning any possible pair of residues. It is possible to choose the matrix to be used depending on the type of sequences one is searching with. For example, BLOSUM 62 matrix is the best matrix for detecting most weak protein similarities, but for long and weak alignments BLOSUM 45 is recommended.

Gap Costs Gapped BLAST and PSI-BLAST use *affine gap costs*, which charge the score $-a$ for the existence of a gap and the score $-b$ for each residue in the gap. Thus, a gap of y residues receives a total score of $-(a + by)$; specifically, a gap of length 1 receives the score $-(a + b)$. For proteins, Table 5.2 is a provisional table of recommended substitution matrices and gap costs for various query lengths.

Query length	**Substitution matrix**	**Gap costs**	
		Existence (creation)	**Extension**
< 35	PAM30	9	1
35–50	PAM70	10	1
50–85	BLOSUM80	10	1
> 85	BLOSUM62	11	1

PSSM Not applicable.

Other advanced same as BLASTn.

PHI pattern Not applicable.

Various format options available under BLAST

Show Same as BLASTn.

Use new formatter Under this head, options for masking character and colour are provided. Characters can be masked either with small letters or the default character, i.e., 'X' for proteins and 'n' for nucleotides. The masking colour can be black, grey, and red.

Descriptions Same as BLASTn. The default value is 500 descriptions.

Alignments Same as BLASTn. The default value is 250.

Alignments views Same as BLASTn.

Format for PSI-BLAST with inclusion threshold E-value cutoff for PSSM. Limits all results to an E-value threshold of 1 or less. The default is set at 0.005.

Limits results by Entrez query Same as BLASTn.

Expect value range Same as BLASTn.

Layout Same as BLASTn.

Formatting options with page with results Same as BLASTn.

Autoformat Same as BLASTn.

The e-mail option Enables users to receive BLAST search results in a convenient form at a user-specified e-mail address. As the default setting, BLAST search results are returned in text format.

Result

BLASTp identifies sequences in a database and predicts the biological significance of the matches.

TASK

Input

Use the protein from alcohol dehydrogenase (EC 1.1.1.1) 4 from yeast (Saccharomyces cerevisiae) or the following sequence:

```
mlgityavns tkqlifcclk yltllgyill snrkkgqrtn mykrvisisg llktgvkrfs
svyckttinn kftfattnsq irkmssvtgf yippisffge galeetadyi knkdykkali
vtdpgiaaig lsgrvqkmle erdlnvaiyd ktqpnpnian vtaglkvlke qnseivvsig
ggsahdnaka iallatngge igdyegvnqs kkaalplfai nttagtasem trftiisnee
kkikmaiidn nvtpavavnd pstmfglppa ltaatgldal thcieayvst asnpitdaca
lkgidlines lvaaykdgkd kkartdmcya eylagmafnn aslgyvhala hqlggfyhlp
hgvcnavllp hvqeanmqcp kakkrlgeia lhfgasqedp eetikalhvl nrtmniprnl
kelgvktedf eilaehamhd achltnpvqf tkeqvvaiik kayey
```

Attribute settings

General options

Set subsequence Leave this blank.

Choose databases nr (non-redundant) protein database is selected.

***Do CD*-Search** Yes.

Options for advanced BLAST

Limit by entrez query Leave this blank.

Select from organisms drop down list Select **Homo sapiens**.

Composition-based statistics Yes.

Choose filter Low complexity is checked and the remaining left unchecked.

Expect The threshold of statistical significance for reporting matches against database sequences is set to 10.

Word size Set to 3.

Matrix BLOSUM 62.

Gap Costs Existence: 11, Extension: 1.

PSSM Leave this blank.

Other Advanced Leave this blank.

PHI pattern Leave this blank.

Format options available under BLASTp

Show Formatting options such as Graphical overview, Linkout, Sequence Retrieval, NCBI-gi are checked. The format option of **Alignment** is HTML.

Use new formatter **Masking Character** and Masking Colour are 'n' and Black, respectively.

Descriptions and alignments Default values of 500 and 250 are selected, respectively.

Alignments views is pairwise with identities .

Format for PSI-BLAST with inclusion threshold

Limit results by entrez query Leave this blank **or select from** Homo sapiens

Expect value range <default value>

Layout Two windows

Formatting options on page with results None.

Autoformat Full auto.

Result

The result page for BLASTp is shown in Fig. 5.29.

RID: 1111994805-18821-140973348408.BLASTQ2

Query = (465 letters)

Database: All non-redundant GenBank CDS translations+PDB+Swiss-Prot+PIR+PRF excluding environmental samples 2,408,643 sequences; 815,228,232 total letters

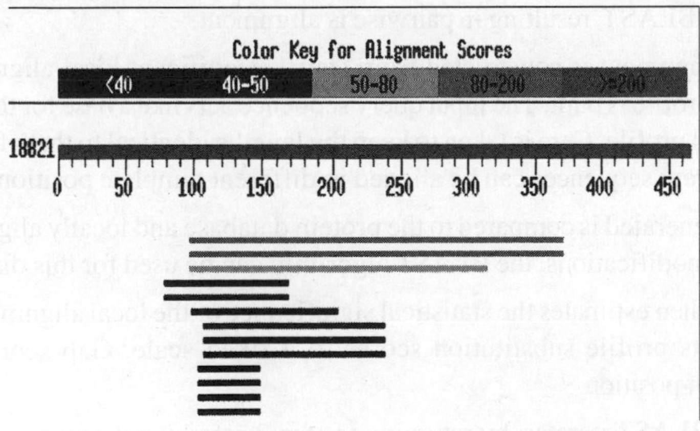

Fig. 5.29 Result page of BLASTp

```
                                                                 Score    E
Sequences producing significant alignments:                      (bits) Value

gi|16552547|dbj|BAB71335.1|  unnamed protein product [Homo s...    92   3e-18  G
gi|25989126|gb|AAK44223.1|  alcohol dehydrogenase 8 [Homo sa...    92   3e-18  G
gi|40352926|gb|AAH64634.1|  ADHFE1 protein [Homo sapiens]          68   3e-11  G
gi|6996447|emb|CAB75426.1|  chaperonin 60, Hsp60 [Homo sapie...    34   0.70   G
gi|306890|gb|AAA36022.1|  chaperonin (HSP60)                       34   0.70   G
gi|40788306|dbj|BAA31598.2|  KIAA0623 protein [Homo sapiens]       33   1.2    G
gi|27754771|ref|NP_002578.2|  protocadherin 1 isoform 1 prec...    31   4.6    G
gi|27754773|ref|NP_115796.2|  protocadherin 1 isoform 2 prec...    31   4.6    G
gi|24212077|sp|Q08174|PCH1_HUMAN  Protocadherin 1 precursor ...    31   4.6    G
gi|56204886|emb|CAI18994.1|  tetratricopeptide repeat domain...    31   6.0    G
gi|30582473|gb|AAP35463.1|  tetratricopeptide repeat domain ...    31   6.0    G
gi|56204887|emb|CAI18995.1|  tetratricopeptide repeat domain...    31   6.0
gi|56204885|emb|CAI18993.1|  tetratricopeptide repeat domain...    31   6.0    G
gi|4507831|ref|NP_003556.1|  unc-51-like kinase 1 [Homo sapi...    31   6.0    G
gi|4406093|gb|AAD19853.1|  tetratricopeptide repeat protein ...    31   6.0    G
gi|20521139|dbj|BAA34442.2|  KIAA0722 protein [Homo sapiens]       31   6.0    G
gi|23241685|gb|AAH34988.1|  ULK2 protein [Homo sapiens]            30   7.8    G
gi|23199985|ref|NP_055498.2|  unc-51-like kinase 2 [Homo sap...    30   7.8    G
```

5.2.2 PSI-BLAST

Tool for Position-specific iterated BLAST (PSI-BLAST) is used for performing BLAST on the set of hits resulting form BLASTp to increase the sensitivity of the result. It computes a position-specific scoring matrix (PSSM) that yields high scores to highly conserved positions and low scores to weakly conserved positions.

Note

PSSM is a profile that is generated by calculating position-specific scores for each position in the alignment.

Algorithm used Iterated profile search methods have led to several biologically important observations, but suffered from drawbacks such as being slow and not providing the means for evaluating the significance of their results. The PSI-BLAST program was designed to overcome drawbacks. The procedure that PSI-BLAST uses is the following.

1. A single protein sequence is taken as an input and compared to a protein database using gapped BLAST, resulting in pairwise is alignment.
2. A multiple alignment is constructed from step 1, significant local alignments are found, and a profile is built. The input query sequence acts like a base for the multiple alignment and profile. Care is taken to keep the lengths identical to that of the query. Several different sequences can be aligned in different template positions.
3. The profile generated is compared to the protein database and locally aligned. After a few minor modifications, the BLAST algorithm can be used for this directly.
4. PSI-BLAST then estimates the statistical significance of the local alignments found and constructs profile substitution scores to a fixed scale. Gap scores remain independent of position.
5. Finally, PSI-BLAST iterates, by returning to step 2 an arbitrary number of times or until convergence is achieved.

Unlike other tools, PSI-BLAST runs as one program, starting with a single protein sequence, and the intermediate steps of multiple alignment and profile construction are invisible to the user.

Tool screenshot Same as BLASTp.

Input data Same as BLASTp.

Attributes

General

Set subsequence Same as BLASTn.

Choose databases The searches are conducted against several protein databases such as nr, Refseq, sp, env_nr, pat, pdb, and month.

Do CD-search Same as BLASTp.

Options for advanced BLAST

Limit by Entrez query Same as BLASTn.

Select from organisms drop down list Same as BLASTn.

Composition-based statistics This option allows the calculation of E-values based on the amino acid composition of the reported sequences that result in alignments. It has two advantages: increase in E-value accuracy and decrease in the number of false positives. The calculation of E-values will not tally with the result of the usual substitution matrices such as BLOSUM62 and PAM70 because of the use of a different scoring system. Therefore it is not surprising that the E-values received after BLAST differ from those of PSI-BLAST even though the query sequence is identical. Furthermore, identical alignments can receive different scores, based upon the compositions of the sequences they involve. Also, two PSI-BLAST parameters get adjusted automatically/internally. The pseudocount constant default (amino acid frequency) is changed from 10 to 7, and the E-value threshold for including matches in the PSI-BLAST model is changed from 0.001 to 0.002.

Choose filter Same as BLASTp.

Expect Same as BLASTn.

Word size Same as BLASTn. The available word sizes in BLASTp are 2 and 3.

Matrix Same as BLASTp.

Gap Costs Same as BLASTp.

PSSM The PSSM is a motif descriptor. The description includes a weight (score, probability, likelihood) for each symbol occurring at each position along the motif. A position-specific scoring matrix is constructed using the output of a BLAST run and the matrix in place of the query sequence. Profiles or motifs are better at detecting weak relationships than database searches using a query sequence. PSSMs enable the scoring of multiple alignments with sequences, structures, etc.

Other advanced Same as BLASTn.

PHI pattern Pattern-hit initiated (PHI) BLAST is a search that combines matching of regular expressions with local alignments surrounding the match. Given a particular protein sequence and a normally associated pattern, it looks for other protein sequences that both have a similar sequence and are homologous to that pattern. The option is advantageous in cases where the pattern occurrence is random and thus indicative of homology.

Format options available under PSI-BLAST are as below

Show Same as BLASTn.

Use new formatter Under this head options for masking character and colour are provided. Character can be either with small letters or the default character i.e. 'X' for proteins and 'n' for nucleotides. Masking colour can be black, grey and red.

Descriptions Same as BLASTn. The default value is 500 descriptions.

Alignments Same as BLASTn. The default value is 250.

Alignments views Same as BLASTn.

Format for PSI-BLAST with inclusion threshold

Limit values by entrez query Same as BLASTp.

Expect Value Range Same as BLASTp.

Layout Same as BLASTp.

Formatting options on page with results Same as BLASTn.

Autoformat Same as BLASTn.

Result

1. Graphical result of hit sequences with colour code.
2. Result of hit sequences with score in bits and E-value.
3. Graphical result of hit sequences with colour code.
4. Result of hit sequences with score in bits and E-value.
5. Statistical report.

TASK

Interpro (http://www.ebi.ac.uk/interpro/DisplayIproEntry?ac=IPR001404) gives the description of Hsp90 as follows.

Prokaryotes and eukaryotes respond to heat shock and other forms of environmental stress by inducing synthesis of heat-shock proteins (Hsp). The 90-kDa heat shock protein, Hsp90, is one of the most abundant proteins in eukaryotic cells, comprising 1%–2 % of cellular proteins under non-stress conditions. Its contribution to various cellular processes including signal transduction, protein folding, protein degradation, and morphological evolution has been extensively studied. The full functional activity of Hsp90 is gained in concert with other co-chaperones, playing an important role in the folding of newly synthesized proteins and the stabilization and refolding of denatured proteins after stress. Apart from its co-chaperones, Hsp90 binds to an array of client proteins, where the co-chaperone requirement varies and depends on the actual client. The sequences of Hsp90s show a distinctive domain structure, with a highly conserved, N-terminal domain separated from a conserved, acidic C-terminal domain by a highly acidic, flexible linker region.

Input

```
LOCUS      XP_611032           112 aa        linear   MAM 25-MAR-2005
DEFINITION PREDICTED: similar to Heat shock protein HSP 90-beta (HSP 84),
partial [Bos taurus].
ACCESSION    XP_611032
VERSION      XP_611032.1  GI:61836621
DBSOURCE     REFSEQ: accession XM 611032.1
KEYWORDS     .
SOURCE       Bos taurus (cow)

firgvvdsed lplnisreml qqskilkvir knivkkclel fselaedken ykkfyeafsk
nlklgiheds tnrrrlsell ryhtsqsgde mtslseyvsr mketqksiyy it
```

Attribute Settings

General options

Set subsequence NA

Choose databases nr (non-redundant) nucleotide database is selected.

Do CD-Search Checked.

Options for advanced BLAST

Limit results by Entrez query NA.

Select from organisms drop down list Saccharomyces cerevisiae.

Composition-based statistics Checked.

Choose filter Low complexity is checked and the remaining are left unchecked.

Expect The threshold of statistical significance for reporting matches against database sequences is set to 1.

Word size To increase the sensitivity of the search, the word size is kept to 7.

Matrix NA.

Gap costs NA.

PSSM NA.

Other advanced NA.

PHI pattern NA.

Format options available under BLAST

Show NA.

Formatting options Graphical Overview, Linkouts, Sequence Retrieval, and NCBI-gi are checked.

Format option of alignment HTML.

Use new formatter Masking character and colour are 'n' and Black, respectively.

Descriptions and alignments Default values of 500 and 250.

Alignment views Pairwise with identities.

Format for PSI-BLAST with threshold value NA.

Limit results by Entrez query NA.

Expect value range NA.

Layout This option enables one to display the result in two windows or in one window.

Formatting options on page with results None.

Autoformat Full auto.

Result

Database All non-redundant GenBank CDS translations+PDB+Swiss-Prot+PIR+PRF excluding environmental samples 2,408,643 sequences; 815,228,232 total letters

1. The graphical result of the hit sequences with the coluor code is shown in Fig. 5.30.

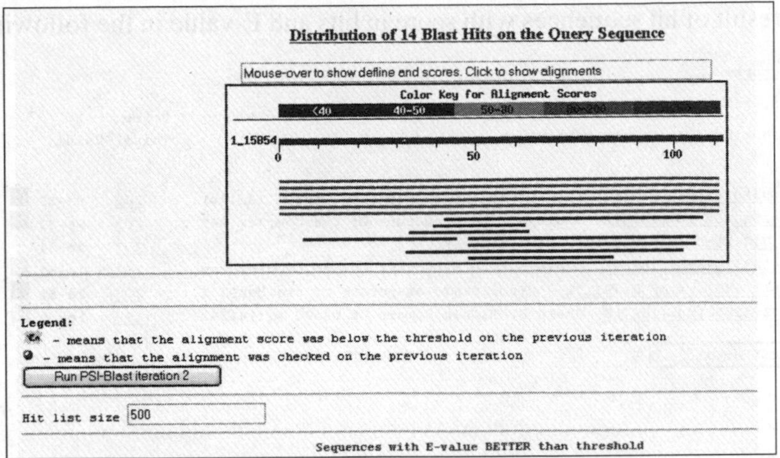

Fig. 5.30 Results of PSI-BLAST iteration 1

2. The result of hit sequences with score in bits and E-value is the following:

	Sequence	Description	Score	E-value	
☐	gi|6320890|ref|NP_010969.1|	Yer049wp [Saccharomyces cerevisiae] ...	27	0.90	G
☐	gi|6323487|ref|NP_013559.1|	Fmp27p [Saccharomyces cerevisiae] >g...	26	1.7	G
☐	gi|51013101|gb|AAT92844.1|	YBR272C [Saccharomyces cerevisiae] >g...	26	2.1	G
☐	gi|37362638|ref|NP_010773.2|	Protein of unknown function, compon...	26	2.4	G
☐	gi|927756|gb|AAB64928.1|	Ydr485cp; CAI: 0.14 [Saccharomyces cere...	26	2.5	
☐	gi|6323078|ref|NP_013150.1|	Ylr049cp [Saccharomyces cerevisiae] ...	24	6.2	G
☐	gi|6322777|ref|NP_012850.1|	Lhs1p [Saccharomyces cerevisiae] >gi...	24	6.8	G
☐	gi|6321012|ref|NP_011091.1|	Chd1p [Saccharomyces cerevisiae] >gi...	24	7.6	G

Run PSI-Blast iteration 2

3. The graphical result of hit sequences with colour code are as follows (Fig. 5.31).

Results of PSI-BLAST iteration 2: No new sequences were found above the 0.005 threshold!

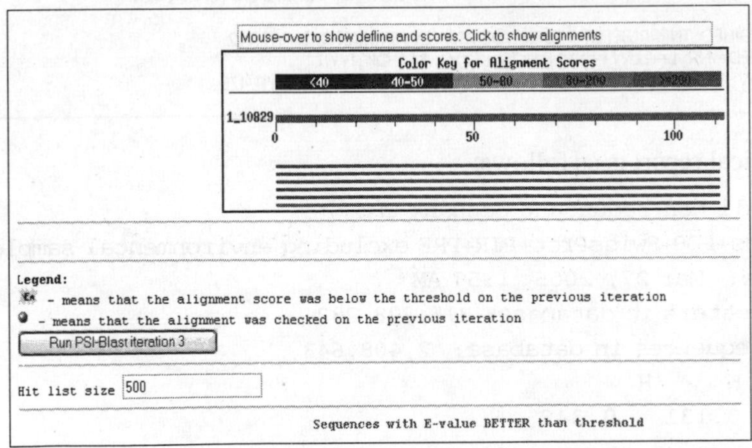

Fig. 5.31 Results of PSI-Blast iteration 2

4. The result of hit sequences with score in bits and E-value in the following:

```
Related Structures

                                                                          Score    E
Sequences producing significant alignments:                               (bits) Value

gi|6325016|ref|NP_015084.1|  Cytoplasmic chaperone (Hsp90 family)...       205   4e-52 G
gi|42543756|pdb|1USU|A  Chain A, The Structure Of The Complex Bet...       203   9e-52 S
gi|171723|gb|AAA02813.1|  hsc82 protein                                    203   1e-51
gi|42543764|pdb|1USV|G  Chain G, The Structure Of The Complex Bet...       203   1e-51 S
gi|6323840|ref|NP_013911.1|  Cytoplasmic chaperone of the Hsp90 f...       203   2e-51 G
gi|42542980|pdb|1HK7|B  Chain B, Middle Domain Of Hsp90 >gi|42542...       199   2e-50 S

Run PSI-Blast iteration 3
```

Legend

NEW means that the alignment score was below the threshold on the previous iteration

● means that the alignment was checked on the previous iteration

```
gi|6325016|ref|NP_015084.1| G Cytoplasmic chaperone (Hsp90 family) required for pheromone signaling
and negative regulation of Hsf1p; docks with the mitochondrial import receptor Tom70p for preprotein delivery;
interacts with co-chaperones Cns1p, Cpr6p,Cpr7p, and Sti1p (Saccharomyces cerevisiae)
gi|1370495|emb|CAA97961.1| G HSP82 (Saccharomyces cerevisiae)
gi|1061249|emb|CAA91604.1| G HSP90/HSP82? (Saccharomyces cerevisiae)
gi|72226|pir||HHBY90 G heat shock protein 90 - yeast (Saccharomyces cerevisiae)
gi|123677|sp|P02829|HSP82_YEAST G ATP-dependent molecular chaperone HSP82 (Heat shock protein Hsp90
    heat inducible isoform) (82 kDa heat shock protein)
gi|171725|gb|AAA02743.1| G hsp82 protein
    Length = 709

Score = 205 bits (521), Expect = 4e-52
Identities = 77/112 (68%), Positives = 101/112 (90%)

Query: 1   FIRGVVDSEDLPLNISREMLQQSKILKVIRKNIVKKCLELFSELAEDKENYKKFYEAFSK 60
            F++GVVDSEDLPLN+SREMLQQ+KI+KVIRKNIVKK +E F+E+AED E ++KFY AFSK
Sbjct: 364 FVKGVVDSEDLPLNLSREMLQQNKIMKVIRKNIVKKLIEAFNEIAEDSEQFEKFYSAFSK 423

Query: 61  NLKLGIHEDSTNRRRLSELLRYHTSQSGDEMTSLSEYVSRMKETQKSIYYIT 112
            N+KLG+HED+ NR  L++LLRY++++S DE+TSL++YV+RM E QK+IYYIT
Sbjct: 424 NIKLGVHEDTQNRAALAKLLRYNSTKSVDELTSLTDYVTRMPEHQKNIYYIT 475
```

5. The statistical report is as follows:

```
Database: All non-redundant GenBank CDS
translations+PDB+SwissProt+PIR+PRF excluding environmental samples
Posted date:  Mar 27, 2005  1:57 AM
Number of letters in database: 815,228,232
Number of sequences in database:  2,408,643
Lambda     K      H
   0.315    0.131    0.348
Gapped
```

```
Lambda     K      H
   0.267  0.0453    0.140
Matrix: BLOSUM62
Gap Penalties: Existence: 11, Extension: 1
Number of Hits to DB: 140,419
Number of Sequences: 2408643
Number of extensions: 6133
Number of successful extensions: 24
Number of sequences better than 10.0: 4
Number of HSP's better than 10.0 without gapping: 3
Number of HSP's successfully gapped in prelim test: 1
Number of HSP's that attempted gapping in prelim test: 21
Number of HSP's gapped (non-prelim): 4
length of query: 112
length of database: 4,694,612
effective HSP length: 79
effective length of query: 33
effective length of database: 3,840,622
effective search space: 126740526
effective search space used: 126740526
T: 11
A: 40
X1: 16 ( 7.3 bits)
X2: 38 (14.6 bits)
X3: 64 (24.7 bits)
S1: 41 (21.6 bits)
S2: 50 (23.7 bits)
```

5.2.3 Tool Name: RPS-BLAST

Tool for RPS-BLAST (reverse PSI-BLAST) searches a query sequence against a database of profiles, or score matrices, producing BLAST-like output.

Link http://www.ncbi.nlm.nih.gov/Structure/cdd/wrpsb.cgi and several other sites on the Internet like the Biology Workbench at http://workbench.sdsc.edu/

Type of tool Web-based and downloadable

Institute NCBI. Shavirin and Madden coded RPS-BLAST.

Algorithm used BLAST-like algorithm.

Tool screenshot The screenshot of RPS-BLAST is shown in Fig. 5.32.

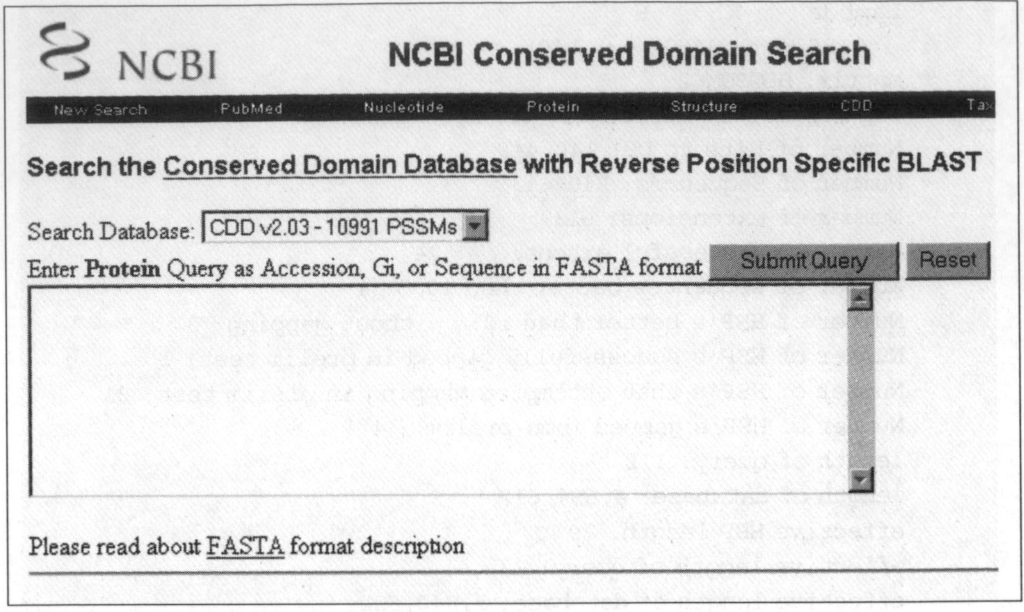

Fig. 5.32 Screenshot of RPS BLAST tool

Input data Same as BLASTp.

Attributes

Search database The query sequence is searched against databases such as Smart, Pfam, COG, KOG, and CDD, or score matrices of these databases.

Advanced options

Expect The various options available are 0.000001/0.0001/0.01/1/10/100. The default **Expect** threshold value setting is 0.01. For more stringent results, settings less than 0.01 are preferred. When results show E-values in the range of 1 and more, the sequences are considered to be putative false positive.

Filter The filter option is available to eliminate low-complexity regions from the searched sequences. It can be either checked or unchecked. If the option is checked the graphical display on the result page will highlight the filtered regions.

Search mode This option allows to select among multiple hits, one-pass, single, hit, one-pass, and single hit, two-pass. The one-pass search offers greater speed than two-pass search but is less sensitive. However, stringency is greater in case of single hit, one pass as compared to multiple hit, one pass.

Output formatting options

Display up to This option is used to set the number of hits displayed. For mid-sized proteins, the default value of 100 is usually sufficient.

With These are options such as **no graphic/condensed graphic/extended graphic overview**. When the **extended graphic overview** option is selected, then the location of domain hits on the query sequence are displayed graphically. In the **condensed graphic** view, only the best scoring hit for a domain is displayed.

In There are three colour scheme options that help to distinguish and align similar and masked regions. These are colour scheme 1, 2, and 3. For example, in colour scheme 3, identical sequence areas are coloured red, similar residues coloured blue, while masked out regions are italicized.

Print graphics using This option offers various resolution options: from high to low, such as 5 pixels per residue/2 pixels per residue/default width/2 residues per pixel/5 residues per pixel/10 residues per pixel. The default width is 1 residue per pixel.

Retrieve previous CD-Search request # ID number (optional):

Result

The result page of RPSBLAST has the following sections.

Graphical summary If in the attributes, graphical display is selected, the result page shows the query sequence as a black bar with divisions indicating the number of amino acids. Below the bar are shown domains from several databases that match the query sequence using CDART colour codes. When the mouse pointer is held over this, the domain accession number, type of domain (which serves as the domain name; may or may not be present), score bits, and E-value in the textbox above the graphics are displayed. Clicking on one of the domains takes one to the domain entry of the corresponding database (not necessarily the same as what was selected during RPS-BLAST). It also displays multiple alignments of the query sequence with the proteins having the same domains. Red bars indicate identical matches; pink indicate probable homologues; and green, blue, and black represent more distantly related homologues. Partial length bars reflect similarity between a portion of the query and a database entry. These regions of similarity may be a shared domain or motif. All domains that are coloured grey belong to the single-domain family. Some CD alignment models may contain multiple domains and would not have been used in neighbouring CDs. These are coloured grey with a black outline. The domain bar with jagged edges may point to truncated query sequences, false-positive hits, or unusual domain architectures involving long insertions.

Show domain relatives This option allows one to click and view the domain relatives of the query sequence in CDART.

CD-search hitlist This displays all significant alignments with score bits and E-values. The values set for **Expect** and **Display up to** decide the number of hits displayed.

Pairwise alignments This displays regions of alignment of the query sequence against the CD hits provided by the search. The bit score and the E-values of the alignment may not always correlate with the values of the hit list, as they are calculated on the basis of the position-specific scoring matrix (PSSM), based on the alignment of all the sequences in the database with the query sequence.

Pink dot A pink dot preceding the entries of the **CD-search hitlist** and the alignments links to the 3D structure. Clicking on any of these allows one to download the **.cn3** file that can be visualized be using Cn3D tools.

Input

```
MSLKEEQVSIKQDPEQEERQHDQFNDVQIKQESQDHDGVDSQYTNGTQNDDSERFEAAESDVKVEPGLGM
GITSSQSEKGQVLPDQPEIKFIRRQINGYVGFANLPKQWHRRSIKNGFSFNLLCVGPDGIGKTTLMKTLF
NNDDIEANLVKDYEEELANDQEEEEGQGEGHENQSQEQRHKVKIKSYESVIEENGVKLNLNVIDTEGFGDKEKYGFAGFPVTAD
GKRNAKLVGAITSRDIQFVEDNSLLVQDVMIKNPVTGAQGITLSEGNEILKKIKKGRLLVVDEKGNLVSMLSRTDLMKNQKYPLASK
SANTKQLLWGASIGTMDADKERLRLLVKAGLDVVILDSSKEKYGFAGFPVTADGKRNAKLVGAITSRDIQFVEDNSLLVQDVMIK
NPVTGAQGITLSEGNEILKKIKKG
RLLVVDEKGNLVSMLSRTDLMKNQKYPLASKSANTKQLLWGASIGTMDADKERLRLLVKAGLDVVILDSS
```

Attribute settings

Search database CDD

Advanced options

Expect The default **Expect** threshold value was set, i.e., 0.01.

Filter Low complexity was checked.

Search mode Multiple hits 1-pass was selected, which is the default option.

Output formatting Options

Display up to 100.

With Extended Graphic Overview.

In Colour scheme Colour scheme 3.

Print graphics using Default width.

Retrieve previous CD-Search request # Leave this blank.

Result and its interpretation

RPS-BLAST 2.2.9 [May-01-2004]

Query= local sequence: lclltmpseq_0 unnamed protein product

490 letters)

Database: cdd.v2.03

Figure 5.33 shows the result of RPS-BLAST.

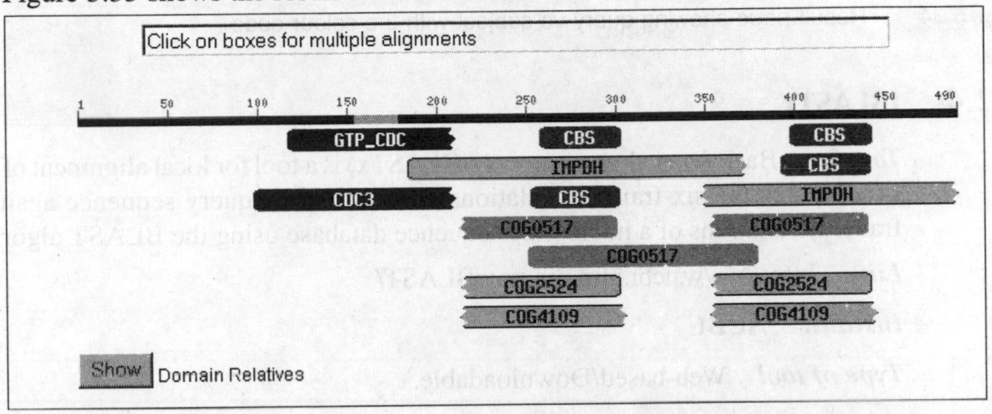

Fig. 5.33 Result Page of RPS-BLAST

PSSMs producing significant alignments:	Score (bits)	E value
● gnl|CDD|24208 smart00116, CBS, Domain in cystathionine beta-synthase and oth...	40.8	4e-04
● gnl|CDD|24208 smart00116, CBS, Domain in cystathionine beta-synthase and oth...	40.8	4e-04
● gnl|CDD|25520 pfam00478, IMPDH, IMP dehydrogenase / GMP reductase domain. Th...	144	3e-35
● gnl|CDD|25520 pfam00478, IMPDH, IMP dehydrogenase / GMP reductase domain. Th...	133	5e-32
gnl|CDD|7746 pfam00735, GTP_CDC, Cell division protein. Members of this fam...	86.4	7e-18
● gnl|CDD|25557 pfam00571, CBS, CBS domain. CBS domains are small intracellula...	46.8	6e-06
● gnl|CDD|25557 pfam00571, CBS, CBS domain. CBS domains are small intracellula...	46.8	6e-06
gnl|CDD|14149 COG5019, CDC3, Septin family protein [Cell division and chromo...	95.7	1e-20
● gnl|CDD|10388 COG0517, COG0517, FOG: CBS domain [General function prediction...	54.0	4e-08
● gnl|CDD|10388 COG0517, COG0517, FOG: CBS domain [General function prediction...	54.0	4e-08
● gnl|CDD|10388 COG0517, COG0517, FOG: CBS domain [General function prediction...	38.2	0.002
gnl|CDD|12118 COG2524, COG2524, Predicted transcriptional regulator, contain...	51.8	2e-07
gnl|CDD|12118 COG2524, COG2524, Predicted transcriptional regulator, contain...	51.8	2e-07
gnl|CDD|13391 COG4109, COG4109, Predicted transcriptional regulator containi...	43.3	7e-05
gnl|CDD|13391 COG4109, COG4109, Predicted transcriptional regulator containi...	43.3	7e-05

Figure 5.34 shows the result page displaying the query versus subject with the colour code.

- gnl|CDD|24208, smart00116, CBS, domain in cystathionine beta-synthase and other proteins. Domain present in all three forms of cellular life. Present in two copies of inosine monophosphate dehydrogenase, of which one is disordered in the crystal structure. A number of disease states are associated with CBS-containing proteins including homocystinuria, Becker's disease, and Thomsen disease.

```
CD-Length = 49 residues, 95.9% aligned
Score = 40.8 bits (96), Expect = 4e-04
Query: 257 NPVTGAQGITLSEGNEILKKIKKGRLLVVDEKGNLVSMLSRTDLMKN 303
Sbjct: 1   DVVTVSPDTTLEEALELLREHGIRRLPVVDEKGRLVGIVTRRDIIKA 47
```

Fig. 5.34 Result page showing query vs subject with the colour code

5.2.4 tBLASTx

Tool for Basic local alignment tool (tBLASTx) is a tool for local alignment of sequences. It compares the six-frame translations of a nucleotide query sequence against the six-frame translations of a nucleotide sequence database using the BLAST algorithm.

Link http://www.ncbi.nlm.nih.gov/BLAST/

Institute NCBI.

Type of tool Web-based/Downloadable.

OS required Windows.

Algorithm used tBLASTx algorithm compares the six-frame translations of a nucleotide query sequence against the six-frame translations of a nucleotide sequence database.

Tool screenshot Figure 5.35 shows a screenshot of the tBLASTx tool.

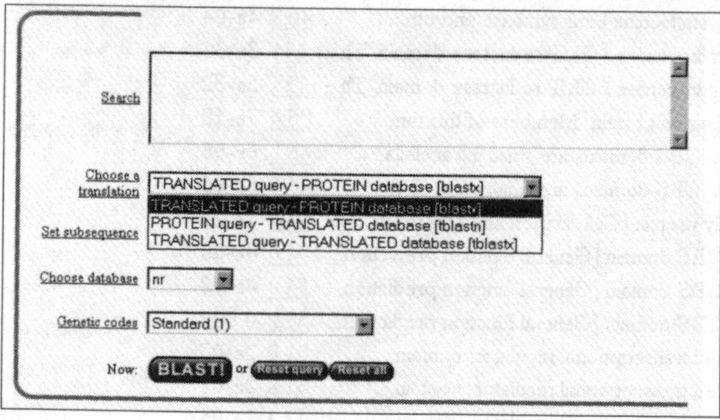

Fig. 5.34 Screenshot of the tBLASTx tool

Input data Nucleotide sequence.

Attributes

Attributes that need to be selected

Choose a translation Among three different options, TRANSLATED query - Protein database [blastx] is selected.

Set subsequence Same as BLASTn.

Choose databases The searches are conducted against several protein databases such as nr, Refseq, Swiss-Prot, pat, pdb, env_nr, month.

Genetic codes There are several genetic codes available as mentioned below:

- Standard Code
- Vertebrate Mitochondrial Code
- Yeast Mitochondrial Code
- Mold, Protozoan, and Coelenterate Mitochondrial Code ; Mycoplasma/Spiroplasma Code
- Invertebrate Mitochondrial Code
- Ciliate, Dasycladacean and Hexamita Nuclear Code
- Echinoderm and Flatworm Mitochondrial Code
- Euplotid Nuclear Code
- Bacterial and Plant Plastid Code
- Alternative Yeast Nuclear Code
- Ascidian Mitochondrial Code
- Alternative Flatworm Mitochondrial Code
- Blepharisma Nuclear Code
- Chlorophycean Mitochondrial Code
- Trematode Mitochondrial Code
- Scenedesmus Obliquus Mitochondrial Code
- Thraustochytrium Mitochondrial Code

Options for advanced BLAST

Limit results by Entrez query Same as BLASTn.

Select from organisms drop down list Same as BLASTn.

Choose filter Same as BLASTp.

Expect Same as BLASTn.

Word size Same as BLASTn. The available word sizes in BLASTp are 2 and 3.

Matrix The various available options are PAM30/PAM70/BLOSUM80/BLOSUM62/ BLOSUM45. Same as BLASTp.

Gap costs Same as BLASTp

Other advanced Same as BLASTn.

Format

Show Same as BLASTn.

Use new formatter Same as BLASTn.

Descriptions Same as BLASTn. The default value is 100 descriptions.

Alignments Same as BLASTn. The default value is 50.

Alignments view Same as BLASTn.

Limit results by entrez query Same as BLASTn.

Select from organisms drop down list Same as BLASTn.

Expect value range Same as BLASTn.

Layout Same as BLASTn.

Formatting options on page with results Same as BLASTn.

Autoformat Same as BLASTn.

Result

1. Graphical result of hit sequences with colour code.
2. Result of hit sequences with score in bits and E-value.
3. Graphical result of hit sequences with colour code.
4. Result of hit sequences with score in bits and E-value.
5. Statistical report.

TASK

Input

>gi|50593113:227733-228944 Saccharomyces cerevisiae chromosome I, complete chromosome

sequence

```
ATGGCCGCCATTAGAGACTACAAGACCGCACTAGATCTTACCAAGAGCCTACCAAGACCGGATGGTT
TGTCAGTGCAGGAACTGATGGACTCCAAGATCAGAGGTGGGTTGGCTTATAACGATTTTTTAATCTTACCA
GGTTTAGTCGATTTTGCGTCCTCTGAAGTTAGCCTACAGACCAAGCTAACCAGGAATATTACTTTAAACATT
CCATTAGTATCCTCTCCAATGGACACTGTGACGGAATCTGAAATGGCCACTTTTATGGCTCTGTTGGATGGT
ATCGGTTTCATTCACCATAACTGTACTCCAGAGGACCAAGCTGACATGGTCAGAAGAGTCAAGAACTAT
GAAAATGGGTTTATTAACAACCCTATAGTGATTTCTCCAACTACGACCGTTGGTGAAGCTAAGAGCATGAA
GGAAAAGTATGGATTTGCAGGCTTCCCTGTCACGGCAGATGGAAAGAGAAATGCAAAGTTGGTGGGTG
CCATCACCTCTCGTGATATACAATTCGTTGAGGACAACTCTTTACTCGTTCAGGATGTCATGACCAAAAAC
CCTGTTACCGGCGCACAAGGTATCACATTATCAGAAGGTAACGAAATTCTAAAGAAAATCAAAAAGGGT
AGGCTACTGGTTGTTGATGAAAAGGGTAACTTAGTTTCTATGCTTTCCCGAACTGATTTAATGAAAAATCAGA
AGTACCCATTAGCGTCCAAATCTGCCAACACCAAGCAACTGTTATGGGGTGCTTCTATTGGGACTATGGA
CGCTGATAAAGAAAGACTAAGATTATTGGTAAAAGCTGGCTTGGATGTCGTCATATTGGATTCCTCTCAAG
GTAACTCTATTTTCCAATTGAACATGATCAAATGGATTAAAGAAACTTTCCCAGATTTGGAAATCATTGCTGGT
AACGTTGTCACCAAGGAACAAGCTGCCAATTTGATTGCTGCCGGTGCGGACGGTTTGAGAATTGGTATG
```

GGAACTGGCTCTATTTGTATTACCCAAAAAGTTATGGCTTGTGGTAGGCCACAAGGTACAGCCGTCTACA
ACGTGTGTGAATTTGCTAACCAATTCGGTGTTCCATGTATGGCTGATGGTGGTGTTCAAAAACATTGGTCATA
TTATTACCAAAGCTTTGGCTCTTGGTTCTTCTACTGTTATGATGGGTGGTATGTTGGCCGGTACTACCGAATCA
CCAGGTGA

Attribute Settings

Choose databases nr (non-redundant) protein database is selected.

Genetic code Standard (1)

Choose filter Filter (Low-complexity) is checked and the remaining left unchecked.

Expect The threshold of statistical significance for reporting matches against database sequences is set to 10.

Word size The word size is kept to 3.

Matrix BLOSOM62

Gap costs Existence: 11, Extention: 1.

Formatting options such as Graphical Overview, Linkout, Sequence Retrieval, NCBI-GI are checked. The format option of alignment is HTML. **Masking Character** and **Masking colour** are 'n' and 'Black', respectively. Default values of 100 and 50 are kept for **Descriptions** and **Alignments**. **Alignment view** is Pairwise.

Layout Two windows.

Formatting option None.

Autoformat Full auto.

Result

Figure 5.36 shows the result page of tBLASTx.

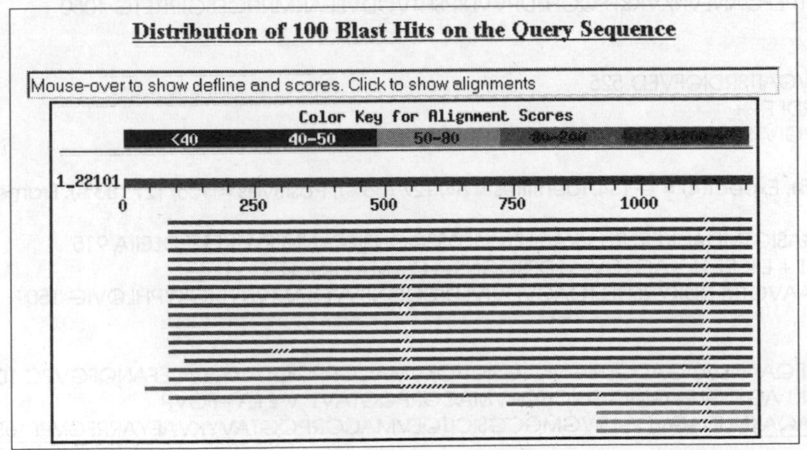

Fig. 5.36 Result page of tBLASTx

Sequences producing significant alignments:		Score (bits)	E Value	N	
gi|34328929|ref|NM_000883.2|	Homo sapiens IMP (inosine ...	211	e-134	4	G U E
gi|34528441|dbj|AK122994.1|	Homo sapiens cDNA FLJ16764 ...	211	e-134	4	G U
gi|21751052|dbj|AK092452.1|	Homo sapiens cDNA FLJ35133 ...	211	e-134	4	G U E
gi|34328927|ref|NM_183243.1|	Homo sapiens IMP (inosine ...	211	e-134	4	G U
gi|21706906|gb|BC033622.1|	Homo sapiens IMP (inosine mo...	211	e-134	4	G U E
gi|16549257|dbj|AK054667.1|	Homo sapiens cDNA FLJ30105 ...	211	e-134	4	G U
gi|57999522|emb|CR933672.1|	Homo sapiens mRNA; cDNA DKF...	211	e-134	4	G
gi|13543972|gb|BC006124.1|	Homo sapiens IMP (inosine mo...	215	e-132	4	G U E
gi|40226019|gb|BC015567.2|	Homo sapiens IMP (inosine mo...	215	e-132	4	G U E
gi|15277479|gb|BC012840.1|	Homo sapiens IMP (inosine mo...	215	e-132	4	G U E
gi|39748488|gb|AY421629.1|	Homo sapiens IMPDH2 gene, VI...	215	e-132	4	
gi|186393|gb|J05272.1|HUMIMPH	Human IMP dehydrogenase t...	211	e-132	4	G U E
gi|4504688|ref|NM_000884.1|	Homo sapiens IMP (inosine m...	215	e-132	4	G U E
gi|186391|gb|J04208.1|HUMIMP	Human inosine-5'-monophosp...	215	e-132	4	G U E
gi|47077067|dbj|AK131294.1|	Homo sapiens cDNA FLJ16255 ...	211	e-130	4	G U
gi|33874351|gb|BC009236.2|	Homo sapiens cDNA clone IMAG...	199	e-128	5	G U
gi|17985929|gb|AC069362.12|	Homo sapiens, clone RP11-74...	197	e-122	4	
gi|50657430|gb|AC091807.4|	Homo sapiens chromosome X cl...	197	e-122	4	

```
>gi|34328929|ref|NM_000883.2| G U E Homo sapiens IMP (inosine monophosphate)
dehydrogenase 1 (IMPDH1), transcript variant 1, mRNA,
Length = 2880, Score = 211 bits (456), Expect(4) = e-134, Identities = 84/140 (60%), Positives = 107/140 (76%),
Frame = +1 / +1

Query: 106  GLAYNDFLILPGLVDFASSEVSLQTKLTRNITLNIPLVSSPMDTVTESEMATFMALLDGI 285
            GL YNDFLILPG +DF + EV L + LTR ITL  PL+SSPMDTVTE++MA  MAL+ GI
Sbjct: 691  GLTYNDFLILPGFIDFIADEVDLTSALTRKITLKTPLISSPMDTVTEADMAIAMALMGGI 870

Query: 286  GFIHHNCTPEDQADMVRRVKNYENGFINNPIVISPTTTVGEAKSMKEKYGFAGFPVTADG 465
            GFIHHNCTPE QA+ VR+VK +E GFI +P+V+SP+ TVG+    K ++GF+G P+T G
Sbjct: 871  GFIHHNCTPEFQANEVRKVKKFEQGFITDPVVLSPSHTVGDVLEAKMRHGFSGIPITETG 1050

Query: 466  KRNAKLVGAITSRDIQFVED 525
            +KLVG +TSRDI F+ +
Sbjct: 1051 TMGSKLVGIVTSRDIDFLAE 1110

Score = 211 bits (455), Expect(4) = e-134, Identities = 84/127 (66%), Positives = 106/127 (83%), Frame = +1 / +1

Query: 736  KQLLWGASIGTMDADKERLRLLVKAGLDVVILDSSQGNSIFQLNMIKWIKETFPDLEIIA 915
            KQLL GA++GT + DK RL LL +AG+DV++LDSSQGNS++Q+ M+ +IK+ +P L++I
Sbjct: 1330 KQLLCGAAVGTREDDKYRLDLLTQAGVDVIVLDSSQGNSVYQIAMVHYIKQKYPHLQVIG 1509

Query: 916  GNVVTKEQAANLIAAGADGLRIGMGTGSICITQKVMACGRPQGTAVYNVCEFANQFGVPC 1095
            GNVVT  QA NLI AG DGLR+GMG GSICITQ+VMACGRPQGTAVY V E+A +FGVP
Sbjct: 1510 GNVVTAAQAKNLIDAGVDGLRVGMGCGSICITQEVMACGRPQGTAVYKVAEYARRFGVPI 1689
```

Statistical report

Database: All GenBank+EMBL+DDBJ+PDB sequences (but no EST, STS, GSS,environmental samples or phase 0, 1 or 2 HTGS sequences)
Posted date: Mar 26, 2005 9:03 PM
Number of letters in database: -309,317,664
Number of sequences in database: 1,134,558

Lambda K H
0.318 0.134 0.401

Matrix: BLOSUM62
Number of Hits to DB: 14,280,690,067
Number of Sequences: 1888534
Number of extensions: 248637816
Number of successful extensions: 756416
Number of sequences better than 10.0: 123
length of database: 1,456,963,265
effective HSP length: 62
effective length of database: 1,431,781,531
effective search space used: 488237502071
frameshift window, decay const: 40, 0.5
T: 13
A: 40
X1: 16 (7.3 bits)
X2: 0 (0.0 bits)
S1: 41 (21.7 bits)

5.2.5 CDART

Tool for NCBI's CDART (conserved domain architecture retrieval tool) is primarily a comparative sequence analysis used to display the functional domains that make up a protein and lists proteins with similar domain architectures when presented with a protein query sequence. Sequences that share one or more similar domains are found and displayed in the form of a graphical summary. It determines the domain architecture of a query protein sequence by comparing it to the database of conserved domain alignments. The search utilizes reverse position specific BLAST. Links to the individual sequences as well as rther information on their domain architectures are also provided. Since protein domains may be considered elementary units of molecular function, proteins related by domain architecture may play similar roles in cellular processes. Thus, CDART serves as a useful tool.

Link http://www.ncbi.nlm.nih.gov/Structure/lexington/lexington.cgi?cmd=rps

Institute NCBI

Type of the tool Web-based /Downloadable.

OS required Windows

Note

Even though RPS-BLAST and PSI-BLAST use similar methods to find the conserved features of a protein family, RPS-BLAST compares a query sequence against a database of profiles prepared from ready-made alignments while PSI-BLAST builds alignments starting from a single protein sequence. There is an essential difference in the utility of the two tools. RPS-BLAST is used to identify conserved domains in a query sequence, whereas PSI-BLAST is used to identify other members of the protein family to which a query sequence belongs.

Algorithm used CDART, developed by Geer, Domrachev, Lipman, and Bryant in 2002, uses an algorithm that finds protein similarities across significant evolutionary distances using sensitive protein domain profiles rather than by direct sequence similarity. Proteins similar to a query protein are grouped and scored. Domain profiles are searched based on those derived from several collections of domain definitions, which include functional annotations. Searches can be further refined by taxonomy and by selecting domains of interest.

Tool screenshot Figure 5.37 shows the screenshot of CDART.

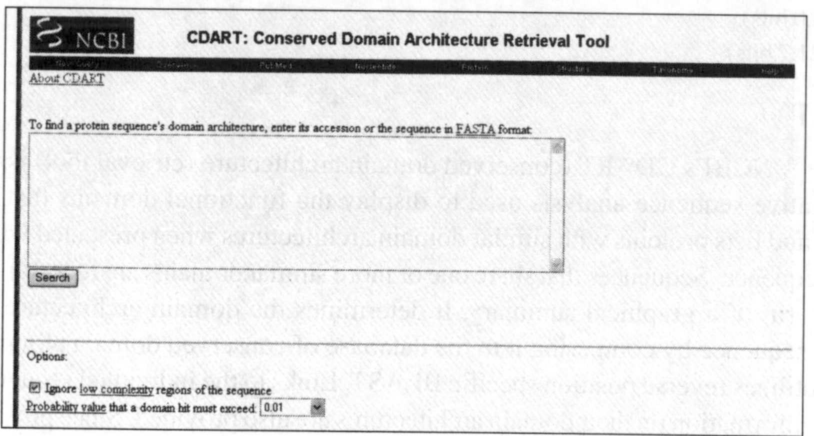

Fig. 5.37 Screenshot of CDART

Input data Protein sequence in FASTA format.

Attributes Probability value of the domain hit should exceed 0.01. Avoid low-complexity regions.

Result Sequences that share one or more similar domains displayed in a graphical format.

TASK

Step 1 Carefully reexamine the data given below:
Cdc3p [*Saccharomyces cerevisiae*]
Other Aliases: YLR314C
Other Designations: Component of the septin ring of the mother-bud neck that is required for cytokinesis; septins recruit proteins to the neck and can act as a barrier to diffusion at the membrane, and they comprise the 10-nm filaments seen with EM
Chromosome: XII
GeneID: 851024

Step 2 Look for the accession number of the corresponding protein in NCBI.

Step 3 Your search will give you three results. Pick up NP_013418 Cdc3p [Saccharomyces cerevisiae] gi|6323346|ref|NP_013418.1|[6323346]

Step 4 Go to the CDART tool at the link given below
http://www.ncbi.nlm.nih.gov/Structure/lexington/lexington.cgi?cmd=rps

Step 5 Your result page will look like the screen shot shown in Fig. 5.38.

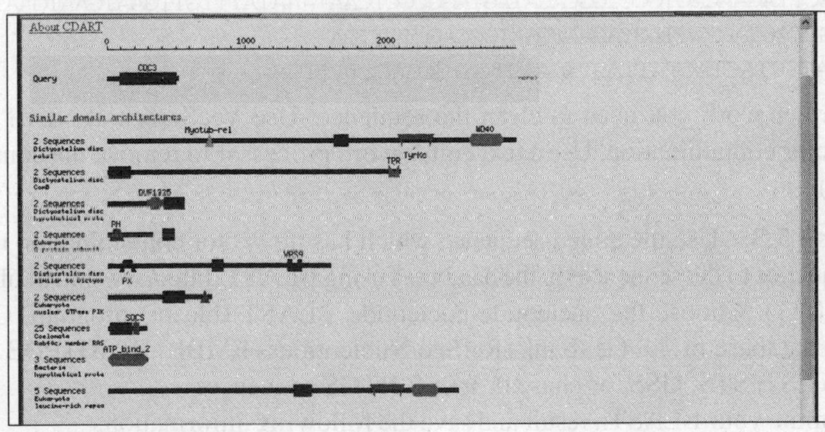

Fig. 5.38 Result page of CDART

Conclusion

The ultimate goal of mining DNA sequences is to discover all genes, and study how they are regulated and how they work together to produce higher function and behaviour levels.

A word of Caution—data mining is a tool and not a magic wand. While it can assist one to find patterns and their relationships with other data, it cannot interpret the value and meaning of these patterns. Thus, the patterns uncovered by data mining must be validated and interpreted on all counts. In other words, data mining cannot automatically discover solutions without intelligent inputs and guidance.

EXERCISES

Exercise 5.1 Go to the NCBI home page (http://www.ncbi.nlm.nih.gov). Select the Saccharomyces cerevisiae Ste23p gene partial cds. Open http://www.ebi.ac.uk/emboss/transeq/ and use the tool Saccharomyces cerevisiae Ste23p gene partial cds to translate the DNA sequence into the corresponding protein sequence. Use the Expasy ProtParam tool to find the following:

(a) Number of amino acids, molecular weight, and theoretical pI.
(b) Total number of positively and negatively charged residues.
(c) The number of tryptophans in the sequence.
(d) Whether or not the protein is stable using the instability index.
(e) The grand average of hydropathicity of this protein.

Exercise 5.2 You have cloned and deduced a sequence to be the following:

```
TATAAATACAAATACGTATACATGTCTATTATAATGAAAAATTGCCAATCTTGTTTAAGCAAATGCATTC
TATCGTTATTATAAATGTTAGTTCTAGCTTTATTTACTTCAAAATCTTAAATCAGAATAAATTAATATTG
TATTGCTGCTGTGCGTGGAAAAAGATGATGTTTATGTTCTTATAGAATAAAAGCTGTGGTTNTTTATTGT
CTGTCTCCTCCACTAGANTGTAAGCTCCATGAGGGCAGGGATTTTGTCTGTYTTGTTCACTGCTGTATCC
CCAGCGCCTAGAACAGTGCCTGGCACATAGTAGGCGCTCAATAAATATTTGTTGAATGAATGAATG
AATACGCTACTACTATTAGTAGAATTGATGCCACCTTTTCA
      ACGCTACTACTATTAGTAGAATTGATGCCACCTTTTCA
```

For further work you need to clean the sequence. Use VecScreen at NCBI to remove any vector contamination. Use a text editor/word processor to remove the contamination (if any).

Exercise 5.3 Use the edited sequence which has no vector contamination to compare this sequence to the sequences in the databases using BLAST (http://www.ncbi.nlm.nih.gov/BLAST/17). Choose the nucleotide-nucleotide BLAST (blastn) program. Choose the default database nr: all GenBank+RefSeq Nucleotides+EMBL+DDBJ+PDB sequences (but no EST, STS, GSS, or phase 0, 1, or 2 HTGS sequences).

Examine your BLASTn result and save the following information:

(a) Header
(b) Graphic overview of alignment
(c) One-line description of matched sequences (bit score, E value)
(d) Pairwise sequence alignment

Exercise 5.4 Use the uncontaminated sequence obtained as an output of Exercise 5.2 and conduct a BLASTx search to compare this sequence against the protein databases at NCBI. Translate your nucleotide sequence into all possible amino acid sequences (in all six reading frames) and compare it to the protein databases. Choose the default database nr: all non-redundant GenBank CDS translations+RefSeq+Proteins+PDB+Swiss-Prot+PIR+PRF.

Examine your BLASTx result and view its GenBank report.

Exercise 5.5 Obtain the Saccharomyces cerevisiae chromosome VI complete DNA sequence (gi:2804269). Use the ORF Finder at NCBI to translate the first 600 bp length and compare the protein sequence to the Swiss-Prot sequences. Compare it to the human protein.

[***Hint:*** Try ORF Finder and GENESCAN (Swiss-Prot/Proteomics tools...)]. What is the result and why?

Exercise 5.6 Copy the following sequence:

```
gccaaaacca aaaccactgg cagagaagaa gcatcgatca attcaagtta ccaagcaact
tggatcttct cgatgtttcc ttgcgaagat gagttcctct tggaccccca agacggagtg
gcccgagcta gtttgccgga tgatcaagga ggccaaggag aagatcaaag cagaccgtcc
agatctcaag attgaggtgg ttccggtagg caccatcgtc actcaagagt tcgacgagaa
tcgcgttcgc atctgggtcg acacagtggc aaagaccccc acaatcggtt aagctgaaac
cccgatatag gcatacatct agcaagtgta cgtccatgca agtatttctg gatatggtcc
agtatgataa ataaataaat aaataaataa aaaataaata aataaataaa taaataaata
aaaaaaaaaa aaaaaaaaaa aaaaaaaaaa aaaaaaaaaa aaaaaa
```

Use BLASTN to run a BLAST search. Examine the results with Fig. 5.39.

(a) What do you believe the query sequence encodes?

(b) How many of the matches do you believe are biologically relevant? Examine the alignments of the 'best' matches. (Recall that you may do this by clicking on the score for each alignment.)

(c) How may bases of your query match each of these sequences?

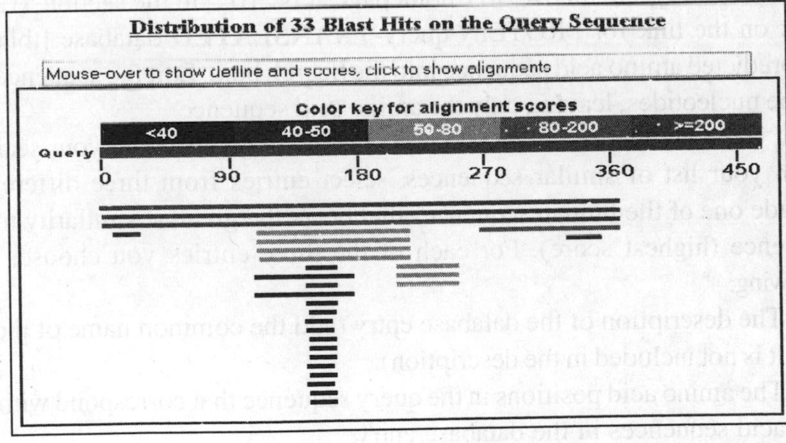

Fig. 5.39 Result page of BLASTn

Exercise 5.7 The structure of Concanavalin B from jack bean (*Canavalia ensiformis*) is found in the Protein Data Bank as 1cnv.

(a) Using Entrez, find the sequence.

(b) What is its oligomeric unit? Determine the number of residues this protein contains.
(c) Determine the function of this protein?
(d) Which proteins is it similar to?
(e) Why is the first amino acid not methionine?

Exercise 5.8 Do a PSI-BLAST search with the following sequence of *Homo sapiens* H+ transporting ATPase located on the lysososme (V1 subunit F).

(a) What are the results of your search?
(b) Run a CD search for the protein sequence with GI 9229839 (A-ATPase A-subunit) from *Thermoplasma acidophilum*. Do you see any unexpected domains in the results of your search?
(c) In the table from your PSI-BLAST iterations, what happens to the total hits, number of new hits in the description lines, and the E-value as the number of iterations is increased?
(d) How do you know when to stop executing further iterations of PSI-BLAST?

Exercise 5.9 Use the DNA sequence given in Exercise 5.6 to search the sequence for the presence of open reading frames, which are likely to be exon sequences that are translated to produce protein products. The results will be presented as a graphic display with six bar graphs. Each bar represents one possible reading frame (three forward and three reverse). Click on the bar that represents the largest ORF. The predicted amino acid sequence (along with the associated DNA sequence) will be presented.

(a) List the first 15 amino acids in the predicted polypeptide sequence.
(b) Copy the entire predicted polypeptide sequence (along with the associated DNA sequence) and go to the BLAST homepage at NCBI . In the section **Translations**, click on the link for PROTEIN query-**TRANSLATED** database [tblastn]. Paste the predicted amino acid sequence into the Search box. Edit the sequence to remove all the nucleotides, leaving only the amino acid sequence.
(c) What is the number of database entries that were identified in your search?
(d) From your list of similar sequences, select entries from three different species. Include one of the human sequences that bears the greatest similarity to the query sequence (highest score). For each of the three entries you choose, record the following:
    - The description of the database entry (add the common name of the species if it is not included in the description).
    - The amino acid positions in the query sequence that correspond with the amino acid sequences in the database entry.
    - The number of amino acids predicted by the query sequence that are identical to each database entry.
    - The number of amino acids predicted by the query sequence that are identical or similar to each database entry.

Exercise 5.10 Duchenne's and Becker's are two forms of classic dystrophies that are defined by progressive deterioration of muscle tissue and resultant weakness. Other types of skeletal muscle dysfunction have been occasionally reported in association with abnormalities of the dystrophin gene. Go to the NCBI site and click on **Structure** and then on CDD (for Conserved Domains Database). Run a CDD search. Feed in NP_732427 (*Drosophila dystrophin*) and run the query. Prepare a report on the following:

(a) Graphic display: location of a variety of conserved domains located along the length of dystrophin. Move your mouse over the coloured domains and watch the display window. Provide information on the different regions.

(b) List of the domains found.

(c) List the Pfam or Smart accession number along with a textual description.

(d) What is the E-value for the match.

(e) Run a CDART search by searching for conserved domains instead of the primary sequence. Prepare a report on the following:

- Architecture of the *Drosophila dystrophin* domain
- Proteins with similar domain architectures
- Similar domain positions to those in our protein
- Two calponin domains to the left
- WW and ZZ domains on the right
- Spectrin repeats are more variable, but present

6 Prediction Tools

Learning Objectives

- To understand and analyse phylogenetic trees
- To learn about gene prediction process
- To learn about protein structure and function prediction

Introduction

Today, efficient methods of prediction have become crucial, and often bottlenecks, as a result of the steady influx of data that is arising from sequencing projects. Therefore, it is not surprising that a number of varied tools have emerged to help assign structures and functions to the sequence data. Several platforms, scattered globally, offer public programs that help researchers.

6.1 Phylogenetic Trees and Phylogenetic Analysis

The similarity of molecular mechanisms of the organisms that have been studied strongly suggests that all organisms at some point in time had a common ancestor. Extrapolating this would mean that any set of species is related and this relationship is known as phylogeny. Phylogenetics is a technique that helps to demonstrate and build a relationship between species. Such an analysis is important in building up information about the relationship and in determining the direction of evolution. This relationship is usually represented by a phylogenetic tree. Figure 6.1 shows a sample phylogenetic tree. The evolutionary distances are measured as lengths of branches or dendrograms. Crudely put, phylogeny helps bioinformaticians to build such a tree by observing the existing organisms.

There are certain terms that we need to famaliarize with before we discuss the programs and packages that help us perform a phylogenetic analysis.

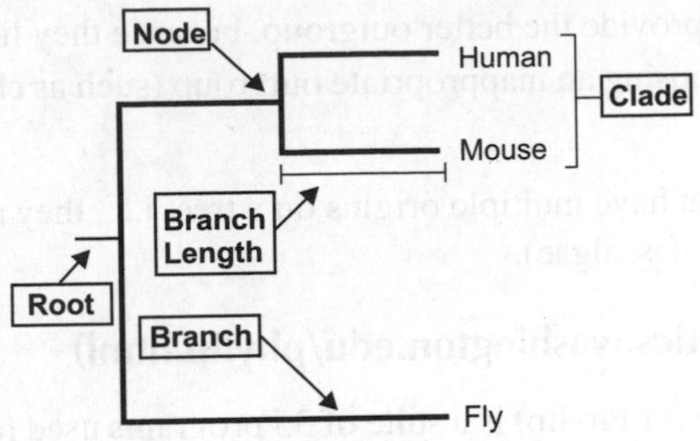

Fig. 6.1 Sample phylogenetic tree (*source*: NCBI)

Node represents a taxonomic unit. It can be either an existing species or an ancestor.

Branch defines the relationship between the taxa in terms of descent and ancestry.

Topology is the branching pattern of a phylogenetic tree.

Branch length is a very significant part of the phylogenetic tree. It represents the number of changes that have occurred in a branch.

Root is the 'common ancestor' of all taxa.

Distance scale is a scale chosen to represent the number of differences between organisms or sequences.

Clade is a group of two or more taxa or DNA sequences that include both their common ancestor and all of their descendents.

Bootstrap is a method for assessing the statistical significance of a particular node on a phylogenetic tree. It is arrived at by randomly resampling subsets of the data.

Characters define the units upon which analysis is made, e.g., amino acids, nucleotides, etc.

Long branch attraction refers to a phenomenon in which relatively long branches in a tree tend to associate simply due to their 'length' rather than their evolutionary relationship. It is an artefact most likely to occur in parsimony analyses, and the method of 'invariants' is the least likely to have it.

Monophyletic refers to a group of taxa that have a single origin on a tree, i.e., all taxa descend from a common ancestor (e.g., apicomplexans, kinetoplasts, nematodes).

Outgroup is one or more taxa that are related to, but outside the group of taxa you are studying. Choosing an appropriate outgroup is very important, because this group defines the root and polarizes the direction of change in your tree. While both frogs and monkeys are outgroups for apes, monkeys provide the better outgroup, because they lie closest to the root of all apes. Of course, choosing an inappropriate outgroup (such as chimps) will grossly disturb your tree.

Polyphyletic is a group of taxa that have multiple origins on a tree, i.e., they arose twice in evolution (e.g., helminths, parasites, algae).

6.1.1 Phylip (http://evolution.genetics.washington.edu/phylip.html)

The **Phyl**ogeny **i**nference **p**ackage (or Phylip) is a suite of 35 programs used for deriving evolutionary trees or phylogeny and analysing how the nucleic and protein sequences might have been derived during evolution. This package has been in distribution since 1980. Often the relationship between two sequences is sought and the evolutionary tree that results has the two sequences placed as the outer branches of the tree. The branching pattern depicts the degree of relationship/diversity between them. The results smay also be displayed as nested parentheses notation such as (A,(B,(C,D))). The tool is available on the internet and its code has been written in C. It is compatible with Windows (95/98/NT/2000/me/xp), MacOS 8 and 9, MacOS X, and Linux systems. Phylip has been compiled successfully on Unix and Linux systems with the GCC C compiler, with proprietary compilers from Sun, HP, Compaq/Digital, SGI, and many other systems.

In the earlier chapters you have learnt to align sequences: such alignments form the basis of the phylogenetic analysis. Even very similar sequences that have a small variability degree can result in a full blown evolutionary tree. Phylip utilises parsimony, distance matrix, and likelihood methods, including bootstrapping and consensus trees, to derive its evolutionary trees. In addition to sequence, the input data can be in the form of gene frequencies, restriction sites and fragments, distance matrices, and discrete characters.

The data is read into the program from a text file that has a flat ASCII or text only format. There are certain sequence analysis programs such as the ClustalW alignment program which can write data files in the Phylip format. Most of the programs look for the data in a file called 'infile'. Output is written onto special files with names such as 'outfile' and 'outtree'.

There are various methods that can be used to predict and construct phylogenetic trees. The flow chart shown in Fig. 6.2 can be used to determine the method to predict and construct phylogenetic trees.

The molecular sequence programs available at Phylip allow estimate phylogenies from protein sequence or nucleic acid sequence data. The input format for the sequence programs uses the symbols A,T,G,C,U. The first line of the input file contains the number of species and the number of sites. Data that belongs to each species starts on a new line. The first ten characters of that line are the species name. Then follows the base sequence of that species. While blanks and numerics are ignored. It allows GenBank and EMBL sequence entries to be read with minimum editing. The input for the protein sequence programs contains a line that holds the number of species and the number of amino acid positions. This first line also includes options such as U (user tree) and W (weights). The U option is used to select that user-defined trees are provided at the end of the input file. The trees in FITCH are regarded as unrooted, and are specified with a trifurcation (three-way split) at their base such as ((A,B),C,(D,E)). It is important to realize that when the U option is selected, and if the tree being constructed requires every branch length to be estimated from the data, it will be possible to solve for the branch lengths and sum of squares when there is some missing data.

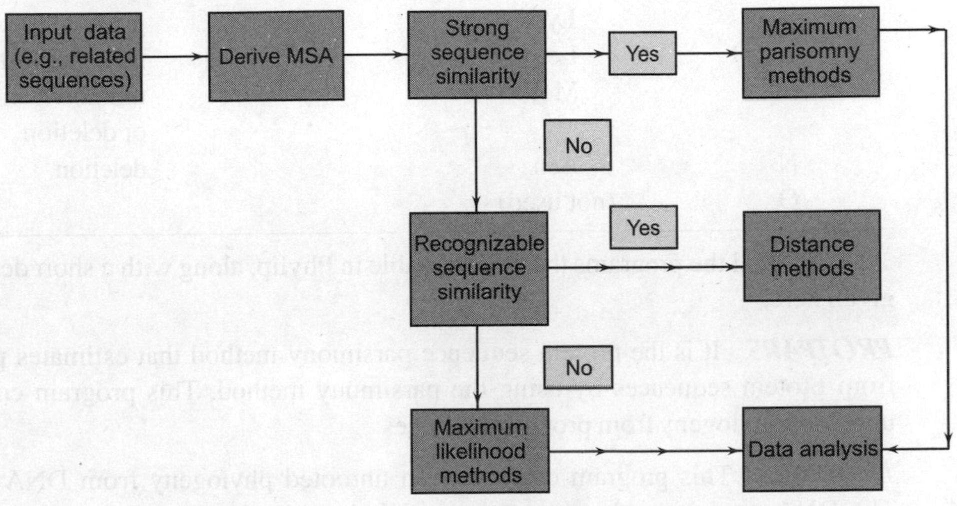

Fig. 6.2 Flows chart to select the method of constructing a phylogenetic tree

The next line contains information on the species data. Each sequence starts on a new line, has a ten-character species name that must be blank-filled to be of that length, followed immediately by the species data in the one-letter code. The sequences must be in either the 'interleaved' or 'sequential' format. Sequences can have internal blanks, but there must be no extra blanks at the end of the terminated line.

Protein sequences are given by the one-letter code used by Dayhoff's group in the *Atlas of Protein Sequences*, and the IUB standard abbreviations. These are shown in Table 6.1.

Table 6.1 One-letter code used by Dayhoff in the *Atlas of Protein Sequences*

Symbol	Stands for	Symbol	Stands for
A	Ala	P	pro
B	Asx	Q	gln
C	Cys	R	arg
D	Asp	S	ser
E	Glu	T	thr

(contd)

Table 6.1 *(contd)*

Symbol	Stands for	Symbol	Stands for
F	Phe	U	(not used)
G	Gly	V	val
H	His	W	trp
I	Ileu	X	unknown amino acid
J	(not used)	Y	tyr
K	Lys	Z	glx
L	Leu	*	nonsense (stop)
M	Met	?	unknown amino acid or deletion
N	Asn	-	deletion
O	(not used)		

A list of all the programs that are available in Phylip, along with a short description, is given here.

PROTPARS It is the protein sequence parsimony method that estimates phylogenies from protein sequences by using the parsimony method. This program constructs an unrooted phylogeny from protein sequences.

DNAPARS This program constructs an unrooted phylogeny from DNA sequences. The DNA parsimony program estimates phylogenies by the parsimony method. The method of Fitch is used to count the number of changes of base needed on a given tree. DNAPARS can construct both bifurcating and multifurcating trees.

DNAMOVE Interactive DNA parsimony is an interactive DNA parsimony program. The program reads in a dataset, which is prepared in almost the same format as DNAPARS. It allows the user to choose an initial tree and view different sites as well as the way the nucleotide states are distributed on that tree. It allows the user to specify how the tree is to be rearranged, rerooted, or written out to a file. The O (outgroup), W (weights), T (threshold), and 0 (graphics type) options are the usual ones and are described in the main documentation file.

DNAPENNY This is a program that will find all of the most parsimonious trees implied in the data when the nucleic acid sequence parsimony criterion is employed. It does so by using the more 'branch and bound' algorithm, a strategy first applied to phylogenetic inference by Hendy and Penny (1982). The program should be slower than the other tree-building programs in Phylip, but can be used comfortably for as many as ten species. The algorithm (branch and bound) finds all the most parsimonious trees without generating all possible trees.

DNACOMP The DNA compatibility program implements the compatibility method for DNA sequence data. For a four-state character without a character-state tree, as in nucleic acid sequences, the usual theorems cannot be applied. Thus this program directly evaluates each tree topology by counting the number of substitutions needed at each site and comparing this to the minimum number that might be needed (one less than the number of bases observed at that site). It then evaluates the number of sites which achieve the minimum number. This is the evaluation of the tree (the number of compatible sites), and the topology is chosen so as to maximize that number. Compatibility is particularly appropriate when sites vary greatly in their rates of evolution.

DNAINVAR This program reads in nucleotide sequences for four species and computes the phylogenetic invariants. This method of deriving evolutionary trees was initiated by Lake and Cavender and by Felsenstein. It examines and tests alternative tree topologies. The invariants are formulas in the expected pattern frequencies, not the observed pattern frequencies.

DNAML The DNA maximum likelihood program for DNA sequences estimates phylogenies. This method is based on an algorithm given by Felsenstein and Churchill. It is based on the substitution model where expected frequencies of the four bases as well as that of transitions and transversions are allowed to be unequal. It has several approaches of allowing different rates of evolution at different sites and one of them is the hidden Markov model.

DNAMLK This program uses the maximum likelihood method for DNA sequences. It estimates the trees constructed to be consistent with a molecular clock. This program is also based on DNAML. The use of the two programs together permits a likelihood ratio test of the molecular clock hypothesis to be made.

PROML The protein maximum likelihood program estimates phylogenies from protein amino acid sequences by maximum likelihood. It uses either the Jones–Taylor–Thornton or the Dayhoff probability model of change between amino acids. It also uses hidden Markov model of rates of evolution of amino acids to determine the rates. This allows different rates of change at known sites to be inferred. The input data is the same as other programs under Phylip.

PROMLK The protein maximum likelihood program is the same as PROML but assumes a molecular clock. The use of the two programs together permits a likelihood ratio test of the molecular clock hypothesis to be made.

DNADIST It is a program to compute the distance matrix from nucleotide sequences. It derives four different distances between species from nucleic acid sequences, which then are used in the distance matrix programs. This program uses the distances from the Jukes–Cantor formula, Kimura's 2- parameter method, and F84 model such as DNAML. The program can also make a table of percentage similarity among sequences.

PROTDIST This program uses protein sequences to compute a table of similarity between the amino acid sequences and a distance matrix, under four different models of amino acid replacement. The distance for each pair of species gives an estimate of the total branch length between the two species, which can be used in the distance matrix programs FITCH, KITSCH, or NEIGHBOR. This program can be used instead of the parsimony program PROTPARS.

RESTDIST Restdist uses and takes in the same restriction sites format as the RESTML and computes a restriction site's distance or a restriction fragment's distance. The original restriction fragments and restriction sites distance methods were introduced by Nei and Li. Their original method for restriction fragments is also available in this program, although its default methods are modifications of the original Nei and Li methods. Distances are calculated from restriction site's data or restriction fragment's data. The restriction site's option is the one to use to estimates distances for randomly amplified polymorplic DNAs (RAPDs) or amplified fragment length polymorphisms AFLPs.

RESTML The restriction site's maximum Likelihood program qualifies as the slowest running program of Phylip. It runs a maximum likelihood method for restriction site's data only. The program utilizes the approach of Smouse and Li (1987) who gave explicit expressions for computing the likelihood for three-species trees. It also allows for multiple restriction enzymes. The algorithm has been described by Felsenstein. The assumptions of the present model are that each restriction site evolves independently, different lineages evolve independently, each site undergoes substitution at an expected rate which is prespecified, and substitutions consist of replacement of a nucleotide by one of the nucleotides chosen randomly.

SEQBOOT Bootstrap, Jackknife, or permutation resampling of molecular sequence, restriction site, gene frequency, or character data is a package that reads in a dataset and produces multiple datasets from it by bootstrap resampling. Bootstrapping is a computational method used to generate confidence intervals and standard errors. The advantages of bootstrapping include the fewer assumptions (data needs not be normally distributed), greater accuracy. SEQBOOT is a general bootstrapping and dataset translation tool. It is intended to allow the user to generate multiple datasets that are resampled versions of the input dataset. More details about bootstrapping in relation to Phylip can be obtained from http://evolution.genetics.washington.edu/phylip/doc/seqboot.html. In addition SEQBOOT allows multiple datasets to be bootstrapped, jackknifed, or permuted. The input to SEQBOOT can be in the form of sequences, binary characters, restriction sites, or gene frequencies. It can also convert datasets between sequential and interleaved formats. In order to carry out a bootstrap, jackknife, or permutation test in Phylip, it is better to use SEQBOOT with the original dataset and produce a large number of secondary bootstrapped or jackknifed datasets. The second step is to find the phylogeny estimate for each of these secondary datasets. This would generate a big output file as well as a treefile with the trees from secondary datasets. This treefile would become the input for CONSENSE. When CONSENSE is run the majority rule consensus tree will result, which shows the outcome of the analysis.

FITCH The Fitch-Margoliash and least-squares distance method estimates phylogenies from distance matrix data under the 'additive tree model', according to which the distances are expected to equal the sums of branch lengths between the species. FITCH can be used to determine distances computed from sequences, restriction site's or fragment's distances, with DNA hybridization measurements, and with genetic distances computed from gene frequencies. But the program does not assume an evolutionary clock. It uses a special input option, called G, the global search option that is not available in KITSCH or NEIGHBOR. This option can be used to remove and re-add a group after the last species is added to the tree, hence giving a better result since the position of every species is reconsidered.

KITSCH This program derives phylogenies from distance matrix data using the additive tree model and an evolutionary clock. Effectively, the total length from the root of the tree to any species is the sum of the lengths of the branches. The input format is the same as in the distance matrix methods. The output is a rooted tree, where actual branch lengths are not exactly proportional to the lengths drawn on the displayed tree so that even the shortest branch is visible and the longest branch fits on the screen.

NEIGHBOR This program uses the neighbor-joining method of Saitou and Nei and the UPGMA method of clustering to construct trees. The branch lengths are not optimized by the least squares criterion, but the methods are very fast. The output consists of a tree which is rooted if UPGMA is elected and unrooted if the neighbor-joining method is used. The lengths of the interior segments of the tree are also displayed. Neighbor joining is a distance matrix method without the assumption of a clock, whereas UPGMA does assume a clock.

CONTML This program estimates phylogenies of a given dataset by using the restricted maximum likelihood method based on the Brownian motion model. The model also assumes that all divergence is due to genetic drift in the absence of new mutations. Users have the option of either using this program or first computing Nei's genetic distance and use one of the distance matrix programs. This program assumes that the characters evolve at equal rates and in an uncorrelated fashion, so that it does not take into account the usual correlations of characters. CONTML is extensively used to analyse microsatellite data. The output gives the topology of the tree as an unrooted tree diagram.

Note

Caution must be exercized in using this program under the following conditions:

- In case gene frequencies are being analysed and the gene frequencies change not only by genetic drift but also by mutation, the model is not correct.
- If continuous characters are being analysed and the characters have not been transformed to new coordinates that evolve independently and at equal rates, then the model is also violated.
- The input data has two species (or populations) with identical transformed gene frequencies: as when the sample sizes are small and/or many loci are monomorphic.

GENDIST This program of Phylip computes genetic distances from gene frequencies using three formulas: Nei's genetic distance, the Cavalli–Sforza chord measure, and the genetic distance of Reynolds et al. These are given as an output file in a format that can be read by the distance matrix phylogeny programs FITCH and KITSCH. All three formulas assume that all differences between populations arise from genetic drift. Nei's distance is used and based on the infinite isoalleles model of mutation, in which there is a rate of neutral mutation. It is assumed that all loci have the same rate of neutral mutation and that the genetic variability initially in the population is at equilibrium between mutation and genetic drift, with the constraint population size of each population remaining constant.

CONTRAST This program contrasts for comparative methods using a dataset of the standard quantitative characters sort and a tree from the treefile. The second step in this program is the contrast between species which according to that tree are statistically independent. This is done for each character. The third step is to standardize the contrasts by branch lengths. It assumes a Brownian motion model.

PARS It is a general parsimony program that uses the parsimony method with multiple states. This accommodates changes among all states. The criterion is to find the tree which requires the minimum number of changes. The model assumes that ancestral states are unknown, different characters evolve independently, different lineages evolve independently, changes to all other states are equally probable (Wagner), all other kinds of evolutionary events such as retention of polymorphism are far less probable than these state changes and the rates of evolution in different lineages are sufficiently low that two changes in a long segment of the tree are far less probable than one change in a short segment.

MIX It is a general parsimony program that carries out the Wagner (pars) and Camin–Sokal parsimony methods together, where each character can have its method specified separately.

MOVE It is an interactive parsimony program that uses a dataset which is prepared in almost the same format as the MIX. The user can choose an initial tree and can look at different characters and the way their states are distributed on that tree. The tree can be rearranged, rerooted, or written out to a file. A manual search for the most parsimonious tree is made by the user by looking at different rearrangements.

PENNY It is a program that constructs all of the most parsimonious trees implied by the input dataset by examining all possible trees using the branch and bound algorithm method of exact search. This program is slower and can be used for up to about ten species only.

DOLLOP This program of Phylip estimates phylogenies by polymorphism parsimony criteria for discrete character data with two states (0 and 1). The method is named after Dollo, who asserted that in evolution it is harder to gain a complex feature than to lose it.

The algorithm explains the presence of the state 1 by allowing up to one forward change 0→1 and as many reversions 1→ 0 as are necessary to explain the pattern of states seen.

DOLMOVE It is an interactive parsimony program that uses the Dollo and polymorphism parsimony criteria. The input data is almost the same as for the Dollo and DOLLOP. This program evaluates parsimony and compatibility criteria for those phylogenies and displays reconstructed states throughout the tree. This program can be used to find parsimony or compatibility estimates by hand.

DOLPENNY This program finds all of the most parsimonious trees implied by input data when the Dollo or polymorphism parsimony criteria are employed. It does so by using the more sophisticated branch and bound algorithm. The program is slower than the other tree-building programs in the package and is useable up to about ten species.

CLIQUE This program finds the largest clique of mutually compatible characters, and the phylogeny which they recommend, for discrete character data with two states (0 and 1). The largest clique (or all cliques within a given size range of the largest one) are found by a very fast branch and bound search method. The method does not allow for missing data.

FACTOR This program uses discrete multistate data with character state trees and produces the corresponding dataset with two states (0 and 1). It then creates a dataset consisting entirely of binary (0,1) characters that, in turn, can be used as input to any of the other discrete character programs in this package, except for PARS.

DRAWGRAM and DRAWTREE These programs interactively plot a cladogram (or phenogram-like rooted tree diagram) with many options including the orientation of tree and branches, style of tree, label sizes and angles, tree depth, margin sizes, stem lengths, and placement of nodes in the tree. Note that DRAWTREE can plot unrooted phylogenies only. You can get additional information on these two programs at http://evolution.genetics.washington.edu/phylip/doc/draw.html

TREEDIST This program computes the branch score distance between trees, which allows for differences in the tree topology and which also makes use of branch lengths. It also computes the Robinson–Foulds symmetric difference distance between trees, which allows for differences in the tree topology but does not use branch lengths.

CONSENSE This program computes consensus trees by the majority-rule consensus tree method. It is not able to compute the Adams consensus tree. The lengths on the tree on the output tree file are not branch lengths but the number of times that each group appeared in the input trees. This number is the sum of the weights of the trees in which it appeared. This program can be used as the final step in doing bootstrap analyses for other methods in Phylip.

RETREE It is an interactive tree rearrangement program. This tree editor reads a tree, allows the user to construct one, and displays the tree on the screen. The user can then specify how the tree is to be rearranged, re-rooted, or written out to a file. The user can re-root a tree, flip branches, change names of species, and change or remove branch lengths. This program is used most commonly to change rootedness of trees so that a rooted tree derived from one program can be fed in as an unrooted tree to another.

Note

In KITSCH tree are to be regarded as rooted and have a bifurcation at the base such as ((A,B),(C,(D,E))). Such user trees are not available in NEIGHBOR. In addition, FITCH allows the user to select branch lengths. This option can be used to avoid having those branches iterated, so that the tree is evaluated with their lengths fixed. More information on the distance matrix programs is available at the website of Pittsburgh Supercomputing Center.

Now that we learnt about the suite of programs found in Phylip, let us try to generate some phylogenetic trees for ourselves. Table 6.2 can serve as a guide to select an appropriate program for a particular phylogeny.

Table 6.2 Programs and their brief descriptions

S. No	Programs	Brief Description
1	CLIQUE	Finds the largest clique of mutually compatible characters, and the phylogeny which they recommend, for discrete character data with two states
2	CONSENSE	Computes consensus trees by the majority-rule consensus tree method
3	DNADIST	Computes four different distances between species from nucleic acid sequences
4	DNAPARS	Estimates phylogenies from nucleic acid sequences using parsimony
5	DOLLOP	Executes Dollo and polymorphism parsimony program
6	DRAWgram	Interactively plots a cladogram or phenogram-like rooted tree diagram
7	DRAWTREE	Interactively plots an unrooted tree diagram
8	FITCH	Estimates phylogenies from distance matrix data under the 'additive tree model' using the Fitch–Margoliash criterion and some related least-squares criteria
9	KITSCH	Estimates phylogenies from distance matrix data under the 'ultrametric' model (same as the additive tree model except that an evolutionary clock is assumed)
10	MIX	Estimates phylogenies by several parsimony methods for discrete character data with two states
11	PARS	Executes the Wagner parsimony method with multiple states

(*contd*)

Table 6.2 *(contd)*

S. No	Programs	Brief Description
12	NEIGHBOR	Estimates a phylogenetic tree using neighbor joining (does not assume a clock) and UPGMA (assumes a clock) methods
13	PROTDIST	Computes a distance measure for protein sequences using maximum likelihood estimates based on one of three metrics
14	PROTPARS	Estimates phylogenies from protein sequences using parsimony
15	SEQBOOT	Produces multiple datasets from a single dataset by bootstrap resampling

TASK 

Step 1 Locate the DDBJ nucleotide database on the Internet.

Step 2 Search with the keywords 'human growth factor' through the SRS (left link).

Step 3 This step demonstrates the **generation of shortest-distance neighbor-joining tree**.

In this exercise, you will use the protein alignment as a basis to construct a distance matrix for all pairwise combinations of sequences under study using the PROTDIST program. This distance matrix will then be used by the NEIGHBOR program to calculate the shortest-distance neighbor-joining tree. The tree will then be plotted with DRAWGRAM and later modified using a graphics program.

Dataset has 4 taxa and 500 aligned amino acids positions. It is a subset of the alignment of the largest subunit of the RNA polymerase II.

Part A: Running PROTDIST

1. Orient your browser to the BioWeb Phylip page at http://bioportal.cgb.indiana.edu/tools/phylip/
2. Go to the PROTDIST page by clicking on **protdist**.
3. Paste your protein alignment as shown in Fig. 6.3 in Phylip format (generated in ClustalW) into the sequence input box.
4. Use all default parameters. Type in your e-mail address; then Click **Run Protdist**.
5. A new page will pop up with your results. To view your protein distance matrix, click **outfile**. Copy and paste the distance matrix into a word file. If your alignment contains a lot of sequences, you may have to reduce the font size and convert it into a proportional font, e.g., courier, so that the matrix is displayed correctly.

```
4   500
Hom.sap   MECPGHFGHI  ELAKPVFHVG  FLVKTMKVLR  CVCFFCSKLL  VDRLTHVYDL
Ara.tha   MECPGHFGYL  ELAKPMYHVG  FMKTVLSIMR  CVCFNCSKIL  ADRLKKILDA
Sac.cer   MECPGHFGHI  DLAKPVFHVG  FIAKIKKVCE  CVCMHCGKLL  LDRFAAIWTL
Vai.nec   MSCPGHFGHI  ELTKPMFHVG  YMTKIKKILE  CVCFYCSRLK  ISDLNFVWNI

          CKGGCGRYQP  RIRRSGLELY  AEWQEKKILL  SPERVHEIFK  RISDEECFVL
          CKNGCGAQQP  KLTIEGMKMI  AEYAERKQTL  GADRVLSVLK  RISDADCQLL
          CKTGCGNTQP  TIRKDGLKLV  GSWPELRVLS  TEEILNIFKH  ISVKD-FTSL
          SKTGCGNKQP  VIKKEGMSLI  A-FSDGKVIL  NGERVHNILK  KIVNEDAVFL

          GMEPRYARPE  WMIVTVLPVP  PLSVRPAVVM  QGARNQDDLT  HKLADIVKIN
          GFNPKFARPD  WMILEVLPIP  PPPVRPSVMM  DASRSEDDLT  HQLAMIIRHN
          GFNEVFSRPE  WMILTCLPVP  PPPVRPSISF  NEQRGEDDLT  FKLADILKAN
          GFDQKFTKPE  WLILTVLLVP  PPSVRPSIVM  EGLRAEDDLT  HKLADIVKAN

          NQLRRNEQNG  AAAHVIAEDV  KLLQFHVATM  VDNELPGLPR  AMQKSGRPLK
          ENLKRQEKNG  APAHIISEFT  QLLQFHIATY  FDNELPGQPR  ATQKSGRPIK
          ISLETLEHNG  APHHAIEEAE  SLLQFHVATY  MDNDIAGQPQ  ALQKSGRPVK
          TYLKKYELEG  APGHVVRDYE  QLLQFHIATM  IDNDISGQPQ  ALQKSGRPLK
          IQLKRQTDRG  AKSHVLQDLC  SLLQFHITTL  FDNDIPGMPI  ATTRSKKPIK
          KRLRAQRSES  STDTAMKETR  TLLQYHITTY  MINDKPSIER  AVTKSGRPLK

          SLKQRLKGKE  GRVRGNLMGK  RVDFSARTVI  TPDPNLSIDQ  VGVPRSIAAN
          SICSRLKAKE  GRIRGNLMGK  RVDFSARTVI  TPDPTINIDE  LGVPWSIALN
          SIRARLKGKE  GRIRGNLMGK  RVDFSARTVI  SGDPNLELDQ  VGVPKSIAKT
          SISARLKGKE  GRVRGNLMGK  RVDFSARSVI  TPDPNISVEE  VGVPSEIAKI

          MTFAEIVTPF  NIDRLQELVR  RGNSQYPGAK  YIIRDNGDRK  VERHMCDGDI
          LTYPETVTPY  NIERLVRLVF  IS---FSETK  YIIRDDGQRK  VERHLQDGDF
          LTYPEVVTPY  NIDRLTQLVR  NGPNEHPGAK  YVIRDSGDRK  VERHIMDNDP
          HTFPEIITPF  NIDRLTKLVS  NGPNEYPGAN  YVIRNDGQRV  VERHMQDGDV

          VIFNRQPTLH  KMSMMGHRVR  ILPWSTFRLN  LSVTTPYNAD  FDGDEMNLHL
          VLFNRQPSLH  KMSIMGHRIR  IMPYSTFRLN  LSVTSPYNAD  FDGDEMNMHV
          VLFNRQPSLH  KMSMMAHRVK  VIPYSTFRLN  LSVTSPYNAD  FDGDEMNLHV
          VLFNRQPSLH  KMSMMAHFVR  VMEGKTFRLN  LSCVSPYNAD  FDGDEMNLHM

          PQSLETRAEI  QELAMVPRMI  VTPQSNRPVM  GIVQDTLTAV  RKFTKRDVFL
          PQSFETRAEV  LELMMVPKCI  VSPQANRPVM  GIVQDTLLGC  RKITKRDTFI
          PQSEETRAEL  SQLCAVPLQI  VSPQSNKPCM  GIVQDTLCGI  RKLTLRDTFI
          PQSYNSKAEL  EELCLVSKQV  LSPQSNKPVM  GIVQDSLTAL  RLFTLRDSFF

          ERGEVMNLLM  FLKVPQPAIL  KPRPLWTGKQ  IFSLIIPDTK  VVVENGELIM
          EKDVFMNTLM  WWKVPAPAIL  KPRPLWTGKQ  VFNLIIPDTQ  VRIERGELLA
          ELDQVLNMLY  WVVIPTPAII  KPKPLWSGKQ  ILSVAIPDNG  MLIIDGQIIF
          DRRETMQLLY  SVILNFPAIS  YPKKLWTGKQ  ILSYILPDSY  VIIRNGEILS

          GILCKKSLGT  SAGSLVHISY  LEMGHDITRL  FYSNIQTVIN  NWLLIEGHTI
          GTLCKKTLGT  SNGSLVHVIW  EEVGPDAARK  FLGHTQWLVN  YWLLQNGFTI
          GVVEKKTN    LYFAINAFSIVGS  SNGGLIHVVT  REKGPQVCAK  LFGNIQKVVN
          FWLLHNGFST  GIIDKKAVGS  TQGGLIHIIA  NDFGPDRVTC  FFDDAQKMM
```

Fig. 6.3 Protein alignment in Phylip format

Part B: Running neighbor joining

1. The outfile generated above in PROTDIST is used as the input to generate the neighbor-joining phylogenetic tree.
2. Choose the **Neighbor** program from the BioWeb Phylip page at http://bioportal.cgb.indiana.edu/tools/phylip/
3. The Neighbor page will pop up. Paste the outfile generated by PROTDIST into the dialogue box.
4. Scroll down, and under **Other Options**, type in the number of the protein you wish to use as an outgroup. Your outgroup should be a protein sequence that you know for sure is ancestral to all others. If your outgroup is the fifth sequence listed on your alignment, then type in the number 5. Type in your e-mail, and click **Run neighbor**.
5. On the next page, click **outfile** to view a text version of the N-J tree. Save this file.
6. Click **treefile**. Save this file as a text file. This file will be used as an input into a tree-plotting program called DRAWGRAM.

Part C: Running DRAWGRAM

1. To generate a graphical output of the tree displayed in the outfile, you will use a program called DRAWGRAM. Orient your browser to the DRAWGRAM page at hppt://biowebpasteur.fr/seqanal/interfaces/drawgram.html
2. Paste the treefile generated by NEIGHBOR into the dialogue box.
3. Under DRAWGRAM options, choose PCX format (or PICT for MAC users).
4. Type in you e-mail, and click **run Drawgram**.
5. On the next page, save the plotfile. It can be viewed using the PAINT program, or it can be imported into Word.

Step 4 This step explains the **generation of a maximum parsimony tree.** In this exercise you will use the protein alignment to infer a maximum parsimony tree. To do this, you will use the alignment file as the input into the program PROTPARS. The tree file generated will then be used as the input into DRAWGRAM to generate a picture of the tree.

Part A: Running PROTPARS

1. Choose the **Protpars** program from the BioWeb Phylip page at http://bioportal.cgb.indiana.edu/tools/phylip/.
2. The **PROTPARS** page will pop up. Paste the alignment in Phylip format into the dialogue box.
3. Scroll down, and under **Other Options**, type in the number of the protein you wish to use as an outgroup. Your outgroup should be a protein sequence that you know for sure is ancestral to all others. If your outgroup is the fifth sequence listed on your alignment, then type in the number 5.
4. Type in your e-mail, and click **Run Protpars**.

5. On the next page, click **outfile** to view a text version of the parsimony tree. Save this file.
6. Click **treefile**. Save this file as a text file. This file will be used as the input into a tree-plotting program called DRAWGRAM.

Part B: Running DRAWGRAM

1. To generate a graphical output of the tree displayed in the PROTPARS outfile, you will use a program called DRAWGRAM. Orient your browser to the DRAWGRAM page at hppt://biowebpasteur.fr/seqanal/interfaces/drawgram.html.
2. Paste the treefile generated by PROTPARS into the dialogue box.
3. Under DRAWGRAM options, choose **Postscript** format (or PICT for MAC users).
4. Type in you e-mail, and click **run Drawgram**.
5. On the next page, save the plotfile. It can be viewed using the PAINT program, or it can be imported into Word.

6.1.2 Phyml (http://atgc.lirmm.fr/phyml/)

The ever-increasing size of homologous sequence datasets and complexity of substitution models stimulate the development of better methods for building phylogenetic trees. Phyml a simple, fast, and accurate algorithm to estimate large phylogenies by maximum likelihood. PHYML Online is a web interface to Phyml, which can generate phylogenies from DNA and protein sequences. This tool provides the user with a number of options, e.g., nonparametric bootstrap and estimation of various evolutionary parameters, in order to perform comprehensive phylogenetic analyses on large datasets.

Algorithm The method is a simple hill-climbing algorithm that adjusts the tree topology and branch lengths simultaneously. Due to this simultaneous adjustment of the topology and branch lengths, a few iterations are sufficient to reach the optimum. The topological accuracy of this method is at least as high as that of the existing maximum-likelihood programs and much higher than the performance of distance-based and parsimony approaches. The program is heuristic and is based on tree-swapping operation, namely 'nearest-neighbour interchange'. This swapping takes three possible topological configurations around each internal branch. For each of these configurations, the length of the internal branch that maximizes the likelihood is estimated using numerical optimization. This algorithm dramatically reduces the computing time and can be used with much larger and more complex datasets. Results are sent to the user by e-mail. The following is the pattern in which the results from this tools are presented:

- The first file presents a summary of the options selected by the user, maximum likelihood estimates of the parameters of the substitution model that were adjusted, and the log likelihood of the model given the data.
- The second file shows the maximum likelihood phylogeny(ies) in NEWICK format.
- Trees can be viewed through an applet available on the PHYML Online server.

This applet runs the program ATV that provides numerous options to display and manipulate large phylogenetic trees.

TASK 

Use Phyml for a dataset of your choice and calculate the tree with the highest likelihood using a model for Among Site Rate Variation that has a proportion of invariant site estimated from the data.

A test file containing vacuolar/archaeal ATPases from the following prokaryotes and eukaryotes can be used: plants such as *Daucus carota, Arabidopsis thaliana, Gossypium hirsutum*; algae such as *Acetabularia acetabulum, Cyanidium caldarium*; mammals such as *Mus, Homo, Bos*; birds such as *Gallus*; insects such as *Aedes*; fungi such as *Saccharomyces, Candida, Schizosaccharomyces*, and *Neurospora, Dictyostelium discoideum*; *protists or protozoa Entamoeba, Plasmodium falciparum, Trypanosoma, Giardia* and *Methanococcus jannaschii* (80oC), *Haloferax volcanii* (37oC), *Halobacterium salinarium Methanobacterium Desulfurococcus* sp, *Thermococcus* sp such as archaea,

Step 1 The interface of the program, as shown in Fig. 6.4, can be accessed from the Internet.

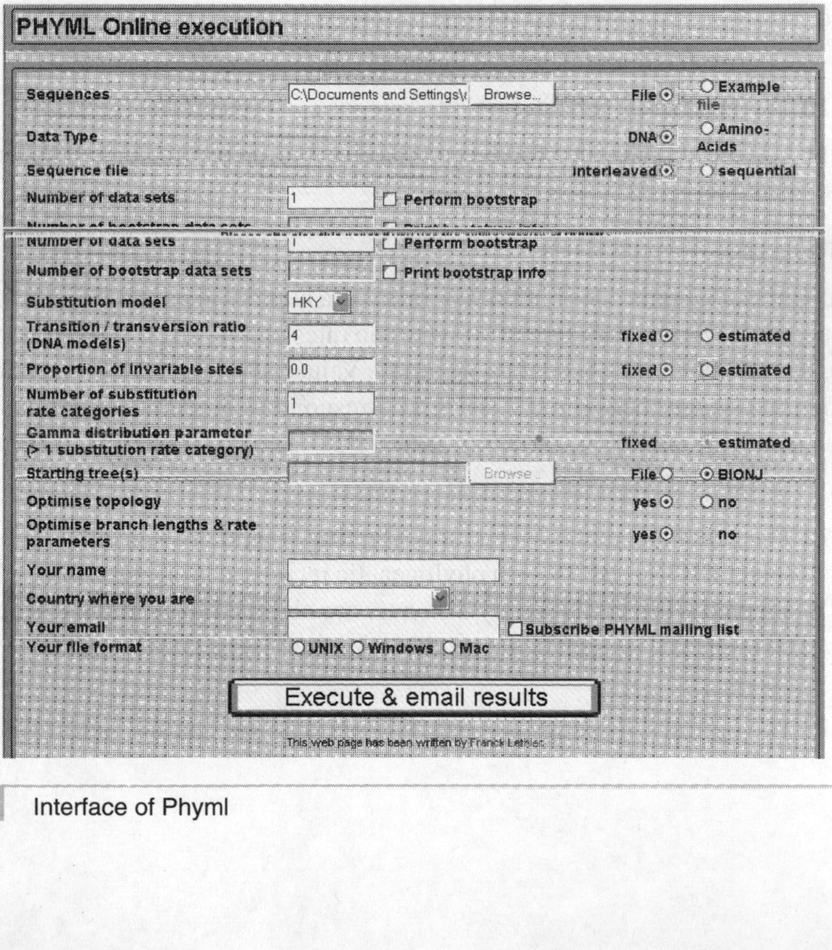

Fig. 6.4 Interface of Phyml

Step 2 The first textbox asks for the sequences that need to be used for constructing the tree.

Step 3 The options available for its treatment are to specify the sequence data type: **DNA** or **amino acid**. Select **DNA**.

Step 4 Sequence format needs to be specified as **interleaved** or **sequential**. Select **interleaved**.

Step 5 Next specify the number of datasets to be used in the construction of the tree. Also specify if you want to bootstrap it or not. Do not use this option initially.

Step 6 The number of bootstrap datasets that are to be used is to be specified next. Leave this option blank initially. The option to print the bootstrap results is also available.

Step 7 The next step is to specify the substitution model that you want to use. This tool gives the following options: HKY, JC69, K2P, F81, F84, TN93, and GTR. (In case you select amino acids in Step 3, the options available would be Dayhoff, JTT, mtREV, WAG, and DCMut.)

Step 8 The transition/transversion ratio (DNA models) can be chosen to be as either **fixed** or **estimated**. Choose the option **fixed** and enter the value as 4.

Step 9 The proportion of invariable sites can be chosen to be as either **fixed** or **estimated**. Choose **fixed** and enter the value as 0.0.

Step 10 The number of substitution rate categories is selected as 1.

Step 11 Use the default value for the gamma distribution parameter (> 1 substitution rate category).

Note

The shape of gamma distribution is defined by a numerical parameter. The higher its value, the lower the variation of substitution rates among sites. The default value is 1.0. It corresponds to a moderate variation. Values less than 0.7 correspond to high variations. Values between 0.7 and 1.5 correspond to moderate variations. Higher values correspond to low variations. The value can also be optimized to maximize the likelihood of phylogeny.

Step 12 The next input that is asked for is **Starting tree(s)**. You an either upload a file or select **BIONJ**. BIONJ is a phylogenetic reconstruction algorithm, which is well adapted when evolutionary distances are obtained from aligned sequences. It is an improved version of the neighbor-joining (NJ) algorithm. Choose the **BIONJ** option.

Step 13 The optimise topology has only two options: **yes** and **no**. Select **yes**.

Step 14 The next input required is **Optimise branch lengths & rate parameters**. This has only two options: **yes** and **no**. Select **yes**.

Step 15 The tool asks for the e-mail address, country, and operating system of the user.

Step 16 The search is executed and the results are sent to the user through e-mail in the format chosen.

Step 17 The tree can be visualized using a Java applet as shown in Fig. 6.5.

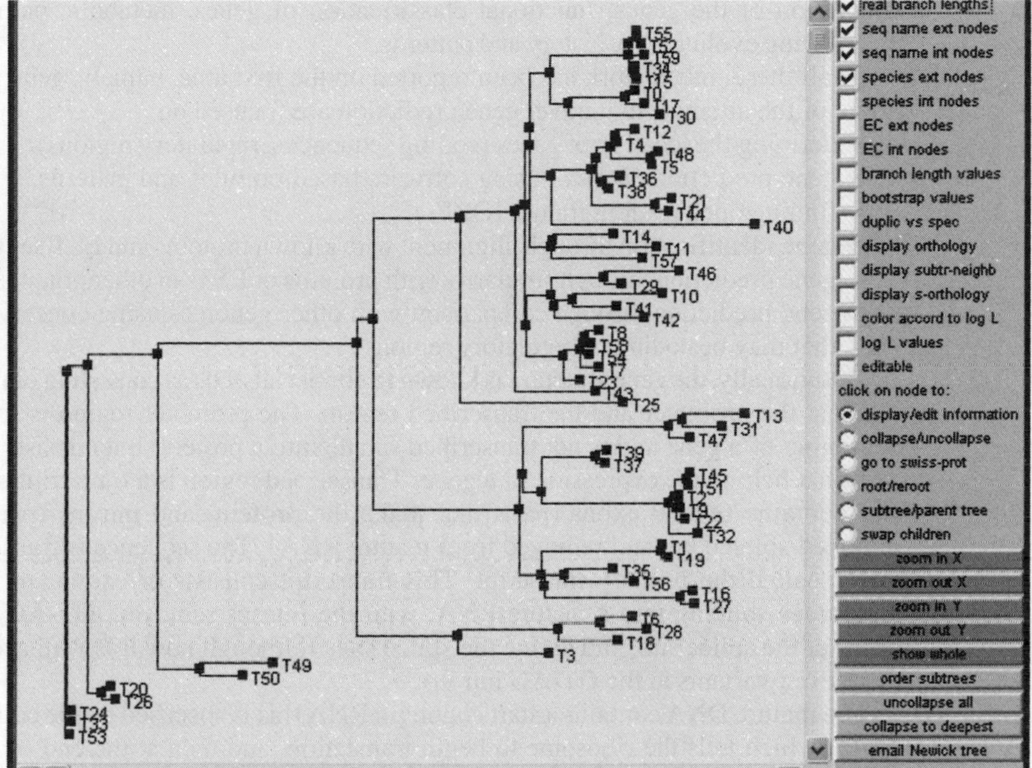

Fig. 6.5 Phylogenetic tree generated from Phyml

Click on nodes of the tree to

- display and edit information of a node. To edit information, **Editable** needs to be checked.
- collapse and uncollapse subtrees.
- go to Swiss-Prot and display the corresponding entry in the browser.
- place a root in the middle of the parent branch.
- display a subtree. To go back to the parent tree, click on the root node of the subtree.
- swap the children of a node (a pure cosmetic operation).

6.2 Gene Prediction

So, what does one do with a complete or near-complete genome such as the human genome? After all, a sequenced genome consists only of so many bases in a defined order. Analysis is obviously necessary in order to obtain biologically interesting information. The analysis of a genome covers many different aspects. The potential for interesting discoveries in the complete genomes is great. While the analysis of the genome is possible in innumerable ways, some of the common tasks undertaken are as follows: gene prediction, annotation of the genes, functional classification of genes, metabolic pathways, and discovering evolutionary history and patterns.

Of all these, much work has been reported on the first area, namely, gene prediction. Some of the attempts to unravel gene prediction are focused on

- defining the location of genes (coding sequences, regulatory regions),
- gene prediction *ab initio* using software based on rules and patterns,
- finding open reading frames (ORFs),
- gene identification through alignment with known proteins and EST sequences,
- gene prediction through similarity with proteins or ESTs in other organisms, and
- gene prediction through comparison with other genomes using conserved regions that may be coding or regulatory regions.

Traditionally, the gene structure is known to almost all of us as consisting of two distinct regions: the promoter and the transcribed region. The promoter region is found at the beginning of a gene and is not transcribed (made into a protein) but consists of binding sites that help in the expression of a gene. Transcribed region is a transcription start site and contains several exons (parts that make the protein) and introns (parts that are removed/spliced out and removed from mature RNA). The sequence is transcribed into RNA, called the primary transcript. This transcript consists of exons and introns. It undergoes splicing into a mature RNA, with the introns removed. GT-AG sequences occur at the splice site, and define the splice sites (although now it is well accepted that there exist variants to the GT-AG intron).

The mature DNA contains a start codon (in RNA this is specified by the codon ATG or AUG) which tells the ribosome to begin translation, and then at the end of the coding region there is a stop codon (TGATAA or TAG) that tells the ribosome to stop translation and release the protein. An untranslated region (UTR) flanks it and promotes its maturing, but it is not translated. In addition, it has a regulatory role in the cell such as directing the RNA within a cell. The mature RNA is translated into protein, which folds into a specific conformation and is functional. See Fig. 6.6 to see the process of transcription, splicing, and translation.

The human genome sequence data and the later discovery of the genome of several organisms have brought us to a point where we ask ourselves: where are the genes in the sequence? Answering this question largely depends on the genome we are looking at. For example, in the yeast genome, about 70% of the sequence codes for proteins, whereas in humans the genome is noncoding (only about 1.5% codes). Hence it is logical to have different approaches to finding genes in these genomes. Yeast genes consist of intergenic regions which are followed by a contiguous coding region and flanked by another intergenic region with no introns in between. It is easier to find the protein coding regions in these genes. In yeast the mean length of a coding region is about 500 codons and creates a protein with 500 amino acids.

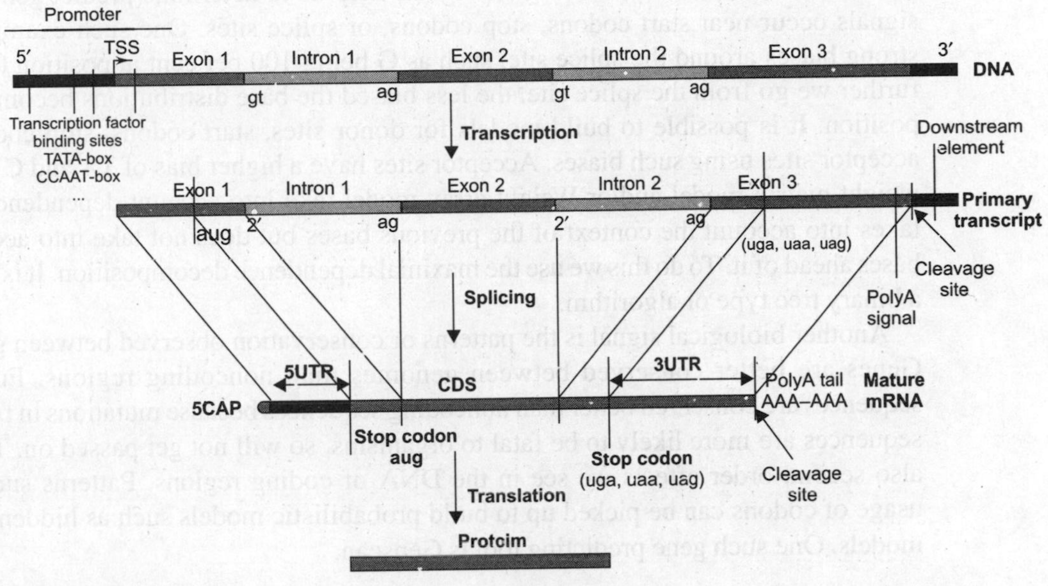

Fig. 6.6 Process of transcription, splicing, and translation

Another way to predict the occurrence of a gene is to look for ORFs (open reading frames). An ORF is a region that starts with a potential start codon and contains no in frame stop codons. It is a region that could theoretically be processed by a ribosome. The idea is to look at all the ORFs and find all the long ones that have the potential to be genes. A long ORF is unlikely to occur just by chance. In yeast a cut-off is around 350–400 codons, and it is safe to assume that any ORF over this length is likely to be a gene.

Unfortunately this is not true for higher organisms because of introns. Since introns are distributed within the coding sequences, all of the ORFs are much shorter and subsequently coding regions can be spread across many separate ORFs. For example, Drosophila has as many as 3.4 introns per gene. The mean intron length is about 475 codons and the mean exon length is about 397. In humans it is even higher (about 8.8 introns per gene). The mean intron length is about 4400 codons and the mean exon length is about 165.

To visualize/predict the genes in a genome/given sequence, programs have been devised that help us predict the presence of a gene based on the following assumptions:

- Genes have regular patterns.
- Genes start with start codons and end with stop codons.
- Introns have characteristic sequences. Introns usually start with GT/GC and end with AG.
- Exons alternate with introns.

The composition of coding DNA and that of the noncoding DNA differ because coding DNA codes for proteins and noncoding DNA has a different function. Because the cell uses different amino acids in different frequencies and patterns in its proteins, this difference in frequency and patterns is reflected in the codons as well. In addition, certain cell types show a bias towards the usage of certain amino acids—a phenomenon called codon bias. In addition to this, several biological signals help us to determine/predict genes. These signals occur near start codons, stop codons, or splice sites. One such example is the strong biases around the splice site, such as G being 100 per cent at position 0. But the further we go from the splice site, the less biased the base distributions become at each position. It is possible to build models for donor sites, start codons, stop codons, and acceptor sites using such biases. Acceptor sites have a higher bias of T's and C's and the weight matrix model and/or Weight array model take into account dependencies. This takes into account the context of the previous bases but does not take into account the bases ahead of it. To do this we use the maximal dependence decomposition. It is basically a binary tree type of algorithm.

Another biological signal is the patterns of conservation observed between genomes. Genes are better conserved between genomes than noncoding regions. Functional sequences are conserved better than noncoding sequences because mutations in functional sequences are more likely to be fatal to organisms, so will not get passed on. There are also second-order effects we see in the DNA of coding regions. Patterns such as the usage of codons can be picked up to build probabilistic models such as hidden Markov models. One such gene predicting tool is Genscan.

6.2.1 GENSCAN

This tool identifies genes from genomic sequences. It predicts gene exons and introns and returns the encoded protein sequence. GENSCAN was created at Stanford by Burge and Karlin in 1997. It was created before the human genome project, so there were no alignments available. Their estimate for the human gene count was about 100,000. The GENSCAN model uses explicit state duration HMM as shown in Fig. 6.7 where the states can have arbitrary length (not just geometric) distributions. If we have maximal state duration of size *n*, then time complexity is *n* times worse than a regular HMM.

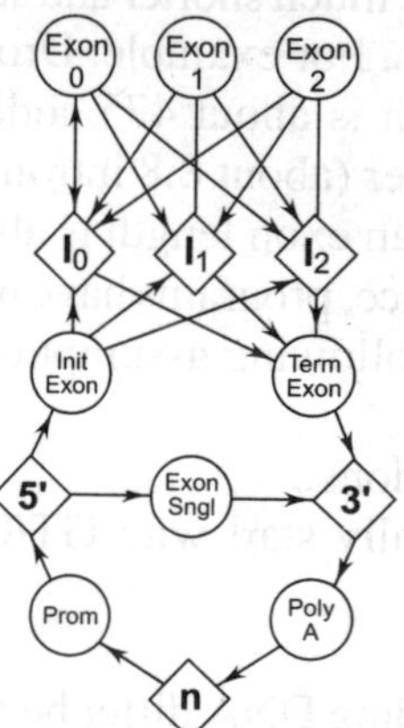

Fig. 6.7 GENSCAN based on *explicit state duration HMM*

GENSCAN finds the optimal —'parse' of a sequence, where a 'parse' is a succession of

- intergenic regions,
- 5′UTR (untranslated regions),
- exons,
- introns,
- 3′UTRs.

The program evaluates alternative parses with the help of

- Weight matrices or similar models for sites: promoters, translation starts, splice donors and acceptors, translation stops, polyadenylation sites.
- Interpolated Markov chains (3-periodic HMMs), and length distributions for exons, introns, 5′ and 3′ UTRs, and intergenic regions.

An example of a GENSCAN parse is shown in Fig. 6.8.

Sequence

Parse

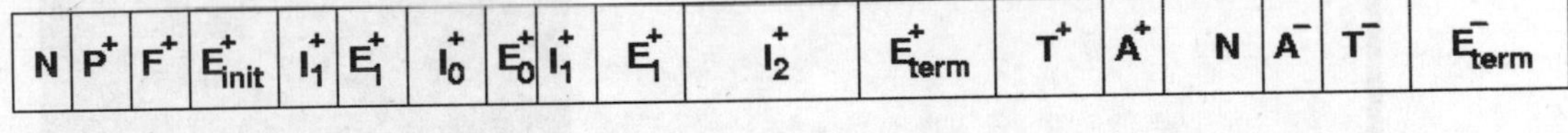

Genes

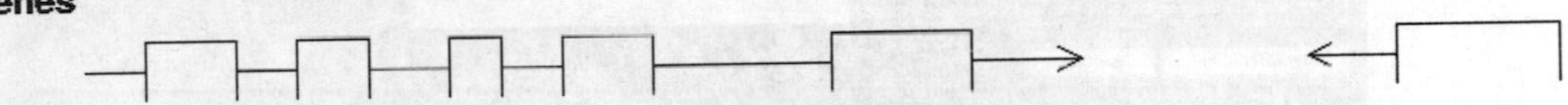

Fig. 6.8 Parsing in GENSCAN

TASK 

Use an unknown sequence to perform this gene prediction exercise.

Step 1 Type in the following url in your browser: http://genes.mit.edu/GENSCAN.html The interface of GENSCAN is shown in Fig. 6.9.

Step 2 The first textbox asks for the organism type, with three parameter options: **Vertebrate** for human/vertebrate sequences (also Drosophila), **Arabidopsis** for Arabidopsis thaliana sequences and **Maize** for Zea mays sequences. Choose the **Vertebrate** option.

Step 3 Suboptimal exon cut-off may be specified in the next textbox. The options available are: 1.00, 0.50, 0.25, 0.10, 0.05, 0.02 and 0.01. Let us select the **1.00** option.

Step 4 The next step is to fill in the sequence name in the textbox provided.

Step 5 Next specify the print options as either **Predicted peptides only** or **Predicted and CDS peptides**. Choose the option **Predicted peptides only**.

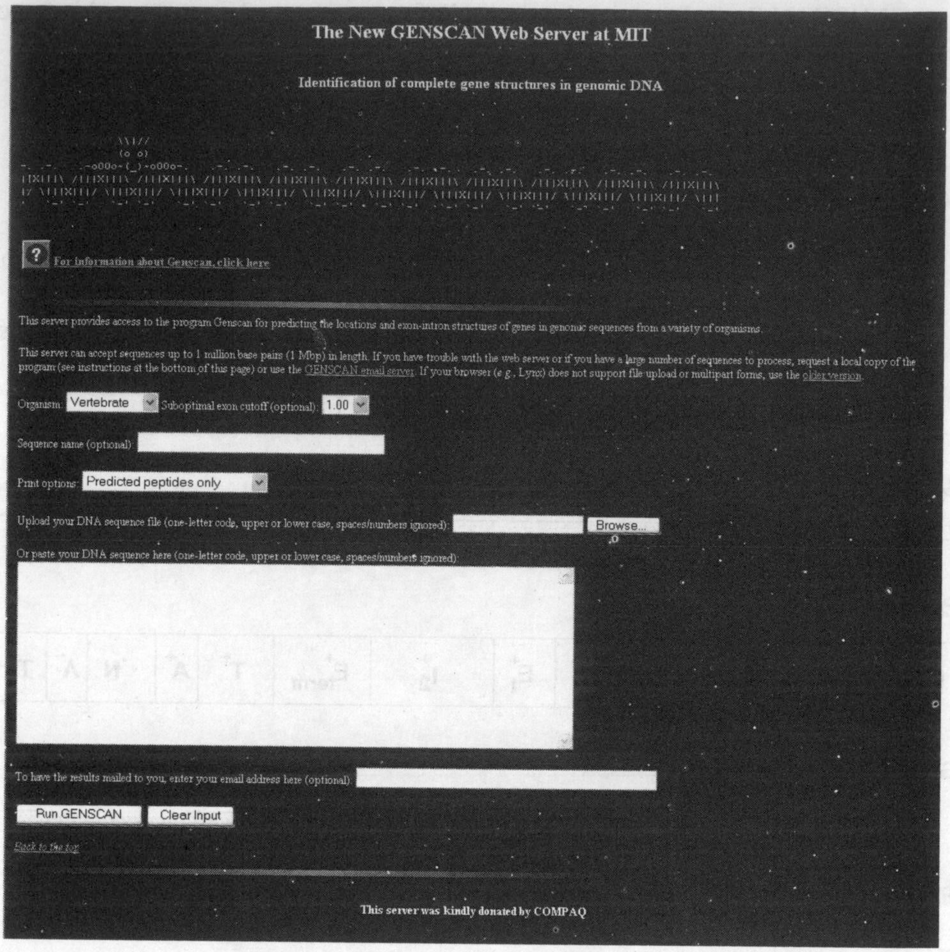

Fig. 6.9 Interface of GENSCAN

Step 6 You can enter or pate data in the fig textbox provided or upload the data file using the **Browse** option.

Step 7 In case you want the results to be mailed to you, enter your e-mail ID in the textbox provided next.

Step 8 The results that are reported back by GENSCAN are shown in Fig. 6.10.

Step 9 Prepare a small report on the following:

- The number of exons predicted in the sequence you submitted.
- The start and end points for each exon.
- Other elements/information given by the program (***Hint:*** Poly A sites, GC content)
- The length of the protein sequence predicted.
- Other elements/information given by the program (***Hint:*** molecular weight, other characteristic attributes, matches in a database, structure).

```
GENSCANW output for sequence genscan

GENSCAN 1.0        Date run: 29-Mar-106        Time: 01:48:57

Sequence genscan : 12360 bp : 50.50% C+G : Isochore 2 (43 - 51 C+G%)

Parameter matrix: HumanIso.smat

Predicted genes/exons:

Gn.Ex Type S .Begin ...End .Len Fr Ph I/Ac Do/T CodRg P.... Tscr..
----- ---- - ------ ------ ---- -- -- ---- ---- ----- ----- ------

 1.04 PlyA -   2539   2534    6                              1.05
 1.03 Term -   7042   6028 1015  0  1  106   49  2037 0.935 193.18
 1.02 Intr -   7946   7820  127  0  1   76   71   143 0.999  11.14
 1.01 Intr -   8402   8304   99  0  0   79  117   141 0.241  16.18

Click here to view a PDF image of the predicted gene(s)

Click here for a PostScript image of the predicted gene(s)

Predicted peptide sequence(s):

>genscan|GENSCAN_predicted_peptide_1|413_aa
XRIQCGKKLLQPSKQLRMFLKAVVLTLALVAVAGARAEVSADQVATVMWDYFSQLSNNAK
EAVEHLQKSELTQQLNALFQDKLGEVNTYAGDLQKKLVPFATELHERLAKDSEKLKEEIG
KELEELRARLLPHANEVSQKIGDNLRELQQRLEPYADQLRTQVNTQAEQLRRQLTPYAQR
MERVLRENADSLQASLRPHADELKAKIDQNVEELKGRLTPYADEFKVKIDQTVEELRRSL
APYAQDTQEKLNHQLEGLTFQMKKNAEELKARISASAEELRQRLAPLAEDVRGNLRGNTE
GLQKSLAELGGHLDQQVEEFRRRVEPYGENFNKALVQQMEQLRQKLGPHAGDVEGHLSFL
EKDLRDKVNSFFSTFKEKESQDKTLSLPELEQQQEQQQEQQQEQVQMLAPLES

Explanation

Gn.Ex : gene number, exon number (for reference)
Type  : Init = Initial exon (ATG to 5' splice site)
        Intr = Internal exon (3' splice site to 5' splice site)
        Term = Terminal exon (3' splice site to stop codon)
        Sngl = Single-exon gene (ATG to stop)
        Prom = Promoter (TATA box / initation site)
        PlyA = poly-A signal (consensus: AATAAA)
S     : DNA strand (+ = input strand; - = opposite strand)
Begin : beginning of exon or signal (numbered on input strand)
End   : end point of exon or signal (numbered on input strand)
Len   : length of exon or signal (bp)
Fr    : reading frame (a forward strand codon ending at x has frame x mod 3)
Ph    : net phase of exon (exon length modulo 3)
I/Ac  : initiation signal or 3' splice site score (tenth bit units)
Do/T  : 5' splice site or termination signal score (tenth bit units)
CodRg : coding region score (tenth bit units)
P     : probability of exon (sum over all parses containing exon)
Tscr  : exon score (depends on length, I/Ac, Do/T and CodRg scores)

Comments

The SCORE of a predicted feature (e.g., exon or splice site) is a
log-odds measure of the quality of the feature based on local sequence
properties. For example, a predicted 5' splice site with
score > 100 is strong; 50-100 is moderate; 0-50 is weak; and
below 0 is poor (more than likely not a real donor site).
```

Fig. 6.10 Result page of GENSCAN

6.2.2 GrailEXP (http://compbio.ornl.gov/grailexp/gxpfaq1.html)

GrailEXP is a suite of tools for locating genes that encode proteins within DNA sequences. This suite can also be used to predict EST/mRNA alignments, certain types of promoters, polyadenylation sites, CpG islands, and repetitive elements. Essentially it functions as a gene finder, an EST alignment utility, an exon prediction program, a promoter/polya recognizer, a CpG island finer, and a repeat masker. GrailEXP is currently used to analyse human and mouse sequences, but other systems such as **Arabidopsis**, **Drosophila**, rice, corn, wheat, and many more are being developed.

GrailEXP v3.0–v3.2 were developed by Doug Hyatt at Oak Ridge National Laboratory. and is used at the Computational Biology Section at Oak Ridge to annotate the entire known portion of the human and mouse genomes, including both finished and draft data. An important differentiator is that this suite has been included in the annotations offered by Celera and DoubleTwist. This program takes many of its components from its earlier versions Grail 1.3 which could predict repeats, polyAs, promoters, exons, genes, and complex repetitive elements.

Presently, GrailEXP features include the following:

- Grail-like exon finder (with improved splice site recognition and other minor changes) adapted from the Grail 1.3 code.
- Gene modelling has been vastly improved by searching a database of known gene messages (complete and partial) and building gene models based on the corresponding alignments. It is these two additional powerful tools (the gene message alignment program and the gene assembly program) that distinguish GrailEXP from Grail. In addition, the Smith–Waterman-like complex repeat Grail finder, which takes forever to run, has been replaced by a much faster BLAST-based method, although admittedly less precise.

GrailEXP runs on all the standard UNIX platforms. This suite consists of three binaries wrapped by a single Perl script which calls the binaries on the user's requests. The suite consists of the following programs:

- **Perceval** (the exon prediction program) It can also be used to find repeats, exon prediction, and CpG island location.
- **Galahad** (the gene message alignment program) It can search against any number of databases, align with a single mRNA/EST, either relying on Grail or Genscan exons as a seed or independently.
- **Gawain** (the gene assembly program) It assembles genes from the EST alignments, recognizes alternative splicing, finds 5' and 3' untranslated regions, and predicts polyA and promoter elements.

Let us try and learn more about these programs.

Perceval (**P**rotein-coding **e**xon, **r**epetitive, and **c**pG-island **eval**uator) This program takes a DNA sequence as its input and gives as the output a list of possible Grail Exon candidates. It also filters the output list against a repetitive element database and also looks for CpG islands. These candidates are arrived at by using neural networks. Each candidate's sequence has a 'begin', 'end', 'strand', and 'frame'. Each begins with either a start codon or an AG acceptor splice site and ends with either a stop codon or a donor splice site. These are then sorted into clusters, with the highest scoring exon in each cluster clearly indicated. These best exons in each cluster are traditionally referred to as 'Grail exons'. To predict and search repetitive elements, the program blasts the input sequence against a database of repetitive elements. A sample output (pretty output) from GrailEXP is given in Fig. 6.11.

```
GrailEXP v3.2
http://compbio.ornl.gov/grailexp/

Authors:  Doug Hyatt, Manesh Shah, Victor Olman, Richard Mural, Ying
Xu, and  Edward C. Uberbacher, 1996-2001

Reference:  "Automated Gene Identification in Large-Scale Genomic
Sequences", Xu, Y. and Uberbacher, E.C., Journal of Computational
Biology, Volume 4,  Number 3, 1997

Sequence:  >GrailEXP Input Sequence (36741 bp)
----------------------------------------------------------------------
PERCEVAL Exon Candidates (15 predicted)

 Index Std   Begin        End     Frm    Type      Len    Scr   Quality

    1 -       200         386   2  Internal    187    79      Good
    2 -       595         693   0  Terminal     99    80      Good
    3 -      9207        9254   0  Internal     48    66  Marginal
    4 -      9910        9986   0  Initial      77    70      Good
    5 -     14287       14794   2  Terminal    508    49  Marginal
    6 +     19230       19291   0  Internal     62    99 Excellent
    7 +     26344       26466   2  Internal    123   100 Excellent
    8 +     28908       29051   2  Internal    144   100 Excellent
    9 +     29823       29938   2  Internal    116   100 Excellent
   10 +     31176       31303   1  Internal    128    98 Excellent
   11 +     32425       32496   0  Internal     72   100 Excellent
   12 +     32573       32674   0  Internal    102   100 Excellent
   13 +     32851       32915   0  Internal     65    60  Marginal
   14 +     34354       34483   2  Internal    130   100 Excellent
   15 +     35100       35202   0  Internal    103    94 Excellent
----------------------------------------------------------------------
```

Fig. 6.11

The output reports only the highest-scoring exon in each cluster. All indexing is from the forward-strand perspective. Another output format is the Genome channel file shown in Fig. 6.12.

```
exon_grailexp_v3=1|f|1|2|19230|19291|19170|19295|0.99|1
exon_grailexp_v3=2|f|1|2|19234|19291|19170|19295|0.91|0
exon_grailexp_v3=3|f|1|1|26291|26466|26267|26470|0.86|0
exon_grailexp_v3=4|f|2|1|26291|26470|26267|26470|0.78|0
exon_grailexp_v3=5|f|1|1|26344|26466|26267|26470|1|1
exon_grailexp_v3=6|f|2|1|26344|26470|26267|26470|0.94|0
exon_grailexp_v3=7|f|0|1|26402|26466|26267|26470|0.84|0
exon_grailexp_v3=8|f|0|0|28837|29051|28750|29055|0.88|0
exon_grailexp_v3=9|f|1|0|28908|28976|28750|29055|0.88|0
exon_grailexp_v3=10|f|1|0|28908|28986|28750|29055|0.82|0
exon_grailexp_v3=11|f|1|0|28908|29022|28750|29055|0.91|0.............................
```

Fig. 6.12

Genome channel output format is always indexed from the target strand's perspective. This means all forward-strand objects are indexed relative to the forward strand, and all reverse-strand objects are indexed relative to the reverse strand.

Galahad **(Gene message alignment program)** Before we move on to learn more about this program, let us be clear about what is meant by the term 'gene message alignment'. A gene message alignment is an alignment between a genomic sequence (in which genes are present in small bits) and a gene message (in which a gene is a single contiguous piece of DNA built from spliced together exons). It differs from a regular global alignment because there is additional information that can help with the alignment process, e.g., the location of the internal 'edges' of each alignment piece and its conformity with standard splice site (GT.... AG) boundaries.

Note

GrailEXP produces 'alignments' but the output displays boundaries and scores; and not the actual bases lined up against each other as in BLAST.

When such as program is executed it aligns a gene message with a genomic sequence so that the exon/intron boundaries become identifiable. The program searches for alignments against a large number of databases such as CBIL/UPenn DOTS, TIGR EGAD Transcript Database, NCBI RefSeq and dbEST databases, Miscellaneous mRNAs from Genbank.

Like Perceval, this program blasts the input sequence against a list of Grail Exon candidates. It replaces all non-exonic portions of the sequence with n's and blasts that sequence against the search database. The program examines the resulting alignments and selects those ESTs/cDNAs that are the best. It then performs a second blast, in which it blasts the entire sequence against the EST/cDNAs of interest. It then optimizes all the edges (left, right, and internal), doing an exhaustive search of all splice sites in the vicinity of each edge. The resulting alignments, with exons, introns, and splice sites calculated, are the final gene message alignments.

The pretty output version for Galahad looks like as shown in Fig. 6.13.

```
GrailEXP v3.2
http://compbio.ornl.gov/grailexp/

Authors:  Doug Hyatt, Manesh Shah, Victor Olman, Richard Mural, Ying Xu,
and dward C. Uberbacher, 1996-2001

Reference:  "Automated Gene Identification in Large-Scale Genomic
Sequences", Xu, Y. and Uberbacher, E.C., Journal of Computational
Biology, Volume 4,   Number 3, 1997

Sequence:  >GrailEXP Input Sequence (36741 bp)
-------------------------------------------------------------------------
GALAHAD Gene Alignments (97 located: 25 displayed, 72 redundant)
```

```
Index Std    Begin    End  Accession     Database  Organism    Length

  1 -         320     702  AA633758.1     dbest     human       383

   1 piece      Seq exons = (320..702)
  99% ident     Ref exons = (383..1)

  2 +        3936    35973 HT36257        tigr      human      1337

  12 pieces     Seq exons =
 (3936..4063,19230..19295,26411..26466,28908..29051,
 29823..29938,31273..31303,32425..32496,32573..32674,

 32851..32915,34354..34483,35100..35202,35651..35973)
  99% ident     Ref exons =
 (1..128,129..195,196..253,254..397,398..515,516..542,
                        543..614,615..716,717..781,782..911,912..1014,
                        1015..1337)

  3 +        3936    35975   HT1292       tigr      human      1498

  12 pieces     Seq exons =
 (3936..4063,19230..19291,26344..26466,28908..29051,

 29823..29938,31176..31303,32425..32496,32573..32674,

 32851..32915,34354..34483,35100..35202,35651..35975)
  99% ident     Ref exons =
 (1..128,129..190,191..313,314..457,458..573,574..701,

 702..773,774..875,876..940,941..1070,1071..1173,
                        1174..1498)

  4 +      3939      31296   BE407945.1     dbest      human       756
..............................................................(truncated)
..............................................................(truncated)
 17 +    32440      32620   ET63508         tigr      mouse      1399
   2 pieces     Seq exons = (32440..32496,32573..32620)
  87% ident     Ref exons = (714..770,771..818)

 18 +     32571      35202  AW631750.1      dbest     other       572
   3 pieces     Seq exons = (34353..34483,35100..35202,35651..35976)
  97% ident     Ref exons = (4..130,131..232,233..553)
------------------------------------------------------------------
```

Fig. 6.13

Note that unlike the GCA and Raw output formats, the pretty output performs a redundancy check and shows only alignments that provide unique information.

Seq exons indicate the bounds of the predicted exons within the genomic sequence.

Ref exons indicate the bounds of the predicted exons within the reference sequence (mRNA, EST, cDNA).

In addition to the pretty output format, this program can give its output as the Genome Channel output format and the Raw output format.

Gawain (gene **a**ssemblies **w**ith **a**lignment **in**formation) This is a gene assembly program. It first designs a gene model from Grail Exon candidates and/or GrailEXP genomic alignments. A good gene model consists of the following elements:

- 5′ untranslated region,
- coding portion,
- 3′ untranslated region,
- exons,
- introns,
- a polyadenylation site, and
- a promoter.

However, it is not necessary to have both exon candidates and alignments. The program can use either of them or use both. The program also outputs the mRNA and protein translation for each gene model.

> **Gawain**
> Gene assembly program that assembles genes from the EST alignments, recognizes alternative splicing, finds 5′ and 3′ untranslated regions, and predicts polyA and promoter elements.

The program uses genomic alignments for clustering and assigns a frame using a recursive frame-scoring function which takes in the Grail coding score for each potential frame, the frames of Grail exons that match the alignment pieces, and splicability among the various pieces. The program then uses an elaborate dynamic programming algorithm to build gene models. False stop codons in the genomic sequence due to sequencing errors or pseudogenes can cause incorrect start and stop codons to be identified. PolyAs and promoters are located using a simple Markov model. The promoter system uses a neural net and looks for specific types of signals (TATA, CAAT, GC-box). The pretty output version for Gawain is shown in Fig. 6.14.

This output format provides information about each gene model, including the FASTA header for the highest scoring reference, a reference path, a complete list of exons and CDS regions, and the mRNA and predicted protein for that gene. The other output formats available are raw and GCA.

```
GrailEXP v3.2                                    http://compbio.ornl.gov/grailexp/

Authors:  Doug Hyatt, Manesh Shah, Victor Olman, Richard Mural, Ying Xu, and   Edward C.
Uberbacher, 1996-2001

Reference:  "Automated Gene Identification in Large-Scale Genomic Sequences", Xu, Y. and
Uberbacher, E.C., Journal of Computational Biology, Volume 4,  Number 3, 1997

Sequence:  >GrailEXP Input Sequence (36741 bp)
---------------------------------------------------------------------
GAWAIN Gene Predictions (2 predicted, 2 with database similarity)


Genes with Database Similarity (2 predicted, 0 with alternative splices)

Gene 1, Variant 1          Strand: -  Bounds: 320-702  Exons: 1
                           Start Codon:  Yes     Stop Codon:  Yes

Top-Scoring Reference:  AA633758.1 (383 bp) (99% id, 320-702)
     >human|AA633758.1|dbest|AA633758 ac27b12.s1 Stratagene ovary (#937217)       Homo
sapiens cDNA clone IMAGE:857663 3'
Reference Path:  AA633758.1 (383 bp) (99%, 320-702)

 ---Index---- --------Exons-------- ---------CDS--------- -Ph- -Fr- -Len- -Scr-
 1.1.1              320        702        595        687    0    0   383   99

Gene 2, Variant 1          Strand: +  Bounds: 3936-35976  Exons: 12
                           Start Codon:  Yes     Stop Codon:  Yes

Top-Scoring Reference:  HT1292 (1498 bp) (99% id, 3936-35975)
     >human|HT1292|tigr|adenosine deaminase, alt. transcript 1, 3'
Reference Path:  HT1292 (1498 bp) (99%, 3936-35975)
                 AV735384.1 (567 bp) (97%, 34354-35976)

 ---Index---- --------Exons-------- ---------CDS--------- -Ph- -Fr- -Len- -Scr-
 Promoter          1978       2055        ...        ...    .    .    78   69
 2.1.1             3936       4063       4031       4063    1    1   128  100
 2.1.2            19230      19291      19230      19291    0    2    62  100
 2.1.3            26344      26466      26344      26466    2    1
.........................................................................
(Truncated)
130  100
 2.1.11           35100      35202      35100      35202    0    2   103  100
 2.1.12           35651      35976      35651      35664    1    0   326  100
 PolyA            35947      35952        ...        ...    .    .     6   89
```

>GrailEXP Gene 1, Var 1 mRNA|Similar to AA633758.1
cctccccagcctctcatgacattcaagtccctcttcaacatcagtcccaccacttcccaccacattgctg
....
ctcttgaacattctctgggaacccaagcttctagaagagaccatcatgctaagagagtccacatatgg
ccattccagttgacagtcccactgagcccaggc

>GrailEXP Gene 1, Var 1 protein|Derived from similarity to AA633758.1
MTFKSLFNISPTTSHHIAEMPINLDELLVA*

>GrailEXP Gene 2, Var 1 mRNA|Similar to HT1292
accgctggccccagggaaagccgagcggccaccgagccggcagagacccaccgagcggcggcggagggag
cagcgccggggcgcacgagggcaccatggcccagacgcccgccttcgacaagcccaaagtagaactgcat
(truncated)
cttgcacatgggcatggttgaatctgaaaccctccttctgtggcaacttgtactgaaaatctggtgctca
ataaagaagcccatggctggtggcatgca

>GrailEXP Gene 2, Var 1 protein|Derived from similarity to HT1292
MAQTPAFDKPKVELHVHLDGSIKPETILYYGRRRGIALPANTAEGLLNVIGMDKPLTLPDFLAKFDYYMP
(truncated)YGMPPSASAGQNL*
Genes with No Database Evidence (0 predicted)

Fig. 6.14

TASK

Use this unknown sequence to perform this gene prediction exercise

Step 1 Go to the EMBL database at http://www.ebi.ac.uk/embl/ and select **Nucleotide sequences** as shown in Fig. 6.15.

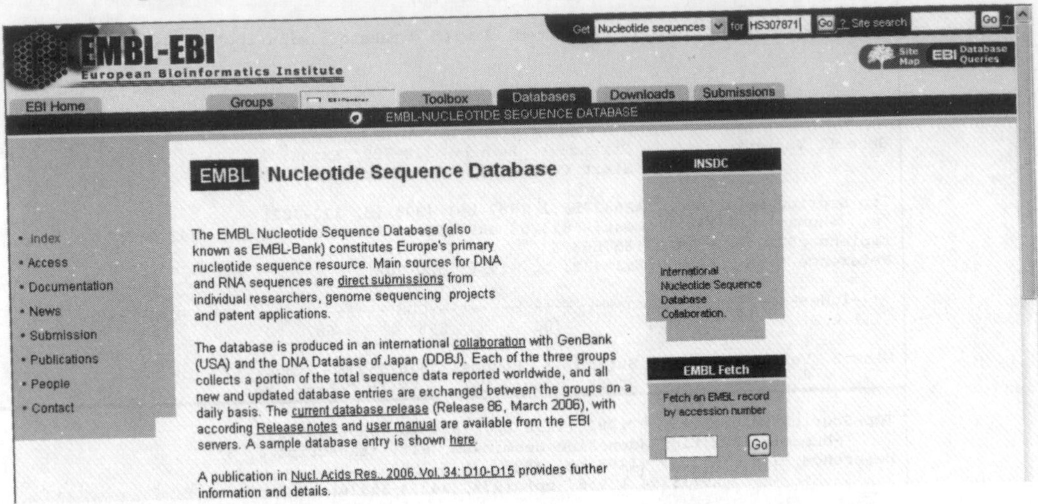

Fig. 6.15 Interface of EMBL Nucleotide Sequence Database

Step 2 Retrieve sequence entry U30787, and detect the mRNA and CDS exons as shown in Fig. 6.16.

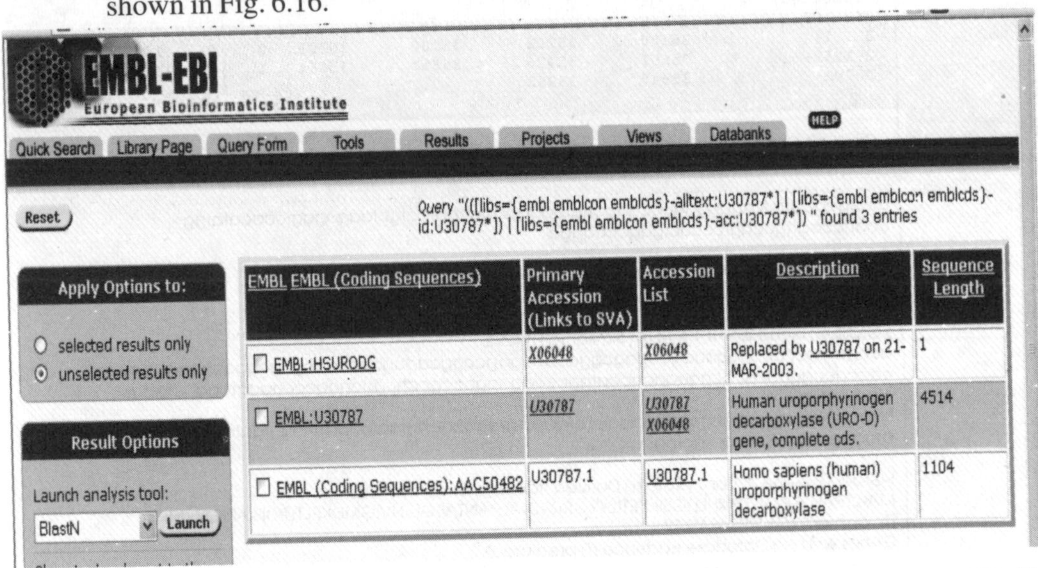

Fig. 6.16 Result page of EMBL search for sequence

Step 3 Retrieve the sequence in the FASTA format (see Fig. 6.17).

```
>embl|U30787|U30787 Human uroporphyrinogen decarboxylase (URO-D)
gene, complete cds. ...
aagcttcgtaagcacctctcgcggcacgaaagccagcgctgcctaggcgccgcccggcgc
gaggctctcacctctgccaagaagcgcaccggcccagcagctgccggggggactccagca
ccgcgccgggccatggacccgccatgagtcagctggcgcgaccgcggacagagcttccca
ccacgcccttccccgcctttggccagcctttgccgtatgttctggactaagcgcacccca
gctctcactgtattggactgtgtactcccacactcaaccatattacttatctctgtgcca
ccctaacccagccgaccaaacccaagattggtgattgctacctgatcaatctccctctct
ccatttccttgtgactaccattttatctctactgctactaccctcattcaagtcaccatt
ctagctagcctgggtcattgccaacagtcattttctggttcttcggcctgctgttttc
ctcccactcccagcgaatctgctggactccctatcctatgggtggtgtgattaaagtgtt
tgagacaatggccccttccctgccactgacaggagtcttgagtcattagggttgagttc
tgtttgacactcctaatcccaaggacactggagatcattattcatttaatgtgattgct
gatttctgtttccccagtcttgtagctccttaaaggctggggtgtcttgagcagagctaa
cctctgcacctactataggtccaggctatagtatggacctggctggataagactgttggt
atcatagttgggacttgcgccaagctccggatacccagactgtcagatgagaacaaattc
ctcatgtcaccgtaagatacatttacagcggagtttctttgggcctttgttgtttcgt
cgctacagcaaactttacggtgaaaaaggtaggggtctacggcagcagcagggcagccc
tggagctgtcgctggagtccgatcatgtgatcttcaacatggcgacgctcttggttccct
acagaaaggggcggagcctggactgggggcaggctcagattcaggttaaattgtggatt
gagctcgcagttacagacagctgaccatggaagcgaatgggttggggtgagttctccaga
gcacgcggtgtggctagccgggcttctaatttgagtcttccaactcaggactctatccct
ctactcccctttccccaccctggagaacctcccaacctgaactccgttagctggatcctg
aatcctaaaccatggattttgagatgttcatcccagggccttaattcaagggatgcct
caggatttccaaccaggatcttcattctgggaccatcaactctgatccctctttatcccc
cagcctgggtatttctcagccctgaaccagcccagtgacatttcccggtttctgaggct
cactagttcgaagaccccaaactatccttagtgggccttcattccctcccccagtccc
tctggttgcttcgagcttggaagagtagagactaagtggagggaagaggccccagggcgg
gcccttctggagtttgtgcactgataggcagagaggaggcggaacgggcggaaagccagg
gtttgggagctggcctggaggaggtaggatagcggtcctggactgaatcggccttatgaa
```

Fig. 6.1

Step 4 Go to the GrailEXP homepage at http://compbio.ornl.gov/grailexp/

Step 5 The page gives the option of selecting from human, Drosophila, Arabidopsis, and mouse. Select **Human** (*Homo sapiens*).

Step 6 Select output type options as **Human Readable Text**.

Step 7 Check the **Perceval Exon Candidates** check box.

Step 8 Either paste or upload the DNA sequence and click **Go**.

Step 9 The result page will look like as is shown in Fig. 6.18.

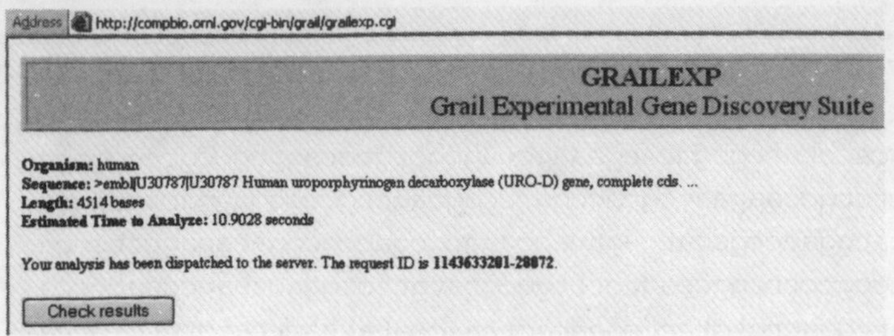

Fig. 6.18 Result page of GrailExp

Step 10 Click the **Check Results** button.

Step 11 The results that are obtained are given in Fig. 6.19.

```
gc_object_start: gene_grailexp --organism human --output pretty --nodb
--noassemble --dbpat grailexp_v3

# Service: gene_grailexp
# Version: 3.3
# Description: GAT GrailEXP Gene Prediction Service
# Last Modified: October, 2001
# Tool:  GrailEXP 3.3 from ORNL.  Last updated:  October, 2001.
# Database:  GrailEXP Database Thu Feb 27 16:15:37 EST 2003 from
NCBI/TIGR/Baylor/Riken (15960696 entries).  Last updated:  Thu Feb 27
16:15:37 2003.
# Sequence Name: >gene_grailexp|PID=14720
# Sequence Length: 4514
# Output_begin: pretty
----------------------------------------------------------------------
---------
GrailEXP v3.31 [March, 2002]
http://compbio.ornl.gov/grailexp/

Authors:  Doug Hyatt, Manesh Shah, Victor Olman, Richard Mural, Ying
Xu, and   Edward C. Uberbacher, 1996-2001

Reference:  "Automated Gene Identification in Large-Scale Genomic
Sequences",  Xu, Y. and Uberbacher, E.C., Journal of Computational
Biology, Volume 4,   Number 3, 1997

Sequence:  >gene_grailexp|PID=14720 (4514 bp)
----------------------------------------------------------------------
PERCEVAL Exon Candidates (6 predicted)
 Index Std   Begin       End      Frm    Type       Len    Scr    Quality

     1 +      1755       1860    0   Internal    106     57    Marginal
     2 +      2434       2631    0   Internal    198    100   Excellent
     3 +      2749       2910    0   Internal    162    100   Excellent
     4 +      3324       3416    0   Internal     93     92   Excellent
     5 +      3576       3676    0   Internal    101    100   Excellent
     6 +      4179       4340    0   Terminal    162    100   Excellent
----------------------------------------------------------------------
# Output_end: pretty

gc_object_end: gene_grailexp --organism human --output pretty --nodb --
noassemble --dbpat grailexp_v3
```

Fig. 6.19

Step 12 Connect to the GENSCAN at http://genes.mit.edu/GENSCAN.html and paste this DNA sequence and run Genscan.

Step 13 Compile a report on the compare annotations and predicted exons as given by both the programs.

6.3 Protein Structure and Function Prediction

Large scale analysis of protein sequences is done for identification of disease associated proteins and designing drug to complement target cells. The drug molecule is expected to complement the diseased protein and should not bind to other protein molecules to avoid side effects. It should pass through various filtration processes such as those defined by ADMET. The need of this hour is well supported by automated structure and function prediction tools instead of time intensive manual processes. Each of these tools follow various approaches where even protein structure prediction is used to gain hints of protein function. Protein function can also be predicted from a sequence or by de novo function assignment. Here we will deal with two kinds of tools that are broadly used for protein structure and function prediction purpose.

6.3.1 Prosite

PROSITE is a database of protein families and domains. It consists of biologically significant sites, patterns, and profiles which help to map unknown sequences to known protein families. The current release of PROSITE contains 1344 documentation entries which describe 1322 patterns, 4 rules, and 515 profiles/matrices. PROSITE was constructed by Amos Bairoch. It is based on the proteins sequences in Swiss-Prot. Certain regions of protein sequences have been conserved in nature, and as a basic thumb rule these regions form the basis of the function of a protein and/or for the maintenance of its three-dimensional structure. The 'patterns' or signatures that these regions of the protein sequences display help in setting them aside from other unrelated sequences. Therefore it is possible to ascertain certain families of proteins to an unknown sequence based on the signature pattern/profile of the particular domain. In fact an often used analogy is that of assigning a fingerprint to an individual. Using the same analogy a protein signature can be used to assign a newly sequenced protein to a specific family of proteins and thus to formulate hypotheses about its function.

The basis of PROSITE is regular expressions describing characteristic subsequences of specific protein families or domains. It has now been extended to contain some profiles, which can be described as probability patterns for specific protein sequence families. PROSITE makes a distinction between patterns and rules, which are both described by regular expressions.

- A pattern is defined as a unique characteristic that belongs to a protein domain family.
- A rule only displays certain features in a protein sequence that may or may not help in assigning it to a specific protein family. For example, a potential phosphorylation site can be found in many protein sequences.

In this section we will just describe the notation for patterns and rules in PROSITE. Patterns and rules are described using the same notation. Unfortunately, the PROSITE notation for sequence patterns is different from the UNIX-type regular expressions.

However, the concepts are the same, and it is not so difficult to translate a PROSITE pattern into a UNIX-type regular expression. As an example, let us use the PROSITE pattern CBD_FUNGAL (accession code PS00562). The original text entry, as is given in the downloadable PROSITE data file, is as follows.

```
ID  CBD_FUNGAL; PATTERN.
AC  PS00562;
DT  DEC-1991 (CREATED); NOV-1997 (DATA UPDATE); JUL-1998 (INFO UPDATE).
DE  Cellulose-binding domain, fungal type.
PA  C-G-G-x(4,7)-G-x(3)-C-x(5)-C-x(3,5)-NHG-x-FYWM-x(2)-Q-C.
NR  /RELEASE=38,80000;
NR  /TOTAL=21(18); /POSITIVE=21(18); /UNKNOWN=0(0); /FALSE_POS=0(0);
NR  /FALSE_NEG=1; /PARTIAL=0;
CC  /TAXO-RANGE=??E??; /MAX-REPEAT=4;
CC  /SITE=1,disulfide; /SITE=7,disulfide; /SITE=9,disulfide;
CC  /SITE=16,disulfide;
DR  Q00023, CEL1_AGABI, T; Q12714, GUN1_TRILO, T; P07981, GUN1_TRIRE, T;
DR  O59843, GUX1_ASPAC, N;
DO  PDOC00486;
//
```

The main line is the PA line, which contains the pattern. Let us go through this pattern step by step.

```
PA  C-G-G-x(4,7)-G-x(3)-C-x(5)-C-x(3,5)-NHG-x-FYWM-x(2)-Q-C.
```

Let us go through the elements in the pattern to see what they mean.

Each non-x letter defines one particular type of amino acid residue in that position in the pattern. Here we must have a tripeptide Cys-Gly-Gly in the beginning of the matching segment of a protein chain. The characters '-' add no information to the pattern and are added to make the pattern slightly easier to read. The notation x(4,7) means that at least 4 and at most 7 residues of any type may occur at this position. This corresponds to the notation. {4,7} in a UNIX-type regular expression. The notation NHG means the same thing as in a UNIX-type regular expression: in this position any of the residues within the brackets may be chosen. One and only one such residue must be at this position. The notation x(2) means that exactly two residues of any type may occur at this position. This corresponds to the notation .. or .{2,2} in a UNIX-type regular expression. The notation {GP} (not shown in this example) means that all residues except Gly and Pro are allowed in this position. The lines marked DR are the protein sequence entries in Swiss-Prot which match (character T) or do not match (character N) the regular expression. In this case, the protein GUX1_ASPAC does not match the PROSITE rule, although it should; it is a false negative.

The interface of PROSITE is shown in Fig. 6.20.

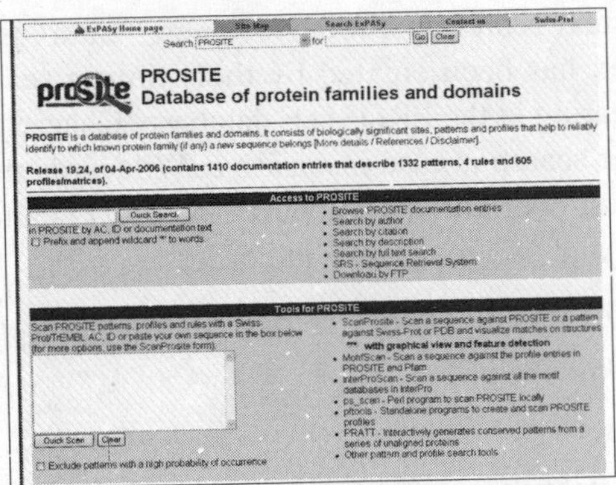

Fig. 6.20 Interface of Prosite

Access to PROSITE

Browse PROSITE documentation entries It is possible to search PROSITE using the following parameters: post-translational modifications, compositional biased regions, domains, DNA or RNA associated proteins, enzymes, electron transport proteins, other transport proteins, structural proteins, receptors, cytokines and growth factors, hormones and active peptides, toxins' inhibitors, protein secretion and chaperones, and others. The entries are alphanumeric and follow a definite structure, e.g.,

PDOC00020 2 Kringle domain signature and profile

- The character in the first column is used to indicate whether a documentation entry is new in this release '+', or has been modified '*'.
- The numerical characters in positions 3 to 7 provide the documentation entry accession number.
- The numerical character in position 9 is used to indicate how many data entries (patterns, rules and profiles/matrices) are described by a documentation entry.

Examples are

- PDOC00001 1 N-glycosylation site
- PDOC00351 2 Disintegrin domain signature and profile
- PDOC00027 2 'Homeobox' domain signature and profile
- PDOC00086 1 Thymidylate synthase active site
- PDOC00160 1 DNA topoisomerase II signature
- PDOC00109 2 Phospholipase A2 active sites signatures

Search by author Searching PROSITE by author name is possible.

Search by citation PROSITE can be searched using this option for journals published after 1965 with options such as journal name, volume, and year.

Search by description It is possible to search the PROSITE database of protein families and domains by the description of sites and patterns by using a keyword. The keyword may be any word or partial word appearing in the description of the documentation entrics.

Search by full text search The text search is a relatively new feature available on PROSITE Release 19.24. The index has been created by the Glimpse search engine. Search terms may be any word appearing in the PROSITE database. Characters allowed as part of a 'word' are _-#'(),./*. The search is case insensitive. Wildcards may be used to search with partial words. Boolean operators (AND, OR, and NOT) can be used to restrict search, with braces (i.e., '{' and '}') to specify the order.

SRS (sequence retrieval system) PROSITE offers two alternative search forms under this category. These are self-explanatory.

Tools available at PROSITE

Prosite offers the following tools on its web interface.

ScanProsite This tool enables a user to scan a sequence against PROSITE or a pattern against Swiss-Prot/TrEMBL or PDB for the occurrence of patterns, profiles, and rules (motifs) stored in the PROSITE database. It also visualizes matches on structures with a graphical view and feature detection. PRATT can be used to generate a user's own patterns.

MotifScan It allows a user to scan a sequence against the profile entries in PROSITE and Pfam. Motif scanning allows search for all known motifs that occur in a given unknown sequence. This tools allows the user to input a protein sequence, select the collections of motifs to scan for, and launch the search.

InterProScan This is a Perl stand-alone InterProScan package that allows the user to query the input/unknown sequence against InterPro. The InterPro database integrates PROSITE, PRINTS, Pfam, ProDom, SMART, TIGRFAMs, PIR superfamily, SUPERFAMILY,Gene3D, and PANTHER databases. This tool combines different protein signature recognition methods into one resource. The program then runs the input sequence against all the motif databases in InterPro.

PRATT This is a tool that interactively generates conserved patterns from a series of unaligned proteins. The tool may be used to look for patterns that are conserved in a set of sequences. The advantage of using this tool is that the user can specify what kind of patterns should be searched for and how many sequences should match a pattern to be reported. The only constraint is that the patterns need to be a subset of the set of patterns that can be described using the PROSITE notation (described earlier).

ps_scan It is a Perl program to scan PROSITE locally.

pftools These are standalone programs to create and scan PROSITE profiles.

TASK 

We will use the following sequence as the Input.

```
>LCLYTHIGRNIYYGSYLYSETWNTGIMLLLITMATAFMGYVLPWGQMSFWGATVITNLFSAIPYIGTNLV
EWIWGGFSVDKATLNRFFAFHFILPFTMVALAGVHLTFLHETGSNNPLGLTSDSDKIPFHPYYTIKDFLG
```

LLILILLLLLLALLSPDMLGDPDNHMPADPLNTPLHIKPEWYFLFAYAILRSVPNKLGGVLALFLSIVIL
GLMPFLHTSKHRSMMLRPLSQALFWTLTMDLLTLTWIGSQPVEYPYTIIGQMASILYFSIILAFLPIAGX
IENY

Step 1 Set your browser to the PROSITE web page at http://www.expasy.org/prosite/

Step 2 Click **ScanProsite**.

Step 3 Place the given unknown sequence in the textbox under **Protein(s) to be scanned**. Deselect the option **Exclude motifs with a high probability of occurrence**.

Step 4 Select the output format as **Graphical rich view** and click **START THE SCAN**.

Step 5 The result viewer that opens in the next page will display your hits (see Fig. 6.20).

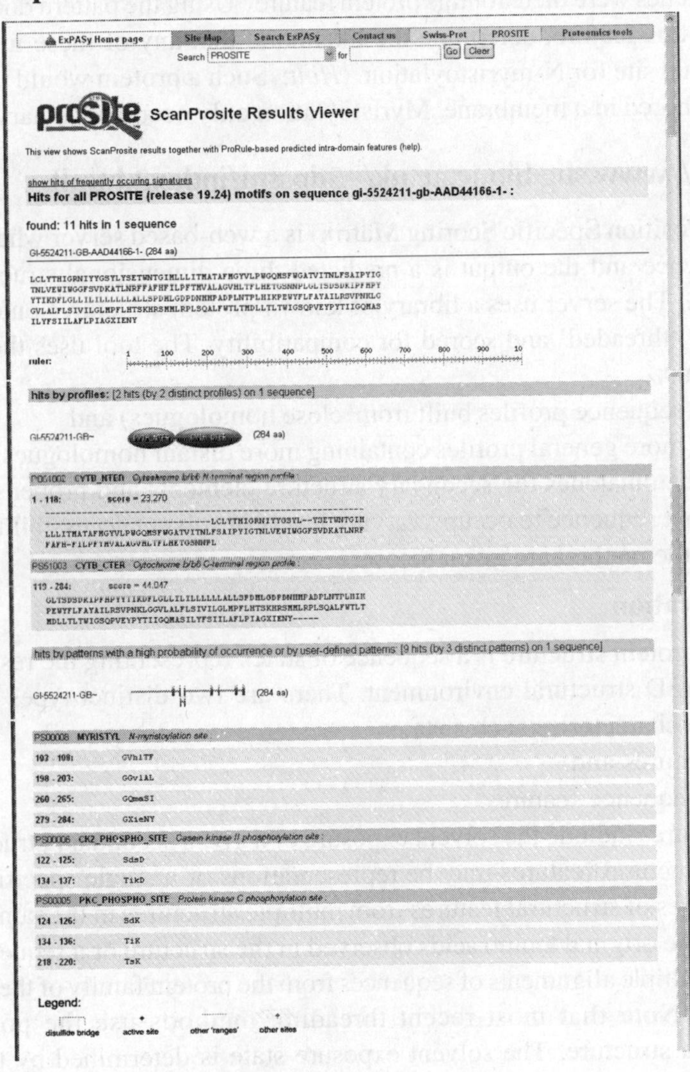

Fig. 6.21 ScanProsite result page

Step 6 Save this output to a file and prepare a report on the following:

(a) Are the results separated into different types: hits by PROSITE profiles, profiles with a high probability of occurrence, patterns, patterns with a high probability of occurrence or by user-defined patterns?

(b) Look at the pattern documentation for one or two of these pattern or profile matches carefully and determine the number of disulphide bridges found.

(c) Explore the details of the first two hits listed: PS51002 CYTB_NTER Cytochrome b/b6 N-terminal region profile and PS51003 CYTB_CTER Cytochrome b/b6 C-terminal region profile.

(d) Scroll until you see the bold heading **MYRISTYL**. How many potential matches were there for this protein feature? Using the pattern documentation for this feature, determine the likelihood that any of these matches is an actual site for N-myristoylation. (*Hint*: Such a protein would not be found anchored in a membrane. Myristoylation only occurs in eukaryotes.)

6.3.2 3DPSSM (http://www.sbg.bio.ic.ac.uk/~3dpssm/index2.html)

The 3D-PSSM (**P**osition **S**pecific **S**coring **M**atrix) is a web-based server where the input is a protein sequence and the output is a predicted three-dimensional structure and its probable function. The server uses a library of known protein structures onto which the input sequence is 'threaded' and scored for compatibility. The tool uses the following scoring components:

- 1D-PSSMs (sequence profiles built from close homologues) and
- 3D-PSSMs (more general profiles containing more distant homologues).

The program then matches the secondary structure elements and propensities of the residues in the input sequence to occupy varying levels of solvent accessibility.

The details of the method are given below.

1D-profile generation

A 1D model of a protein structure is a sequence of states representing the residue as if it were present in a 3D structural environment. There are two distinct types of features frequently used to characterize such a state:

- structural features and
- amino acid sequence features.

The structural features include the solvent exposure of a given secondary structure of the residue. These structural features may be representations of a single specific structure or weighted averages of structural features from multiple structures in the same family. It may also include the original amino acids observed in the structure or a sequence profile representing the multiple alignments of sequences from the protein family of the structure's native sequence. Note that most recent threading methods use the poly-alanine representation of a structure. The solvent exposure state is determined by the solvent exposure of an alanine placed at each residue position. Some approaches vary the radii of the solvent molecule and the beta carbon.

To generate a 1D-profile the following sequence of steps are undertaken.

Step 1 Take the sequence of the domain from the master protein (A0) of the known structure in a superfamily

Step 2 Search the master sequence against NRPROT using 20 iterations of PSI-BLAST [expectation for including a sequence in the iteration (H) of 0.0005 and a theoretical expectation value (ET) of a hit < 0.0005]

Step 3 Remove sequences that contain 'X' characters, overlap less than 75% of the query, and are > 80% identical to other sequences in the alignment.

Step 4 Use multiple alignment to generate a 1D-PSSM using the method described for PSI-BLAST.

Step 5 Repeat for all master proteins in the fold library

3D-profile generation

After the 1D profile has been generated, the next step is to conduct a three-dimensional structural superposition (using the SAP program of Orengo et al. 1992; Taylor & Orengo 1989; http://mathbio.nimr.mrc.ac.uk/) between the domain of the master structure A0 and all other proteins within the same superfamily. The next step is the selection of structures that 'match'. Structures that superpose with a weighted RMS deviation less than 6.0 Å to the master structure are selected. In the beginning the closest fitting (lowest RMS) structure is added to the alignment (A0 and X0). A search is performed for the next candidate alignment. The alignment with the lowest RMS to either A0 or X0 is used. Hierarchic multiple structural alignments are built by progressively adding alignments that are closest to an existing member of the alignment. This ensures that at all times the alignment with the most confident available structural alignment is considered. Only residues with a SAP equivalence score greater than 0 are considered in the alignment.

Use the residue equivalences from the structural alignment to augment the 1D profile of A0 with 1D-profiles from X0, Y0, ... Note that these are residues at a profile level. This yields a profile with sequences (A0, A1, A2,…AnA, X0, X1, X2,…XnX, Y0, Y1,Y2,…YnY). This step is repeated for all master proteins in the PSSM fold library.

Secondary structure matching

A library of three-state (Coil, Helix, Strand) secondary structures is built. STRIDE (Frishman & Argos 1995) is used for this purpose. STRIDE is a program to recognize secondary structural elements in proteins from their atomic coordinates. It performs the same task as the DSSP but utilizes hydrogen bond energy and main chain dihedral angles rather than hydrogen bonds alone. It relies on database-derived recognition parameters with the crystallographers' secondary structure definitions as a standard of truth.

Once a three-state secondary structure assignment is made for each library entry, they are grouped as

- 310 helices with alpha helices
- bridges with beta-strands
- turns and pi-helices with coil

The secondary structure of query sequences is predicted by PSI-Pred (Q3 77%), which uses the double dynamic programming algorithm proposed by Jones. Here an input query sequence with backbone coordinates is aligned to a given template protein structure, taking into account the detailed pairwise interactions. A simple scoring function for matching the secondary structure types between two residues is implemented where matching of identical secondary structure types gives a score of +1, and otherwise –1.

Solvation potential

Solvation potential is modelled using the approach of Jones et al. (1992). The potential is a term for scoring the preference of an amino acid to occupy a specific structure position with a given exposure. For a given pair of atoms, a given residue sequence separation and a given interaction distance provide a measure of energy, which relates to the probability of such an interaction being present in the native protein. The degree of burial of a residue is defined as the ratio between its solvent accessible surface area (as calculated by DSSP) and its overall surface area. 21 bins in 5% accessibility increments are used, ranging from 0% (buried) to 100% (exposed). The coarseness of this potential means cross-validation is unnecessary.

Bi-directional scoring

It is known that matching a query sequence to a template PSSM is not the same as matching a template sequence with a query PSSM. Often homologies can be detected in one direction and not in the other. To account for this, each query sequence is scanned against the sequence library using PSI-Blast. A 1D profile is generated in exactly the same way as the 1D profiles for the library sequences.

Searching the probe against the 3D-PSSM library

For each probe, the 3D-PSSM library is scanned using the global dynamic programming algorithm that was developed for our fold recognition algorithm FOLDFIT (Russell et al. 1998). The score for a match between a residue in the probe and a residue in the library sequence is calculated as the sum of the secondary structure, solvation potential, and PSSM scores. Three passes of dynamic programming are performed for each query-library sequence match. Each pass differs in the PSSM used for the scoring, with the secondary structure and solvation being held constant.

Pass 1 Library sequence is matched to the query PSSM.

Pass 2 Query sequence is matched to the library 1D PSSM.

Pass 3 Query sequence is matched to the library 3D PSSM.

The final score is simply the maximum of the scores from the three passes. An affine gap penalty of 10 to open and 1 per gap extension is used based on preliminary trials. End gaps are also penalized.

The significance of a match is evaluated by fitting a linear relationship between log (number of hits up to a score) against log (total score). Only the top end of the distribution is used and the possibility of the correct hit contributing to the tail of the distribution considered by removing the top scoring hit and all consecutive entries belonging to the same super family. The top end of the distribution is defined using a penalty function algorithm as described by (Kelley and Sutcliffe 1996). The probability of obtaining a match with that score by chance is converted into a theoretical error rate per query (ET).

TASK

Step 1 Type the following url in your browser: http://www.sbg.bio.ic.ac.uk/~3dpssm/index2.html and click **Go**.

Step 2 Select **Recognize a Fold** from the left links.

Step 3 The first textbox asks for your e-mail ID. The next textbox asks for a single line protein descriptor.

Step 4 Next specify the format of your submission. You can choose from **Single sequence** (a string of amino acids in one letter code), **Multiple Sequence alignment** (Clustal format), and **Probe** (in Block format). Choose **Multiple Sequence alignment** (Clustal format).

Step 5 Paste/upload an unknown sequence in the textbox shown in Fig. 6.22 and perform this protein structure prediction exercise.

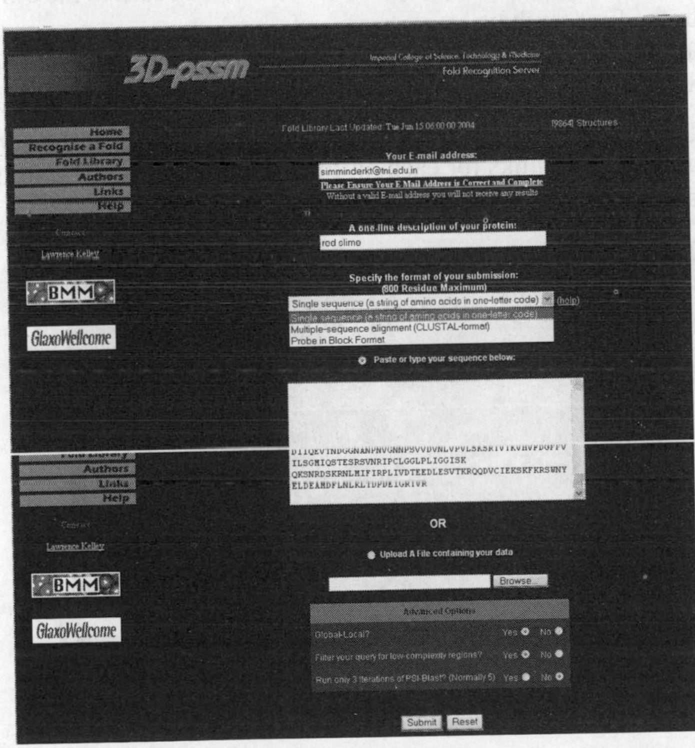

Fig. 6.22 Interface of 3D PSSM

Step 6 There are several advanced options available. Select **Global-Local** and **Filter your query for low complexity regions**. Do not select the **Run only 3 iterations of PSI-Blast** option as shown in Fig. 6.23.

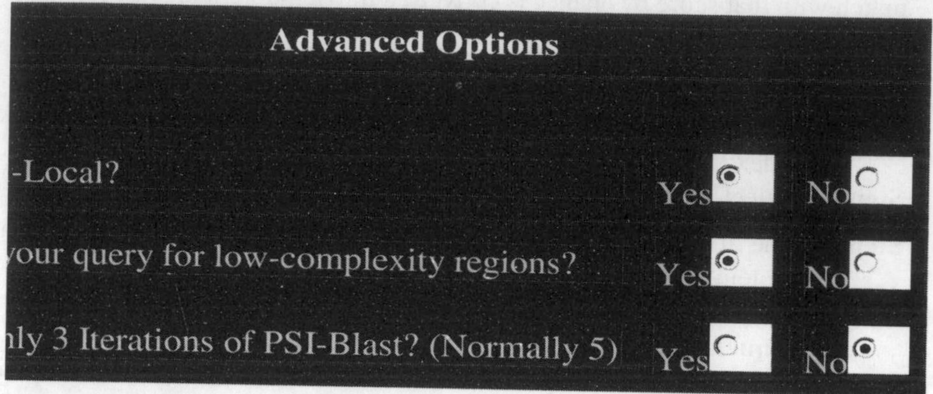

Fig. 6.23 Advanced options of 3D PSSM

Step 7 Click Submit to submit your query. The results will be sent to the e-mail ID your entered in the first textbox.

Step 8 The results will be mailed back to you as a zipped tar file along with a link by the 3D-PSSM web server. At this url you will be able to view models built for each of the top 20 hits, colour-coded alignments indicating regions of good and bad alignment, and hyperlinks for each hit to the Scop database. For this particular query the results page is shown in Fig. 6.24.

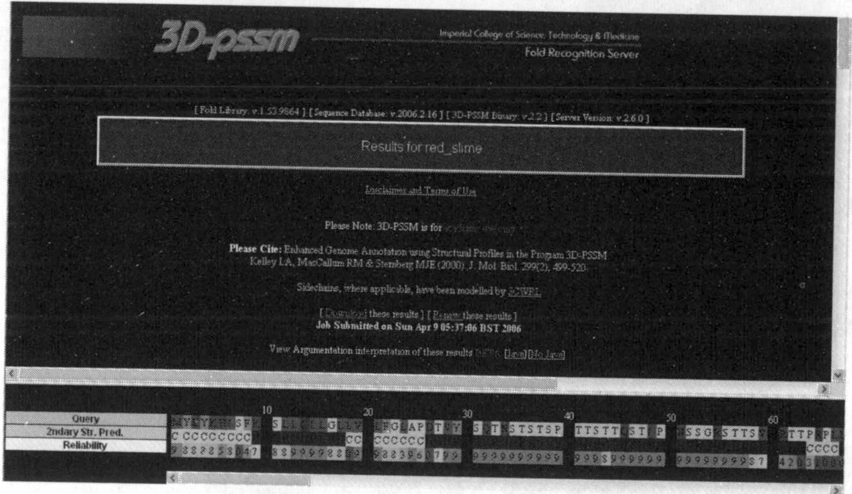

Fig. 6.24 Results page of 3D PSSM

Step 9 The results are given under the flowing heads in a tabular format: View Alignment, Fold Library, Template Length, Model, PSSM E_value, SAWTED E-value, Biotext, Class, Fold, SuperFamily, Family, Protein, Species, ID, Region, CS, and PSSM Type. The alignments with their scores are shown in Fig. 6.25.

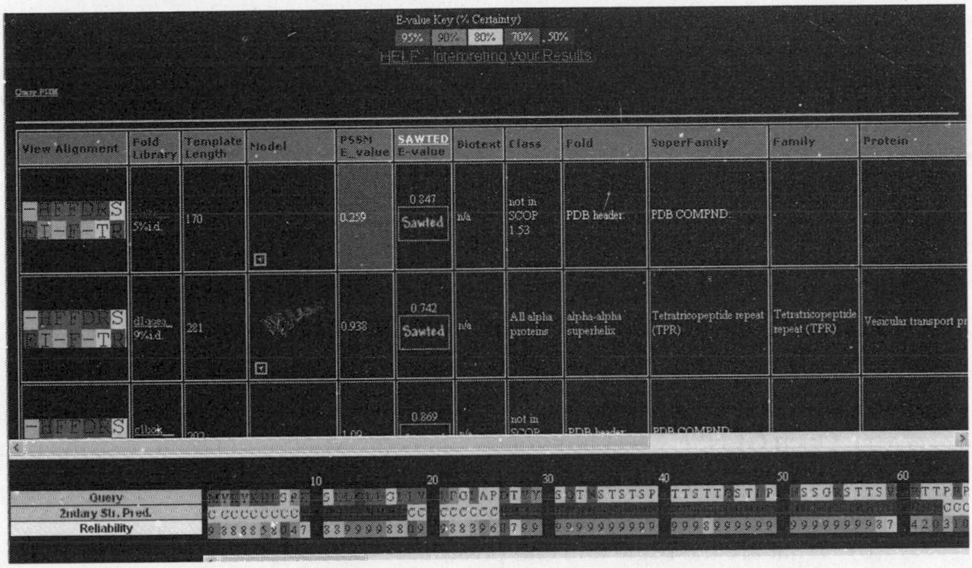

Fig. 6.25 Result alignments with scores

Conclusion

Like bioinformatics data analysis, there is no end to prediction. In this chapter you got acquainted with some kinds of broad predictions such as phylogeny, gene, protein structure, and function. There are many finer predictions as well. Like within protein structure prediction, there are tools for predicting secondary structure, transmembrane protein topology, protein folds, domains, etc. Again each kind of prediction may have different approaches or methods, to predict the same. Therefore, each prediction may have different tools with different methods, thereby resulting in different results. One needs to decide the approach and select the tool depending on what one is ready to sacrifice and what one cannot, depending on the kind of research one is into. In Chapter 7 we will learn about modelling tools for 2D and 3D protein structure as well as for molecules that we tried to predict in this chapter.

EXERCISES

Exercise 6.1 Describe PSSM, emphasizing the following aspects:

(a) scoring

(b) training sets
(c) comparison with dynamic programming

Exercise 6.2

(a) Draw all possible unrooted trees for four taxa A, B, C and D
(b) You are given a sequence. Name three methods that can be used to determine phylogenetic trees. Give a sentence about each one of them.
(c) For each of the rooted trees depicted here what are the possible positions at which a new leaf may be added?

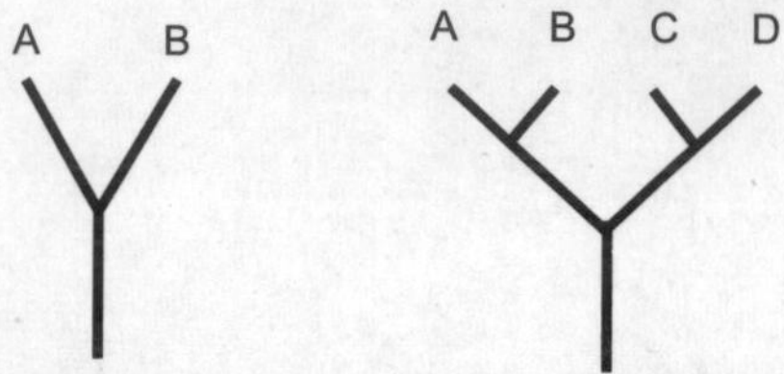

Exercise 6.3 Input sequences from two sources, human and mouse, are given to you. Open GENSCAN at http://genes.mit.edu/GENSCAN.html. Use the input sequence given above and get the output. Analyse the output to answer the following:

(a) How many exons are predicted in the sequence?
(b) What are the begin and end positions?
(c) What is the probability of the possible exons?
(d) On which strand (+ or –) is the gene located?
(e) Write down the first six amino acids and the total length of the predicted protein sequence.

Exercise 6.4 Use the same input sequence as in Exercise 6.3 and feed it into another gene prediction tool GrailEXP Perceval (http://compbio.ornl.gov/grailexp/gxpfaq1.html).

(a) How many exons are predicted in the sequence? Rate them into three groups.
(b) What are the begin and end positions?
(c) What is the probability of the possible exons? Which is the largest of them?
(d) On which strand (+ or –) is the gene located?
(e) Write down the first six amino acids and the total length of the predicted protein sequence
(f) Compare the output with that in Exercise 6.3.

Exercise 6.5 Use the input data given and submit the same to PROSITE (http://www.expasy.ch/prosite). Identify domains within the sequence.

(a) What is the PROSITE identifier number PSxxxxx for this domain.
(b) What is the nature of these domains?
(c) Do any other proteins also have this domain?
(d) What is interesting about the domain architecture for these domains?

Exercise 6.6 You are given PDB files from *Salmonella* (accession number: P02906) and *E. coli* (accession number: P238610) as your input. They are binding proteins. Go to the 3DPSSM server at http://www.sbg.bio.ic.ac.uk/~3dpssm/index2.html. Choose **Recognize a Fold** from the left panel and go to the **advanced submissions page**. Paste the protein sequence and submit to the server. Store the results obtained in two separate files.

(a) Which folds are selected by the sequence of the sulfate binding protein (P02906)?
(b) Does it recognize the fold of the polyamine binding protein?
(c) Click **View Alignment** and note the template length of the best hit.
(d) How does the sorted value help?

Exercise 6.7 The human Alpha-N-acetylglucosaminidase protein is given as input for this exercise. To know more about the protein and its clinical significance, visit PubMEd and OMIM databases. Use the 3D PSSM server to determine its structure prediction. Do the same using PHD. Store the results separately. Pick up the structures in Rasmol and analyse the predictions given by the two methods in light of the threading approach using 1D and 3D profiles coupled with secondary structure and solvation potential used by PSSM. Will these methods work for proteins for which there are currently no known structures that have high enough sequence homology?

Exercise 6.8 Obtain the following sequences in their PDB formats: 1ak5, 3cox, 1gcu, 1bgl, 1efv, and 1p04. Go to the predictProtein sever at http://www.predictprotein.org/. Submit your PDB files and store and analyse the results. Start by locating the actual structure information for each of the six proteins in PDB. Now you can compare the real (pdb) secondary structure with what was returned by the prediction method. Each of these pairs of sequences contains a short region that has a different secondary structure in the two proteins. Take a close look at the following:

(a) RVPALV 1ak5 starting at position 125 3cox starting at position 7.
(b) LITTAHA 1gcu starting at position 121 1bgl starting at position 833.
(c) LLPRVA 1efv starting at position 100 1p04 starting at position 75.

Recompare the actual secondary structure for these short sequences and the predicted secondary structure. How well were these regions predicted?

7 Modelling Tools

Learning Objectives

- To learn and locate various 2D protein modelling tools
- To learn and locate various 3D protein modelling tools
- To set the attributes for each tool and process structure data
- To analyse results obtained from each tool

Introduction

The aim of this chapter is to give a practical introduction to the use of the available modelling tools that use structure data from various databases. The chapter presents an introductory analysis of the model from the output of the tools used. In this chapter we will also look at protein modelling tools. For ease of learning, each tool is introduced in a predefined format under several subheadings. There are various visualization and modelling tools for 2D or 3D structures of proteins. These tools are available on Internet for online viewing, e.g., protein explorer, quickpdb, webmol, etc. There may be browser plug-in applications for both online and offline viewing, i.e., Chime (Chemscape), Cn3D, MAGE, and Rasmol. For better and clear understanding, it is better to download these tools and work on these tools while you work through this chapter.

7.1 Tools for 2D Protein Modelling

There are various tools available for 2D protein modelling. Most of them are available freely on Internet and many are downloadable also. For the ease of work and also to maintain consistency, we have mostly picked up myoglobin data from source *Physeter catodon*, commonly known as sperm whale. Only in tools where the length of the sequence is a constraint, we have picked up genes of smaller sizes from the same organism. Also the objective is that when all the tools from this section are considered with the do-it-along TASKs, you should end up doing a miniproject on 'modelling tools for protein'.

7.1.1 Rasmol 2.6

Tool for Rasmol is a molecular graphics program used for the visualization of proteins, nucleic acids, and small molecules. It reads a coordinate file of a molecule and interactively displays the molecule on the screen in various colour schemes and molecule representations.

Web link http://www.mw-software.com/software/rasmol/rasmol.html

Institute Raswin Molecular Graphics

Type of the tool Downloadable

OS required Rasmol runs on a wide range of architectures and operating systems including Microsoft Windows, Apple Macintosh, UNIX, and VMS systems.

Tool snapshot

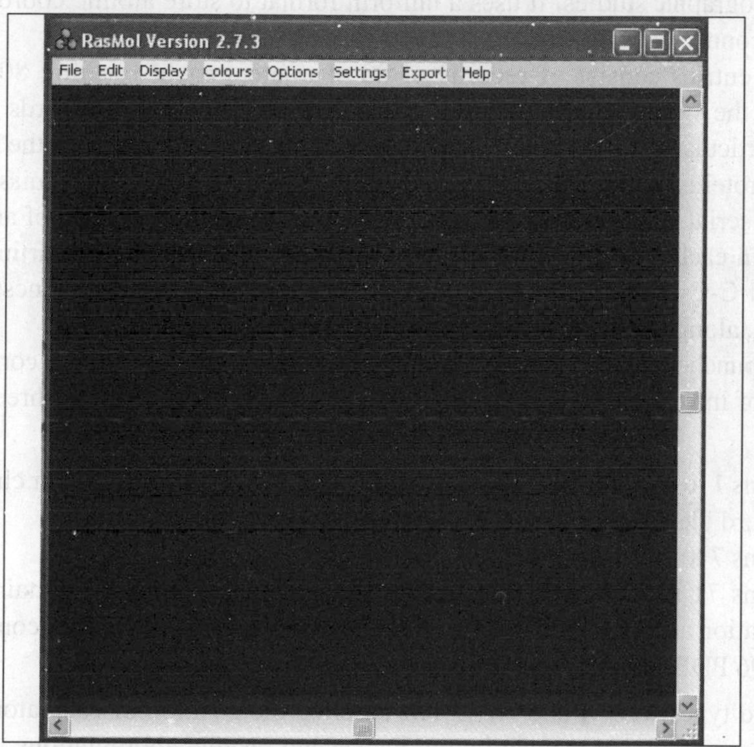

Fig. 7.1 Rasmol snapshot

Input data The input data to fed to the tool in the format discussed later. The data is fed using the **File Open** option, which will be discussed in detail next.

Attributes/menus This tool does not support any attribute. But there are menus and submenus one must know how to deal with.

File There are six submenus under this menu.

Open You can open files with this option. It will allow you to open upto five files. So you can load and display five molecules together in one screen. Any one or all of the molecules may be rotated and translated. This tool can read files in multiple formats. They are

- BrookHaven databank
- Alchemy file format
- Sybyl MOL2 format
- MDL MOL file format
- MSC (XMOL) XYZ format
- CHARMm file format
- MOPAC file format

Let us discuss the BrookHaven databank in detail. The bank uses data that is derived from crystallographic studies. It uses a uniform format to store atomic coordinates and partial bond connectivities.

PDB file entries consist of records of 80 characters each, forming 80 columns. Interestingly the syntax of each record is independent of the order of records within any entry for a particular macromolecule. Residues occur in order, starting from the N-terminal residue for proteins. If the residue sequence is known and there are some missing atoms, certain atom serial numbers may be omitted to allow for future insertion of any missing atoms. Within each residue, atoms are ordered in a standard manner, starting with the backbone (N-C-C-O for proteins) and proceeding in increasing remoteness from the alpha carbon, along the side chain.

Using the punched card analogy, those 80 columns are grouped into 16 sets corresponding to 16 kinds of information. Broadly, these 16 sets can be grouped into three as shown below:

1. Columns 1 to 6 contain a record-type identifier, out of which first four characters of the record identifier can uniquely identify the type of record.
2. Columns 7 to 70 contain data.
3. Columns 71 to 80 is normally blank in older entries, but may contain sequence information added by library management programs. In new entries conforming to the 1996 PDB format, there is other information in those columns.

The record types ATOM and HETATM describe the position of each atom. ATOM/HETATM records contain standard atom names and residue abbreviations, along with sequence identifiers, coordinates in Angstrom units, occupancies, and thermal motion factors. HETATM records are used to define post-translational modifications and cofactors associated with the main molecule. Let us look at each of those sets for these record types, the number of columns in each set, the information that each of the sets carries and its use in the PDB formats (in the 1992 and earlier formats or in the 1996 and later formats or all) under heading **Column**, **Content,** and **Fmt** as shown below:

Column	**Content**	**Fmt**
1–6	ATOM or HETATM	all
7–11	Atom serial number (may have gaps)	all
13–16	Atom name, in IUPAC standard format	all
17	Alternate location indicator indicated by A, B or C	all
18–20	Residue name, in IUPAC standard format	all
23–26	Residue sequence number	all
27	Code for insertions of residues (i.e., 66A & 66B)	all
31–38	X coordinate	all
39–46	Y coordinate	all
47–54	Z coordinate	all
55–60	Occupancy	all
61–66	Temperature factor	all
68–70	Footnote number	92
73–76	Segment Identifier (left-justified)	96
77–78	Element Symbol (right-justified)	96
79–80	Charge on the Atom	96

Other record types provide information as below:

HEADER: The file name and date.

COMPND: The name of the protein.

SOURCE: The organism from which the protein was obtained.

AUTHOR: The person(s) who placed this data in the Protein Data Bank.

REVDAT: All revision dates for data on this protein.

JRNL: Journal articles about the structure of this protein.

REMARK: General information about the contents of this file.

SPRSDE: Older coordinate files of this same structure.

SEQRES: The amino acid sequence of the protein.

FTNOTE: Lines which contain footnotes.

HET and FORMUL: Lines which list the cofactors, prosthetic groups, inhibitors, or other nonprotein substances present in the structure.

HELIX, SHEET and TURN records provide initial secondary structure assignments.

CRYST1, ORIG, and SCALE: Lines which contain some general information about the protein crystals from which this structure was obtained by the technique of x-ray crystallography.

ATOM and HETATM: Lines which contain the atomic coordinate data needed to display the structure of the protein.

TER records are interpreted as breaks in the main molecule's backbone.

CONECT records provide information about the bonds between nonprotein atoms in the file.

END record indicates the end of the file.

Information It gives information about the number of groups, atoms, and bonds in a molecule.

Close In case one file is open, it will close the open file. In case more than one file is open, it will close the file that is active.

Print It will print the model.

Print setup It will allow you to adjust the print setup.

Exit It will close the application.

Edit There are five submenus under this menu.

Select all It will select the whole model.

Cut It will cut the selected portion of the model.

Copy It will copy the selected portion of the model.

Paste It will paste the copied portion of the model.

Delete It will delete the selected portion of the model.

Display There are eight kinds of views or displays under this menu.

1. *Wireframe view* This view represents each bond within the selected region of the molecule as a cylinder, a line, or a depth-cued vector (Fig. 72). One can specify the cylinder radius in angstrom or Rasmol units, colour, etc. By default, non-bonded atoms become invisible and can be marked by a special command.

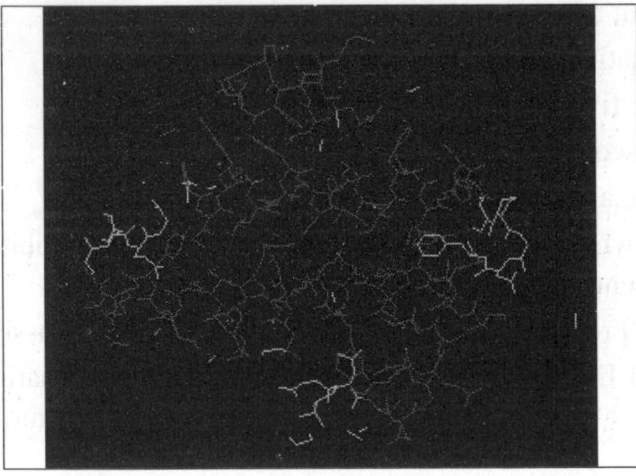

Fig. 7.2 Wireframe view of myoglobin

2. *Backbone view* This representation displays the polypeptide backbone as a series of bonds connecting the adjacent alpha carbons of each amino acid in a chain (Fig. 7.3). Like the wireframe view, one can specify the cylinder radius, colour, etc. One can also render smoothen backbone or a backbone displayed with dashed lines.

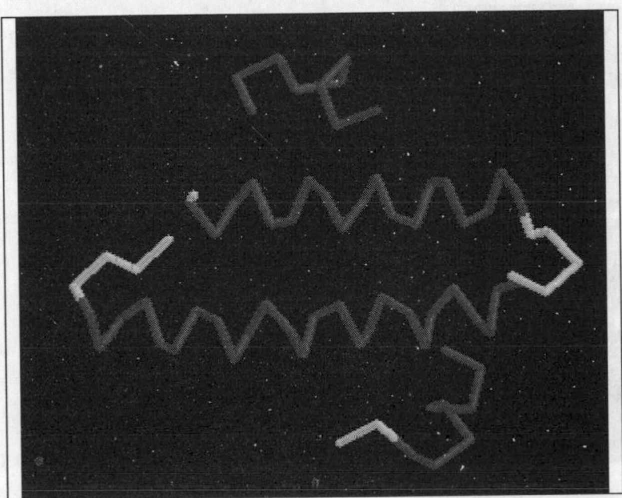

Fig. 7.3 Backbone view of myoglobin

3. *Sticks view* In this view the bonds are displayed as sticks (Fig. 7.4). One can specify the colour and width of the sticks. Different bonds are represented in different colours.

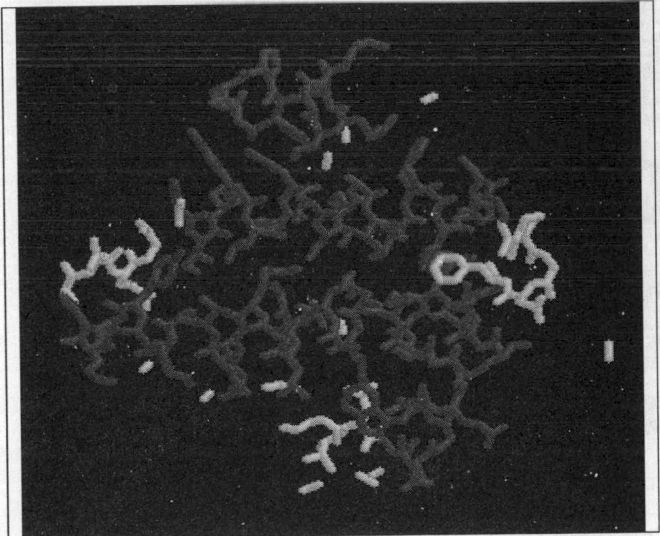

Fig. 7.4 Sticks view of myoglobin

4. *Spacefill view* This view is used to represent all of the currently selected atoms as solid spheres (Fig. 7.5). It can also display both union-of-spheres and ball-and-stick models of a molecule. One can provide the sphere radius in Rasmol units (1/250th an angstrom) or in decimal numbers.

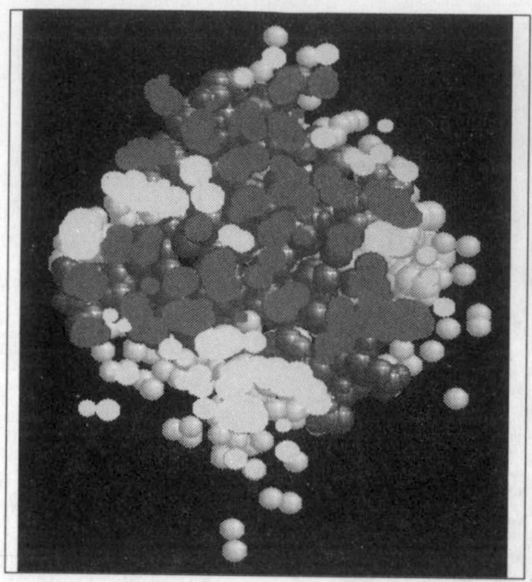

Fig. 7.5 Spacefill view of myoglobin

5. *Ball and stick view* In this kind of display the atoms are displayed as small spheres or balls and bonds are displayed as sticks (Fig. 7.6). One can specify the colour and width or radius of the sticks or balls. Different atoms are represented in different colours.

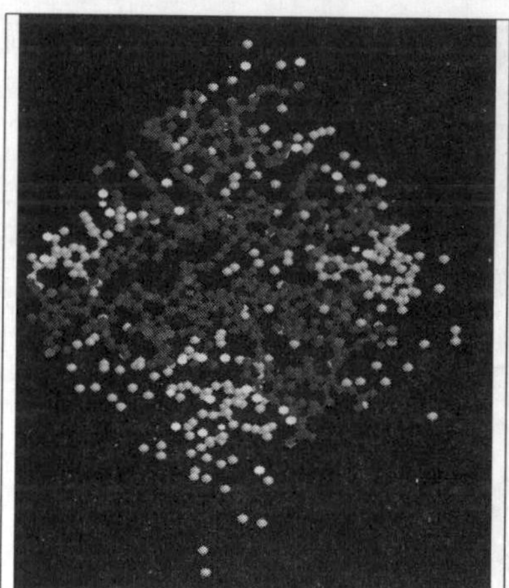

Fig. 7.6 Ball and stick view of myoglobin

6. *Ribbons view* This view displays the selected molecule of protein or nucleic acid as a smooth solid 'ribbon' surface passing along the backbone of the protein (Fig. 7.7). The ribbon is drawn between each amino acid whose alpha carbon is currently selected. One can specity the width and colour of the ribbon.

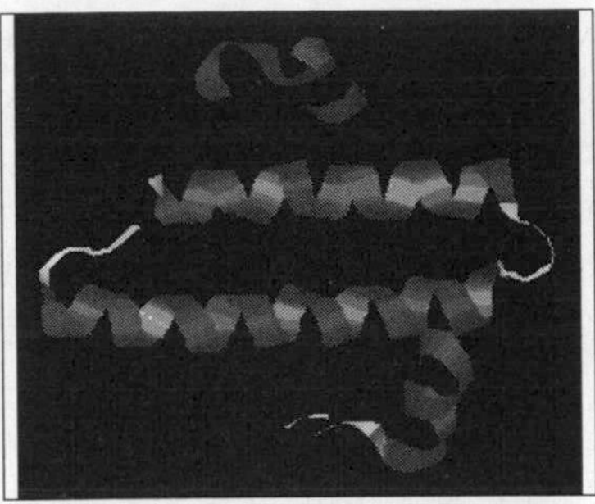

Fig. 7.7 Ribbons view of myoglobin

7. ***Strands view*** This representation is similar to the ribbon view (Fig. 7.8). The ribbon is composed of a number of strands that run parallel to one another along the peptide plane of each residue. Here also the width and colour of the strand can be specified.

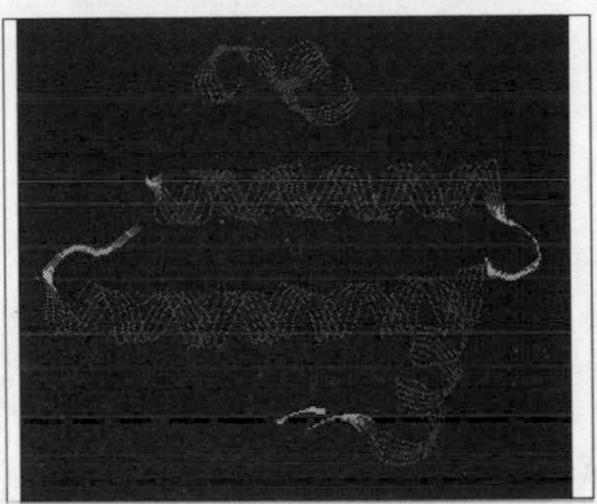

Fig. 7.8 Strands view of myoglobin

8. ***Cartoons view*** Here the view of a molecule is represented as thick, deep ribbons Richardson (MolScript) style protein 'cartoons' (Fig. 7.9). By default, the C-termini of beta-sheets are displayed as arrow heads that can be disabled if one wish to. One can also set the depth of the ribbon.

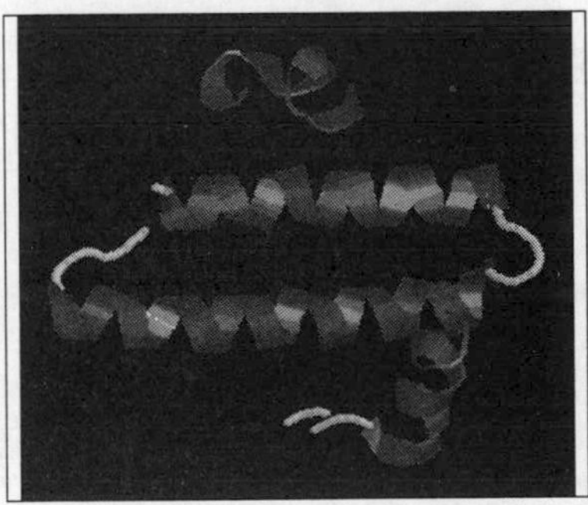

Fig. 7.9 Cartoons view of myoglobin

9. *Molecular surface view* The surface view of the molecule renders a Lee–Richards molecular surface resulting from rolling a probe atom on the selected atoms (Fig. 7.10). One can specify the radius of the probe. If the radius is given in the first form, the evolute of the surface of the probe is displayed, i.e., the solvent excluded surface. If the radius is given in the second form, the envelope of the positions of the centre of the probe is displayed, i.e., the solvent accessible surface.

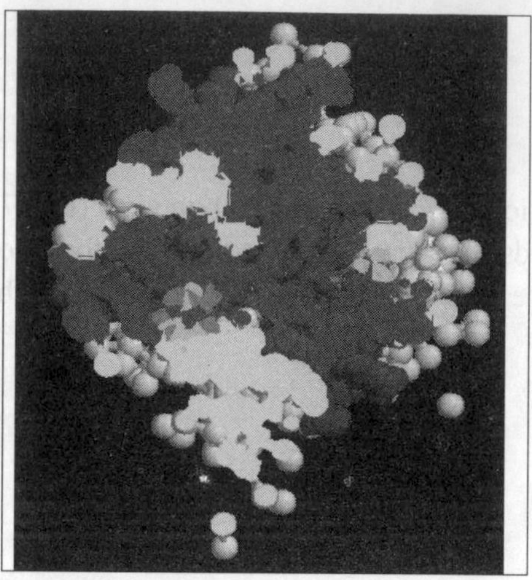

Fig. 7.10 Molecular surface view of myoglobin

Colours There are five submenus under this menu.

Monochrome This option will draw the model in one colour, preferably whitish grey on a black background.

CPK This option provides colour schemes that were developed by Corey, Pauling, and later improved by Kultun. It is based upon the colours of the popular plastic spacefilling models, which colours 'atom' objects by the atom (element) type. This scheme is conventionally used by chemists.

Colours of the most commonly used element types

Carbon	Light grey	[200,200,200]
Oxygen	Red	[240,0,0]
Hydrogen	White	[255,255,255]
Nitrogen	Sky blue	[143,143,255]
Sulphur	Yellow	[255,200,50]
Phosphorous	Orange	[255,165,0]
Chlorine	Green	[0,255,0]
Bromine, Zinc	Brown	[165,42,42]

Sodium	Blue	[0,0,255]
Iron	Orange	[255,165,0]
Magnesium	Forest green	[34,139,34]
Calcium	Dark grey	[128,128,144]
Unknown	Deep pink	[255,20,147]

Shapely This option follows the colour scheme that colours the residues by their amino acid property. This scheme is based upon Bob Fletterick's 'Shapely Models'.

Group This option provides a colour scheme that colour codes residues by their position in a macromolecular chain. Each chain is drawn as a smooth spectrum from blue through green, yellow, and orange to red. Hence the N terminus of proteins are coloured red and the C terminus of proteins are coloured blue. If a chain has a large number of heterogeneous molecules associated with it, Rasmol performs group colouring by using the residue numbering given in the PDB file. Hence the lowest residue number is displayed in blue and the highest residue number is displayed in red.

Note

If a PDB file contains a large number of heteroatoms, the protein is displayed in the blue-green end of the spectrum and the waters are displayed in the yellow-red end of the spectrum. For example, if a PDB file contains a large number of water molecules, those being heteroatoms occupy high residue numbers, and therefore the colour of the model gets biased as there are typically many more water molecules than amino acid residues. The solution to this problem is to deselect **Hetero Atoms** on the **Options** menu before selecting **Group** on the **Colour** menu.

Chain This option assigns each macromolecular chain a unique colour. This colour scheme is useful for distinguishing individual strands of a DNA chain or identifying the parts of multimeric structure.

Temperature This option colour codes each atom according to the anisotropic temperature, i.e., the beta value stored in the PDB file. It is often used to associate a 'colour scale' value with each atom in a PDB file and to colour the molecule appropriately. It gives a measure of the mobility/uncertainty of a given atom's position. High values are coloured in the spectrum of red colour and lower values in the spectrum of blue colour.

Note

The difference between the 'temperature' and 'charge' colour schemes is that increasing temperature values proceed from blue to red, whereas increasing charge values go from red to blue.

Structure This option colours the molecule by the protein secondary structure. The secondary structure is either read from the PDB file (HELIX, SHEET, and TURN records), or determined using Kabsch and Sander's DSSP algorithm. This option may be used to force DSSP's structure assignment to be used. The colours used corresponding to various secondary structures are shown below.

- Alpha helices are coloured magenta [240, 0,128].
- Beta sheets are coloured yellow [255, 255, 0].
- Turns are coloured pale blue [96,128, 255].
- All other residues are coloured white.

User This option allows Rasmol to use the colour scheme stored in the PDB file. The colours for each atom are stored in COLO records placed in the PDB data file. This convention was introduced by David Bacon's Raster3D program.

Model This option codes each NMR model with a distinct colour. The NMR model number is taken as a numeric value. High values are coloured blue and lower values are coloured red. Instead of using a fixed scale, this colour scheme determines the maximum value of the NMR model number and accordingly interpolates from red to blue.

Alt The 'alt' (alternate conformer) colour scheme codes the base structure with one colour and applies a limited number of colours to each alternate conformer. In a Rasmol model for 8-bit colour systems, four colours are allowed for alternate conformers. Otherwise, eight colours are available.

Options There are five submenus under this menu.

Slab mode

Syntax: `slab {<boolean>}`
`slab <value>`

This command enables, disables, or positions the z-clipping plane of the molecule. The program only draws those portions of the molecule that are further than the slabbing plane from the viewer. Integer values range from zero at the very back of the molecule to 100, which is completely in front of the molecule. Intermediate values determine the percentage of the molecule to be drawn. This command interacts with the **depth <value>** command, which clips to the rear of a given z-clipping plane.

Hydrogens

Syntax: `set hydrogen <boolean>`

This parameter is used to modify the **default** behaviour of the **select** command, i.e., the behaviour of **select** without any parameters. When this value is **false**, the default **select** region does not include any hydrogen, deuterium, or tritium atoms. When this value is **true**, the default **select** region may contain hydrogen atoms.

Hetero atoms

Syntax: `set hetero <boolean>`

This parameter is used to modify the **default** behaviour of the **select** command, i.e., the behaviour of **select** without any parameters. When this value is **false**, the default **select** region does not include any heterogeneous atoms. When this value is **true**, the default **select** region may contain heterogeneous atoms. This parameter is similar to the **hydrogen** parameter that determines whether hydrogen atoms should be included in the default set. If both **hetero** and **hydrogen** are **true**, **select** without any parameters is equivalent to **select all**. This command also instructs Rasmol to only use non-hetero residues in the group colour scaling.

Specular

Syntax: `set specular <boolean>`

This command enables and disables the display of specular highlights on solid objects drawn by Rasmol. Specular highlights appear as white reflections of the light source on the surface of the object. The current Rasmol implementation uses an approximation function to generate this highlight. The specular highlights on the surfaces of solid objects may be altered by using the specular reflection coefficient, which can be altered using the Rasmol 'set specpower' command.

Shadows

Syntax: `set shadow <boolean>`

This command enables and disables ray tracing of the currently rendered image. Currently only the space-filling representation is shadowed or can cast shadows. Enabling shadowing will automatically disable the *z*-clipping (slabbing) plane using the **slab off** command.

> **Note**
>
> It is recommended that shadowing be normally disabled while the molecule is being transformed or manipulated, and only enabled once an appropriate point of view is selected, to provide a greater impression of depth.

Stereo

Syntax:
```
stereo on
stereo [-] <number>
stereo off
```

This stereo command provides side-by-side stereo display of images. Stereo viewing of a molecule may be turned on (and off) either by selecting **Stereo** from the **Options** menu, or by typing the commands **stereo on** or **stereo off**. Turning stereo on and off does not reposition the centre of the molecule.

Set stereo

Syntax:
```
set stereo <boolean>
set stereo [-] <number>
```

This parameter controls the separation between the left and right images. The positive values result in a crossed-eye viewing and negative values in relaxed (wall-eyed) viewing.

The inclusion of [-] <number> in the **stereo** command, as, for example, in stereo 3 or stereo -5, also controls angle and direction.

Labels

Syntax: label {<string>}
label <boolean>

This label command allows an arbitrary formatted text string to be associated with each currently selected atom. This string may contain embedded 'expansion specifiers', which display properties of the atom being labelled. An expansion specifier consists of a % character followed by a single alphabet specifying the property to be displayed (similar to C's printf syntax). An actual % character may be displayed by using the expansion specifier %%. Atom labelling for the currently selected atoms may be turned off with the command **label off**. By default, if no string is given as a parameter, Rasmol uses labels appropriate for the current molecule. Rasmol uses the label %n%r:%c.%a if the molecule contains more than one chain; %e%i if the molecule has only a single residue (a small molecule); and %n%r.%a otherwise. The following table lists the current expansion specifiers.

%a		atom name
%b	%t	B-factor/temperature
%c	%s	chain identifier
%e		element atomic symbol
%i		atom serial number
%n		residue name
%r		residue number
%M		NMR model number (with leading /)
%A		alternate conformation identifier (with leading ;)

Settings

Pick

This **set picking** series of commands affect how a user may interact with a molecule displayed on the screen in Rasmol.

(a) *Enabling/disabling atom identification picking*: Clicking on an atom with the mouse results in identification and the display of its residue name, residue number, atom name, atom serial number, and chain in the command window. This behaviour may be disabled with the command **set picking none** and restored with the command **set picking ident**. The command **set picking coord** adds the atomic coordinates of the atom to the display.

(b) *Measuring distances, angles, and torsions*: Interactive measurement of distances, angles, and torsions is achieved using the commands *set picking distance*, *set picking* **monitor**, **set picking** angle, and **set picking torsion**, respectively. In addition, every atom picked increments a modulo counter such that in distance mode, every second atom displays the distance (or the distance monitor) between this atom and the previous one. In the angle mode, every third atom displays the angle between the previous three atoms. In the torsion mode every fourth atom displays the torsion between the last four atoms.

(c) *Labelling atoms with the mouse*: The mouse may also be used to toggle the display of an atom label on a given atom. The command **set picking label** removes a label from a picked atom if it already has one, otherwise it displays a concise label at that atom position.

(d) *Centring rotation with the mouse*: A molecule may be centred on a specified atom position using **set picking centre** or **set picking center** commands. In this mode, picking an atom causes all further rotations about that point.

(e) *Picking a bond as a rotation axis*: Any bond may be picked as an axis of rotation for the portion of the molecule. This feature changes the conformation of the molecule. The bond cannot be used for rotation if it is part of a ring of any size. The most recently selected bond may be actively rotated.

(f) *Enabling atom/group/chain selection picking*: Atoms, groups, and chains can get selected with the **select** command with the **set picking atom**, **set picking group**, **set picking chain** commands.

Rotate

Syntax:
```
rotate <axis> {-} <value>
rotate bond {<boolean>}
rotate molecule {<boolean>}
rotate all {<boolean>}
```

This option rotates the molecule about the specified axis. Permitted values for the axis parameter are `x`, `y`, `z`, and `b bond`. The integer parameter states the angle in degrees for the structure to be rotated. For the X- and Y-axes, positive values move the closest point up and right, and negative values move it down and left, respectively. For the Z-axis, a positive rotation acts clockwise and a negative angle acts anticlockwise. If **rotate bond true** is selected, the horizontal scroll bar will control rotation around the axis selected by the **bond src dst pick** command. If **rotate all true** is selected and multiple molecules have been loaded, then all molecules will rotate together. In all other cases, the mouse and dials control the rotation of the molecule selected by the **molecule n** command.

Export You can export the model in any of the formats as mentioned below:

(a) BMP
(b) GIF
(c) EPSF
(d) PPM
(e) RAST

Result

On selection of various menus/submenus the result will be a model that can be exported in the formats mentioned under **Export** menu.

7.2 Tools for 3D Protein Modelling

Insights into the three-dimensional structure of a protein are of great help when planning experiments aimed at the understanding of protein function and during the drug design process. We all know that proteins from different sources and even with diverse biological functions can have similar sequences. At the same time, it is also accepted that high sequence similarity is reflected by distinct structure similarity. Indeed, the relative mean square deviation (rmsd) of the alpha-carbon coordinates for protein cores sharing 50% residue identity is expected to be around 1 Å. This fact served as the premise for the development of comparative protein modelling (also often called modelling by homology or knowledge-based modelling), which is presently the most reliable method. Comparative model building consists of the extrapolation of the structure for a new (target) sequence from the known 3D structure of related family members (templates). It is a fact that the high precision structures can only be obtained experimentally; but on the other hand, theoretical low resolution protein models provide enough essential information about the spatial arrangement of important residues to guide the design of experiments.

7.2.1 Deep View 3.7

Tool for Deep View was previously known as Swiss-PdbViewer. It got this new name for the ease of pronunciation. It is a user friendly, yet powerful molecular graphics program. It can be used to view the structure files of several proteins simultaneously. It can also build models from scratch, even from an amino acid sequence. For proteins of a known sequence but unknown structure, the tool submits the amino acid sequence to ExPASy to find homologous proteins, where the Swiss model server builds a final model, called a homology model, and returns the result by e-mail. It is possible to thread a protein primary sequence into a 3D template model with an immediate feedback of how well the input protein is accepted by the reference structure. After that, a subsequent request can be generated to build missing loops and refine side chain packing.

This tool is also preferred as an analytical tool that can be used to compare protein structures, manipulate them, and analyse several proteins at the same time. As an analytical tool it does the following:

- It can find hydrogen bonds within proteins and between proteins and ligands.
- It allows examining electron-density maps from crystallographic structure.
- It allows superimposing multiple proteins and comparing their structures, e.g., like deducing structural alignments, comparing active sites, or any other relevant parts.
- It computes electrostatic potentials and molecular surfaces, and carries out energy minimization. It can also read electron density maps, and provides various tools to build into the density.
- It also computes amino acid mutations, H-bonds, angles and distances between atoms with the help of graphic and menu interfaces.

In order to make good ray-traced quality images, Deep View can also generate POV-Ray scenes from the current view.

Web link http://www.expasy.org/spdbv/

Institute The tool was developed within the Swiss Institute of Bioinformatics (SIB) in collaboration between GlaxoSmithKline R & D and the Structural Bioinformatics Group at the Biozentrum in Basel.

Type of the tool Downloadable

OS required Swiss-PdbViewer (Deep View) is available for Macintosh, Windows, Linux, and SGI.

Tool snapshot

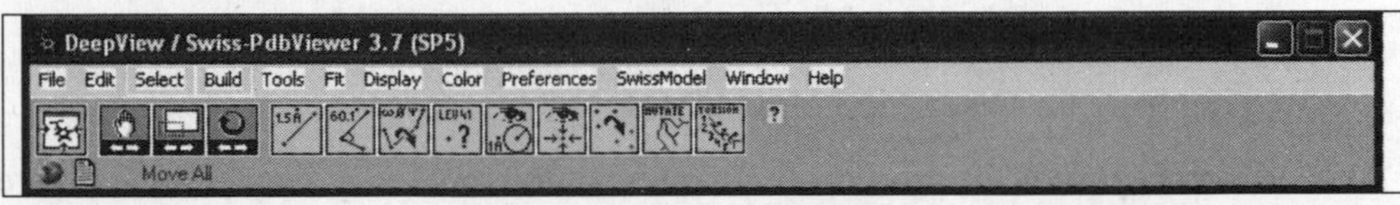

Fig. 7.11 Toolbar of the Deep View Tool

Input data The input data is fed to the tool in different formats. The data is fed using the **File Open** option that will be discussed later in the subsection *Learn some file operations*.

Menus In this tool there are menus and submenus that one must explore. In the earlier section you were taken through each menu and submenu and repeating the same for this tool is not needed. Here, let us learn the tool with a different approach.

(a) Get yourself acquainted with the workspace and related windows

When you execute the Deep View application, unlike other common applications like Rasmol, it opens only the toolbar and not the blank workspace display window where the molecule (s) get displayed. As mentioned earlier, the Deep View tool can load and display several molecules into separate layers in the same display windows simultaneously. You can link or remove the link between the toolbar and the display window(s).

Other than the display window, the tool can allow you to open several other windows for an open file as windows for the following:

- Control panel
- Alignment
- Layers info
- Ramachandran plot
- Electron density map
- Cavity
- Text

Open one or more PDB file(s) and try to open these windows from the menu **Window**. In case of multiple open files, you can only control one molecule at the same time, and to switch between the currently loaded molecules you can use the TAB key or the pop-up located at the top of the Control Panel window or click on the name of the layer you want to activate in other kind of windows.

You will find options for electron density map, cavities, and text are disabled. Click **Tools** and **Compute Electrostatic Potential** to enable the option for electron density maps and view the window. Similarly, click **Tools** and **Compute Molecular Surface** to enable the option for cavities and view the window. To enable text window, you will have to open a text window, using **Open text file** from the **File** menu.

Note

There is a question mark icon in every window that you open like the one shown in the toolbar of the Deep View tool. This will provide help for the respective window.

Let us look at each kind of windows.

Control panel It provides a unique, natural, and easy way to manipulate individual groups of any currently loaded molecule. Each molecule is composed of groups such as amino acids, nucleotides, etc., which in turn are composed of atoms, whose coordinates are taken directly from a PDB file. The control panel has two sections: one header that helps to toggle between layers and a list that contains all the groups of the currently active layer.

Header The top of the control panel gives you the name of the current molecule. Two checkboxes, **visible** and **can move**, are present, which are initially enabled. If **visible** is disabled, the molecule will be removed from the view but not from the memory. This will help to speed up the rendering of other molecules. The **can move** checkbox is quite explicit. If it is disabled, the current molecule is frozen. It allows other molecules to be moved or rotated without affecting the current molecule. This feature helps to place a substrate.

List In the leftmost column of the control panel, the name of each group appears, immediately preceded by the chain it belongs to (if any). A group can be either selected (in red) or not (in black). A group is displayed only if a check (v) is present in the second column (the 'show') column. Similarly, its sidechain, label, Vander Waals dots, spheres, and ribbon are visible only if a check is present in their respective columns.

The colour of each group is displayed in the last column. A '–' in the box means default atom colours defined in the preferences are applied for this group. This colour reflects the backbone colour, the side chain colour, the ribbon colour or the backbone + side chain colour according to the status of the pop-up menu located between the color button and the first color box. As a reminder of the current mode, a `B` for backbone, a `S` for side chain, a `R` for ribbon or a `BS` for backbone + side chain is written after the 'colour' button.

Not only amino acids, other groups such as nucleotides and HETATM are also listed in this window. In the case of HETATM the side chain column is ignored. It displays or hides the base part of nucleotides, leaving the ribose chain untouched. To select groups in a column, the SHIFT and CTRL keys work like that of Microsoft applications. You may also select **All items** from the **select** menu or use CTRL key to do discontinuous selections.

- *Alignment*: It shows the alignment of the molecules open in different layers. It is an alternate way of selecting groups, and is useful to compare multiple proteins. It shows the alignment of the proteins, gives some feedback information, and permits to thread a sequence onto a reference in order to submit a homology modelling request to the Swiss model. A third window containing the primary sequence in one-letter amino acid code is also available.
- *Layer Infos*: This window will let you alter the displayed features of a layer (Hydrogen, H-bonds, H2O, etc.). You can also control which layers will be allowed to 'blink' which is a convenient way to compare proteins by displaying only one layer at a time and quickly switching among the layers.
- *Ramachandran plot*: It shows the Ramachandran plot of the currently selected amino acids.
- *Electron density map*: This lists the currently loaded electron density maps and lets you alter what is displayed (unit cell, coarse contouring, sigma contouring value, colour, dottet or plain lines). Holding down **Control** while clicking on the name brings a more complete dialog box.
- *Cavities*: It lists the surface and cavities in terms of area and volume with colour coding.
- *Text window*: You can open this window from the **File** menu. A text window can contain a PDB file, an on-line help, or any other text file that you want to open with the **Open Text File** submenu.

(b) Learn some file operations

You can perform file operations such as loading one or more molecules on PDB, mmcif, or MOL format in different layers, running a script file to open a file and perform certain operations on it, importing files from the disk or a server by providing the IP address and other relevant information and closing one or all the layers. The tool also gives you the liberty to save as well as reopen files in the form of the surface, electron density map, and electrostatic potential. Unlike any other similar tools it provides the option to save the molecule in either the original orientation or in the modified orientation. While doing so it changes the coordinate values in the PDB file. While working on this tool you can do all these from the **File** menu. You can also close the open file or 'close all' the layers with multiple open files.

If you open a protein file, you will find that options for discarding the surface, electron density map, and electrostatic potential are disabled. Click on the menu **Tools** and **Compute Electrostatic Potential** to enable the discard options for **Electrostatic Potential** and **Electron Density Maps**. Similarly, click on the menu **Tools** and **Compute Molecular Surface** to enable the discard option for **Surface**.

(c) Manipulate a model or its view

Broadly, you can centre, resize, translate, zoom, or rotate a molecule around any three axes with the help of the first four icons in the tool bar. The four icons are shown in Fig. 7.12:

Fig. 7.12 Icons to manipulate model in Deep View

There is a toggle between a 'little earth icon' and a 'little protein icon'. In the first case, the rotation takes place in absolute coordinates allowing to rotate the molecule around any atom, provided this atom has previously been centered (translated to the (0,0,0) coordinate) and in the second case molecules are rotated around their centroid.

You can also select part of a molecule using submenus under the **Select** tool, switch to another layer with the help of control panel, and merge molecules from the **Create Merged Layer** submenu of the **Edit** menu.

You can explore all the options under the Display menu to view different aspects of the model, view the model from a different perspective, or save any view that you have manually done using translate, zoom, or rotate icon. Let us look at each option under the **Display** menu.

- *Label kind*: It selects the kind of label to draw on each group. By default, it is the group name that appears on the carbon alpha, but you can choose to display atom names or type. Labels can appear only if this option is checked in the control panel list before a group under column **labl**.
- *Slab*: It enables two Z clipping planes. It allows you to look inside a protein without having to see all peripheral atoms. By default, the visible section depth (i.e., the distance between the two clipping planes) is 10 Å and the section is centered on Z = 0.
- *Show axis*: When checked, the axis orientation of the first protein loaded will appear at the top left of the screen. Note that it is different from the axis that can be displayed from the Layer Infos window that always appears at the (0,0,0) coordinate of each concerned layer.
- *Show backbone oxygen*: For very large molecules, it is sometimes useful to see only the backbone trace. (This menu can also be accessed from the Layer Infos window).
- *Show side chains even when backbone is hidden*: When checked, this menu allows the display of side chains even when no mark is present in the show column of the control panel. This can be useful for certain kinds of final renderings with a ribbon representation of the protein. (This menu can also be accessed from the **Layer Infos** window.)

- *Show dots surface*: When checked, this menu allows the display of van der Waals surfaces. It supersedes the setting found in the **dot** column of the control panel. This is because the dot surface takes long to compute, and you may want to temporarily hide them without loosing your individual settings.
- *Show hydrogens*: It toggles the hydrogen display status. It is sometimes easier to see something if H are temporarily hidden. This menu can also be accessed from the Layer Infos window.
- *Show H-bonds*: It toggles the H-bonds. To compute H-bonds of the *current layer*, choose the appropriate item under the **Edit** menu.
- *Show H-bonds distances*: Draw the distance between the donor and the acceptor at the middle of each H-bond. When several H-bonds are very close, some screen garbage can occur, and you may need to rotate the molecule a little to be able to read the distance. (This menu can also be accessed from the Layer Infos window.)
- *Show only H-bonds from selection*: Show only H-bonds whose one extremity belongs to the selected groups (groups appearing in red in the control panel).
- *Show only groups with visible H-bonds*: This option allows to focus on important things. This is very useful in conjunction with the previous function. Imagine that you want to display only groups that make H-bonds with an enzymatic cofactor (NAD, ATP,...), click on **NAD** in the control panel (it is now the only selected group); choose **show only H-bonds from selection**, and choose show only groups with visible H-bonds, which will clean up the view.
- *Render with OpenGL (only PC)*: This common will generate a solid image withOpenGL. The colours, lights, appearance, and kind of the renderer can be altered in the two appropriate preferences dialogs.

(d) Measure and compute

Let us learn different measurements and computation required for analysing a protein molecule. Look at the general usage tools. All these functions will ask you to pick atoms in the display window. Follow the instructions that appear in red under the tools; the results will be given at the same place as well as directly on the molecule. If CAPS LOCK is down, you can measure several distances or angles successively. To exit the 'repeated' measurement mode, you can either press CAPS LOCK or hit ESC. We will look at each of them to learn the measurement that can be done.

Fig. 7.13 Icons to measure and compute in Deep View

- The first icon helps to calculate the distance between two atoms.
- The second icon helps to calculate the angle between three atoms.
- The third icon by default gives a measure of omega, phi, and psi angles of the picked

amino acid (you can pick any atom of the aa). But when this tool is invoked with the CONTROL key, it will prompt you to pick 4 atoms, which will allow measuring the torsion angle of any specific bond.

- Provenance of an atom (will give the name of the molecule, group chain, and atom).
- Display/Undisplay groups that are/are not at a certain distance from an atom.
- Centre the molecule on one atom (using this tool will automatically switch to world coordinates).
- Fit a molecule onto another (available only when two or more molecules are loaded). In this case, you will be prompted to pick three corresponding atoms of the molecule to fit. Note that there are other better ways to superpose two proteins available from the **Tool** menu
- Mutation tool.
- Torrion tool.

TASK

In this example we will look at the active site of the lactate dehydrogenase (PDB entry 1ldm).

Step 1 Open the PDB file 1LDM (it is included in the tutorial package).

Step 2 Select the **Compute H-bonds** item from the Tools menu.

Step 3 Click on the control panel window, scroll to the bottom (or hit the PAGE DOWN key), and select the groups **NAD1** and **OXM2**. They should become red in the control panel. If they do not, it is a good idea to read the **Control Panel Selection** section).

Step 4 Choose the item **Show Only H-bonds from selection** of the **Display** menu. Then choose the item **Show only groups with visible H-bond** of the same menu. Finally hit the "=" key of the numerical keypad (right mouse button for PC) to rescale and recentre the view. At this stage, you should be in this situation as shown in Fig. 7.14.

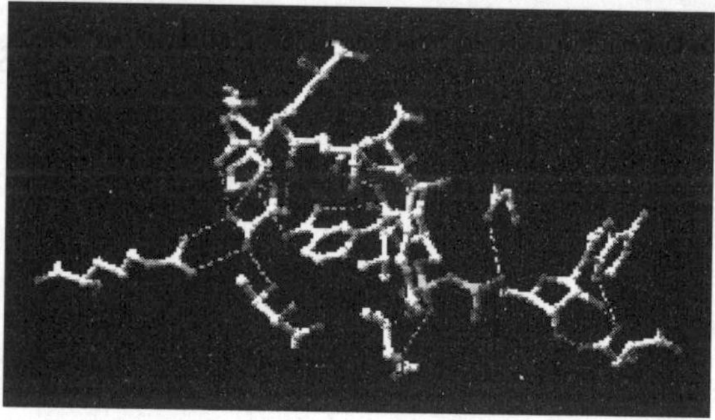

Fig. 7.14 Model with H bonds

Step 5 Now you can measure distances using the tools located at the top of the display window, or show the H-bonds distances with the appropriate item from the **Display** menu.

TASK

In this example we will learn to use the building loop tools.

Step 1 Open three copies of the PDB file 1CRN provided with the tutorial package. Use the Display menu to bring up the Layer Infos window. Hide the second and third copies of the protein.

Step 2 Now use the **Build Loop** option from the **Build** menu. You will be prompted to pick two residues which will serve as 'anchors ones'. Pick L18 and A24 (the blue and the green residue) After a while, a window containing a list of possible loops appears. The currently selected loop appears in red.

```
clash score: 1
PP:-10.56
FF:200.2
 C-N+   CA-C-N+  C-N+-CA+
----------------------------
 0.04    -1.48    -7.09
 0.06     0.00     0.00
-0.11     0.00     0.00
-0.11     0.00     0.00
 0.36     0.00     0.00
 0.33     0.00     0.00
 0.47     0.00     0.00
 0.36     0.00     0.00
```

The first column gives the diviation in angstroms to the ideal closure bond length, while the next two columns give the deviation (in degrees) to the ideal angle closure. As you can see, the first loop is the only one with the values computed. Simply click on a line to see how well it fits the next loop. Note that you can also use the UP and DOWN arrow keys to browse among the solutions. To help you find the best loop, a count of the clashes (bad contacts or H-bonds) is displayed at the top of the window. There is also an energy information (computed with a partial implementation of the GROMOS force-Field) that appears after the `FF` text, and a mean force potential value (PP) computed from a 'Sippl-like' mean force potential [ref. 6]. You can click on either of those lines of the header to sort the results accordingly to this specific criterion. Play a little with the various loops proposed, and then select one that seems good enough.

Step 3 Now make the second copy of 1crn (the one in the second layer) visible again and compare your best solution with the actual one. Hide the first copy of 1crn (the one that contains the rebuilt loop) and make the second layer active by clicking on the second protein listed in the Layer Infosd window. Now use the

Scan Loop item from the **Build** menu. As before, you will be prompted to pick two residues which will serve as 'anchors ones'. Again, pick L18 and A24 (the blue and the green residue). After a while, a window containing a list of possible loops appears. The currently selected loop appears in red. The window content is slightly different, and gives you the name of PDB files that contain a suitable loop, the chain identifyer, the starting residue, the sequence of the possible fragment, and the resolution (in Å) at which the structure has been solved. Note that a resolution of 0.0 Å means that the structure has been solved by NMR, whereas a resolution of 9.99 Å means that the source structure is a model.

```
PGTPE          clash score:1     bad G->X: 0
. ...  (-3)     bad Phi/Psi:2     bad X->P: 1
HLEHK           PP:-10.46         bad X->P: 2
FF:34753.9      access:0.00       rms:0.00
-----------------------------------------------
2STV    10 HLEHK  2.50
2STV    92 VLNTA  2.50
1PYP   113 NNPID  3.00
4PTI    35 GCRAK  1.50
1EST    61 NQNNG  2.50
5ABP   130 KESAV  1.80
8ADH   140 GTSTF  2.40
8ADH   212 AGAAR  2.40
```

The header gives you the sequence of the loop you want to build, aligned with the sequence of a fragment selected from a database of folds. The similarity score for the fragment appears within parenthese and is computed from the PAM200 matrix. In addition to this information, you have, as before, a force field score, a mean force potential score, a clash score, and the number of residues from the source loop that have bad phi/psi angles (in other words, residues that would have phi/psi angles lying beyond the allowed zones of the ramachandran plot). Also, you have to consider that Gly and Pro are somehow special residues that adopt special phi/psi combinations. Gly can accept any kind of combination, whereas Pro have phi angles constrained around –60 degrees. Therefore, the number of 'bad' transversions (Gly in the source loop that would not readily become something else in the loop you want to build; or residues that would not readily become a Pro) are also summarized in the header. As for the previous building loop tool, you can sort the loops by energies or clashes to ease the process of identifying the best loop. Select a suitable loop and make the third layer visible to compare with the actual solution. After having constructed your loops, an energy minimization is mandatory. Swiss-PdbViewer provides an energy miminization facility using the GROMOS96 force field.

TASK

In this example we will learn how to apply non-crystallographic symmetries and build a full tetramer from a PDB file containing only a monomer of a protein.

Step 1 Open the PDB file 1LDM (provided with the tutorial package), have a look at it, and rotate it. Did you ever wonder how it was possible that the first six N-terminal amino acids were out in the solvent and not properly folded around the protein? Have not you learnt that proteins were globular and tightly folded?

Step 2 Now reset the orientation (**Edit** menu), and open three more copies of the PDB file. (You can use the **1LDM** option at the bottom of the **File** menu, as Swiss-PdbViewer remembers files recently opened.) Each copy is loaded into a new layer.

Step 3 Now select all residues in all layers (hold down the SHIFT key while invoking the **Select All** option of the **Select** menu).

Step 4 Make the second layer active (use the control panel pop-up menu to switch among layers). Click on the little text icon in the main display window. This will let you consult the PDB file as a text file, which is useful to have a look at the annotations. Scroll down the PDB file until you find MTRIX lines (they are just before the ATOM lines). You can see 9 lines MTRIX. They represent three transformation matrices, and allow you to build the non-crystallographic symmetries of the protein.

```
MTRIX1   1 -1.000000  0.000000  0.000000        0.00000
MTRIX2   1  0.000000 -1.000000  0.000000        0.00000
MTRIX3   1  0.000000  0.000000  1.000000        0.00000

MTRIX1   2 -1.000000  0.000000  0.000000        0.00000
MTRIX2   2  0.000000  1.000000  0.000000        0.00000
MTRIX3   2  0.000000  0.000000 -1.000000        0.00000

MTRIX1   3  1.000000  0.000000  0.000000        0.00000
MTRIX2   3  0.000000 -1.000000  0.000000        0.00000
MTRIX3   3  0.000000  0.000000 -1.000000        0.00000
```

Step 5 Click on the first line of the first MTRIX record. This will load the transformation matrix into a dialog. Simply click **OK** and the transformation will be applied on the current layer.

Step 6 Now make the third layer active with the control panel pop-up menu, and click on any line of the second group of matrix, and apply the transformation.

Step 7 Now make the fourth layer active with the control panel pop-up menu, and click on any line of the third group of matrix, and apply the transformation.

Step 8 Now color by layer (in the color menu) and observe your full functional unit. As you can see from Fig. 7.15, the first 6 N-terminal amino acids are no longer isolated in the solvent.

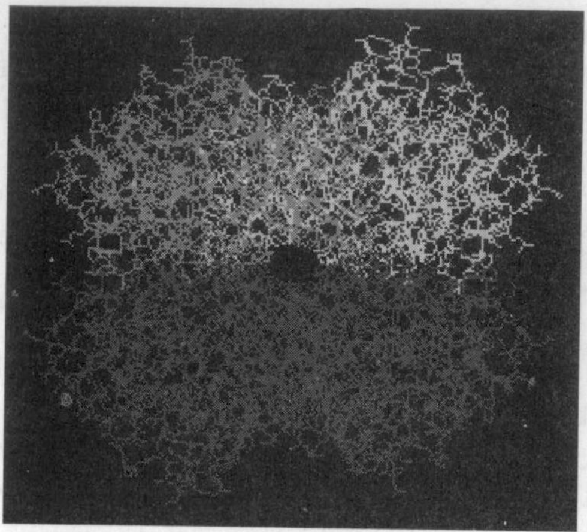

Fig. 7.15 Model with non-isolated first 6 N-terminals

We will now observe which residues are making non-crystallographic contacts. Make the first layer active (the yellow one).

Step 9 Now use the **Groups close to another Layer** option from the **select** menu, choose the **Display only Groups that are within** radio button, set 5 Å as the cut-off value, and check the box **act on all layers**. This lets you observe which residues are involved in protein–protein contacts as in Fig. 7.16.

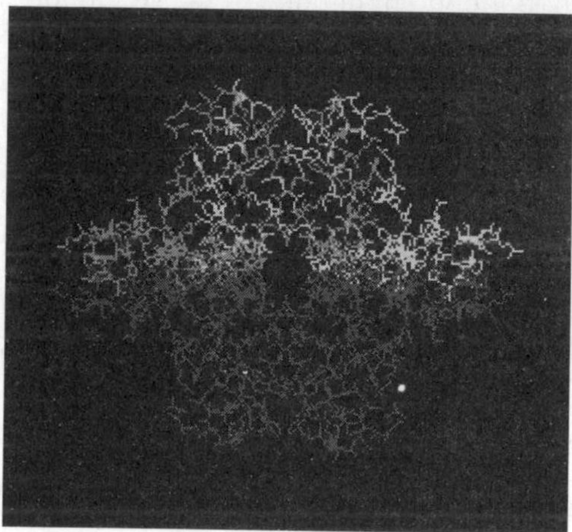

Fig. 7.16 Model showing residues involved in protein–protein contact

Step 10 Now repeat the operation but uncheck **act on all layers**. Every residue within 5 Å of any residue of the current layer only will be selected in this case, only residues close to a 'yellow' residue. Change the active layer to the second protein (the blue one) and do it again, to fully understand how this tool works (see Fig. 7.17).

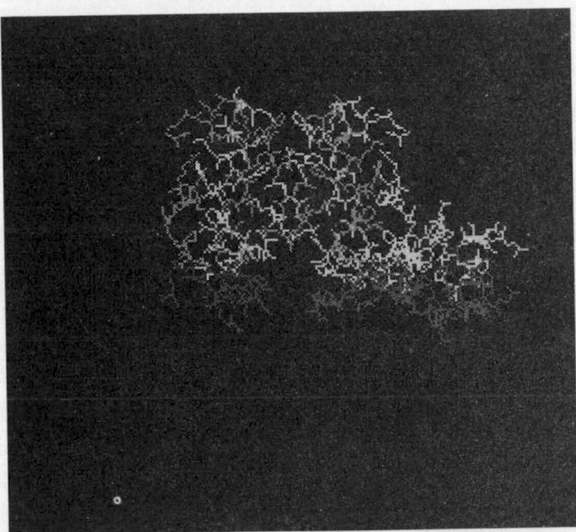

Fig. 7.17 Model showing residues involved in protein–protein contact for the current layer

Conclusion

By now you got exposed to various kinds of tools but that is not the end of bioinformatics tools exploring session. There are still many that exist now and that we have not covered this book. In fact it is not even possible for a single book to portray all the tools that are available for bioinformatics analysis. There will be many more as you read this book or work in this field. Your aim should be to identify what you want to do and hunt for the right tool to do that job. As you all know that there are great difficulties in obtaining sufficient protein diffracting crystals and there are many other technical aspects that hamper the experimental elucidation of the 3D structure of proteins. This is the reason why the number of solved 3D structures increases slowly as compared to the rate of sequencing of novel cDNAs. In this context the predictive methods including prediction, visualization, or modelling have gained much interest. Therefore, like any other specific areas, in the field of prediction and modelling also you may have to use more than one tool to do the same job and collate the results at the end. If you are conversant with programming languages you are free to develop your own tools too.

EXERCISES

Exercise 7.1

(a) Install Rasmol and Chime on your computer(s).
(b) List some molecules you would like to visualize in your class(es).
(c) Search the Brookhaven PDB for suitable atomic coordinate files and download them. (Structures have not been determined for many molecules!)
(d) With Rasmol, explore the structures in these PDB files.
(e) Select structural features and views to show your class which can be effectively visualized with the available PDB's.

(f) Gather the following:
   i. On diskettes, the PDB files you have downloaded for your molecules. (If they are large, consider compressing them with WinZip).
   ii. A plan specifying the molecules, structural features, views.

Exercise 7.2

(a) Prepare small notes for the following steps of comparative protein modelling
- identification of modelling templates
- aligning the target sequence with the template sequence
- building the model
- framework construction
- building non-conserved loops
- completing the backbone
- adding side chains
- model refinement

(b) Take an unknown sequence and do the above steps and paste your result including graphics.

Exercise 7.3 Describe each of the following images from your understanding of protein modelling

(a)

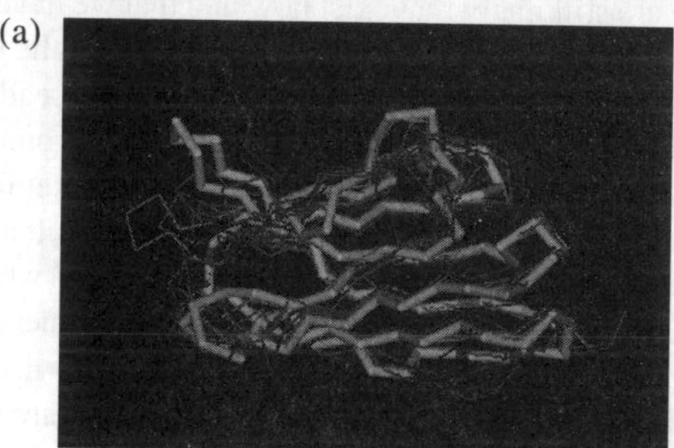

(b)

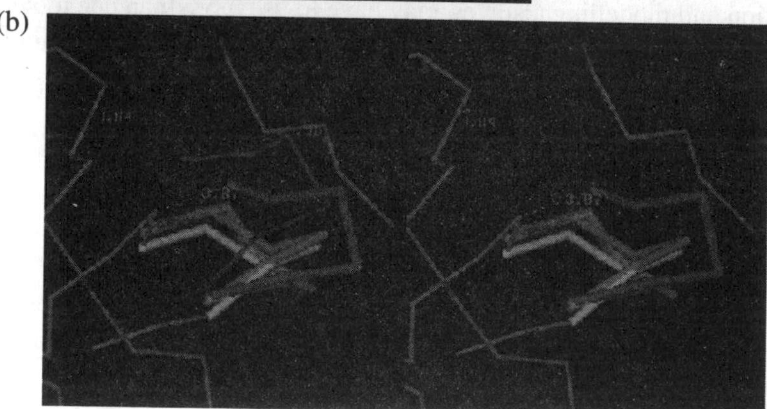

(c)

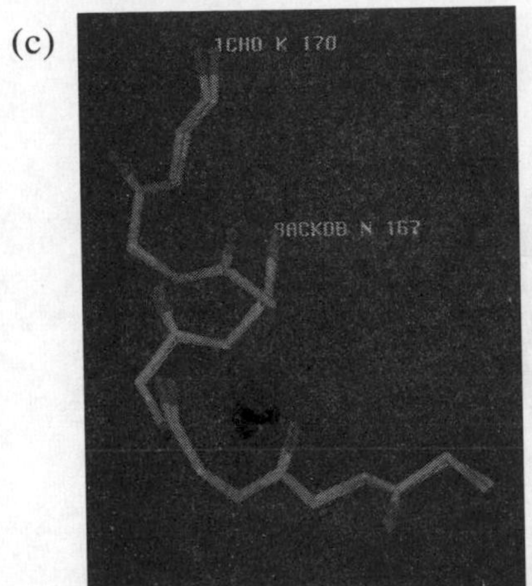

(d)

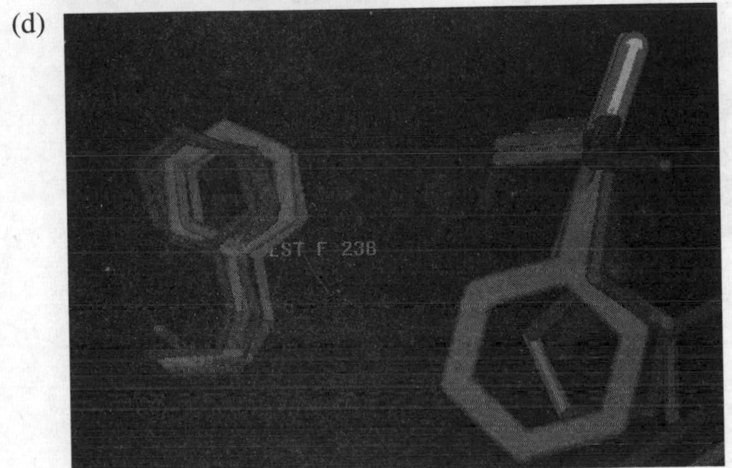

(e)

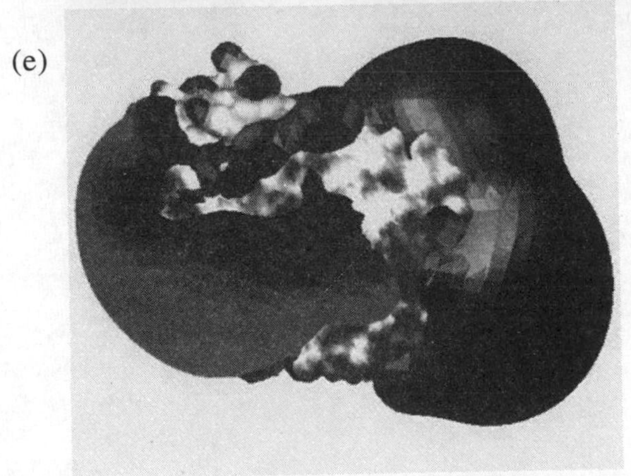

(f)

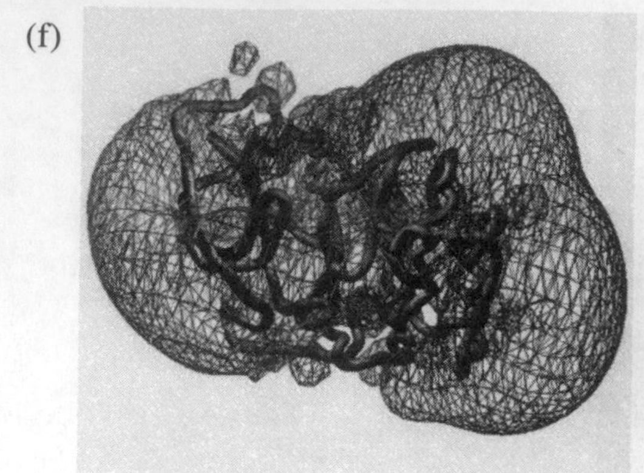

(g)

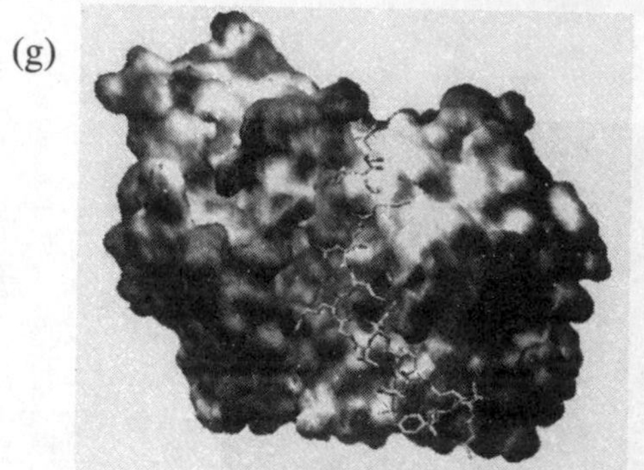

PART D
Algorithms

Chapter 8: Data Analysis Algorithms

Chapter 9: Prediction Algorithms

D Algorithms

Learning Objectives

- To define and understand an algorithm and its associate data structure
- To classify algorithms on the basis of their design and implementation
- To understand implementation of algorithms
- To map various bioinformatics tasks and the corresponding algorithms
- To map various algorithms and the corresponding software

Introduction

The term 'algorithm' (pronounced al-go-rith-um) is derived from the name of the nineth century Persian mathematician Al-Khwarizmi. Originally it was referred to the rules of performing arithmetic using Arabic numerals. Later the word evolved into *algorithm* by the eighteenth century. It includes procedures or formulae for solving problems or performing tasks. It can be understood as a finite set of well-defined instructions for accomplishing some task. Other than steps in series, it also has steps that repeat (that implement iteration) or compare (that implement logic) until the task is completed. It starts with an initial state and continues till it reaches an end state. The end state is the objective of the task. All these should be defined so that the steps are applied in any possible circumstances that could arise. For example, when a conditional step is encountered, the criteria of the result that can lead to one of the possible cases must be clear and computable. Here it should be noted that we are referring to the algorithm in terms of *imperative programming*. Thus, one can think of algorithms as a set of well-defined operations to solve some problem in a finite amount of time. These operations are performed on some data representation. Some common examples of algorithms are as follows: recipes, instructions for assembling a machine, and a set of instructions for solving a problem via a computer.

Analogy

Let us consider the task of boiling potatoes. Its initial state is unboiled potatoes and the final state is boiled potatoes. There can be various ways of performing this task. Each of those ways of performing the task are algorithms. It is the user's decision to select the algorithm based on the resource availability as well as other specific needs of the task such as effort, time, etc.

(i) *Now one algorithm for this task can be the following:*

1. *Take a pan.*
2. *Put some water into the pan.*
3. *Boil the potatoes.*

This algorithm is cost effective to implement because a pan is cheaper than a pressure cooker. But it will take more time to finish the task.

(ii) *Another algorithm for this task can be the following:*

1. *Take a pressure cooker.*
2. *Put some water into the cooker.*
3. *Boil the potatoes.*

This algorithm is not cost effective (if boiling potatoes is to be done once) but is efficient in terms of time. If potatoes boiling is to be done regularly, this algorithm becomes cost effective. Now if the pressure cooker is not there in the resource list, user has no choice other than implementing the first algorithm. But if both the pan and pressure cooker are available in the resource list and the need of the task is to do a quick job, the user will obviously choose the second algorithm.

Algorithms are generally represented with *pseudocodes*. But as the algorithm is a list of precise steps, the order of computation is critical to the functioning of the algorithm. Instructions are listed explicitly, and are described from the top going down to the bottom. Such ideas are better described by *flowcharts*. If there is flaw in the algorithm or, to be more precise, in the instructions of the algorithm, performing each step correctly will not result in the desired end state. Also, different algorithms may complete the same task with a different set of instructions using more or less time, space, or effort.

The data that algorithms work on are represented as *data structures*, which are nothing but high-level ways of representing data. Some common data structures are strings, arrays, lists, stacks, queues, trees, and graphs. However, the way we represent things has a large influence on what we can do with them. It also influences how efficiently data can be manipulated. Some fundamental algorithms, associated with a variety of data structures and hence common to a variety of applications, are those that add, modify, delete, arrange, count, compare, test for existence, search, count, and compare.

Note

Algorithm efficiency is measured as the worst-case efficiency, which is the largest amount of time an algorithm can take given the worst possible input of a given size. Big-O notation describes the running time of an algorithm. If the running time of an algorithm is $O(n^2)$, it means the running time of the algorithm on an input size n is limited by a quadratic function of n.

D.1 Classification of Algorithms

There are many ways to classify algorithms. Each way has its own merit. An algorithm can be correct or incorrect, fast or slow and *iterative* or *recursive*. It can be a combination of these types such as fast, correct, and recursive algorithm. One way of classifying algorithms is by their design methodology or paradigm. The other way is based on implementation. There is a certain number of paradigms, each different from the other. Furthermore, each of these categories include many different types of algorithms. Some commonly found paradigms are discussed in the following subsections.

D.1.1 Algorithm Design

Algorithms can be classified on the basis of their design. Think of a very common problem that we face such as losing a pen. When you realise that you lost your pen, what you do is quickly plan how you want to search that. Sometimes you just search every room of your house and every nook and corner of each room, one after another in a systematic manner, in your house from study table drawers to the refrigerator ice box. Sometime you search only places where you visited since last you saw your pen. Next we will discuss the various kinds of algorithms that are grouped in terms of their design or approach towards solving the problem.

(a) Divide and conquer

Sometimes one big problem is difficult to solve. Such a problem is divided into two smaller problems to make it easier. A divide-and-conquer algorithm reduces a problem to two or smaller subproblems and sometimes reduces those subproblems to further smaller problems till these became small enough to solve easily. The solutions to subproblems are combined to create a solution to the original problem. This is also called the process of reduction. This process of splitting and merging takes a longer time.

Example D.1 The binary search from a given ordered array of n elements is an example of divide-and-conquer paradigm. The basic idea of binary search is to search a given element. To do that, first we need to 'explore' the middle element of the array. The middle element is checked against the given element that is to be searched. The search is continued in either the lower or the upper segment of the array, depending on the outcome until we reach the required (given) element.

Algorithm

```
binarySearch(a, value, left, right)
while left <= right
       mid : = floor((left+right)/2)
        if a[mid] = value
            return mid
        else if value < a[mid]
right := mid-1
        else if value > a[mid]
```

```
  left  := mid+1
return not found
```

Divide and conquer algorithms have been used also in structure-based protein designing. While Global Minimum Energy Conformation (GMEC) is the key approach, Dead End Elimination (DEE) algorithms guarantee the selection of only rotomers that belong to GMEC. In this approach the conformational space is partitioned into several sections such that no single competition rotomer outperforms the designed conformation for every conformation and there exists a different dominant competitor for each partition (Georgeiv et al 2006)

(b) Dynamic programming

There are cases when a problem is solved with the optimal solution constructed from optimal subsolutions to subproblems. These subproblems are found to be overlapping, i.e., the same subproblems are used to solve many different problems of the same type. Such problems are known as *optimal substructures*. These problems are solved effectively using dynamic programming, where solutions that have already been computed are not computed again.

Example D.2 The shortest path to a goal from a vertex in a weighted graph can be found by using the shortest path to the goal from all adjacent vertices. Some of the algorithms that use dynamic programming are Viterbi, Levenshtein distance or edit distance, Needleman-Wunsch, Smith Watermann, etc. Let us do a global sequence alignment using Needleman/Wunsch techniques. The two sequences to be globally aligned are

```
GAATTCAGTTA (sequence 1)
GGATCGA (sequence 2)
```

It will give two alignments with the same score:

Alignment 1

```
GAATTCAGTTA (sequence 1)
GGATCGA (sequence 2)
```

Giving an alignment of

```
G A A T T C A G T T A
| |   | |   |       |
G G A _ T C _ G _ _ A
```

Giving an alignment of

```
G _ A A T T C A G T T A
|   |   | |   |       |
G G _ A _ T C _ G _ _ A
```

Dynamic programming is the most fundamental approach in bioinformatics which has been extended to sequence comparison (Needleman–Wunsch global alignment), gene prediction (gene models for XGrail 1.3), gene chip recognition (local high dimensional segment alignment), RNA secondary structure prediction (including pseudoknots) (Giegrich 2000). You will learn more about the algorithm in Chapter 8.

(c) Greedy algorithms

A greedy algorithm is similar to a dynamic programming algorithm. In this case, at each stage, you need not have the optimal solutions to the subproblems, you can make a 'greedy' choice from the subsolutions of what looks best for the moment. These algorithms are 'shortsighted' in their approach, i.e., they take decisions on the basis of information at hand without worrying about the effect these decisions may have in the future. They are easy to both invent and implement, and most of the time quite efficient since the approach generally is the most obvious one; however, it may be subtly wrong. Several bioinformatics problems have been attempted using these algorithms, such as selection of primer sets (G-primers), multiple aligment (FOLDALIGN).

The greedy algorithm consists of four functions:

- A function that checks whether the chosen set of items provides a solution.
- A function that checks the feasibility of a set.
- The selection function tells which of the candidates hold high prospects.
- An objective function, which does not appear explicitly, gives the value of a solution.

Example D.3 The Make Change is a classical example of a greedy algorithm. Let us assume some coins that are available to you as

- dollars (100 cents)
- quarters (25 cents)
- dimes (10 cents)
- nickels (5 cents)
- pennies (1 cent)

You need to take a change of a given amount using the smallest possible number of coins.

The pseudocode can be given as follows:

- Start with nothing.
- At every stage without passing the given amount.
- Add the largest to the coins already chosen.

Algorithm

```
MAKE-CHANGE (n)
        C ← {100, 25, 10, 5, 1}      // constant.
        Sol ← {};                   // set that will hold the solution set.
        Sum ← 0 sum of item in solution set
        WHILE sum not = n
            x = largest item in set C such that sum + x ≤ n
            IF no such item THEN
                RETURN    "No Solution"
            S ← S {value of x}
            sum ← sum + x
RETURN S
```

(d) Linear programming

Often you will find practical problems around you that include a number of varying quantities. Some common examples may be network flow problems or scheduling task problems. In such cases it is possible to develop mathematical equations that capture the interactions involved in the problem and convert the problem into a mathematical problem. This process of deriving a set of mathematical equations whose solution implies the solution to the given practical problem is called *mathematical programming*. The term 'programming' refers to the process of choosing the variables and setting up the equations. If all the equations involved are linear combinations of the variables, we identify the problem as linear programming. Such algorithms are better applied to infer the strength of protein–protein interaction, solve fixed-backbone homology modelling, multiple sequence alignments, and protein design problems (RAPTOR).

Algorithm

Maximize $x_{AB} + x_{AD}$

Subject to the constraints

$x_{AB} \leq 6$ $\quad$ $x_{CD} \leq 3$

$x_{AC} \leq 8$ $\quad$ $x_{CE} \leq 3$

$x_{BD} \leq 6$ $\quad$ $x_{DF} \leq 8$

$x_{BE} \leq 3$ $\quad$ $x_{EF} \leq 6$

$x_{BD} + x_{BE} = x_{AB}$

$x_{CD} + x_{CE} = x_{AC}$

$x_{BD} + x_{CD} = x_{DF}$

$x_{BE} + x_{CE} = x_{EF}$

$x_{AB}, x_{AC}, x_{BD}, x_{BE}, x_{CD}, x_{CE}, x_{DF}, x_{EF} \geq 0$

(e) Search and enumeration

There are many problems that can be modelled as graphs. Here the word 'graph' is different from that we know from the scientific perspective as a list of data represented in a coordinate system. Here by 'graph' we mean a collection of points connected by lines, where points are called *vertices* and lines are called *edges*. A graph can be described as follows

- Vertex set: V = {a, b, c, d, e}
- Edge set: E = {(a,b), (a,c), (b,c), (b,d), (c,d), (c,e)}

Here the position of the vertices is not important, instead the connection between the vertices is important. Therefore two graphs with vertices positioned differently but with edges connected similarly can be said to be equivalent even if they look different. A graph exploration algorithm or search-and-backtracking algorithm specifies rules for moving around a graph. Common examples are chess, DNA sequencing, protein structure prediction, etc. Look at the algorithm that checks whether a graph is connected or not.

Algorithm

```
test-connected(G)
 {
choose a vertex x
make a list L of vertices reachable from x,
and another list K of vertices to be explored.
initially, L = K = x.
while K is nonempty
   find and remove some vertex y in K
   for each edge (y,z)
        if (z is not in L)
       add z to both L and K
if L has fewer than n items
   return disconnected
else return connected
 }
```

(f) Probabilistic and heuristic algorithms

Algorithms belonging to this class loosely comply to the definition of algorithm.

- ***Probabilistic algorithms*** solve a problem by selecting solutions randomly. They are also known as randomized algorithms since they take certain decisions randomly—one could say they sometimes flip a coin in order to decide how to proceed. Any algorithm that works for all practical purposes but has a theoretical chance of being wrong belongs to this class. It has been proved to provide the fastest solutions. Other advantages include the simulistic approach. They may sometimes be able to break patterns that we are not aware of. This kind of an algorithm is generally used in difficult problems where an exact polynomial-time algorithm is not known. The main advantage is that no input can reliably produce worst-case results because the algorithm runs differently each time. These algorithms have been used to compare genomes (PAGEC), find motifs in orthologous sets (PhyME), identify non-coding RNA (MFOLD), and to characterize intact open-reading frames for pseudogene filtering.
- ***Genetic algorithms*** find solutions to problems by imitating the biological process of evolution. The solutions are generated by random mutations. Thus, they emulate reproduction and 'survival of the fittest'. Unlike linear programming, genetic algorithms begin with a set of solutions (represented by chromosomes) and are called *population*. Solutions from one population are taken and used to form a new population. The new population is assumed to be better than the old one. Solutions that are then selected from the set of new solutions (offspring) are selected according to their fitness—the more suitable they are the more chances they have to reproduce.

Algorithm

- **[Start]** Generate a random population of n chromosomes (x) (suitable solutions for the problem)
- **[Fitness]** Evaluate the fitness $f(x)$ of each chromosome in the population
- **[New population]** Create a new population by repeating the substeps until the new population is complete
  - **[Selection]** Select two parent chromosomes from a population according to their fitness
  - **[Crossover]** With a crossover probability cross over the parents to form new offspring (children) otherwise offsprings are exact replicas of parents.
  - **[Mutation]** With a mutation probability mutate new offspring at each locus (position in chromosome).
  - **[Accepting]** Place new offspring in the new population
- **[Replace]** Use new generated population for a further run of the algorithm
- **[Test]** If the end condition is satisfied, **stop**, and return the best solution in current population
- **[Loop]** Evaluate the fitness

New approaches to finding potential motifs in DNA sequences have been adopted in the recent past. These approaches are based on genetic algorithms (GA), where crossover is implemented with gap penalties to produce optimal child patterns. GAs have also been used for protein structure function assignment, semiautomatic drug design, and sequence analysis.

- ***Heuristic algorithms*** do not find an optimal solution but instead find an approximate solution where the time or resources to find a perfect solution are not practical. The purpose of the random variance is to find close to globally optimal solutions rather than simply locally optimal ones. In such case the 'random' set of solutions decreases as the algorithm approaches to a solution. Some common heuristic algorithms are BLAST and FASTA. The Travelling Salesman problem is a classical example of this kind of algorithm. The problem consists in a salesman who must visit n cities, passing through each city only once, beginning from one of them which is considered as his base, and returning to it. The cost of the transportation among the cities (whichever combination possible) is given. The program of the journey that is the order of visiting the cities in such a way that the cost is the minimum is requested.

Algorithm

```
Input: Number of cities n and array of costs c(i,j) i,j=1,..n
(We begin from city number 1)
Output: Vector of cities and total cost.
```

- (* starting values *)
- C=0
- cost=0
- visits=0
- e=1 (*e=pointer of the visited city)
- (* determination of round and cost)
- for r=1 to n-1 do
    - o choose of pointer j with
    - o minimum=c(e,j)=min{c(e,k);visits(k)=0 and k=1,..,n}
    - o cost=cost+minimum
    - o e=j
    - o C(r)=j
- end r-loop
- C(n)=1
- cost=cost+c(e,1)

Heuristic approach has been adopted to solve the fragment assembly problem, CDHIT, and clustering for gene expression data analysis.

D.1.2 Algorithm Implementation

Algorithms can also be classified on the basis of their implementation. Let us again think of the problem of losing a pen. You have already planned the approach you want to take to search your pen. For example, you want to search every possible place where you can keep your pen. To do that you can either look at those places one after another or you may appoint your younger sister to look at some place while you loot at other. This is what is meant by categorizing algorithms as per their implementation. Here you will learn various kinds of algorithms that are grouped based on how they are implemented to solve the problem.

(a) Recursive algorithm

A recursive algorithm is one that invokes the parent procedure repeatedly by referencing it from the calling procedure until a certain condition matches. A *recursive algorithm* is an algorithm which calls itself with smaller or simpler input values, and which obtains the result for the current input by applying simple operations to the returned value for the smaller (or simpler) input. The famous Tower of Hanoi puzzle is an example of a recursive algorithm. Generating Fibonacci numbers is also an example of a recursive algorithm.

$$F(n) = \begin{cases} 0 & \text{if } n = 0; \\ 1 & \text{if } n = 1; \\ F(n-1) + F(n-2) & \text{if } n > 1. \end{cases}$$

The point to be noted is that the recursive algorithm is different from the iterative algorithm. Iteration is simple repetition of a task till a condition is satisfied.

(b) Serial algorithm

Serial algorithms are designed with the assumption that computers will execute one instruction at a time. Those computers are called as *serial computers*. Algorithms designed for such an environment are called serial algorithms.

Analogy

Consider there is one burner in your kitchen and you are the only person to cook. Your task is to cook your dinner. Here you have no other choice than to cook alone on a single burner one by one.

(c) Parallel algorithms

Such algorithms take the advantage of *parallel computer architectures*. Here several processors can work on a problem at the same time. The various heuristic algorithms such as genetic algorithm fall into this category.

Analogy

Consider there are four burners in your kitchen and you are the only person to cook. Your task is to cook your dinner. Here you have multiple choices as follows:

- *Cooking one after another using one burner.*
- *Cooking all the meals one by one using two burners one after another.*
- *Cooking all the meals one by one using three burners one after another.*
- *Cooking all the meals one by one using all the four burners one after another.*
- *Cooking all the meals at a time using all the burners.*

D.2 Implementing Algorithms

In practice and also in the context of bioinformatics, algorithms are implemented in a software program development process using computer programs. Typically, when an algorithm is associated with a tool (software), data is read from an input device, written to an output device, and/or stored for further use. Stored data is regarded as part of the internal state or intermediate state of the entity performing the algorithm. There are cases where algorithms are implemented as a biological neural network such as human brain, or an insect locating food, or in electric circuits, or in a mechanical device.

The implementation of an algorithm involves execution of various *procedures* in a program which eventually finish. Sometimes procedures can run forever without stopping, based on some entity that is required to carry out such permanent tasks. In the latter case, success can no longer be defined in terms of halting with a meaningful output. Instead, terms of success that allow for unbounded output sequences must be defined.

> **Note**
>
> The analysis and study of algorithms is one discipline of computer science and is often practised abstractly without the use of a specific programming language or other implementation.

D.3 Biological Algorithms

Biological algorithms have been in use by Nature for millions of years. Some of the results of biological algorithms that we see around us are blooming of a flower, growing of a tree, healing of a wound, birth of a child, and many more which we can see in any living being from unicellular amoeba to animals, birds, fish, reptiles, and human beings. Let us look at one of the most simple and ubiquitous example of a biological algorithm, i.e., the replication of DNA in a cell. The replication process requires intricate and complex coordination between different parts of a cell and even molecules! The machine that undertakes the process is nothing but a protein complex known as DNA helicase. Once the helicase separates the two strands of DNA, a replication fork is created as shown in Fig D.1. The strands of DNA, of course, are antiparallel to each other.

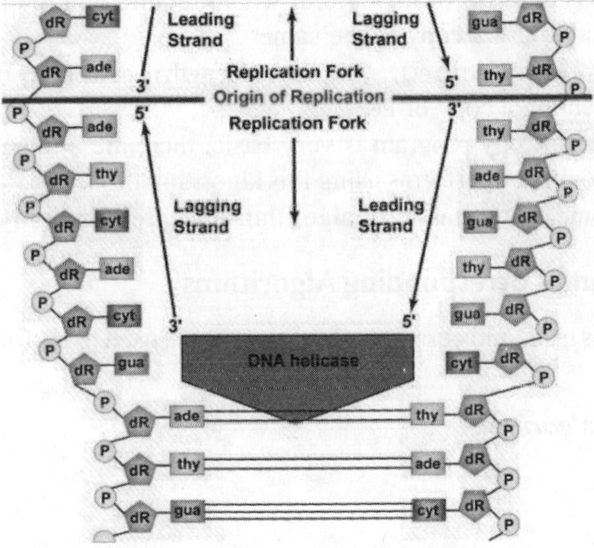

Fig. D.1 Replication forks are the actual site of DNA replication

At this moment two other molecular machines get operational. These are the topoisomerase and the single strand binding protein. Primers (short RNA strands), which are position specific, anchor and stabilize the open strands. Now a fourth molecular machine, DNA polymerase, binds to the open strand in the $3' \rightarrow 5'$ direction. The result is the formation of Okazaki fragments as shown in Fig. D.2.

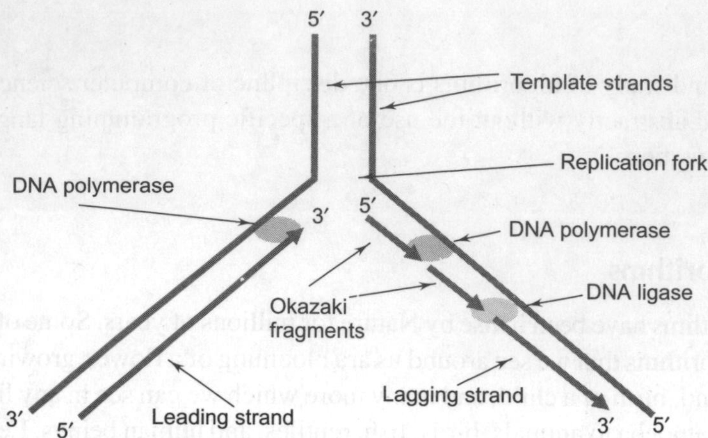

Fig. D.2 Formation of Okazaki fragments at a replication fork

Now another molecular machine, DNA ligase, starts working and seals all the gaps between the Okazaki fragments and forms a single, long DNA strand.

The complexity in this above biological algorithm cannot be reemphasized; however, if you were to put the process in terms of its logic, it would require you to only take the DNA strand as a string and return a copy of the same.

Input: A string D = (`baseA, baseT, baseG,…, baseC`) of length n, as an array of characters
Output: A string representing a copy of D

While this kind of a StringCopy program is very basic, there are a large number of operations that a normal computer performs using this algorithm.

Let us now examine some of the classes of algorithms that are commonly used.

D.4 Bioinformatics Tasks and Corresponding Algorithms

The various bioinformatics tasks and corresponding algorithms used to perform the tasks are as follows:

Sequence comparison algorithms

- Dot matrix
- Hamming distance
- Levenshtein distance

Substitution matrices algorithms

- PAM matrix construction algorithm
- BLOSUM matrix construction algorithm

Sequence alignment optimal algorithms

- Smith–Waterman algorithm
- Needleman–Wunsch algorithm

Sequence alignment heuristic algorithms
- BLAST algorithm
- FASTA algorithm

Prediction algorithm
- Gene/SNP/Promoter prediction
- Phylogenetic prediction
- Protein structure and function prediction

Multiple alignment algorithms
- Progressive algorithm
- Iterative algorithm

Machine learning
- Neural networks
- Hidden Markov models
- Nearest-neighbours method

D.5 Algorithms and Bioinformatics Software

Various bioinformatics software that implement different algorithms for different bioinformatics tasks are as follows:

Sequence comparison algorithms
- Dot matrix: Dotmatcher, Dotpath, Dotup, Polydot

Substitution matrices algorithms
- PAM matrix construction algorithm: PAM n Matrix; n = 100, 120, 160, 200, 250
- BLOSUM matrix construction algorithm: BLOSUM n; n = 90, 80, 60, 52, 45

Sequence alignment optimal algorithms
- Smith-Waterman algorithm: water
- Needleman-Wunsch algorithm: needle

Multiple sequence heuristic alignment algorithms
- BLAST algorithm: BLAST
- FASTA algorithm: FASTA

Prediction algorithm
- Gene/SNP/Promoter prediction: Genscan, GrailEXP
- Phylogenetic prediction: Phylip, Phyml
- Protein structure and function prediction: Prosite, 3DPSSM

In this part you will find chapters where you can learn algorithms for bioinformatics data analysis as well as various kinds of prediction algorithms involved in the field of bioinformatics.

8 Data Analysis Algorithms

Learning Objectives

- To understand and explore data analysis algorithms
- To learn about various types of data analysis algorithms
- To learn how to use data analysis algorithms for data analysis
- To learn about the use and application of data analysis algorithms

Introduction

In the *Algorithm* part you have learnt two basic approaches to classify algorithms. Here you will learn to classify algorithms on the basis of their applications to various bioinformatics tasks. The design of the algorithms depends on the need and complexity of the bioinformatics task on hand. The design methodology is the same as that for any other algorithm like divide and conquer, dynamic programming, and so on. Also on the basis of the machine on which the algorithm will get implemented, it can be designed to be a serial or parallel one. In this chapter we will discuss about different data analysis algorithms. If you recall, in *Tools* part we have learnt various data analysis tools. This chapter will describe the algorithms working behind those tools.

8.1 Sequence Comparison Algorithms

Sequence comparison algorithms deal with two sequences and the similarities between them. Sequences are compared to assign function to a new sequence, predict and construct model protein structures, and design and analyse gene expression experiments. The most common pairwise sequence comparison is done by dot plots.

8.1.1 Dot Plots

Dot plots have proved to be easy yet powerful means of sequence comparison. They have been generally employed to search for regions of similarity in two sequences and repeats within a single sequence. A dot plot is a visual representation of the similarities between two sequences and is likely to be the oldest way of doing so. The principle used to generate dot plots is simple. The top most X and the left most Y-axes of a rectangular array are used to represent the two sequences to be compared. One sequence is placed on the X-axis of a rectangular graph, and the other sequence is placed on the Y-axis. A dot is plotted at every coordinate where there is similarity between the bases. Adjacent regions of identity between the two sequences give rise to diagonal lines of dots in the plot. So, when two sequences are similar over a long stretch, a diagonal line will extend from the beginning of the region of similarity till the end of this region. By the same logic if two sequences are identical a diagonal line will stretch from one corner of the dot plot to the diagonally opposite corner.

Figure 8.1 shows a simple dot plot for the words CORRELATIONS and RELATIONSHIP.

	C	O	R	R	E	L	A	T	I	O	N	S
R			●	●								
E					●							
L						●						
A							●					
T								●				
I									●			
O		●								●		
N											●	
S												●
H												
I									●			
P												

Fig. 8.1 Dot plot of two English words

TASK

Input Try and construct a dot plot as in Fig. 8.2 for the following pair of sequences:

ACTCTAGGAGTCTCGAATTATC

GATAATTCGAGACTCCTAGAGT

Result

	A	C	T	C	T	A	G	G	A	G	T	C	T	C	G	A	A	T	T	A	T	C
G							•	•		•					•							
A	•					•			•							•	•			•		
T			•		•						•		•					•	•		•	
A	•					•			•							•	•			•		
A	•					•			•							•	•			•		
T			•		•						•		•					•	•		•	
T			•		•						•		•					•	•		•	
C		•		•								•		•								•
G							•	•		•					•							
A	•					•			•							•	•			•		
G							•	•		•					•							
A	•					•			•							•	•			•		
C		•		•								•		•								•
T			•		•						•		•					•	•		•	
C		•		•								•		•								•
C		•		•								•		•								•
T			•		•						•		•					•	•		•	
A	•					•			•							•	•			•		
G							•	•		•					•							
A	•					•			•							•	•			•		
G							•	•		•					•							
T			•		•						•		•					•	•		•	

Fig. 8.2 Dot plot of a pair of sequence

The *signal* is a line or otherwise visible pattern, which can be detected visually corresponding to a region of similarity. Our nucleic acid alphabet is a four-letter alphabet (as we treat T and U as equal). Thus implying that the *noise* or *accidental match* or *random chance* of an identical symbol at any given position is 1/4 = 0.25. The numbers of possible dots are too high when applied to weak similarities in nucleic acid or amino acid sequences. Therefore the formula for the number of possible dots is given as

$$\text{Number of possible dots (P)} = (\text{probability of pair}) \times (\text{length of sequence A}) \times (\text{length of sequence B})$$

Consider two sequences of lengths 20 and 18. If they are totally unrelated, we expect $0.25 \times 20 \times 18 = 90$ dots. To tone the so called *signal-to-noise ratio* we need to devise methodologies.

An improvement for the construction of dot plots with a lower signal-to-noise ratio is the usage of word method. For example, on a length of 28 base pairs, if there are 15 matches, the signal is relatively weak. To construct a dot plot, now instead of matching each nucleotide let us try to match words. Here instead of matching and putting a point wherever two letters match, we now can put a dot only when words (called *oligomers*) match. This reduces the chance of a random match. If we use bi-nucleotides, noise will be (1/4)*(1/4) = 1/16 = 6.25% which is already much lower than 25% obtained earlier. A popularly used program suite called GCG uses the default word size of 6—this is (1/4) to the power of 6, which results in a random choice probability of 0.025%. A natural disadvantage of using word match is that the longer the word size, lower the probability that a given word matches in between two different sequences. This sensitivity problem may however be overcome with the permission of mismatches in a word. To do this, you will need to apply the *windows technique*. To derive meaningful dot plots with the windows technique, you must set two parameters: window size and mismatch limit. The GCG program suite (http://www.accelrys.com) calls this *stringency*. Using this window/stringency algorithm, dots can be plotted in the middle of a window rather than a word. Note that you can set window size according your goal of analysis. They can be as follows:

- size of average exon
- size of average protein structural element
- size of gene promoter
- size of enzyme active site

Window size It is the number of bases that is considered and compared to generate a single data point on the plot. The bases under consideration are said to be in a sliding window, which is slid along the sequence during the comparison.

Mismatch limit It determines how similar the two sequences in a window must be to 'match'. For example, if the window size is 9 and the mismatch limit is 2, then up to 2 mismatches in a 9 base window will still be classified as a match.

Method of windows technique A window of predetermined length is usually fixed, together with other parameters, when a sequence window in the first sequence is said to be similar with the other window in the second sequence. Whenever one window in one sequence satisfies the similarity criteria, a dot or short diagonal is drawn at the corresponding position in the array. For example, when a window size of 8 and similarity cut-off 5 is considered, first the nucleotides 1-8 of the X-axis sequence are compared with nucleotides 1-8 of the sequence along the Y-axis. Out of 8 bases if 5 or more bases are identical, a dot is placed in the (1,1) position. Next the window is considered from 2-9 by moving it by 1 base along the X-axis and compared with 1-8 nucleotides of the Y-axis. This is repeated till all the bases of the X-axis are compared. Then the window along the Y-axis is advanced by one base and the process is repeated.

The sliding window reduces noise in the dot plot and improves the visibility of the pattern, showing significant similarity between the two sequences. There is no predecided value that one can set for the window size and cut-off score. It varies depending on the compared pair of sequences and can be set with a trial-and-error approach. It is thus possible to adjust the stringency of the match by adjusting the window size and mismatch limit. A large window of comparison and low mismatch limit yields a stringent comparison.

Example 8.1 Let us construct a dot plot as in Fig. 8.3 for the following pair of sequences using a sliding window of size 4 and a similarity cut-off of 3 nucleotides.

GCTAGTCAGATCTGACGCTA

GATGGTCACATCTGCCGC

	G	C	T	A	G	T	C	A	G	A	T	C	T	G	A	C	G	C	T	A
G									•					•						
A																				
T			•																	
G				•					•											
G					•															
T						•							•							
C							•													
A								•												
C									•											
A										•										
T						•					•									
C												•								
T													•							
G														•						
C															•					
C																•				
G																				
C																				

Fig. 8.3 Dot plot of a pair of sequences

Does this plot reveal any regions of similarity between the two sequences?

A region of strong identity is revealed starting at position 3 of each sequence and continuing until nearly the end of the two sequences.

TASK

Let us try to construct a dot plot for the following pair of sequences:

GCTAGTCAGATCTGACGCTA

GATGGTCACATCTGCCGC - -

	G	C	T	A	G	T	C	A	G	A	T	C	T	G	A	C	G	C	T	A
G	•				•				•					•			•			
A				•				•		•					•					•
T			•			•					•		•						•	
G	•				•				•					•			•			
G	•				•				•					•			•			
T			•			•					•		•						•	
C		•					•					•				•		•		
A				•				•		•					•					•
C		•					•					•				•		•		
A				•				•		•					•					•
T			•			•					•		•						•	
C		•					•					•				•		•		
T			•			•					•		•						•	
G	•				•				•					•			•			
C		•					•					•				•		•		
C		•					•					•				•		•		
G	•				•				•					•			•			
C		•					•					•				•		•		

Fig. 8.4 Dot plot of a pair of sequences

The dot plot looks like as shown in Fig. 8.4. Does this dot plot reveal any regions of similarity?

Looking at the dot plot above, we can infer the following:

- It is possible to display the relationship between two sequences in a graphic fashion.
- The result is rather chaotic.
- Regions of similarity appear as diagonal runs of dots.
- No direct sequence homologies can be derived from this kind of plot.
- Reverse diagonals (perpendicular to diagonal) indicate inversions.
- Reverse diagonals crossing diagonals (X's) indicate palindromes.

(a) Uses of dot plots

Dot plots have many uses. Inversions, duplications, and palindromes have unique 'signatures' in dot matrices. Dot plots can be used

- to align two proteins or two nucleic acid sequences.
- to find amino acid/nucleic acid repeats within a sequence by comparing a sequence to itself. Repeats appear as a set of diagonal runs stacked vertically and/or horizontally.
- to find self-base pairing of RNA (e.g., tRNA) by comparing a sequence to itself complemented and reversed.
- as an excellent approach for finding sequence transpositions.
- to find location of genes between two genomes.
- to find non-sequential alignments.

(b) Drawbacks of Dot Plots

Dot plots have the following drawbacks:

- A problem with dot matrices for long sequences is that they can be very noisy due to lots of insignificant matches.
- Dot plots are not alignments:

Program	Features
Dotmatcher	Uses a threshold, calculated from a substitution matrix (e.g., Blosum) to define whether or not a match is plotted. A window of a specified length moves up all possible diagonals and a score is calculated within each window for each position. The score is the sum of the comparison of the two sequences using the given similarity matrix along the window. If the score is above the threshold, then a linc is plotted on the image over the position of the window (2).
Dotpath	Displays a non-overlapping word-match dotplot of two sequences. It is very similar to Dottup. Both programs look for places where words, called *tuples*, of a chosen length have an exact match, which is represented by a diagonal line over the position of these words. Using a longer word size displays less random noise, runs extremely quickly, but is less sensitive.
Dottup	It is nearly identical to Dotpath. It looks for places where words (tuples) of a specified length have an exact match in both sequences.

It draws a diagonal line over the position of these words. Unlike Dotpath, overlaps are not significant disqualifiers. It is fast, but is not a sensitive way of creating dotplots.

Polydot — It compares all sequences in a set of sequences and draws a dot plot for each pair of sequences by marking where tuples of a certain length have an exact match in both sequences. It should be used when sequences are to be compared exactly one to one.

(c) Some weblinks for construction of dot plots

The Swiss Institute for Bioinformatics provides a JAVA applet that performs interactive dot Plots. The applet to calculate a dot plot from two amino acid sequences is also available at
http://www.isrec.isb-sib.ch/java/dotlet/Dotlet.html
Dotmatcher is available at
http://bioweb.pasteur.fr/seqanal/interfaces/dotmatcher.html.

Alignments

Dot plots are good for visual representation and are applicable for short sequences. But for very long sequence, sequence similarity is evaluated based on a numeric scoring system to determine optimal alignment. Similarity between two sequences is described with this scoring value, which determines whether or not the sequences share a common ancestor. True alignment of a nucleotide or amino acid sequence is one that reflects the evolutionary relationship between two or more homologues (sequences that share a common ancestor).

In a sequence there are three kinds of changes that occur at any position: mutation that replaces the unit of the sequence, an insertion that adds one or more positions, or deletion that deletes one or more position. Insertion and deletion occur at a lower frequency than mutations. Also compared sequences are not homologues when they have inserted or deleted nucleotides.

(i) When such sequences are aligned where one is shorter in length than the other, it can align in a number of ways by changing the position of the shorter sequence. For example, two sequences TTACGAGA and TTCAGA can be aligned in three different ways as shown below:

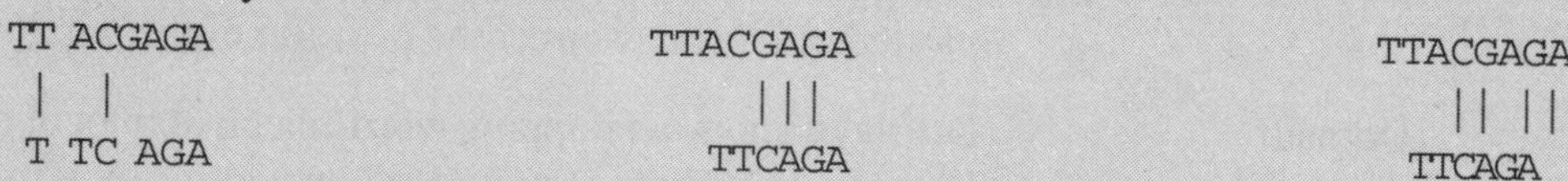

For such an alignment the scoring function is calculated by the total amount of credits earned for aligned pairs of identical residues (i.e., the match score) and the total amount of penalties for aligned pairs of non-identical residues (i.e., the mismatch score).

$$\text{Score} = \sum_{i=1}^{n} \begin{cases} \text{match score; if } seq1_i = seq2_i \\ \text{mismatch score; if } seq1_i \neq seq2_i \end{cases}$$

Let us assume that the match score is 1 and mismatch score is 0. Then the scores for above alignments are 2, 3, and 4 respectively.

(ii) To cater to the occurrences of insertion and deletion, gaps are introduced in sequences during alignment. Gaps increase the number of possible alignments, out of which three possible alignments are shown below:

TTACGAGA	TTACGAGA	TTACGAGA
TTCAGA - -	TTC--AGA	TT-C-AGA

For such an alignment, the scoring function is calculated by the total amount of credits earned for aligned pairs of identical residues (i.e., the match score), the total amount of penalties for aligned pairs of non-identical residues (i.e., the mismatch score), and the total amount of penalties for aligned pairs of a residue and a gap (i.e., the gap penalty).

$$\text{Score} = \sum_{i=1}^{n} \begin{cases} \text{match score; if } seq1_i = seq2_i \\ \text{mismatch score; if } seq1_i \neq seq2_i \\ \text{gap penalty; if } seq1_i = \text{'-' or } seq2i = \text{'-'} \end{cases}$$

Let us assume that the match score is 1, mismatch score is 0, and gap penalty is –1: The scores for above alignments will be 2, 3, and 4, respectively.

(iii) Sometimes gap penalties as used above give optimal alignments with the same score. In that case gaps can be differentiated as isolated gaps and gaps in a longer sequence. When two sequences of unequal lengths are considered for alignment, the difference in lengths can be due to insertion in the longer sequence, deletion in the shorter sequence, or a combination of these two. Such events are commonly referred to as insertion/deletion or indel events.

Statistically the alignment is optimal when the difference is due to a single n-unit indel than multiple insertion and deletions. Here the gap penalty is assigned in two ways: an origination penalty with a higher value and a length penalty with a lower value. The origination penalty is the penalty for starting any new gap and the length penalty for subsequent gaps. Thus considering the previous example, let us assume that the match score is 1, mismatch score is 0, origination penalty is –2, and length penalty is –1. The scores for the alignment are –1, 2, and 2, respectively. Thus in the previous example, though the third alignment was preferred with a high score of 4, here both the second and the third have a similar score of 2 and as per the rule of preference, the second alignment is preferred over third.

(iv) Depending upon the mutation where bases and amino acids undergo substitution, the mismatch penalty can be different. In the case of a conservative substitution where a mutated sequence behaves functionally similar to that of the ancestor sequence, the mismatch penalty is low. On the other hand, in the case of a substitution that change the functionality, the mismatch penalty is high. The values for such substitutions are looked up in the scoring matrices.

- *In the case of a DNA sequence*: Here the scoring matrices are quite simple. The one we already used in our example before is Identity Matrix, where identical residues are scored as 1 and non-identical residues are scored as 0. In the case of BLAST Matrix, the match score is 5 and the mismatch score is –4. In the case of transition –transversion matrix, the match score is 1, mismatch score is –1 in the case of transition, i.e, purine–purine or pyrimidine–pyrimidine substitution and –5 in the case of purine–pyrimidine or vice versa.
- *In the case of an amino acid sequence*: Here the scoring matrices are based on several criteria. Thus common criteria are physical/chemical similarities such as hydrophobicity, charge, electronegativity, size, observed substitution frequency, genetic code, etc.

8.2 Substitution Matrices Algorithms

Scoring matrices are also known as substitution matrices. These are tools that are extensively used to model mutations in sequences as discussed before. The most common method of creating a substitution matrix in the case of amino acids is to observe the actual substitution rates among the various amino acid residues in nature. The score is favourable if substitution is observed frequently. On the other hand, the alignment for a pair of residues is penalized if substitution is not observed frequently. Consider an alignment of protein sequences where aligned amino acid residues that are identical are scored as 1 and that are non-identical are scored as 0. The score using this rule only looks at the degree of the identical nature of the residues. It does not show how similar those proteins are with respect to their structure and function. Thus to make the amino acid alignment and scoring more significant, matrices are developed that score mutation among amino acids with similar physicochemical properties as shown in Fig. 8.5.

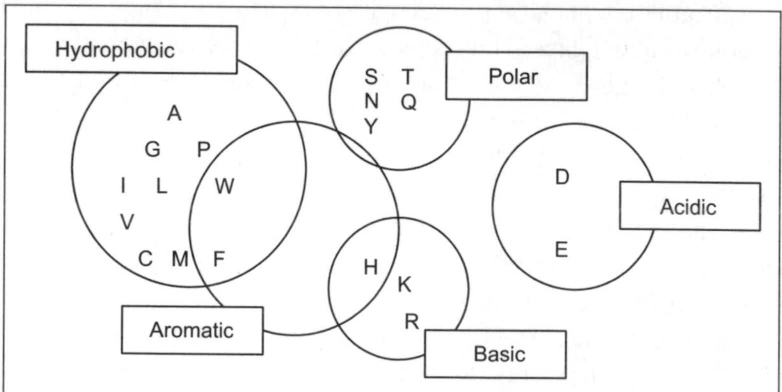

Fig. 8.5 Relationships between physicochemical properties of amino acids

Two most commonly used amino acid substitution matrices are PAM and BLOSUM.

8.2.1 PAM Matrix Construction Algorithm

PAM (point or percent accepted mutations) matrices came into being largely due to the effort of Margaret Dayhoff and colleagues. The scores in a PAM matrix are computed by observing the mutations that are accepted by natural selection that occurred in alignment between similar sequences. PAM is roughly defined as amino acid mutations observed in three major protein families (cytochrome c, hemoglobin, myoglobin, etc.) with sequences of at least 85% similarity. The mutated amino acids that are accepted in a particular sequence during evolution must have similar physical and chemical properties so as to preserve the function of the protein(s).

PAM matrices are based on the evolutionary model from alignments of closely related sequences. These are used to predict the probability of an amino acid to mutate to another amino acid over a given time period. One PAM unit represents an evolutionary distance equivalent to an average change in 1% of all amino acid positions. PAM1 is a scoring system for sequences in which 1% of the residues have undergone mutation. Therefore the PAM256 matrix represents 256% mutations, i.e., an average of 2.5 accepted mutations per residue, a high divergence of very distant relationship. The PAM matrix that is appropriate for a given sequence alignment depends on the length of the sequences and on how closely the sequences are believed to be related.

Note

- PAM1 is used to compare sequences that are closely related and represent shorter evolutionary distance.
- PAM1000 is used to compare sequences with distant relations.
- PAM256 is commonly used in practice as it is not possible to know the evolutionary distance.

Dayhoff and her coworkers used 814 accepted amino acid mutations of the protein sequences to derive the PAM matrix shown in Fig. 8.6.

The log odds amtrix for PAM 256 is as shown in Fig. 8.7.

In the figure you can note that the score for aligning similar amino acids such as L and I or D and E is high, 2 and 3, respectively. These show a high likelihood of substitution as an evolutionary process. On the other hand, the score for D and C is –5, for D and K is 0. You can also note that scores for identical amino acids are different W and W is 17, N and N is 2. This is due to the frequency of occurrence of the amino acids in natural protein sequences. We can thus conclude that the alignment of an uncommon acid, i.e., W is higher than the alignment of a common acid, i.e., S. However, PAM matrices suffer from several drawbacks, which have led to a decline in their use. Some of these are as follows.

- PAM matrices assume that all the amino acids mutate at the same rate.
- The dataset that PAM matrices use was constructed in 1978 and is dated and makes it a rather a 'restrained dataset'. Some of the later substitution matrices such as BLOSUM use larger and use more relevant datasets such as BLOCKS.

- The majority of the protein sequences used to construct PAM matrices are biased, i.e., they represent small globular proteins.

Yet PAM matrices are still used where the emphasis is on understanding how they work.

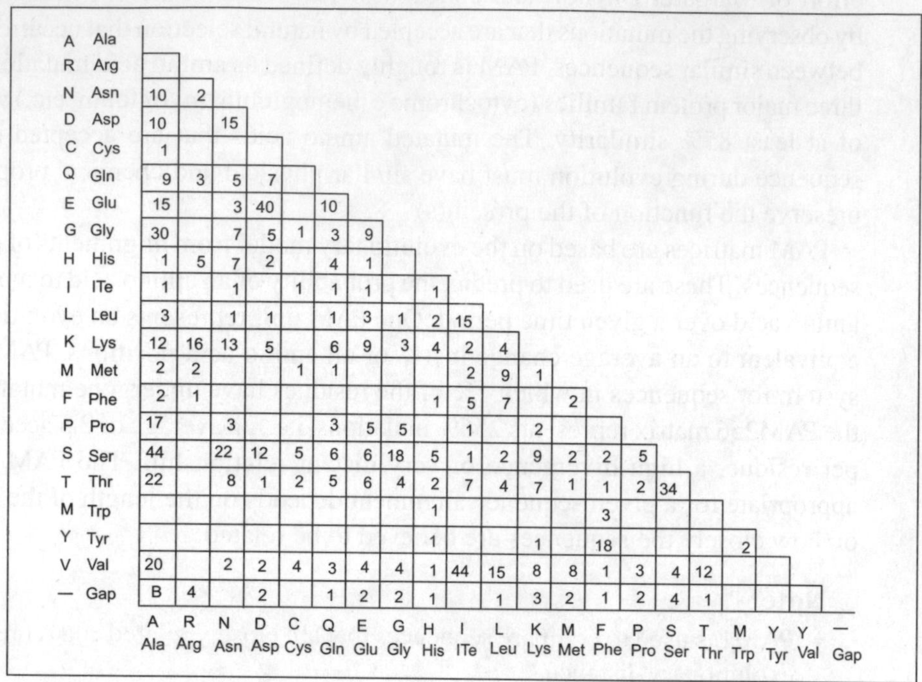

	A Ala	R Arg	N Asn	D Asp	C Cys	Q Gln	E Glu	G Gly	H His	I ITe	L Leu	K Lys	M Met	F Phe	P Pro	S Ser	T Thr	M Trp	Y Tyr	Y Val	— Gap
A Ala																					
R Arg																					
N Asn	10	2																			
D Asp	10		15																		
C Cys	1																				
Q Gln	9	3	5	7																	
E Glu	15		3	40		10															
G Gly	30		7	5	1	3	9														
H His	1	5	7	2		4	1														
I ITe	1		1		1	1	1		1												
L Leu	3		1	1		1	3	1	2	15											
K Lys	12	16	13	5		6	9	3	4	2	4										
M Met	2	2			1	1				2	9	1									
F Phe	2	1							1	5	7		2								
P Pro	17		3			3	5	5		1		2									
S Ser	44		22	12	5	6	6	18	5	1	2	9	2	2	5						
T Thr	22		8	1	2	5	6	4	2	7	4	7	1	1	2	34					
M Trp									1					3							
Y Tyr									1			1		18				2			
V Val	20		2	2	4	3	4	4	1	44	15	8	8	1	3	4	12				
— Gap	B	4		2		1	2	2	1		1	3	2	1	2	4	1				

Fig. 8.6 PAM matrix developed by Dayhoff

	A	C	D	E	F	G	H	I	K	L	M	N	P	Q	R	S	T	V	W	Y
A	2	−2	0	0	−4	1	−1	−1	−1	−2	−1	0	1	0	−2	1	1	0	−6	−3
C		12	−5	−5	−4	−3	−3	−2	−5	−6	−5	−4	−3	−5	−4	0	−2	−2	−8	0
D			4	3	−6	1	1	−2	0	−4	−3	−2	−1	2	−1	0	0	−2	−7	−4
E				4	−5	0	1	−2	0	−3	−2	1	−1	2	−1	0	0	−2	−7	−4
F					9	−5	−2	1	−5	2	0	−4	−5	−5	−4	−3	−3	−1	0	7
G						5	−2	−3	−2	−4	−3	0	−1	−1	−3	1	0	−1	−7	−5
H							6	−2	0	−2	−2	2	0	3	2	−1	−2	−2	−3	0
I								5	−2	2	2	−2	−2	−2	−2	−1	0	4	−5	−1
K									5	−3	0	1	−1	1	3	0	0	−2	−3	−4
L										6	4	−3	−3	−2	−3	−3	−2	2	−2	−1
M											6	−2	−2	−1	0	−2	−1	2	−4	−2
N												2	−1	1	0	1	0	−2	−4	−2
P													6	0	0	1	0	−1	−6	−5
Q														4	1	−1	−1	−2	−5	−4
R															6	0	−1	−2	2	−4
S																2	1	−1	−2	−3
T																	3	0	−5	−3
V																		4	−6	−2
W																			17	0
Y																				10

Fig. 8.7 The log odds matrix for PAM256

> **Note**
>
> To obtain matrices of different PAM numbers, for example, a PAM2 matrix, simply multiply a PAM1 by itself. Similarly, to obtain a PAM256 matrix, simply multiply PAM1 against itself 256 times. A PAMX is obtained by multiplying PAM1 X times.

Creation of a PAM matrix

Step 1 Construct multiple sequence alignments between sequences with a high similarity score (i.e., > 85%).

Step 2 Construct a phylogenetic tree from the aligned sequences that show the order of various substitutions.

Step 3 Compute the relative mutability (m_j) for each amino acid.

Step 4 Compute the relative mutability divided by the total number of mutations and multiplied by the frequency of amino acid and a scaling factor of 100.

Step 5 Compute substitution tally $A_{i,j}$.

Step 6 Calculate mutation probability $M_{i,j}$ for each pair of amino acid.

$$M_{i,j} = m_j A_{i,j} / \sum_i A_{i,j}$$

$\Sigma A_{i,j}$ is the total number of substitutions involving the residue in the phylogenetic tree.

Step 7 Compute mutation probability $M_{i,j}$ divided by the frequency of occurrence f_i of residue (i) and then calculate $R_{i,j}$ by taking the log of the resulting value for each entry of the PAM matrix. This is repeated to compute values for non-diagonal entries in the PAM matrix.

Step 8 Diagonal entries in the PAM matrix are computed by taking $M_{i,j} = 1 - m_j$ and then following step 6.

> **Note**
>
> - Relative mutability is equal to the number of times the amino acid is substituted by any other amino acid in the phylogenetic tree.
> - The total number of mutations affecting the residue is the total number of substitutions across the entire tree multiplied by 2.
> - Scaling factor is 100 to indicate PAM1 will represent 1 substitution per 100 residues.
> - Substitution tally $A_{i,j}$ is the number of times the amino acid j is replaced by other amino acid i. This is tallied for each amino acid pair i,j. It is assumed that substitutions are equally likely in each direction.
> - Mutation probability is defined as the probability of an amino acid to mutate over a certain time interval. Higher the relative mutability, more probable is the mutation of an amino acid.
> - Frequency of occurrence is the number of occurrences of a residue in the multiple alignment divided by the total number of residues.

However, biological sequences are rather unpredictable. Amino acids have different biochemical and physical properties that influence their relative replaceability in evolution. There are certain 'hot spots' that are more mutable. Some amino acids that are very amenable to mutation are Asn, Asp, Glu, and Ser; whereas Cys and Trp are the least mutable. The familiarity with the pattern and probability of changes that are likely to occur during molecular evolution allows bioinformaticians to build amino acid scoring matrices and produce adequate sequence alignments too.

Since the entries of the matrix are build using a logarithm, we can conclude the following:

- Amino acids with values greater than 0 mutate more often than expected by chance.
- Values lesser than 0 mutate less often than predicted by chance.
- Values equal to 0 mutate as predicted by chance.

TASK

Work out the probability of substitution of an isoleucine into a valine using a PAM256 substitution matrix.

Results

S(isoleucine,valine) = 10 × log at the base 10(0.07/0.035) = 3.01 ~ = 3.

In this example, the mutation is two times more likely than expected by chance:

log at the base 10(x) = 0.3.

10^(0.3) = x

x = 2

Going by the same logic:

- 60PAMS ~60% sequence similarity
- 80PAMS ~50% sequence similarity
- 120PAMS ~40% sequence similarity
- 250PAMS ~20% sequence similarity

8.2.2 BLOSUM Matrix Construction Algorithm

PAM-like matrices derive their initial substitution frequencies from global alignments of very similar sequences. Henikoff and Henikoff adopted an alternative approach. They used local multiple alignments of more distantly related sequences to create BLOSUM matrices that will help to model protein sequences having a lesser degree of divergence.

BLOSUM matrices are derived from PROSITE signatures of the BLOCKS database. The BLOCKS database is a set of ungapped local multiple alignments of sequence regions from families of related proteins. It is thus a database of tightly conserved regions at all evolutionary distances. BLOSUM matrices are based on 2000 BLOCKS coming from more than 500 protein families. BLOCKS are conserved regions of protein families that do not contain insertions or deletions. They are located in the BLOCKS database. It uses a clustering approach that sorts the sequences in each block into closely related groups. The clustering was done wherever sequences that were similar at some threshold value of percentage were identified. Substitution frequencies for all pairs of amino acids are calculated between the groups. These frequencies are used to calculate a log odds BLOSUM (BLOcks SUbstitution Matrix) matrix. Different matrices are obtained by varying the clustering threshold. For example, the BLOSUM80 matrix is derived using a threshold of 80% identity.

BLOSUM is used to get a better measure of differences between two proteins specifically for more distantly related proteins. Thus though BLOSUM is widely used, it has a narrower range than the PAM matrix. While this bias limits the use of BLOSUM matrices for some purposes, for other well used programs such as FASTA, and BLAST it is a perfect approach since an accurate measure of distance is not really required when peptides are closely related.

Different levels of the BLOSUM matrix can be created by differentially weighting the degree of similarity between sequences. BLOSUM62 matrix is calculated from protein blocks such that if two sequences are more than 62% identical, then the contribution of these sequences is weighted to sum to 1. In this way the contribution of multiple entries of closely related sequences is reduced. Refer BLOSUM62 in Fig. 8.8.

> **Note**
>
> When the BLOSUM62 matrix is compared to PAM160, it is found that it is less tolerant of substitutions to or from hydrophilic amino acids, while being more tolerant of hydrophobic changes and of cysteine and tryptophan mismatches.

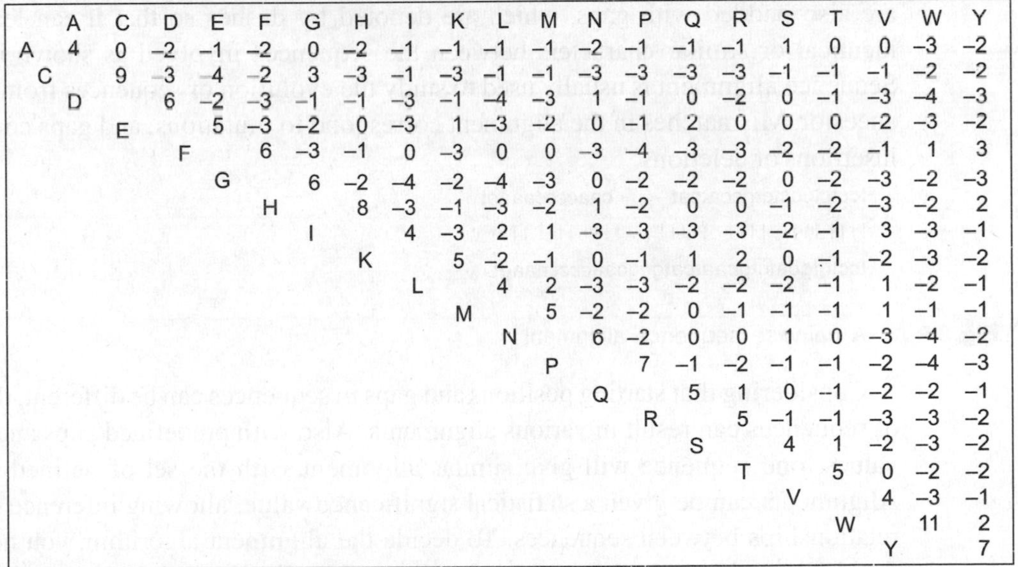

	A	C	D	E	F	G	H	I	K	L	M	N	P	Q	R	S	T	V	W	Y
A	4	0	−2	−1	−2	0	−2	−1	−1	−1	−1	−2	−1	−1	−1	1	0	0	−3	−2
C		9	−3	−4	−2	3	−3	−1	−3	−1	−1	−3	−3	−3	−3	−1	−1	−1	−2	−2
D			6	−2	−3	−1	−1	−3	−1	−4	−3	1	−1	0	−2	0	−1	−3	−4	−3
E				5	−3	−2	0	−3	1	−3	−2	0	−1	2	0	0	0	−2	−3	−2
F					6	−3	−1	0	−3	0	0	−3	−4	−3	−3	−2	−2	−1	1	3
G						6	−2	−4	−2	−4	−3	0	−2	−2	−2	0	−2	−3	−2	−3
H							8	−3	−1	−3	−2	1	−2	0	0	−1	−2	−3	−2	2
I								4	−3	2	1	−3	−3	−3	−3	−2	−1	3	−3	−1
K									5	−2	−1	0	−1	1	2	0	−1	−2	−3	−2
L										4	2	−3	−3	−2	−2	−2	−1	1	−2	−1
M											5	−2	−2	0	−1	−1	−1	1	−1	−1
N												6	−2	0	0	1	0	−3	−4	−2
P													7	−1	−2	−1	−1	−2	−4	−3
Q														5	1	0	−1	−2	−2	−1
R															5	−1	−1	−3	−3	−2
S																4	1	−2	−3	−2
T																	5	0	−2	−2
V																		4	−3	−1
W																			11	2
Y																				7

Fig. 8.8 The log odds matrix for BLOSUM 62

A Comparison of PAM and BLOSUM matrices is given below.

BLOSUM	**PAM**
No model	Explicit evolutionary model
Larger dataset	
Alignment of conserved domain	Alignment of entire protein
BLOSUMX	PAMX
Bigger x, less diverged	Bigger x, more diverged

Tips on choosing a matrix

- Generally, BLOSUM matrices perform better than PAM matrices for local similarity searches.
- When comparing closely related, proteins, one should use lower PAM or higher BLOSUM matrices; for comparing distantly related proteins, use higher PAM or lower BLOSUM matrices.
- The commonly used matrix for database searching is BLOSUM62.

8.3 Sequence Alignment Optimal Algorithms

Sequence alignment is the most common way of comparing biomolecules such as DNA, RNA, or amino acids. The process involves constructing as well as finding significant alignments in a database of potentially unrelated sequences. Here non-exact sequences of biomolecules are matched and similarity is highlighted. These sequences are strings of bases and amino acids which are made to line up against each other. These sequences are also padded with gaps, which are denoted by dashes so that it can lead to more identical or similar characters between the sequences involved as shown in Fig. 8.9. Sequence alignment is usually used to study the evolution of sequences from a common ancestor. Mismatches in the alignment correspond to mutations, and gaps correspond to insertions or deletions.

```
tcctctgcctctgccatcat - - - caaccccaaagt
|||| |||| |||| |||||    |||| |||| ||||
tcctgtgcatctgcaatcatgggcaaccccaaagt
```

Fig. 8.9 A pairwise sequence alignment

Considering that starting positions and gaps in sequences can be different, the same set of sequences can result in various alignments. Also with predefined gaps and mismatch values, one sequence will give similar alignment with the set of defined sequences. Alignments can be given a statistical significance value, allowing inference of possible relationships between sequences. To decide the alignment algorithm, you need to first decide about the method. Searching for all possible alignments is impractical. With DNA sequences of sizes 100 bp and 95 bp, we can have 5 gaps in the shorter sequence at any place. This combination will give 55 million possible alignments! As the length of the sequence grows, the number of possible alignments will be impossible to compute. The problem can be solved using dynamic programming, which employs a method of breaking a problem into reasonable subproblems and using these partial problems to compute the first answer. Usually, alignment methods fall within one of the two categories, local and global sequence alignments. Also both of these categories can be of pairwise or multiple type.

8.3.1 Pairwise Alignment

Pairwise sequence alignment methods are concerned with finding the best matching in local or global alignments of protein (amino acid) or DNA/RNA (nucleic acid) sequences. The most important application of pairwise alignment is the identification of sequences of unknown structure or function. Another important use is the study of molecular evolution. Here the purpose is to find related, i.e., homologues, of a gene or gene product in a database of known examples. Here we will consider pairwise alignment in both global and local alignments.

(a) Global sequence alignment

The goal in this type of sequence alignment is to discover the overall relationship between sequences. To do this, global sequence alignments span the entire length of the sequences being compared. Here, all the characters in both sequences participate in the alignment. Global alignments are useful mostly for finding closely related sequences where sequences are expected to be similar across their entire lengths. Though these sequences are easily identified by local alignment methods, here it helps to locate only highly similar sequences. The algorithm used in global alignment is the Needleman–Wunsch algorithm. It uses the following tools for the purpose of global alignment of two sequences:

- LALIGN
- EMBOSS align

Needleman–Wunsch algorithm The Needleman–Wunsch algorithm uses a dynamic programming approach to obtain an optimal global alignment of two sequences. There are three steps in dynamic programming using the Needleman–Wunsch algorithm. These are discussed next.

Initialization step In this step, two sequences of lengths x and y are considered for alignment as an input. The program creates a matrix with (x + 1) columns and (y + 1) rows. For example, let us do a global sequence alignment using the Needleman–Wunsch technique with the sequences as shown below:

Sequence 1: G A A T T C A G T T A of length 11

`Sequence 2: G G A T C G A` of length 7

The scoring scheme assumed is

- $S_{i,j} = 1$ for the match score if the residue at position i of sequence 1 is the same as the residue at position j of the sequence

- $S_{i,j} = 0$ for the mismatch score
- $w = 0$ for the gap penalty

Matrix fill step The next step in the program is to fill the matrix. The value for each position of the matrix position $M_{i,j}$ is the information about the maximum global alignment score for that position computed with the help of a recurrence relation specific to this algorithm. The positions of the first row (i.e., $M_{0,j}$) and the first column (i.e. $M_{i,0}$) are filled with 0's, as in these cases each residue in the sequence is actually compared with nothing as shown in Fig 8.10. Therefore actual scoring is done starting from $M_{1,1}$, the upper left hand corner of the matrix M and then the row and column corresponding to that position is filled. This way the maximal score $M_{i,j}$ for each position in the matrix is filled. To find $M_{i,j}$ you need to know the score for the left $M_{i-1,j}$, above $M_{i,j-1}$, and the top left diagonal $M_{i-1,\, j-1}$ matrix positions to i,j. Here the example is colour coded for your convenience. You will find $M_{i-1,j-1}$ will be red, $M_{i,j-1}$ will be green, and $M_{i-1,j}$ will be blue. Thus the maximum score $M_{i,j}$ at position i,j is as shown below:

$$
\begin{aligned}
M_{i,j} = \text{MAXIMUM}[\;& \\
& M_{i-1,\, j-1} + S_{xi,yj} \text{ (match/mismatch in the diagonal)}, \\
& M_{i,j-1} + w \text{ (gap in sequence 1)}, \\
& M_{i-1,j} + w \text{ (gap in sequence 2)}]
\end{aligned}
$$

where $M_{i,j}$ represents the value at the (i,j) position of the matrix and $S_{xi,yj}$ represents the value obtained from the substitution matrix for the amino acids x and y, corresponding to the (i,j) position of the 2D matrix. It is also denoted as $S_{i,j}$.
w represents the value of the *gap penalty.*

Note

Sequences can contain gaps, which are penalized during an alignment. The affine gap penalty function is normally given as

$$f(n) = w + e(n - 1)$$

where n is the length of the gap, w is the opening gap penalty, and e is the extension gap penalty.

Thus, a gap of length 3 (n = 3) having –10 as a gap opening penalty and –2 as a gap extension penalty will have an overall penalty score of

$$f(3) = -10 + e(2) = -10 - 4 = -14.$$

The recurrence relation written above fills every (i,j) position of the 2D matrix with values computed using the function. If the scoring scheme assumed before is negative, the maximum score of a matrix position can also be negative. Here we have assumed 1 or 0 for match, mismatch, and gaps. Using this information, the score $M_{1,1}$ at position 1,1 in the matrix can be calculated. Since the first residue is G in both the sequences, $S_{1,1} = 1$. Also there is no gap in both the sequences, so w = 0. Thus, $M_{1,1}$ = MAX[$M_{0,0}$ + 1, $M_{1,0}$

+ 0, $M_{0,1}$ + 0] = MAX[1, 0, 0] = 1. Since w is 0, the values for the rest of the positions in row 1 and column 1 are 1 as shown in Fig. 8.11.

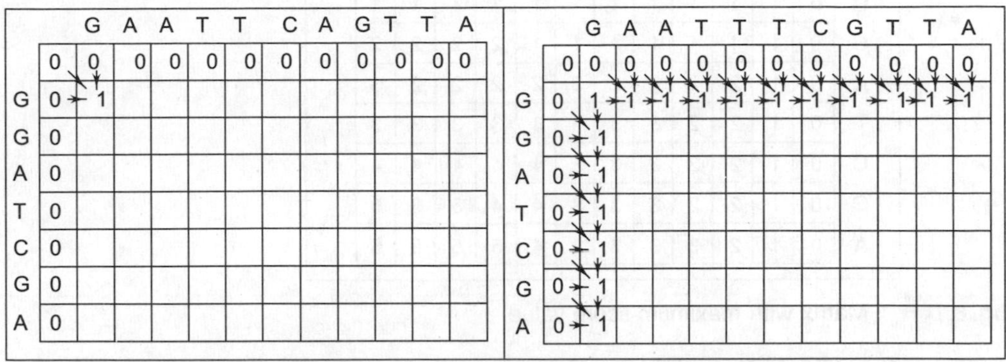

Fig. 8.10 Matrix with values filled in the 0th row and column as well as position 1,1

Fig. 8.11 Matrix with values filled in the 1st row and column

Now let us consider column 2:

$$M_{2,2} = \text{MAX}[M_{1,1} + 0, M_{2,1} + 0, M_{1,2} + 0] = \text{MAX}[1, 1, 1] = 1$$

$$M_{3,2} = \text{MAX}[M_{2,1} + 1, M_{3,1} + 0, M_{2,2} + 0] = \text{MAX}[2, 1, 1] = 2$$

Using the same method you can fill all the positions for columns 2 and 3 as shown in Figs 8.12 and 8.13, respectively.

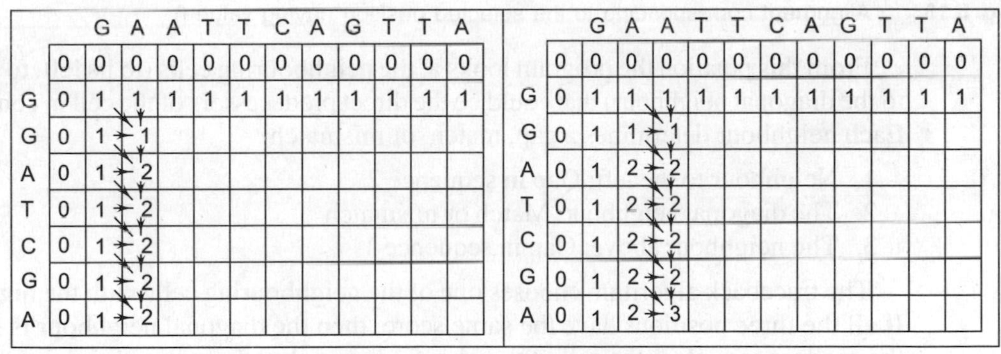

Fig. 8.12 Matrix with values filled in the 2nd column

Fig 8.13 Matrix with values filled in the 3rd column

Likewise, other positions of the matrix are computed. The filled matrix is as shown in Fig. 8.14. You can see that the maximum alignment score is 6 for the two test sequences.

		G	A	A	T	T	C	A	G	T	T	A
	0	0	0	0	0	0	0	0	0	0	0	0
G	0	1	1	1	1	1	1	1	1	1	1	1
G	0	1	1	1	1	1	1	1	2	2	2	2
A	0	1	2	2	2	2	2	2	2	2	2	3
T	0	1	2	2	3	3	3	3	3	3	3	3
C	0	1	2	2	3	3	3	4	4	4	4	4
G	0	1	2	2	3	3	3	4	4	5	5	5
A	0	1	2	3	3	3	3	4	5	5	5	6

Fig. 8.14 Matrix with maximum score value

Traceback step This step determines the actual alignment(s) that result in the maximum score. It begins in the $M_{(x+1),(y+1)}$ position in the matrix that has the maximal score. For example, here $M_{11,7}$ is the position with the maximum score value of 6. In Fig. 10 the neighbours are marked in red. Current alignment corresponding to the selected cell is shown in Fig. 8.15.

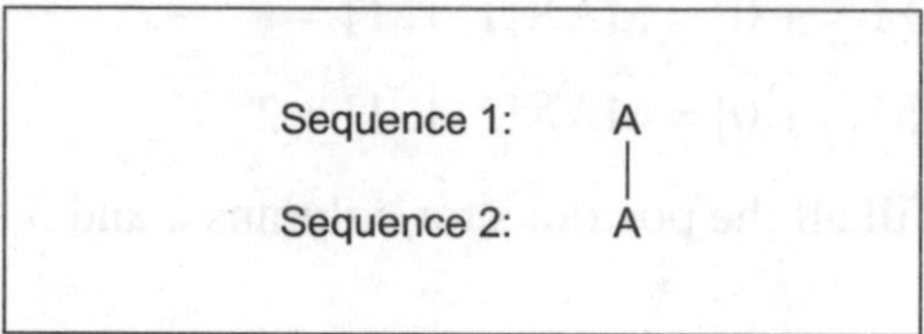

Fig. 8.15 Alignment corresponding to the selected position having value 6

From this position the program looks at the neighbouring cells (to its left, to above, and to the diagonal neighbour) that could be the direct predecessors of the cell in consideration. Each neighbour determines a gap, match, or mismatch:

1. Neighbour to the left: Gap in sequence 2
2. The diagonal neighbour: Match or mismatch
3. The neighbour above: Gap in sequence 1

The traceback algorithm chooses one of the neighbouring cells with the highest score. If all the three positions have the same score, then the diagonal neighbour is selected as the predecessor. Here the cell with value 5 (diagonal to the cell with value 6) is selected as shown in Fig. 8.16 and its alignment is shown in Fig. 8.17.

Note

Since the current cell has a value of 6 and the scores are 1 for a match and 0 for anything else, the only possible predecessor is the diagonal match/mismatch neighbour. If more than one possible predecessor exists, any one can be chosen. Thus, with a simple scoring algorithm such as the one that is used here, there are likely to be multiple maximal alignments.

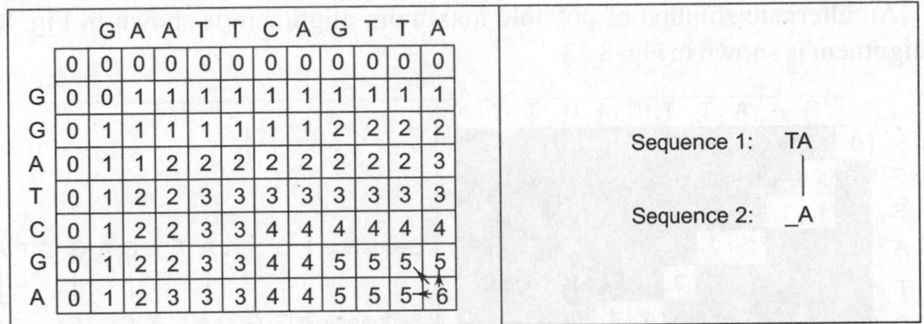

Fig. 8.16 Matrix with the maximum score value and its neighbours

Fig. 8.17 Alignment corresponding to the selected position with values 6, 5

Next the algorithm chooses 5 twice more as the neighbouring cells based on the highest score selection, as shown in Fig. 8.18 and its alignment is shown in Fig. 8.19.

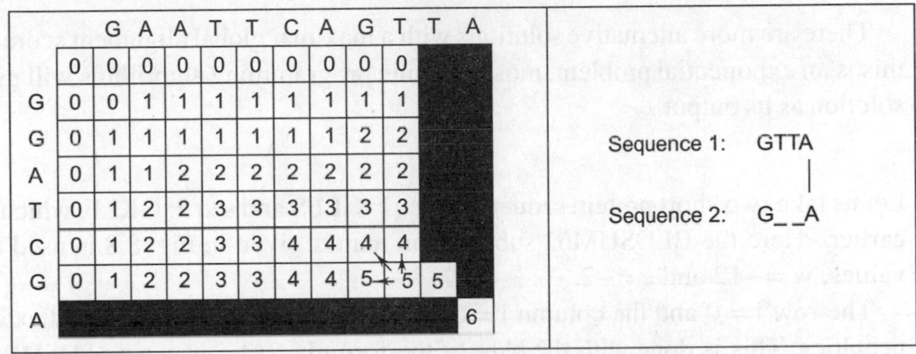

Fig. 8.18 Matrix with maximum score value and its neighbours

Fig. 8.19 Alignment corresponding to the selected position having value 6

The algorithm continues with the traceback step till it reaches the position $M_{0,0}$, which means that traceback is completed, as shown in Fig. 8.20. Its alignment is shown in Fig. 8.21.

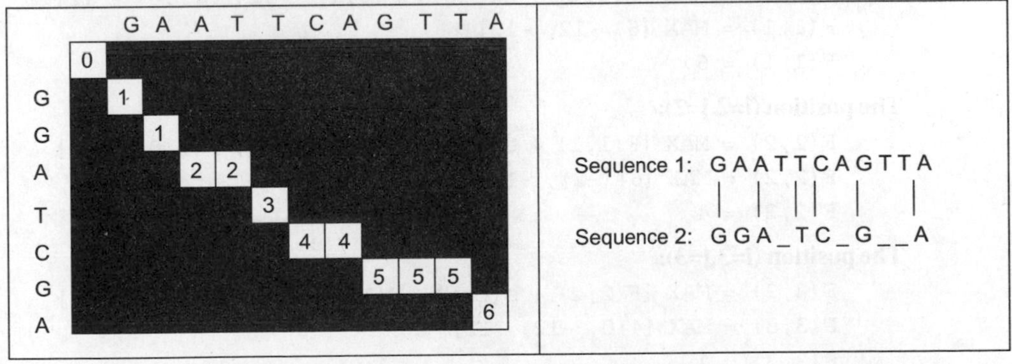

Fig. 8.20 Matrix with maximum score value and its neighbours

Fig. 8.21 Alignment corresponding to the selected position having value 6

An alternate solution of possible maximum alignment is shown in Fig. 8.22 and its alignment is shown in Fig. 8.23:

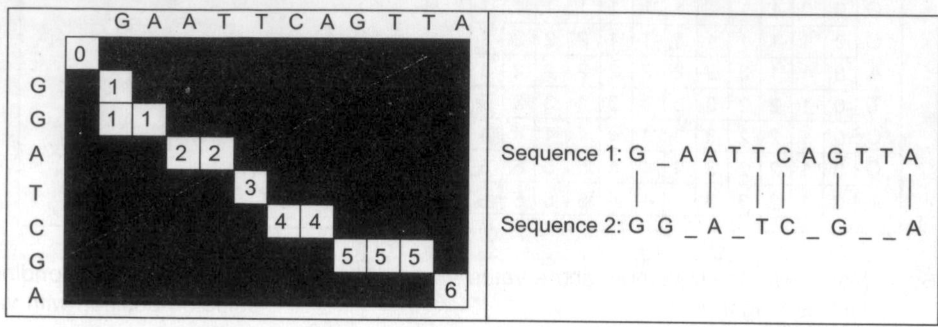

Fig. 8.22 Matrix with maximum score value and its neighbours

Fig. 8.23 Alignment corresponding to the selected position having value 6

There are more alternative solutions with a maximal global alignment score of 6. Since this is an exponential problem, most dynamic programming algorithms will give a single solution as its output.

TASK

Let us take two short protein sequences, seq1: GLFS and seq 2: GKLF, which were used earlier. Here the BLOSUM62 substitution matrix given in Fig. 8.8 is used for s(xi,yj) values. w = –12 and e = –2.

The row i = 0 and the column j = 0 are initialized to the opening and extension gap penalties. This is done with the help of the formula f(n) = w + e(n – 1). Here negative values are allowed.

The remaining positions are filled according to the recurrence relation. The below mentioned procedure is carried out for every position of the 2D matrix.

The position (i=1,j=1):

```
F(1,1) = MAX [F(0,0) + s(G,G), F(0,1) - 12, F(1,0) - 12]
F(1,1) = MAX [6, -12, -12]
F(1,1) = 6.
```

The position (i=2,j=2):

```
F(2,2) = MAX [F(1,1) + s(K,L), F(0,1) - 12, F(1,0) - 12]
F(2,2) = MAX [6+(-2), -12, -12]
F(2,2) = 4
```

The position (i=3,j=3):

```
F(3,3) = MAX [F(2,2) + s(L,F), F(1,2) - 12, F(2,1) - 12]
F(3,3) = MAX [4+0, -12, -12]
F(3,3) = 4
```

The position (i=4,j=4):

```
F(4,4) = MAX [F(3,3) + s(F,S), F(2,3) - 12, F(3,2) - 12]
F(4,4) = MAX [4+-2, -12, 4-12]
F(4,4) = 2
```

	J=0	G	L	F	S
I=0	0	-12	-14	-16	-18
G	-12	**6**	**-6**	**-8**	**-10**
K	-14	**-6**	**4**	**-8**	**-8**
L	-16	**-8**	**-2**	**4**	**-8**
F	-18	**-10**	**-8**	**4**	**2**

The values in bold entries indicate where the values come from and the shaded highlights indicate the optimal alignment. To find the optimal global alignment, the maximum value located at position F(n,m) serves as the starting point and the alignment is found by retracing the appropriate steps until i,j = 0.

The optimal global alignment from the example is obtained by starting from the F(n,m) position, in this case F(4,4) which contains the value 2, and by following back the appropriate values in bold text. The optimal global alignment of the two sequences at F(4,4) with score equal to 2 is

```
GLFS
:
GKLF
```

Local sequence alignment

Local alignment is used to detect regions of high similarity between sequences. This is a more flexible technique, as the alignment score can be high bacause fragments of sequences are considered. For example, all bases from 20–40 positions of sequence A can align with all the bases from 50–70 positions of sequence B. Also the advantage local alignment over global alignment is that related regions, which appear in a different order in the two protein sequences, can be identified as being related. This is known as *domain shuffling*.

Local alignment is used for the following cases.

- Sequences of different lengths are compared.
- Long sequences containing both coding and non-coding regions are compared.
- Proteins from different protein families are compared to find conserved domains.
- Sequence comparison using global alignment does not give the expected score, but there are clues that let you think that the sequences have similar parts.

Local alignment is first described by the Smith–Waterman algorithm. It uses following tools to find out local alignment between two sequences:

- BLAST sequence alignment against a database,
- FASTA sequence alignment against a database,
- LALIGN alignment of two sequences, and
- EMBOSS align alignment of two sequences.

Smith–Waterman algorithm This algorithm is based on dynamic programming. The algorithm was first proposed by Smith and Waterman in 1981. It is widely used to find local similarity regions in two sequences. This algorithm has the required feature to find the optimal local alignment with respect to the scoring system used. The scoring system comprises the substitution matrix and the gap-scoring scheme. However, the Smith–Waterman algorithm is fairly demanding of time and memory resources. In order to align two sequences of lengths m and n, O(mn) time and space are required. This algorithm differs from the global alignment algorithm, i.e., Needleman–Wunsch algorithm in the following points:

- Here i and j are initialized with 0 as the gap penalty values.
- The recurrence function initialize to 0 if no positive values are found.

To find the optimal local alignment of two sequences, consider a two-dimensional matrix of dimension i + 1 and j + 1 corresponding to the lengths of sequences 1 and two, respectively. Each position of the matrix is filled with the information about that position of the sequence with the help of a recurrence relation.

Thus the maximum score $M_{i,j}$ at position i,j is as shown below

```
Mi,j = MAXIMUM[
      0
     Mi-1, j-1 + Si,j (match/mismatch in the diagonal),
     Mi,j-1 + w (gap in sequence 1),
     Mi-1,j + w (gap in sequence 2)]
```

Here, the recurrence relation written above fills every (i,j) position of the 2D matrix with values that are higher or equal to 0. It is then possible to retrace the optimal alignment using these values. The values are thus always positive. We have already learnt the detailed steps of a dynamic programming algorithm. Let us work out an example to get a feel of the Smith–Waterman algorithm.

TASK

Let us take two short protein sequences, seq1: GLFS and seq 2: GKLF. Here the BLOSUM62 substitution matrix given in Fig. 8.8 is used for s(xi,yj) values. w = –12 and e = –2.

The row i = 0 and the column j = 0 are initialized to value 0. This is done with the help of the formula f(n) = w + e(n – 1). As the values obtained from this formula are all negative, 0 values are used. The remaining positions are filled according to the recurrence relation. The below mentioned procedure is carried out for every position of the 2D matrix.

The position (i=1, j=1):

```
F(1,1) = MAX [0, F(0,0) + s(G,G), F(0,1) - 12, F(1,0) - 12]
F(1,1) = MAX [0, 6, -12, -12]
F(1,1) = 6.
```

The position (i=2, j=2):

```
F(2,2) = MAX [0, F(1,1) + s(K,L), F(0,1) - 12, F(1,0) - 12]
F(2,2) = MAX [0, 6+(-2), -12, -12]
```

```
F(2,2) = 4
```

The position (i=3, j=3):

```
F(3,3) = MAX [0, F(2,2) + s(L,F), F(1,2) - 12, F(2,1) - 12]
F(3,3) = MAX [0, 4+0, -12, -12]
    F(3,3) = 4
```

The position (i=4, j=4):

```
F(4,4) = MAX [0, F(3,3) + s(F,S), F(2,3) -12, F(3,2) - 12]
F(4,4) = MAX [0, 4+-2, -12, 4-12]
F(4,4) = 2
```

	J=0	G	L	F	S
I=0	**0**	0	0	0	0
G	0	**6**	0	0	0
K	0	0	**4**	0	0
L	0	0	**4**	4	0
F	0	0	0	**10**	**2**

The bold entries indicated in the matrix (usually represented as an arrow) are where the values come from and the shaded highlights indicate the optimal alignment. The direction of arrow is from top left to right bottom. To find the optimal alignment(s), the maximum value is found, which is usually towards the right bottom position of the matrix. It is then traced back until the value of F(i,j) is equal to 0. When a value comes from two or three different places, then there are two or three different optimal alignments.

The optimal alignment of this example is obtained by starting from the position that has the maximum value F(i = 4,j = 3) = 10 and by following back the values in bold text. The optimal local alignment of the two sequences is

```
LF
::
LF
```

Till now, we have studied about substitution matrices and the dynamic programming algorithm that helped to find solutions with an optimal score according to the specified scoring scheme. But they are not the fastest method of sequence alignment. Dynamic algorithms have time complexity of the order O(n,m), the product of the length of the sequences, which is very high. To overcome this time-related problem, heuristic algorithms come handy.

8.3.2 Multiple Alignment

Multiple alignment is an extension of pairwise alignment. Here an unknown sequence is matched with several known sequences. Multiple alignment methods find common regions between the sequences at once, without making pairwise alignments first. There are several approaches, one of the most popular being the progressive alignment strategy used by the Clustal family of programs. This is used to build phylogenetic trees as well as to build sequence profiles which are used to search sequence databases for more distant relatives. The two most popular methods for detecting remote homologues–PSI-BLAST and hidden Markov model- (HMM-)based methods—both work on this principle. Here we will learn multiple alignment with heuristic algorithms. Phylogenetic trees and hidden Markov Model will be taken up in Chapter 9.

Heuristic algorithms are faster algorithms that are based on assumptions and approximations. Unlike dynamic programming, these algorithms do not make all possible pairwise comparisons to all of the database sequences, and thus they are not so expensive. The process of knowing, i.e., learning to solve a solution by trying rather than by following some pre-established formula is the approach of such algorithms. The algorithms learn by experiences as done by 'rule-of-thumb' or by 'trial-and-error' methods. Thus based on successive approximations, heuristic algorithms solve similarity search and alignment problems. These are methods that are devised to search a small fraction of a dynamic programming matrix by looking at all the high scoring alignments. But heuristic algorithms compromise on sensitivity. There are cases where such algorithms sometimes miss the best scoring alignment. Even the selectivity of these algorithms is comparable to the searches by dynamic programs. Both sensitivity and selectivity of heuristic algorithms are due to the incorporation of certain statistical parameters into the following two programs.

The two best known heuristic algorithms are BLAST and FASTA. These are two most commonly employed computational tools for scanning protein and DNA databases for similarity to a query sequence. Both programs do the following using different approach:

1. Identify very short, exact matches between the query sequence and the database sequence(s).
2. Extend the best short hits from the first step to look for longer stretches of similarity.
3. Optimize the best hits with some form of dynamic programming.

FASTA

It is a heuristic sequence searching and local alignment algorithm found by Pearson and Lipman in 1988. It has restrictions on the word size and window size. Like dot plots, FASTA compares two sequences at a time. The algorithm has a multistep approach that does the following:

1. It uses a look-up table to search for exact, short word matches of length *ktup* between the two sequences. For a protein, ktup is 1 or 2; for DNA it may be 4 or 6.
2. The program matches all identical words from the two sequences and creates diagonals by joining adjacent matches which are non-overlapping as shown in Fig. 8.24.
3. Next, the highest scoring regions are rescored using a substitution matrix such as the PAM250 and the best of these scores is called 'init1' as shown in Fig. 8.25.

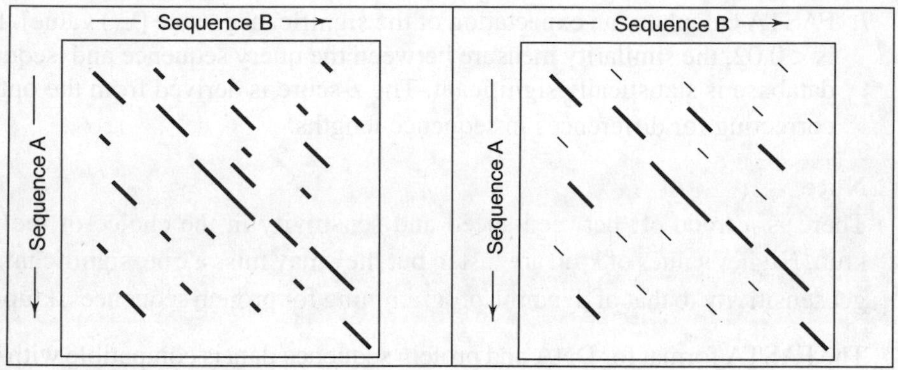

Fig. 8.24 Find runs of identical words

Fig. 8.25 Rescore using the PAM matrix. Keep top scoring segments

4. This step is analogous to the hit extension step in the BLAST algorithm. It extends the exact word matches to find maximal scoring ungapped regions. In the process it possibly joins multiple small matches.
5. In this step the algorithm identifies gapped alignments. The high-scoring diagonals are joined together by gapped regions between them, thus allowing for gap costs. The diagonal with the best score gets the value of 'initn' as shown in Fig. 8.26.
6. This is the last step in which the highest scoring candidate matches in a database search are realigned using the full dynamic programming algorithm. An optimal local alignment is performed between the query sequence and a limited number of database 'hit' sequences with high initn values. But unlike the dynamic programming algorithm, it restricts to a subregion of the dynamic programming matrix, forming a band around the candidate heuristic match. In the final step, for this alignment, the Smith–Waterman dynamic programming approach is employed. This optimal alignment score is labelled 'opt' as shown in Fig. 8.27.

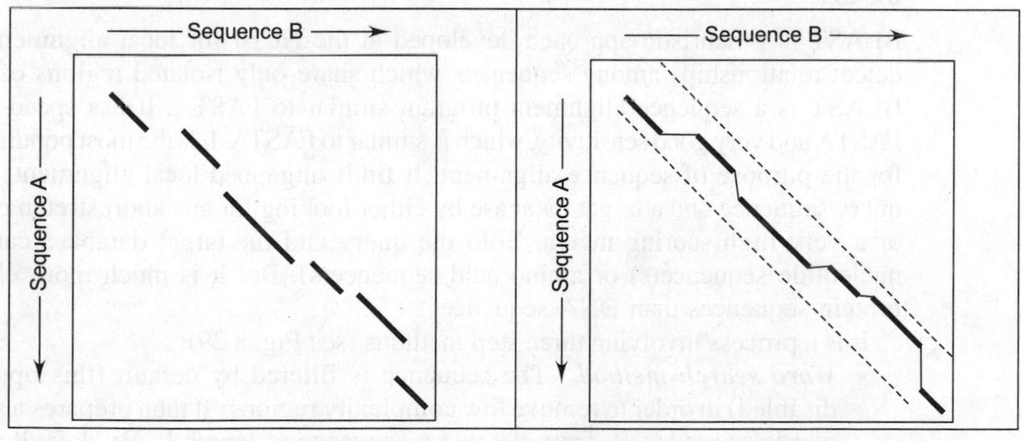

Fig. 8.26 Join segments using gaps and eliminate other segments

Fig. 8.27 Use dynamic programming using optimal alignment

7. FASTA calculates an expectation of the significance value [E() value]. If this value is < 0.02, the similarity measure between the query sequence and sequence(s) in a database is statistically significant. The z-score is derived from the opt score after correcting for differences in sequence lengths.

Note
There is a trade-off between speed and sensitivity in the choice of the parameter ktup. Higher values of ktup are faster, but they may miss a true, significant match. To get sensitivity to that of dynamic programming for protein sequences, ktup is set to 1.

The FASTA format for DNA and protein sequence data is compatible with virtually all molecular biology programs. A sequence in this format has a single-line description (header), which is followed by lines of sequence data. The header line is distinguished from the data lines by a greater than (>) symbol as its first character. An example of a sequence in the FASTA format is as shown in Fig. 8.28.

```
>gi|129369|sp|P04637|P53_HUMAN CELLULAR TUMOR ANTIGEN P53 (PHOSPHOPROTEIN
P53)MEEPQSDPSVEPPLSQETFSDLWKLLPENNVLSPLPSQAMDDLMLSPDDIEQWFTEDPGPDEAPRMPEAA
PPVAPAPAAPTPAAPAPAPSWPLSSSVPSQKTYQGSYGFRLGFLHSGTAKSVTCTYSPALNKMFCQLAKT
CPVQLWVDSTPPPGTRVRAMAIYKQSQHMTEVVRRCPHHERCSDSDGLAPPQHLIRVEGNLRVEYLDDRN
TFRHSVVVPYEPPEVGSDCTTIHYNYMCNSSCMGGMNRRPILTIITLEDSSGNLLGRNSFEVRVCACPGR
DRRTEEENLRKKGEPHHELPPGSTKRALPNNTSSSPQPKKKPLDGEYFTLQIRGRERFEMFRELNEALEL
KDAQAGKEPGGSRAHSSHLKSKKGQSTSRHKKLMFKTEGPDSD
```

Fig. 8.28 A sequence in FASTA format

FASTA does not guarantee an optimal solution at every execution. It is accessible on the web at http://www.ebi.ac.uk/fasta33/. You can also log on to the FTP site of the latest releases of FASTA at ftp://ftp.virginia.edu/pub/fasta/.

BLAST

BLAST is a heuristic approach developed at the NCBI for local alignments that can detect relationships among sequences which share only isolated regions of similarity. BLAST is a sequence alignment program similar to FASTA. It has speed faster than FASTA and very good sensitivity, which is similar to FASTA. It is the most popular algorithm for the purpose of sequence alignment. It finds ungapped local alignments between a query sequence and a target database by either looking for any short stretch of identities or a very high scoring match. Both the query and the target database can be either nucleotide sequence(s) or amino acid sequence(s). But it is much more effective for protein sequences than DNA sequences.

It is a process involving three step methods (see Fig. 8.29):

- ***Word search method*** The sequence is filtered by default (this option can be disabled) in order to remove low complexity regions. It then prepares a set of query words or seeds (W) from the query sequence of length L. By default these seeds are of a fixed length, 11 for nucleic acids and 3 for protein sequences. The number of seeds is L – W + 1.

- ***Identification of exact word match method*** The algorithm next searches the database to finds neighbourhood words. These words are aligned with the query word having the score value equal to or more than neighbourhood score threshold (T). These alignments are conserved and are called *hits*. The scoring is done with a substitution matrix.

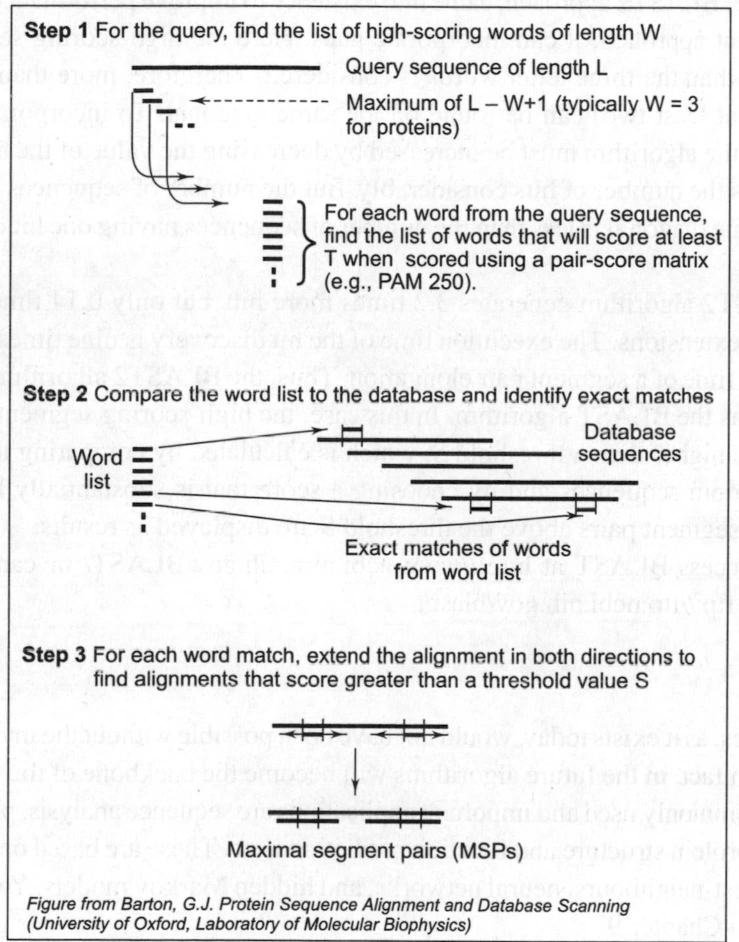

Step 1 For the query, find the list of high-scoring words of length W
Query sequence of length L
Maximum of L – W+1 (typically W = 3 for proteins)
For each word from the query sequence, find the list of words that will score at least T when scored using a pair-score matrix (e.g., PAM 250).
Step 2 Compare the word list to the database and identify exact matches
Word list
Database sequences
Exact matches of words from word list
Step 3 For each word match, extend the alignment in both directions to find alignments that score greater than a threshold value S
Maximal segment pairs (MSPs)
Figure from Barton, G.J. Protein Sequence Alignment and Database Scanning (University of Oxford, Laboratory of Molecular Biophysics)

Fig. 8.29 Sequence alignment using BLAST algorithm

- ***Maximal segment pair alignment method*** The algorithm starts a hit extension process at these matches. In this process it extends the possible match as an ungapped alignment in both directions that stops at the maximum score. This way it looks for local optimal ungapped alignment or HSP (high-scoring pairs) or MSP (maximal segment pairs) with a score of at least S or an E value lower than the specified threshold. This will retrieve most biologically significant similarities, but will miss a few and will include some chance similarities. Nucleotide neighbours are mostly used to build contigs, whereas protein neighbours discern biological function.

This procedure is responsible for 90% of the BLAST execution time. Therefore, a new technique, BLAST2 approach, came into existence to improve performance. BLAST2 uses a different approach; it can incorporate gaps. Here the high-scoring segment pair that is longer than the three letter words is considered. Therefore, more than one three-letter words (at least two) can be found on the same diagonal. To incorporate this, the sensitivity of the algorithm must be increased by decreasing the value of the threshold T. This increases the number of hits considerably. But the number of sequences having two hits on the same diagonal is less than the number of sequences having one hit on the same diagonal.

The BLAST2 algorithm generates 3.2 times more hits but only 0.14 times as many segment pair extensions. The execution time of the hit discovery is nine times faster than the execution time of a segment pair elongation. Thus, the BLAST2 algorithm is roughly twice as fast as the BLAST algorithm. In this case, the high-scoring segment pairs have a score that is higher than a threshold S, which is calculated by comparing the range of scores of random sequences and by choosing a score that is substantially higher. The high-scoring segment pairs above the threshold S are displayed as results.

You can access BLAST at http://www.ncbi.nlm.nih.gov/BLAST/ or can download BLAST from ftp://ftp.ncbi.nih.gov/blast/.

Conclusion

Bioinformatics, as it exists today, would not have been possible without the intervention of algorithms. In fact, in the future algorithms will become the backbone of the field. Some of the most commonly used and important applications are sequence analysis, phylogenetic predictions, protein structure and function predictions, etc. These are based on algorithms such as, nearest-neighbours, neural networks, and hidden Markov models. You will learn about these in Chapter 9.

EXERCISES

Exercise 8.1 Construct a dot plot on an spread sheet or a graph paper for the sequences GCTAGTCAGATCTGACGCTA and GATGGTCACATCTGCCGC. In case you are using a spread sheet, reduce the width of the columns to 2 to see a clear pattern. Does your dot plot reveal any regions of similarity?

	G	*C*	*T*	*A*	*G*	*T*	*C*	*A*	*G*	*A*	*T*	*C*	*T*	*G*	*A*	*C*	*G*	*C*	*T*	*A*
G	*X*				*X*				*X*					*X*			*X*			*A*
	X				*X*		*X*					*X*			*X*					
	T			*X*			*X*					*X*		*X*						*X*
G	*X*				*X*				*X*					*X*			*X*			
G	*X*				*X*				*X*					*X*			*X*			
T			*X*			*X*					*X*		*X*						*X*	
C		*X*					*X*					*X*				*X*		*X*		
A				*X*				*X*		*X*					*X*					*X*
C		*X*					*X*					*X*				*X*		*X*		
A				*X*				*X*		*X*					*X*					*X*
T			*X*			*X*					*X*		*X*						*X*	
C		*X*					*X*					*X*				*X*		*X*		
T			*X*			*X*					*X*		*X*						*X*	
G	*X*				*X*				*X*					*X*			*X*			
C		*X*					*X*					*X*				*X*		*X*		
C		*X*					*X*					*X*				*X*		*X*		
G	*X*				*X*				*X*					*X*			*X*			
C		*X*					*X*					*X*				*X*		*X*		

Exercise 8.2 Construct a dot plot using an spread sheet or a graph paper for the sequences GCTCGTCAGCTCTGCCGCTC and GATGGTCACATCTGCCGC. In case you are using a spread sheet, reduce the width of the columns to 2 to see a clear pattern. Does your dot plot reveal any pattern? What does it indicate?

	G	*C*	*T*	*C*	*G*	*T*	*C*	*A*	*G*	*C*	*T*	*C*	*T*	*G*	*C*	*C*	*G*	*C*	*T*	*C*
G	*X*				*X*				*X*					*X*			*X*			
A				*X*				*X*	*X*						*X*					*X*
T			*X*			*X*					*X*		*X*						*X*	
G	*X*				*X*				*X*					*X*			*X*			
G	*X*				*X*				*X*					*X*			*X*			
T			*X*			*X*					*X*		*X*						*X*	
C		*X*		*X*			*X*			*X*		*X*			*X*	*X*		*X*		*X*
A								*X*												
A								*X*												
A								*X*												
A								*X*												
A								*X*												
A								*X*												
A								*X*												
A								*X*												
C		*X*		*X*			*X*			*X*		*X*			*X*	*X*		*X*		*X*
G	*X*				*X*				*X*					*X*			*X*			
C		*X*		*X*			*X*			*X*		*X*			*X*	*X*		*X*		*X*

Exercise 8.3 Construct a dot plot using an spread sheet or a graph paper for the sequences GCTAGTCACTGATCGGCTAA and GATGGTCACATCTGCCGC. In case you are using a spread sheet, reduce the width of the columns to 2 to see a clear pattern. Does your dot plot reveal any pattern? What does it indicate?

	G	C	T	A	G	T	C	A	C	T	G	A	T	C	G	G	C	T	A	A
G	X				X						X				X	X				
A				X				X				X							X	X
T			X			X				X			X					X		
G	X				X						X				X	X				
G	X				X						X				X	X				
T			X			X				X			X					X		
C		X					X		X					X			X			
A				X				X				X							X	X
C		X					X		X					X			X			
A				X				X				X							X	X
T			X			X				X			X					X		
C		X					X		X					X			X			
T			X			X				X			X					X		
G	X				X						X				X	X				
C		X					X		X					X			X			
C		X					X		X					X			X			
G	X				X						X				X	X				
C		X					X		X					X			X			

Exercise 8.4 Calculate the number of possible alignments for the following strings:

(a) abcde and bdefg

(b) abcde, bdefg, and cdedf

Note that you cannot use gaps for both the sequences at the same position. When there is a gap at a position in one sequence there has to be a residue at that position in the second sequence.

Exercise 8.5 Calculate the score for the following alignments:

GATGGTGTCACGTCTG
GCTAGTCACATCTG

(a) Assume that the match score is 1 and mismatch score is 0.

(b) Assume that the match score is 1, mismatch score is 0, and gap penalty is –1

Exercise 8.6 Calculate the score for the following alignments:

GATGGTGTCACGTCTG
GCTAGTCACATCTG

(a) Assume that the match score is 1, mismatch score is 0, origination penalty is –2, and length penalty is –1.

(b) Assume that the purine–purine or pyrimidine-pyrimidine match score is –1, purine–pyrimidine or pyrimidine–purine match score is –5, origination penalty is –2, and length penalty is –1.

Exercise 8.7 Consider the following pair of sequences:

GCTAGTCACTGATCGGCTAA

GATGGTCACATCTGCCGC

Using the Needleman–Wunch method, do the following:

(a) Compute the values for each position of the matrix.

(b) Traceback the path on the matrix from the position with the highest score to the position of $M_{0,0}$. Each position along the path should traceback the neighbouring cells with the highest score.

(c) Plot the alignment corresponding to the traceback path you have plotted on the matrix.

Exercise 8.8 Consider the following pair of sequences

TTGACACCCTCCCAATTGTA

ACCCCAGGCTTTACACAT

Using the Smith–Watermann method, do the following:

(a) Compute the values for each position of the matrix.

(b) Traceback the path on the matrix from the position with the highest score to the position of $M_{0,0}$. Each position along the path should traceback the neighbouring cells with the highest score.

(c) Plot the alignment corresponding to the traceback path you have plotted on the matrix.

9

Prediction Algorithms

Learning Objectives

- To understand prediction algorithms
- To learn about gene prediction
- To learn about phylogenetic prediction
- To learn about protein structure prediction

Introduction

If certainty is to mechanical then prediction is to natural. One can be certain about mechanical processes where the efficiency of machines can be used to calculate the production output. But when it comes to natural processes, one can only predict the result. From weather forecasting to birth of a child, one can never be definite and has no other choice than to rely on prediction. Similarly, if you are provided with a sequence of 0's and 1's, you can predict the next best element of the sequence. Here the word 'best' means the best possible predicted element, keeping the historical context of the sequence of 0's and 1's that had preceded the prediction. This is also what we can call as finding the most likely that is inferred from the past behaviour. Tasks of bioinformatics include analysis and 'broadly' prediction. The subject is its intancy. It has yet a lot to explore and that includes predicting genes, finding signals, identifying proteins, assuming protein structures, determining protein functions and many more. Thus most bioinformatics algorithms are mostly predictive in nature. Of course, the task is tougher than just predicting next best element in sequence of 0s and 1s.

Analogy

It happens on almost everybody that we predict something or the other. It can be an sms from somebody, a phone call from someone, a mail from a friend, or a visit by some distant relative(s). These predictions are made on the basis of the past experience of the predictor. Our brain records each and every experience that we go through and utilises each piece of information to analyse and conclude. It is definitely an adaptive mechanism where every new experience adds to the knowledge base of the brain and helps lead to a different prediction from the previous experience(s). Say, for example, we have the following facts:

- *Every Sunday morning your uncle joins you for tea.*
- *He misses those Sundays when it rains heavily.*
- *Many a times he gets those brown crispy 'jalebis'.*
- *You love 'jalebis'.*
- *You met him on the way yesterday afternoon after his 15-day official tour.*
- *Your uncle loves to see you happy.*
- *Today it is raining.*
- *Today is thursday but a political party has called for a 'Bandh'.*

With all these facts in mind, your brain will predict that your Uncle will come with jalebis and join you for tea today. Now how does this happen? You observe those past experiences and after filtering the relevant facts with various conditions you try to predict things.

In this chapter we will focus on certain predictions involving tasks of bioinformatics that will throw light on how the prediction mechanism works and how is it implemented. There are many types of prediction algorithms and covering all of them is not possible here.

9.1 Gene Prediction Algorithm

To understand the algorithm behind gene prediction, let us revisit some concepts of genes. Gene prediction is the task of locating genes in a genomic sequence. Prokaryotic organisms do not have split genes, hence gene prediction is simple in such case. Eukaryotic organisms have genes with the intron-exon model where nucleotide segments, called *exons*, are separated by junk segments, called *introns*. Each codon (triplet of nucleotides) of all these exons codes for one amino acid of the corresponding protein. Deletion of one or two consecutive nucleotides effects major changes in the protein, whereas deletion of a codon surprisingly results in minor changes in protein.

Example 9.1 Let us take the phrase 'THE FAT CAT AND THE SHY RAT'. If we remove one alphabet from this phrase say 'F', the phrase will become 'THE ATC ATA NDT HES HYR AT' and if we remove two alphabets from this phrase say 'FA', the phrase will become 'THE TCA TAN DTH ESH YRA T'. In both the cases the phrase will make no sense. But if we remove three alphabets from the phrase say 'FAT', the phrase will become 'THE CAT AND THE SHY RAT' which definitely make some sense.

Therefore, following this experiment, it was proved that a gene and its protein are collinear, whether or not they are broken into several segments like that in eukaryotes or present in a fragment as a continuous segment like that in prokaryotes. There are two approaches to predict the gene location.

The statistical approach This approach is based on detection of statistical variations between coding and non-coding regions by locating features that appear frequently in genes and infrequently elsewhere. For example, splicing signals at exon–intron junctions, i.e., AG and GT on the left and right sides of an exon are lightly conserved. Unfortunately, using profiles to detect splice sites in an eukaryote is not so successful, as these profiles also tend to match frequently in the genome at non-splice sites.

The similarity based approach This approach has an improved accuracy of gene prediction. It is based on the similarity of a newly sequenced gene with a known gene. For example, similarity of mouse genes with that of human genes, where a concatenated set of substrings (putative exons) in the mouse genomic sequence matches with a known human protein.

9.1.1 Statistical Approaches

Consider a genome of length n having a sequence of n/3 codons. The genome is a sequence of four kinds of nucleotides A, T, C, and G. A codon is a combination of three nucleotides that result to 4^3, i.e., 64 combinations and hence 64 different codons each coding for one among 20 amino acids. Refer Table 9.1 for the Codon table where you will find that in many cases more than one kind of codon codes for the same amino acid.

Table 9.1 20 amino acids, their single-letter database codes (SLC), three-letter codes (TLC), and their corresponding DNA codons

Amino Acid	SLC	TLC	DNA Codons
Isoleucine	I	Ile	ATT, ATC, ATA
Leucine	L	Leu	CTT, CTC, CTA, CTG, TTA, TTG
Valine	V	Val	GTT, GTC, GTA, GTG
Phenylalanine	F	Phe	TTT, TTC
Methionine	M	Met	ATG
Cysteine	C	Cys	TGT, TGC
Alanine	A	Ala	GCT, GCC, GCA, GCG
Glycine	G	Gly	GGT, GGC, GGA, GGG
Proline	P	Pro	CCT, CCC, CCA, CCG
Threonine	T	Thr	ACT, ACC, ACA, ACG
Serine	S	Ser	TCT, TCC, TCA, TCG, AGT, AGC
Tyrosine	Y	Tyr	TAT, TAC
Tryptophan	W	Try	TGG
Glutamine	Q	Gln	CAA, CAG
Asparagine	N	Asn	AAT, AAC
Histidine	H	His	CAT, CAC
Glutamic acid	E	Glu	GAA, GAG
Aspartic acid	D	Asp	GAT, GAC
Lysine	K	Lys	AAA, AAG
Arginine	R	Arg	CGT, CGC, CGA, CGG, AGA, AGG
Stop codons	Stop	Stp	TAA, TAG, TGA

To look for potential coding regions in the genomic sequence of your interest, you can look at the segment of sequences called open reading frames (or ORFs) which starts with a start codon (ATG) and ends with a stop codon (TAA or TAG or TGA). There can be overlaps in ORFs within a single genomic sequence as there are six reading frames (three reading frames at positions 1, 2, and 3 on a single 3′ to 5′ strand and another three reading frames at positions 1, 2, and 3 on the reverse 5′ to 3′ strand).

In a random DNA sequence, the average number of codons between two consecutive stop codons is 64/3, i.e., roughly 21. But in an average protein, the number of codons is approximately 300. Therefore those short exons are ignored and ORFs with length longer than a threshold length indicate potential genes. But this fails to detect short genes or genes with short exons. In this approach of gene prediction, subtle statistical variations between coding (exons) and noncoding (introns) regions are recorded by looking at the codon usage biases and preparing an codon usage array for both coding and non coding regions. Also, different organisms often show particular preferences for one of the several codons that encode the same given amino acid. For example, let us look at the codon bias of arginine in species such as Human, Drosophila and E. coli, given in Table 9.2 that gives the percentages of arginine amino acids that are encoded by each of the six codons in various numbers of genes in these species.

You will find in the human genes, codons CGU and AGG code for the same amino acid arginine, but AGG is more likely to be used as genes than CGU. Therefore an ORF that prefers AGG while coding for Arg than CGU will have a likely candidate gene. Similarly, CGC and CGU are preferred in the cases of Drosophila and E. coli.

Table 9.2 Frequencies of six arginine codons in the DNA of Human, Drosophita, and E. coli

Codon	Human	Drosophila	E. coli
Arginine			
AGA	22%	10%	1%
AGG	23%	6%	1%
CGA	10%	8%	4%
CGC	22%	49%	30%
CGG	14%	9%	4%
CGU	9%	18%	49%
Total number of argine codons	2403	506	149
Total number of genes	195	46	149

One can use the likelihood ratio approach that allows the testing of the applicability of these two hypotheses. If the likelihood ratio is large, the first hypothesis is more likely true than the second one. Another better coding sensor is the in-frame hexamer count that reflects frequencies of pairs of consecutive codons. In prokaryote genomes there are several conserved sequence motifs found in the regions around the start of transcription. In the case of eukaryote genomes there are splicing signals, i.e., conserved sequence eight and four nucleotides at exon–intron (donor splice site) and intron–exon (acceptor splice site) boundaries, respectively. As splice site profiles are weak, this approach is complemented by the Hidden Markov Model to capture the statistical dependencies between sites.

A well known tool applying this approach is GENSCAN which was developed in 1977. This approach is based on a probabilistic model of gene structure in the human genomic sequence. The overall model is similar to generalized HMM where it combines the coding region and splicing signal predictions into a single framework. Potential genes on both strands are analysed simultaneously using the explicitly double-stranded genomic sequence model. It covers cases where the input sequence contains no gene, partial gene, complete gene, or multiple genes. It does not address alternative splicing. The accuracy of GENSCAN decreases for genes with many short exons or with unusual codon usage.

9.1.2 Similarity Based Approaches to Gene Prediction

In this approach a gene is predicted in an unknown DNA fragment using previously sequenced genes and their protein products as templates. Here the task is to either find a substring in case of prokaryotes or a set of substrings (candidate exons) in case of eukaryotes from the unknown genomic sequence whose concatenation (splicing) best fits the target. Here the local sequence similarity between the genomic sequence and the target protein sequence is looked into. The substrings of the genomic sequence that show similarity to the target protein sequence are considered as putative exons. These exons may lack canonical exon flanking and can be either extended or shorten so that they are flanked by AG and GT. Each putative exon is assigned a weight, which can be either the score for local alignment or the strength of the flanking acceptor and donor sites reflecting the likelihood of the string to be an exon. The set of substrings is then filtered to avoid overlaps, because exons in real genes do not overlap and result into a large chain with maximum total weight. This problem is called the *exon chaining problem*. A limitation of this approach is that the endpoints of putative exons are not well defined. For example, if a set of substrings contain two putative exons, then the first exon may correspond to the end region of the target protein and the last exon may correspond to the beginning region of the target protein. Thus the putative exons corresponding to the valid optimal chain of intervals cannot be combined into a valid alignment. This shortfall can be overcome by spliced alignment where the global alignment score is maximum for the chain of candidate exons.

Let us look at the hidden Markov model (HMM) that has best of both the worlds, statistical as well as similarity based approach. HMM is a statistical model where it is assumed that the system that is being modelled is a Markov process with unknown parameters. In a regular Markov model, the state is directly visible to the observer, and therefore the state transition probabilities are the only parameters. But in the case of

HMM the state is not visible to the observer. The task of this algorithm is to determine the hidden parameters of the states from the observable parameters of output tokens. The extracted model parameters can then be used to perform further analysis like that in pattern-recognition applications. Thus each state has a probability distribution over the possible output tokens. Therefore, looking at a sequence of tokens generated by an HMM does not directly indicate the sequence of states. Let us look at Fig. 9.1 to understand the state transition in an HMM. In this figure, x1, x2, and x3 are states of the Markov model with a_{12}, a_{21}, and a_{23} as transition probabilities and b1, b2, and b3 as output probabilities giving y1, y2, and y3 as observable outputs. Its HMM can be represented in triple (a, b, x).

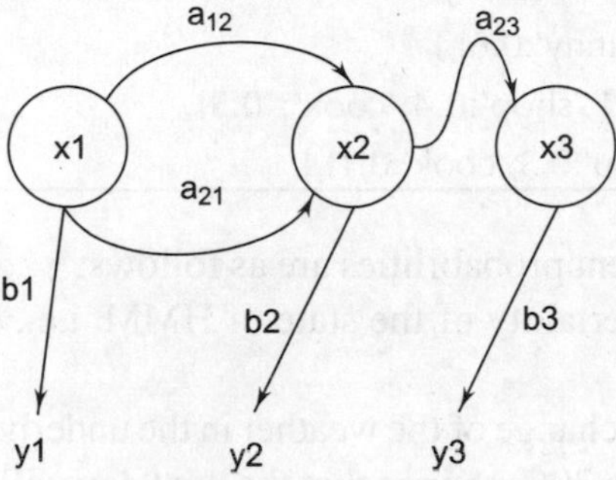

Fig. 9.1 State transitions in a hidden Markov model

To represent state transitions to explicitly represent the evolution of the model over time, Fig. 9.2 shows the states at different times t – 1, t, and t + 1 are represented by different states variables, x(t – 1), x(t), and x(t + 1) giving y(t – 1), y(t), and y(t + 1) as observable outputs .

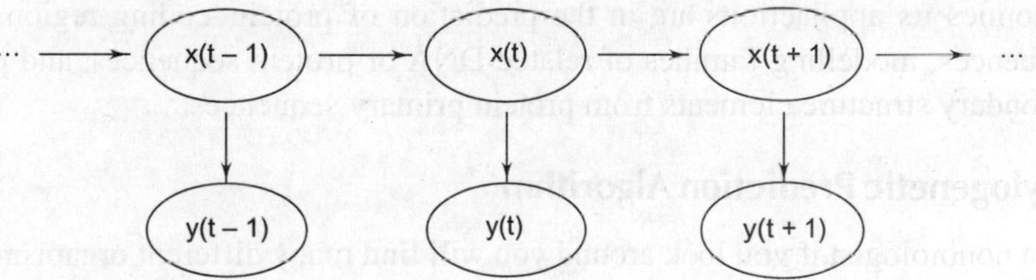

Fig. 9.2 Evolution of a Markov model

Example 9.2 Assume your cousin lives far away and every night both of you talk about what each of you did after office that day. Your cousin's interest is limited and she does either of the three things, going out for a walk, shopping, or cooking a new dish. The choice of what she will do is based on that day's weather. You do not have any information about the weather of the place where she lives. But being her close cousin you know her general trend. So every night when you hear what she did that day after office you try to guess that day's weather. Here you try to model the weather as HMM with its two states, 'Rainy' and 'Sunny' hidden from you. Your cousin does one of the activities 'walk,' 'shop,' or 'cook,' depending on the weather. These activities are observations. Let us represent the parameters 'weather' and your cousin's 'activity' as a fragment algorithm as shown in Table 9.3:

Table 9.3 Fragment algorithm representing HMM parameters

```
states = ('Rainy', 'Sunny')
observations = ('walk', 'shop', 'cook')
start_probability = {'Rainy': 0.6, 'Sunny': 0.4}
transition_probability = {'Rainy': {'Rainy':0.7,'Sunny':0.3},
                          'Sunny': {'Rainy':0.4,'Sunny':0.6},}
emission_probability = {'Rainy': {'walk':0.1,'shop':0.4,'cook': 0.5},
                        'Sunny' : {'walk':0.6,'shop':0.3,'cook':0.1},}
```

In this fragment algorithm, three different probabilities are as follows:

(a) **Start_probability** refers to the uncertainty of the state of HMM, i.e., the weather when your cousin first calls you.

(b) **Transition_probability** refers to the change of the weather in the underlying Markov chain. In this example, there is only a 30% chance that the next day will be sunny if today is rainy.

(c) **Eission_probability** tells you likelihood of your cousin performing activity on each day. If it is rainy, there is a 50% chance that she is cooking. If it is sunny, there is a 60% chance that she will go out for a walk.

There are various applications of hidden Markov models such as in speech recognition, optical character recognition, natural language processing, etc. In bioinformatics and genomics its appliactions are in the prediction of protein–coding regions in genome sequences, modelling families of related DNA or protein sequences, and prediction of secondary structure elements from protein primary sequences.

9.2 Phylogenetic Prediction Algorithm

As a nonbiologist if you look around you will find many different organisms, when you travel, you will find more. And as a biologist you will find so many that you will develop anurge to reconstruct the evolutionary history of all organisms that are there on our Earth. To do so you can either choose to study the fossil records (available in few) or opt for a phylogenetic analysis. We all know that evolution is defined as genetic change. Therefore organisms with a high degree of molecular similarities are expected to be closely related than those that are dissimilar. Earlier, when molecular data was not available in plenty, taxonomist had no other choice than to study phenotypes to infer their genotypes. Such studies have their own limitations. Phenotypic similarities can evolve in distantly related organisms with genetic dissimilarities. Some organisms may not have suitable phenotypic features for comparison. Analyses of molecular data, DNA, or protein sequences are free from such problems. So it is reliable to go for molecular phylogenies. Here we will learn some phylogenetic terminology and the various methods of reconstructing phylogenetic trees from molecular data.

Phylogenetic analysis helps to understand the evolutionary relationships among different species (i.e., the origin of life) and thus can be used to trace the evolutionary history. Such analyses are done by conducting two major activities:

Character analysis Here operational taxonomic units (OTU) are categorized by traits or characters and analysed to discover how different characters have evolved.

Phylogenetic inference This is also known as tree building. The branching orders (topology), and in some cases the branching length (timing), infer the evolutionary history which helps to understand the implications.

One of the important and beneficial application areas of a phylogenetic analysis is in medical science research, which can lead to effective drug development.

Phylogenies are graphs that are usually represented as trees with nodes and branches as shown in Fig. 9.3. Nodes can be of three types.

Terminal nodes The nodes at the tip of the branches are called *terminal nodes*. These nodes represent different taxonomic units corresponding to the gene or organism for which sequence data are available. These units can be species (or higher taxa), populations, individuals, or genes. These are also referred to as *operational taxonomic units* (OTUs).

Internal nodes These are nodes that lie in the middle of the root and the terminal nodes representing inferred ancestors for which empirical data are not available. These nodes represent the divergence point of the ancestral unit like in the case of duplication and speciation. These are also referred to as *hypothetical taxonomic units* (HTUs). When internal nodes have two lineages descending from them, they are said to be bifurcating. When they have multiple lineages descending from them, they are said to be multifurcating.

Root The node that represents the ultimate ancestor of all the taxa is called the *root*.

Branches are lineages or edges that join two nodes. Each edge of the tree is associated with some divergence that is defined by a measure of similarities or differences, i.e., the distance between sequences. Thus branches represent the evolutionary relationship among different nodes.

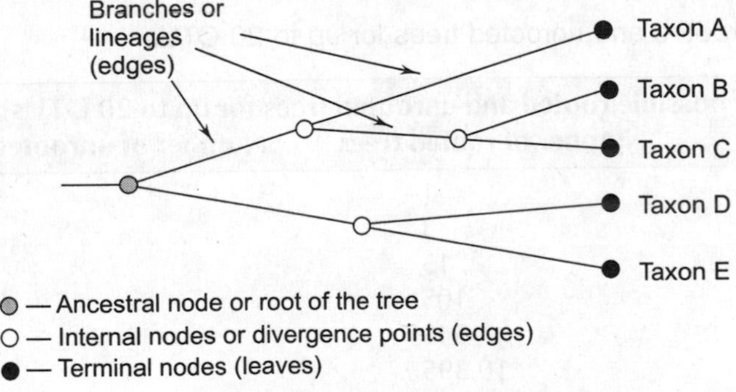

Fig. 9.3 Phylogeny with nodes and branches

Phylogenetic trees can be of various types. We will learn about each type next.

Type A A tree with five taxa can be rooted or unrooted as shown in Fig. 9.4.

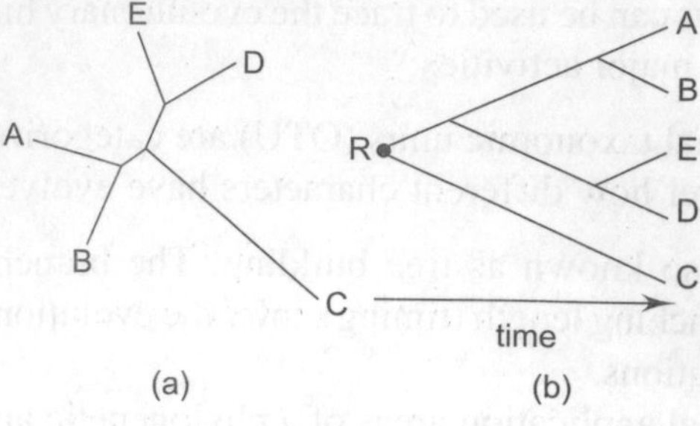

Fig. 9.4 (a) An unrooted and (b) a rooted tree

Rooted tree In this type of tree the root represents the common ancestor of all the taxonomic units of the tree. The direction and the length of each branch correspond to the evolution and time taken for evolution. Such trees are used to study evolutionary relationships.

Unrooted tree In this case the tree specifies the relationships among taxonomic units, but it does not show the evolution path. It also does not identify the common ancestor. You can assign a root in an unrooted tree by using an outgroup. Outgroup is an obvious choice within species under study that can be separated at the earliest.

The number of possible tree topologies increases as the number of OTU increases. In general, the number of possible topologies for a bifurcating rooted tree and an unrooted tree of n taxa are given by

```
Nr = (2n-3)! / 2(n-2)!          Nu = (2n-5)! / 2(n-3)!
```

where Nr denotes the number of rooted trees and Nu denotes the number of unrooted trees.

Table 9.4 shows the number of possible rooted and unrooted trees for up to 20 OTUs. The numbers are calculated by using the above equations.

Table 9.4 Number of rooted and unrooted trees for up to 20 OTUs

Number of possible rooted and unrooted trees for up to 20 OTUs		
Number of OTUs	**Number of rooted trees**	**Number of unrooted trees**
2	1	1
3	3	1
4	15	3
5	105	15
6	945	105
7	10,395	954

(contd)

Table 9.4 *(contd)*

Number of possible rooted and unrooted trees for up to 20 OTUs		
Number of OTUs	Number of rooted trees	Number of unrooted trees
8	135,135	10,395
9	2,027,025	135,135
10	34,459,425	2027,025
11	654,729,075	34,459,425
12	13,749,310,575	654,729,075
13	316,234,143,225	13,749,310,575
14	7905,853,580,625	316,234,143,225
15	213,458,046,676,875	7905,853,580,625
16	6190,283,353,629,375	213,453,046,676,875
17	191,898,783,962,510,625	6190,283,353,629,375
18	6332,659,870,762,850,625	191,893,783,962,510,625
19	221,643,095,476,699,771,875	6332,659,870,762,850,625
20	8200,794,532,637,891,559,375	221,643,095,476,699,771,875

In the case of more number of OTUs, there is a large number of rooted and unrooted trees but it is only one tree that represents the true phylogenetic relation. The other trees are referred to as *inferred trees*. When there is less number of possible tree topologies, it is easy to identify the true tree that represents the true evolutionary relationships among the OTUs. This can be done by eliminating the trees with obvious unlikely evolutionary relationships. But this is difficult when there number of OTUs greater.

Type B A tree with six nodes as shown in Fig. 9.5 is a gene tree that can be broadly grouped into two nodes to represent it as a species tree.

Gene tree Here the branching pattern represents the evolutionary relationships among genes. An internal node in a gene tree indicates the divergence of an ancestral gene into two genes (called *gene splitting event*) with different DNA sequences resulting from mutation. Phylogenetic inference cannot be made when all the sequences are obtained from the same species. In such cases the occurrences of mutation are similar throughout the whole genome. Thus a reliable phylogeny cannot be obtained from insufficient polymorphism.

Species tree In this case the branching pattern represents the evolutionary relationship among species. An internal node in a species tree represents a speciation event, whereby the population of the ancestral species splits into two different species that cannot interbreed. In this case, all the data are obtained from different species, resulting in a tree with each taxon having a distinct sequence. Thus such a tree can yield more dependable phylogenetic trees from which evolutionary relationships can be inferred.

However, the gene-splitting event (mutation) and speciation do not always occur at the same time. In Fig. 9.5 although the first gene divergent time is at G1, which is earlier than the species divergent time at S, the complete time of the gene-splitting event is much longer than the speciation event.

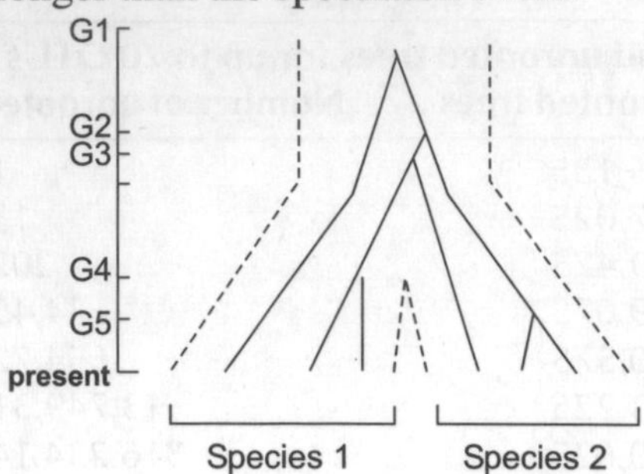

Fig. 9.5 A tree showing gene-splitting and speciation events

There are three ways of representing rooted trees. These are as follows

Cladogram It is a branching diagram representing the most parsimonious distribution of derived characters within a set of taxa. It is shown in Fig. 9.6. The branching pattern of a cladogram is intended to show the relative relationships among taxa. It is not a true 'evolutionary tree' of how those relationships came to be.

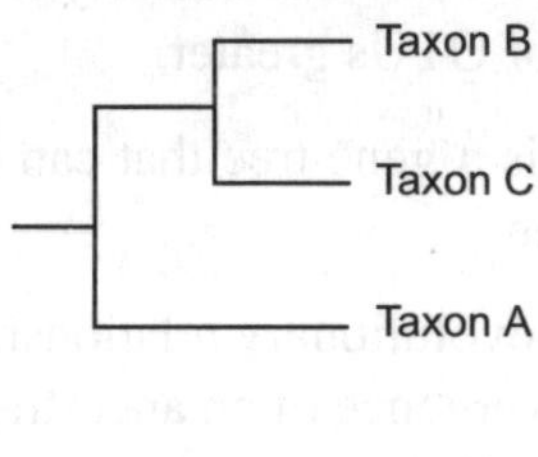

Fig. 9.6 Cladogram

Phylogram It is a phylogenetic tree that indicates the relationships among the taxa. It also shows evolution and the rate of evolution. It is shown in Fig. 9.7.

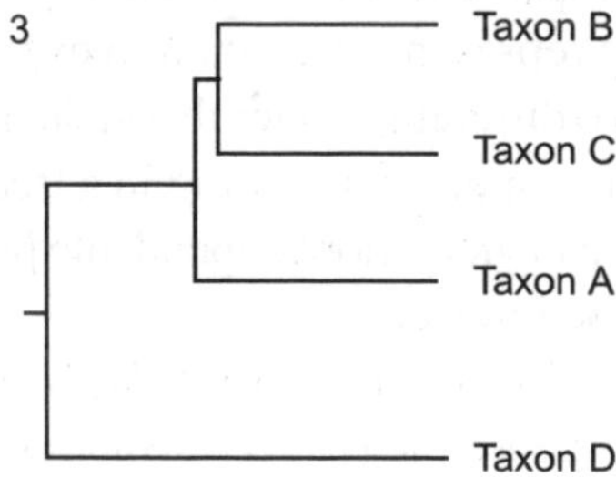

Fig. 9.7 Phylogram

Ultrametric tree It is a rooted tree where each internal node is labelled with a number. Each internal node has at least two offsprings and the labels decrease along the path from any root to leaf. The branch lengths of an ultrametric tree are proportional to the divergent time. An ultrametric tree is shown in Fig. 9.8.

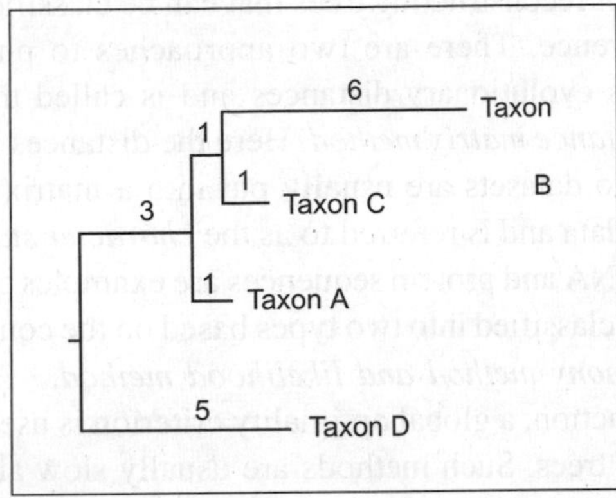

Fig. 9.8 Ultrametric tree

Type C A tree can be scaled and unscaled depending on the branch lengths getting reflected in the tree.

Scaled tree Here the branch lengths of the tree are proportional to the lengths between the pair of the neighbouring nodes.

Unscaled tree Here the branch length does not convey the number of changes between the nodes and instead just represents the relation between them.

Classification of tree reconstruction methods

Inferring a phylogeny is done on the basis of incomplete information because we do not have information about the past. This process involves performing the best estimate of the evolutionary history, so it is an estimation procedure. There are two processes involved in the inference estimation of topology and estimation of branch lengths for a given topology. One is when topology is known, estimation of branch lengths is done. The other is to estimate topology. We have learnt before that the number of possible topologies increases dramatically with an increase in the number of sequences, and it is very hard find true topolgy. We also know that from a set of OTUs different phylogenetic trees can be produced. Therefore, for comparing alternative phylogenies and to select one or few trees to represent the best estimate of the true evolutionary history, we must specify some conditions or criteria. Most phylogenetic methods seek to accomplish this goal by defining a criterion for and deciding which one is the best.

A phylogenetic reconstruction, therefore, consists in two steps:

Defining an optimality criterion or objective function: This step assigns a value to a tree and is subsequently used for comparing the other trees.

Developing specific algorithms: This step is used to develop an algorithm to compute the objective function values. This helps to identify the tree or a set of trees that have the best values according to this criterion.

We have learnt before that phylogenetic trees are reconstructed from molecular data. There are many statistical methods of reconstructing trees that can be classified by types of data used in phylogenetic inference. There are two approaches to phylogenetic reconstructions. One approach uses evolutionary distances and is called the *distance method*, or, more commonly, the *distance matrix method*. Here the distances in the form of pairwise differences between two datasets are usually put into a matrix form. The other approach uses character state data and is referred to as the *character state method*. Anatomic and behavioral traits of DNA and protein sequences are examples of character state data. This method can again be classified into two types based on the computational methods used. They are the *parsimony method and likelihood method.*

In some methods of tree reconstruction, a global optimality criterion is used to search for the best tree among all possible trees. Such methods are usually slow and demand more computing time. The maximum parsimony method uses character state data (e.g., the nucleotide or amino acid sequences at a site) with the global optimality criterion. The smaller the number of evolutionary changes required by a tree, the better the tree. Thus the shortest pathway leading to these character states is chosen as the best tree.

In other methods, OTUs are clustered sequentially according to a local optimality criterion. These then choose the optimal cluster among the possible local clusters during the process of tree building. Such methods are mostly fast, but may not find the globally optimal tree. The maximum likelihood method uses both character state configuration data (e.g., the pattern of nucleotide or amino acid sequence differences at that site) among the sequences under study and evolutionary distances for all possible trees to choose the best tree with the highest maximum likelihood value.

A number of methods that have been frequently used or are convenient for illustrating basic principles are discussed below. These methods can be classified into three types: distance matrix methods, maximum parsimony methods, and maximum likelihood methods. A description of methods is given in the following sections and a comparison of the strengths and weaknesses of methods is also presented.

9.2.1 Distance Matrix Methods

In a distance matrix method, the evolutionary distances are computed for all pairs of taxa, and a phylogenetic tree is constructed by using an algorithm based on some functional relationships among the distance values. The methods described below illustrate different types of functional relationships.

One of the most popular distance approach methods is the unweighted pair-group method with arithmetic mean (UPGMA). This is also the simplest method for tree reconstruction. It assumes that the rates of evolution are approximately constant among the different evolutionary lineages under study, so that an approximately linear relation exists between the evolutionary distance and the divergence time. As this assumption usually does not hold well, the method often does not perform well. There now exist many better methods, but this method is presented here because it is good for explaining some basic concepts and principles in tree reconstruction.

The UPGMA method employs a sequential clustering algorithm, in which local topological relationships are inferred in order of their decreasing similarity and a phylogenetic tree is built in a stepwise manner. That is, we first identify the two OTUs that are most similar to each other (i.e., have the shortest distance) and treat them as a new single OTU. Such an OTU is referred to as a *composite OTU*. Subsequently, from among the new group of OTUs, we identify the pair with the highest similarity, and so on, until only two OTUs are left.

Example 9.3 Let us assume there are four taxa a, b, c, and d with pairwise distances between each of the taxa as shown in the following matrix:

Species	a	b	c
b	Dab	-	-
c	Dac	Dbc	-
d	Dad	Dbd	Dcd

Here Dab, Dac, Dad, Dbc, Dbd, and Dcd, represent the distances between ab, ac, ad, bc, bd, and cd taxa, respectively.

If value of Dab is least among all the values, then a and b are grouped into a composite group and a new distance matrix is computed with the help of the following formulas:

```
D(ab)c = ½ (Dac+Dbc)
```

and

```
D(ab)d =  ½(Dad + Dbd)
```

The taxa separated by the smallest distance in the new matrix are clustered to make another new composite group and this process is repeated till all the taxa are grouped.

The branch length between any two taxa a and b is the matrix value Dab. Therefore the branch length of a to the ancestor internal node is Dab/2 and of b to its ancestor internal node is Dab/2.

9.2.2 Character State Methods

(a) Maximum parsimony

This method uses character state data. The principle of maximum parsimony is to search for a tree that requires the smallest number of evolutionary changes to explain the differences observed among the OTUs under study. Such a tree is called a *maximum parsimony tree*. Often more than one tree with the same minimum number of changes is found, so that no unique tree can be inferred. The method discussed below was first developed for amino acid sequence data by Eck and Dayhoff and was later modified for use on nucleotide sequences by Fitch.

A nucleotide site is phylogenetically informative only if it favours some trees over the others. To illustrate the distinction between informative and noninformative sites, consider the four hypothetical sequences given in Table 9.5.

Table 9.5 Four hypothetical sequences

Sequence	Sites
1	A A G A G T G C A
2	A G C C G T G C G
3	A G A T A T C C A
4	A G A G A T C C G

There are 15 possible rooted trees and three possible unrooted trees for these four OTUs as per Table 9.4. Site 1 is not informative because all sequences at this site have A, so that no change is required in any of the three possible trees. At site 2, only sequence 1 has A, while all other sequences have G. So a simple assumption is made that the nucleotide has changed from G to A in the lineage leading to sequence 1. Thus, this site is also not informative, because each of the three possible trees requires 1 change. As shown in Fig. 9.9 for site 3 each of the three possible trees requires 2 changes and so it is also not informative. Even if we assume that the nucleotide at the node connecting OTUs 1 and 2 in tree I is C instead of G, the number of changes required for the tree remains 2. For site 4, each of the three trees requires 3 changes and thus site 4 is also noninformative. For site 5, tree I requires only 1 change, whereas trees II and III require 2 changes each. Therefore, this site is informative.

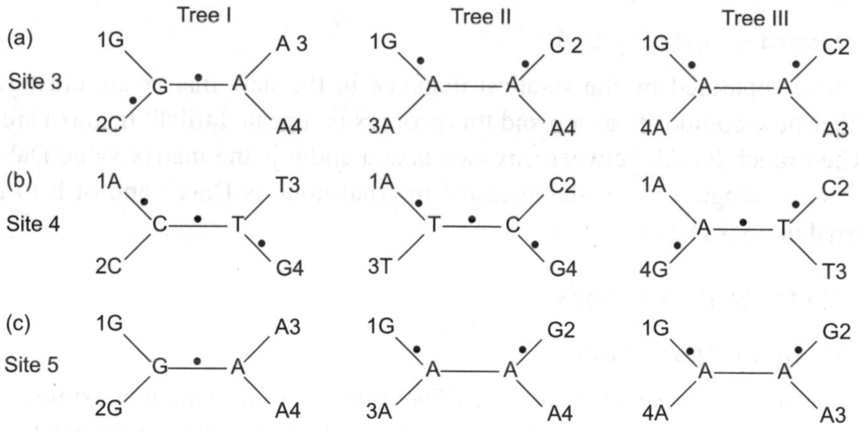

Fig. 9.9 Three possible trees for sites 3, 4, and 5

From these examples, we see that a site is informative only when there are at least two different kinds of nucleotides at the site, each of which is represented in at least two of the sequences under study. For these informative sites, tree I requires 1, 1, and 2 changes, respectively; tree II requires 2, 2, and 1 changes; and tree III requires 2, 2, and 2 changes. Thus, tree I is chosen because it requires the smallest number of changes at the informative sites. When the number of OTUs under study is larger than four, the situation becomes more complicated because there are many more possible trees to consider and because inferring the number of substitutions for each alternative tree becomes more tedious. However, the basic principle remains the same, i.e., to infer the minimum number of substitutions required for a given tree.

(b) Maximum likelihood

The maximum likelihood (ML) method is a well-established method in statistics. The first application of this method to tree reconstruction was made by Cavalli-Sforza and Edwards, who used gene frequency data. Later, Felsenstein developed ML algorithms for amino acid or nucleotide sequence data. This method is more complicated than the maximum parsimony method. We give only a brief description here. The ML method requires a probabilistic model for the process of nucleotide substitution; that is, we must specify the transition probability from one nucleotide state to another in a time interval in each branch.

The likelihood for all sites is the product of the likelihoods for individual sites if all the nucleotide sites evolve independently. For a given set of data, in order to compute the maximum likelihood value for each tree topology, you have to find the branch lengths that give the largest value for the likelihood function. This function will vary with the transition/transversion ratio, base composition, and substitution rate differences among lineages variation across sites. Finally, you can choose the topology with the highest maximum likelihood value as the best tree, which is called the maximum likelihood tree. Since the maximum likelihood method computes the probabilities for all possible combinations of ancestral states, it is computationally very time consuming.

In the past decade, one of the most important topics in evolutionary sequence analysis was the development of methods for statistical testing of phylogenetic hypotheses. These developments are available almost exclusively within the likelihood framework. They help assess which model provides the best fit for a given dataset, which is vital for the selection of the optimal model with which to perform phylogenetic inference. Statistical tests in phylogenetics also permit the assessment of the degree of confidence we have in any given tree topology being the true topology; in summary, they are responsible for the mutual feedforward between our abilities to estimate better trees and to create more realistic models of evolution.

9.3 Protein Structure Prediction

If you look at the atomic structure of even the smallest protein you will be amazed to see the complex convoluted path of the polypeptide main chain and the complex packing arrangements of the main and side chain atoms. Thus an attempt to predict the theoretical structure of a protein from its sequence has limited achievement as there are environmental and other extrinsic factors associated with it. Prediction is thus a pressing problem for many biologists as the discrepancy continues to increase between the number of known protein sequences and the number of experimentally determined protein structures. To deal with the prediction problem, we need to understand some of the dominant effects of protein structure such as the net protein stability, hydrophobic effect, atomic packing, conformational entropy, electrostatic effects due to ion pairs and hydrogen bonds, disulfide bridges, etc. After having known these the next thing that you must understand is the principles of protein architecture that form the basis for the prediction algorithm. They are residue conformation, periodic and non-periodic secondary structure, residue burial and interactions, associations of secondary structures, folds of protein domain, and protein evolution. Figure 9.10 shows a general approach to protein prediction. You will come across various approaches, starting with sequence analysis to secondary and tertiary structure prediction and to modelling with protein structures simulations.

Sequence Analysis

SEARCH for <homologous sequences> in databases
IF <homologous> FOUND
  GENERATE <multiple alignments>
IDENTIFY <known functional motifs>

Secondary Structure Prediction

PREDICT <secondary structure>
IDENTIFY <transmembrane segments, if any >

Tertiary Structure Prediction

IF there is an <experimental structure for a homologue>
PREDICT by <comparative modelling>
IF <no obvious homologue>
SEARCH for <distant homologue> or <an analogous fold> USING <fold recognition>
IF <no homologue or analogue>
IF <small protein>
DOCK <secondary structure>
ELSE
ATTEMPT <de novo folding> USING <simplified representation or lattice simulations>
IF <transmembrane segments>
ASSIGN <topology>

Modelling of Protein Structures

Refine accurately predicted structures and model local conformational changes in experimental structures USING <energy calculation>
Docking <protein-protein> AND <protein-ligand>
USING <experimental structures> AND <accurately predicted structures>

Verification

TEST <predictions> AGAINST <experimental structures> AND <other experimental data>
USE <human expertise> to INTERVENE AND MODIFY <all or parts of these procedures>

Fig. 9.10 Protein prediction flow chart

Other than the documentation produced by commercial companies, the Internet helps even the nonexpert to implement the approaches of protein structure prediction. The problem for the new comers is to choose the algorithm and the tool for the prediction task. Here we will learn about two algorithms for secondary structure prediction.

9.3.1 Chou–Fasman Method

This is a statistical method of secondary prediction that collects frequencies of all amino acid types from elements of known secondary structures, makes a table for each amino acid, and scans a peptide sequence for short segments that have high probability. It is based on the calculation of statistical propensities of each residue forming either an alpha helix or a beta strand. The Chou–Fasman method of secondary structure prediction depends on assigning a set of prediction values to a residue and then applying a simple algorithm to those numbers, conformational parameters, and positional frequencies. Conformational parameters for each amino acid are calculated using three criteria:

- the relative frequency of a given amino acid within a protein,
- its occurrence in a given type of the secondary structure, and
- the fraction of residues occurring in that type of structure.

These three parameters can be taken as measures of a given amino acid's preference to be found in a helix, sheet, or coil. Using these conformational parameters, it is possible to find nucleation sites within the sequence. This is extended until as sequence stretch is found that usually is not found in that type of the structure or until a stretch is found that is more stable in another type of the structure. At that point, the structure is terminated. This process is repeated throughout the sequence until the entire sequence is predicted.

Note

The conformational parameters for each amino acid are calculated by considering the relative frequency of a given amino acid within a protein, its occurrence in a given type of secondary structure, and the fraction of residues occurring in that type of the structure. These parameters are measures of a given amino acid's preference to be found in a helix, sheet, or coil. Using these conformational parameters, one finds nucleation sites within the sequence and extends them until a stretch of amino acids is encountered that is not disposed to occur in that type of the structure or until a stretch is encountered that has a greater disposition for another type of the structure. At that point, the structure is terminated. This process is repeated throughout the sequence until the entire sequence is predicted.

The values for conformational parameters and positional frequencies of 20 amino acids are given in Table 9.6

Table 9.6 Values for conformational parameters and positional frequencies of 20 amino acids

Name	P(a)	P(b)	P(turn)	f(i)	f(i+1)	f(i+2)	f(i+3)
Alanine	142	83	66	0.06	0.076	0.035	0.058
Arginine	98	93	95	0.070	0.106	0.099	0.085
Aspartic Acid	101	54	146	0.147	0.110	0.179	0.081

(contd)

Table 9.6 (*contd*)

Name	P(a)	P(b)	P(turn)	f(i)	f(i+1)	f(i+2)	f(i+3)
Asparagine	67	89	156	0.161	0.083	0.191	0.091
Cysteine	70	119	119	0.149	0.050	0.117	0.128
Glutamic Acid	151	037	74	0.056	0.060	0.077	0.064
Glutamine	111	110	98	0.074	0.098	0.037	0.098
Glycine	57	75	156	0.102	0.085	0.190	0.152
Histidine	100	87	95	0.140	0.047	0.093	0.054
Isoleucine	108	160	47	0.043	0.034	0.013	0.056
Leucine	121	130	59	0.061	0.025	0.036	0.070
Lysine	114	74	101	0.055	0.115	0.072	0.095
Methionine	145	105	60	0.068	0.082	0.014	0.055
Phenylalanine	113	138	60	0.059	0.041	0.065	0.065
Proline	57	55	152	0.102	0.301	0.034	0.068
Serine	77	75	143	0.120	0.139	0.125	0.106
Threonine	83	119	96	0.086	0.108	0.065	0.079
Tryptophan	108	137	96	0.077	0.013	0.064	0.167
Tyrosine	69	147	114	0.082	0.065	0.114	0.125
Valine	106	170	50	0.062	0.048	0.028	0.053

The actual algorithm contains a few simple steps:

1. Assign all of the residues in the peptide the appropriate set of parameters.
2. Scan through the peptide and identify regions where 4 out of 6 contiguous residues have P(a-helix) greater than 100. This is the nucleation criterion for the helix, and this region is declared an alpha-helix. Extend the helix in both directions until a set of four contiguous residues that have an average P(a-helix) less than 100 is reached. This is declared as the end of the helix. If the segment defined by this procedure is longer than 5 residues and the average P(a-helix) is greater than P(b-sheet) for that segment, the segment can be assigned as a helix.
3. Repeat this procedure to locate all of the helical regions in the sequence.
4. Scan through the peptide and identify a region where 3 out of 5 of the residues have a value of P(b-sheet) greater than 100. This is the nucleation criterion for the strand or sheet, and this region is declared as a beta-sheet. Extend the sheet in both directions until a set of four contiguous residues that have an average P(b-sheet) less than100 is reached. This is declared as the end of the beta-sheet. Any segment of the region located by this procedure is declared a beta-sheet if the average P(b-sheet) greater than 105 and the average P(b-sheet) greater than P(a-helix) for that region.
5. Any region containing overlapping alpha-helical and beta-sheet assignments are taken to be helical if the average P(a-helix) greater than P(b-sheet) for that region. It is a beta-sheet if the average P(b-sheet) greater than P(a-helix) for that region. If both the helix and strand are predicted, higher prediction is taken.
6. To identify a bend at residue number j, compute the following
   $p(t) = f(j)f(j + 1)f(j + 2)f(j + 3)$

Here values f(j), f(j + 1), f(j + 2), f(j + 3) are are bend frequencies in the four positions on the beta-turn and are used for j, j + 1, j + 2, and j + 3, respectively. Beta-turn is predicted at a location if the following conditions are satisfied:

- p(t) > 0.000075,
- the average value for P(turn) is greater than 1.00 in the tetrapeptide where P(turn) is the conformational parameter for ß-turn, and
- the averages for the tetrapeptide obey the inequality P(a-helix) < P(turn) > P(b-sheet) where P(a-helix), P(b-sheet), and P(turn) are the conformational parameters for the helix, sheet, and turn, respectively.

Note

P(helix), P(sheet), and P(turn) are the conformational parameters for the helix, sheet, and turn, respectively with values between 50 and 150.
f(j), f(j + 1), f(j + 2), and f(j + 3) are bend frequencies in the four positions on the beta-turn.

This method is one of the most widely used predictive schemes because it is simple to use and easy to understand. The drawback of this algorithm is that it neglects influences of any surrounding amino acids and the result is 50–60% accurate where predictions of alpha-helices and beta-sheets are accurate in about 60% of cases while the reverse turns are accurate in about 47% of cases. The Chou–Fasman method can be improved by taking into account the positions of the amino acid residues. For example, the acidic residues predominate at the N-termini of helices and the basic residues predominate at the C-termini and this results in favourable interactions with the helical macrodipole. In addition, proline is rare in the interiors of helices but is common at the N-termini and glycine is found at the C-terminus of a third of all known helices. Other approaches to secondary structure prediction take into account factors such as stereochemistry, hydrophobi` city, hydrophilicity, and electrostatic properties.

TASK

You are given the following protein sequence that belongs to the mouse hemoglobin beta-2 chain:

```
>gi|122526|sp|P02089|HBB2_MOUSE Hemoglobin beta-2 chain (B2) (Hemoglobin
beta-minor chain)
MVHLTDAEKSAVSCLWAKVNPDEVGGEALGRLLVVYPWTQRYFDSFGDLSSASAIMGNPKVKAHGKKVIT
AFNEGLKNLDNLKGTFASLSELHCDKLHVDPENFRLLGNAIVIVLGHHLGKDFTPAAQAAFQKVVAGVAT
ALAHKYH
```

Let us try and determine its secondary structure using the Chou–Fasman method of prediction.

Step 1 Go to the following url: http://fasta.bioch.virginia.edu/fasta_www/chofas.htm. The interface will look like as shown in Fig. 9.11.

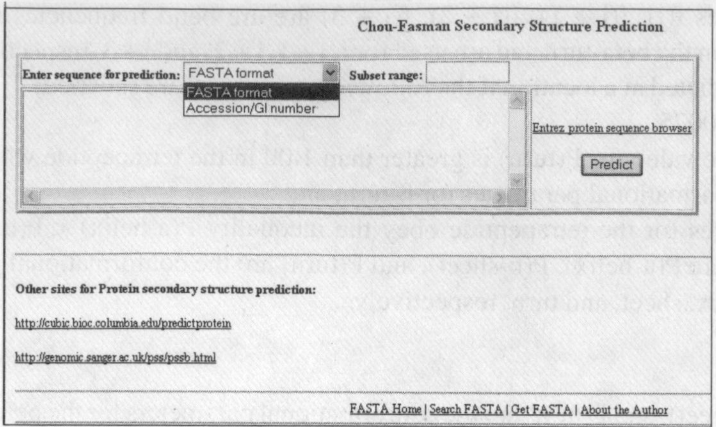

Fig. 9.11

Step 2 From the drop-down, select the FASTA format option.

Step 3 Paste the given sequence in FASTA format in the textbox provided. Click the **Predict button**. The result page will look like this as shown in Fig. 9.12.

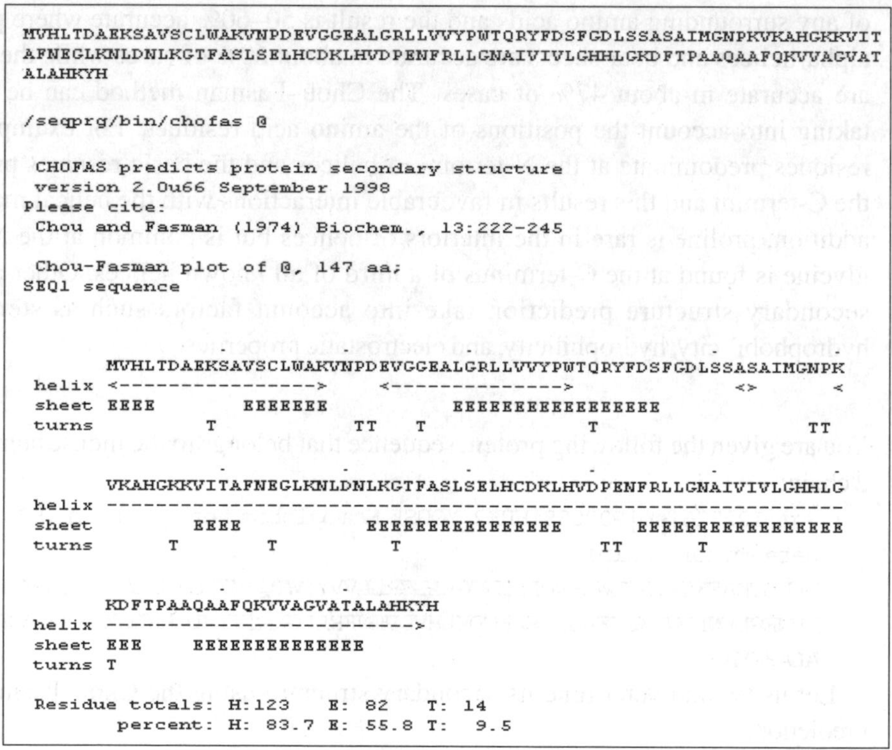

```
MVHLTDAEKSAVSCLWAKVNPDEVGGEALGRLLVVYPWTQRYFDSFGDLSSASAIMGNPKVKAHGKKVIT
AFNEGLKNLDNLKGTFASLSELHCDKLHVDPENFRLLGNAIVIVLGHHLGKDFTPAAQAAFQKVVAGVAT
ALAHKYH

/seqprg/bin/chofas @

 CHOFAS predicts protein secondary structure
 version 2.0u66 September 1998
Please cite:
 Chou and Fasman (1974) Biochem., 13:222-245

 Chou-Fasman plot of @, 147 aa;
SEQ1 sequence

               .         .         .         .         .         .
      MVHLTDAEKSAVSCLWAKVNPDEVGGEALGRLLVVYPWTQRYFDSFGDLSSASAIMGNPK
helix <----------------->   <-------------->             <>      <
sheet EEEE       EEEEEEE          EEEEEEEEEEEEEEEEE
turns         T           TT   T              T                TT

               .         .         .         .         .         .
      VKAHGKKVITAFNEGLKNLDNLKGTFASLSELHCDKLHVDPENFRLLGNAIVIVLGHHLG
helix ------------------------------------------------------------
sheet        EEEE          EEEEEEEEEEEEEEEE      EEEEEEEEEEEEEEEEE
turns      T       T         T                TT      T

               .         .
      KDFTPAAQAAFQKVVAGVATALAHKYH
helix ------------------------->
sheet EEE    EEEEEEEEEEEEEE
turns T

Residue totals: H:123   E: 82   T: 14
       percent: H: 83.7 E: 55.8 T:  9.5
```

Fig. 9.12

> **Note**
> Helical regions are depicted by HHH, sheets as EEE, and turns as TT.

Step 4 Now answer the following:

a. What are the main alpha-helix forming and breaking residues?
b. What are the main beta-sheet forming and breaking residues?
c. Why is proline a breaking residue for both alpha and beta structures?
d. What are the preferred residues at the central two positions of a turn? Are turns simply hydrophilic regions ? Explain.

9.3.2 GOR

The GOR method of secondary structure prediction was developed by **G**arnier, **O**sguthorpe, and **R**obson. It is one of the most widely used methods that are currently being used. Like the Chou–Fasman method, this method also is based on statistical calculation. It is robust and theoretically sound. It interprets protein folding as message translation and method allows the separation and evaluation of different types of information such as intraresidue, directional and pair interaction, etc. involved in protein folding. This method uses both information theory and Bayesian statistics for predicting the secondary structure of proteins, which works on the principal of maximizing 'information'.

```
I(x,y) = information that event y carries about the occurrences of x
I(x,y) = log(P(x|y)/P(x))
I = 0 for no information
I > 0 if A favors helix
I < 0 if disfavors
```

Here x is one of the three states helix (H), extended (E) and coil (C) and y is one of the 20 possible amino acids.

Information function is defined as the logarithm of the ratio of the conditional probability P(x,y) of observing conformation x and the probability P(x)of the occurrence of conformation x.

The GOR method is based on the idea of treating the primary sequence and the secondary structure sequence as two messages related by a translation process, which is examined using the information theory.

There are various versions of GOR that take a sequence in FASTA format and calculate the information function I(x,y) to predict its secondary structure as better accuracy.

- GOR I used a small database of 26 proteins with about 4500 residues.
- GOR II used a database of 75 proteins containing 12757 residues.
- Both GOR I and II predicted four conformations (helix, extended sheets, coil, and turns) and were using singlet frequency information called as directional information within the window.
- GOR III used the same database as GOR II. Apart from the information used earlier, GOR III also used information about the frequencies of pairs of residues within the window.

- GOR IV uses a database of well determined structures with crystallographic resolution of 2.5 A of 267 protein chains containing 63566 residues. It uses all possible pair frequencies within a window of 17 amino acid residues that comprise 8 N-terminal and 8 C-terminal residues plus one central residue for each of the three structural states. GOR IV has a mean accuracy of 64.4% for a three-state prediction (Q3) allows cross-validation within the database. Its drawback is that the method is not related to physico-chemical principles. It has 63% overall accuracy and serious underprediction on strands (46%).
- The GOR V method was written by Jean-Francois Gibrat and was modified to increase its efficiency by Stephen Pheiffer. It combines information theory, Bayesian statistics, and evolutionary information. The major change in the algorithm is the inclusion of evolutionary information using Psi-BLAST. The system is trained on 513 proteins that reside in a server. The secondary structure prediction for a sequence of 100 amino acids takes 1 min using this method. It has an accuracy of prediction, Q3, of 73.5%. It works as mentioned in the following steps:
  1. A peptide is taken and the system calculates the helix, sheet, and coil probabilities at each residue position. Based on the structural states having highest probabilities, it makes an initial prediction.
  2. After this initial prediction, heuristic rules are applied. These rules include converting to coils from helices that are shorter than five residues and sheets that are shorter than two residues.
  3. It gives a user friendly output, displaying the sequence and the predicted secondary structure in rows, where H = helix, E = extended or beta strand, and C = coil.
  4. It also gives another output that displays the probability values for each secondary structure at each amino acid position. The predicted secondary structure with the highest probability is compatible with a predicted helix segment of at least four residues and a predicted extended segment of at least two residues.

TASK

You are given the following protein sequence that belongs to the mouse hemoglobin beta-2 chain:

```
>gi|122526|sp|P02089|HBB2_MOUSE Hemoglobin beta-2 chain (B2) (Hemoglobin
beta-minor chain)
MVHLTDAEKSAVSCLWAKVNPDEVGGEALGRLLVVYPWTQRYFDSFGDLSSASAIMGNPKVKAHGKKVIT
AFNEGLKNLDNLKGTFASLSELHCDKLHVDPENFRLLGNAIVIVLGHHLGKDFTPAAQAAFQKVVAGVAT
ALAHKYH
```

Let us try and determine its secondary structure using

- GOR I and
- the improved GOR V server.

Structure prediction using GOR I

Step 1 Go to the following url: http://abs.cit.nih.gov/gor/. The interface will look like as shown in Fig. 9.13.

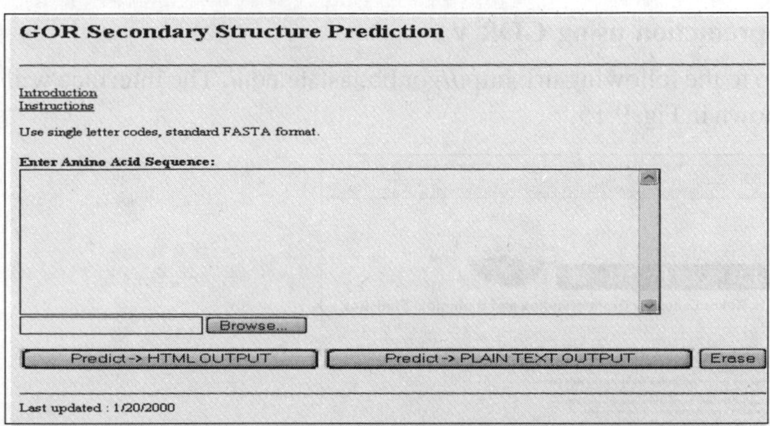

Fig. 9.13

Step 2 Paste the given sequence in FASTA format in the textbox provided. Click **Predict → plain text output button**. The result page looks like as shown in Fig. 9.14.

```
*****************Your Data****************
GI|122526|SP|P02089|HBB2_MOUSE HEMOGLOBIN BETA-2 CHAIN (B2) (HEMOGLOBIN
BETA-MINOR CHAIN)
MVHLTDAEKS AVSCLWAKVN PDEVGGEALG RLLVVYPWTQ RYFDSFGDLS     Sequence
CCCCCCCCCH HHHHHHHCCC CCCCCHHHHH EEEEECCCCC EEECCCCCCC     Predicted
Sec. Struct.
SASAIMGNPK VKAHGKKVIT AFNEGLKNLD NLKGTFASLS ELHCDKLHVD     Sequence
HHHHHHCCCC CCCCCCEEEE HHHHHHHHHC CCCCCHHHHH HHCCCCCCCC     Predicted
Sec. Struct.

PENFRLLGNA IVIVLGHHLG KDFTPAAQAA FQKVVAGVAT ALAHKYH        Sequence
CCHHHHHCCC EEEEECCCCC CCCCHHHHHH HHHHHHHHHH HHCCEEC        Predicted
Sec. Struct.

*************Column Output************
GI|122526|SP|P02089|HBB2_MOUSE HEMOGLOBIN BETA-2 CHAIN (B2) (HEMOGLOBIN
BETA-MINOR CHAIN)
147
SEQ PRD   H     E     C
 M   C  0.000 0.009 0.991
 V   C  0.000 0.035 0.965
 H   C  0.000 0.038 0.962
```

Fig. 9.14

> **Note**
> Coil regions are depicted by CCC, sheets as EEE, and turns (neither an-helix nor a B-sheet) as TT.

Structure prediction using GOR V

Step 1 Go to the following url: http://gor.bb.iastate.edu/. The interface will look like as shown in Fig. 9.15.

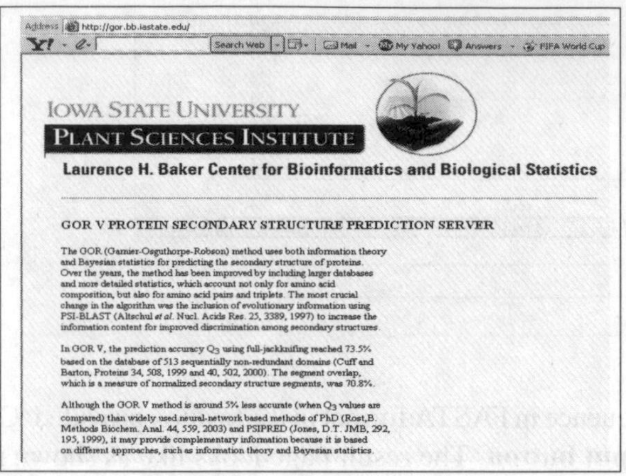

Fig. 9.15

Step 2 Paste the given sequence in FASTA format in the textbox provided. Click the **SUBMIT** button. While processing the web page looks like as shown in Fig. 9.16.

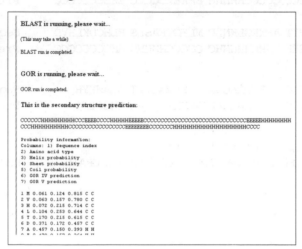

BLAST is running, please wait...

(This may take a while)

BLAST run is completed.

GOR is running, please wait...

GOR run is completed.

This is the secondary structure prediction:

CCCCCCHHHHHHHHHCCCEEEECCCHHHHHEEEEECCCCCCCCCCCCCCCCCCCCCCCCCCCCCEEEEEHHHHHHHH
CCCHHHHHHHHHHCCCCCCCCCCCCCCCCCEEEEEEEECCCCCCHHHHHHHHHHHHHHHHHHHHCCCC

```
Probability information:
Columns: 1) Sequence index
2) Amino acid type
3) Helix probability
4) Sheet probability
5) Coil probability
6) GOR IV prediction
7) GOR V prediction

1 M 0.061 0.124 0.815 C C
2 V 0.063 0.157 0.780 C C
3 H 0.072 0.215 0.714 C C
4 L 0.104 0.253 0.644 C C
5 T 0.170 0.215 0.615 C C
6 D 0.371 0.172 0.457 C C
7 A 0.457 0.150 0.393 H H
```

Fig. 9.16

Step 3 The result (sent by e-mail) looks like as shown in Fig. 9.17.

```
This is the GOR V prediction:
 CCCCCCHHHHHHHHHCCCEEEECCCCHHHHHEEEEECCCCCCCCCCCCCCCCCCCCCCCCCCEE
EEEHHHHHHHHH
 CCCHHHHHHHHHHCCCCCCCCCCCCCCCCCEEEEEEEEECCCCCCCHHHHHHHHHHHHHHHHHHCCCC
 Probability information:
 Columns: 1) Sequence index
          2) Amino acid type
          3) Helix probability
          4) Sheet probability
          5) Coil probability
          6) GOR IV prediction
          7) GOR V prediction

 1 M 0.061 0.124 0.815  C C
 2 V 0.063 0.157 0.780  C C
 3 H 0.072 0.215 0.714  C C
 4 L 0.104 0.253 0.644  C C
 5 T 0.170 0.215 0.615  C C
 6 D 0.371 0.172 0.457  C C
 7 A 0.457 0.150 0.393  H H
 8 E 0.479 0.157 0.364  H H

 Thank you for your interest. Please use the references provided at http://gor.bb.iastate.edu
```

Fig. 9.17

Step 4 Compare the results obtained by the two versions GOR I and GOR V. Answer the following:

(a) Are there any regions where GOR failed to predict the helical (H) regions?
(b) Are there any regions where GOR overpredicted the sheet (E) regions?
(c) Are there any regions where GOR over-predicted the coil (C) regions?

Conclusion

You have studied various types of prediction algorithms in this chapter. By now, you also must have some exposure to applications that use these algorithms. Algorithms are not something that you can read and learn. You need to learn them by way of drawing flowcharts, writing simple pseudocodes, and doing dry runs to get the result. So it is for you to first think of a prediction task and then design the algorithm. The other tasks involve understanding already designed algorithms that you will come across while you work. You should also try to find out the algorithm used behind any applications you need to work with. This will give you a clear picture why you are using the tool and what result is expected.

EXERCISES

Exercise 9.1 Consider three species X, Y, and Z. Draw the following for these species.

(a) All possible rooted trees.
(b) All possible unrooted trees.

Exercise 9.2 Find out the informative and noninformative site between the following sequences

(a) TAACGTAAG
(b) ATATCGGGA
(c) GGGTTAACG

Exercise 9.3 Use UPGMA to reconstruct a phylogenetic tree using the following distance matrix:

Species	M	N	O	P
N	3	-	-	-
O	5	4	-	-
P	8	8	9	-
Q	11	10	12	8

Exercise 9.4 Assume you have to contact your professor only when he is in good mood. S/He does not carry a mobile! You know that when s/he is at his workstation he is online via Yahoo messenger. So the best way to contact her/him is to login to your Yahoo ID and connect with her/him online. It s/he is not online at Yahoo, s/he is either in library or in his lab. There is only a 40% chance that the next day her/his mood is good if today s/he is off. If s/he is in good mood, then there is a 50% chance that s/he is in lab. If s/he is in off mood, there is a 60% chance that s/he is in library. Write a code snippet modelling HMM around this problem.

Exercise 9.5 Calculate the number of rooted and unrooted trees for 10 operational taxonomic units.

Exercise 9.6 Write to-the-point answers for the following:

(a) What are the main alpha-helix forming and breaking residues in the Chou–Fasman secondary structure prediction method?
(b) What are the main beta-sheet forming and breaking residues in the Chou–Fasman secondary structure prediction method?
(c) Why is proline a breaking residue for both alpha and beta structures?
(d) What are the preferred residues at the central two positions of a turn? Are turns simply hydrophilic regions? Explain, basing your answer on either the Chou–Fasman or the GOR prediction method.

Appendix

A1 Biology for Bioinformatics

Cells

Cells are the basic units building blocks of living organisms, with the exception of viruses. Even though, at the first glance, an animal cell looks very different from a plant cell, both have a similar organization. On the basis of the presence or absence of a single organelle the nucleus, all cells that we know of today can be classified as either prokaryotic cells or eukaryotic cells. The *eukaryotes* include protista, fungi, animals, and plants. *Prokaryotes* include archaebacteria and eubacteria, which are essentially single-cell organisms.

Archaebacteria

The archaebacteria live in extreme environments. They have been of great interest to biotechnologists and molecular biologists as several of the commercially viable products in this field available today use these systems or their genes. They belong to three major classes:

- ***Methanogens,*** which live in an anaerobic environment such as swamps and rice fields. These are methane producers and are intolerant to oxygen.
- ***Halophiles,*** which live in very high concentrations of salt (NaCl), e.g., in the Dead Sea and the Great Salt Lake.
- ***Extreme thermophiles,*** which live in a hot, sulphur rich and low pH environment, such as in hot springs, geysers, and fumaroles in the Yellowstone National Park, Kulu Manali valley.

Biomembranes

All biological membranes, including plasma membranes and all organelle membranes, contain lipids and proteins. The lipids found in biomembranes are mainly phospholipids and cholesterol. In the plasma membrane and some of organelle membranes, proteins and phospholipids are attached to carbohydrates, forming glycoproteins and glycolipids, respectively (see Fig. A1.1). To a bioinformatician, membranes are extremely important locations since these provide a microenvironment for proteins to fold and function.

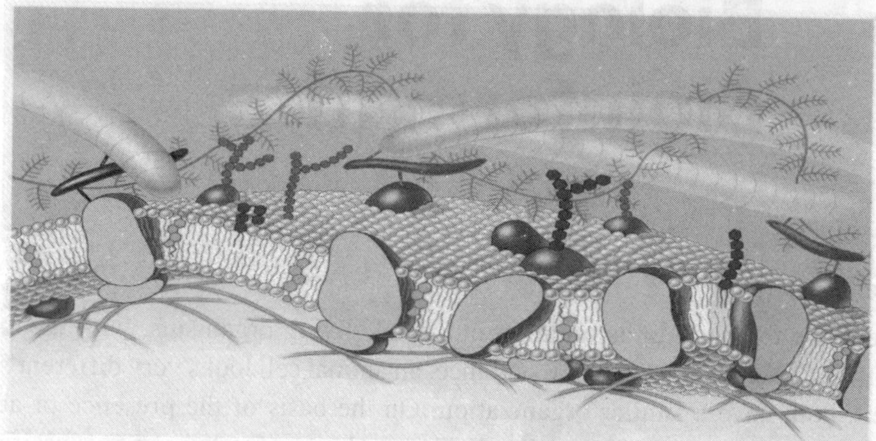

Fig. A1.1 A typical biomembrane showing the organization of lipid bilayers and embedded proteins and carbohydrates

Nucleus

The cell nucleus consists of a nuclear envelope, nucleolus, and nucleoplasm. Most chromosomes are located in the nucleoplasm, but portions of several chromosomes containing clusters of rRNA genes may get together in the nucleolus, forming the nucleolar organizing region. The major role of the nucleolus is to produce rRNA. This single organelle that houses most of the DNA in the cell, has been the focus of the entire Human Genome Project that has given birth to bioinformatics and its siblings, namely, computational biology, proteomics, metabolomics, microarrays, and personalized medicine.

Organelles

Organelles is a term that is used to refer to all membrane-bound structures in a cell (see Fig. A1.2). The nucleus is an example. Others that are found in the cytoplasm are mitochondria, chloroplasts, endoplasmic reticulum, Golgi apparatus, peroxisomes, lysosomes, vacuoles, and glyoxisomes.

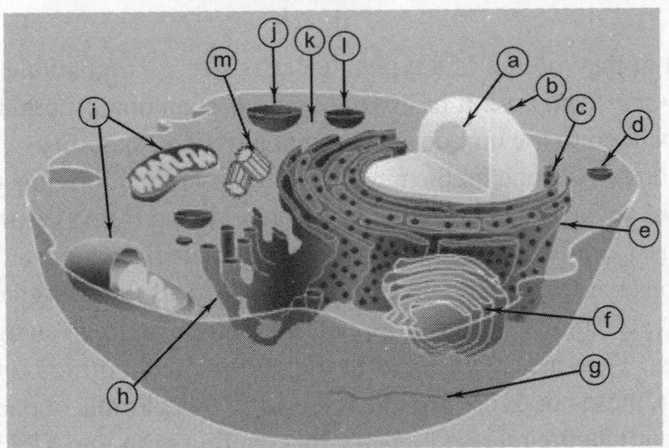

Fig. A2.1 Schematic of a typical animal cell, showing subcellular components, i.e., organelles: (a) nucleolus, (b) nucleus, (c) ribosome, (d) vesicle, (e) rough endoplasmic, reticulum (ER), (f) Golgi apparatus, (g) cytoskeleton, (h) smooth ER, (i) mitochondria, (j) vacuole, (k) cytoplasm, (l) lysosome, and (m) centrioles

Table A1.1 shows gives an overview of organelles.

Table A1.1 Organelles found in eukaryotes and prokaryotes

Organelle	Main Function	Organisms
Mitochondria	ATP synthesis	All eukaryotes
Nucleus	Store house of genetic information	All eukaryotes
Chloroplast	Synthesis of carbohydrates and harvesting of sunlight	All higher plants
Acrosome	Helps spermatozoa fuse with ovum	Many animals
Centriole	Anchor for cytoskeleton	Animals
Cilium	Movement in or of external medium	Animals, protists, few plants
Glyoxysome	Conversion of fat into sugars	Plants
Hydrogenome	Energy and hydrogen production	A few unicellular eukaryotes
Lysosome	Breakdown of large molecules	Most eukaryotes
Melanosome	Pigment storage	Animals
Nucleolus	Ribosome production	Most eukaryotes
Peroxisome	Oxidation of protein	All eukaryotes
Ribosome	Translation of RNA into proteins	Eukaryotes and prokaryotes
Vesicle	Miscellaneous	All eukaryotes
Endoplasmic reticulum	Modification and folding of new proteins and lipids	All eukaryotes
Golgi apparatus	Sorting and modification of proteins	Most eukaryotes
Vacuole	Storage and homeostasis	Eukaryotes

Mitochondria

It occupies one fourth of the volume of a typical cell. Its size is comparable to that of a common bacterium *Escherichia coli*. It has two membranes: an outer membrane and an inner membrane. Mitochondria are unique entities that have the capability to make more of their own kind and do not depend on the nucleus for their replications since they have their own DNA (mtDNA), which encodes some of the proteins and RNAs in mitochondria. However, most proteins operating in mitochondria still originate from nuclear DNA. Also known as the powerhouse of the cell, the mitochondria produces the energy currency of the living cells, namely, ATP (adenosine triphosphate), which carries high energy to power most cellular processes. This energy is stored in the phospho-anhydride bonds of ATP. During ATP hydrolysis, these bonds are broken, releasing 7.3 kcal/mole of energy. Many cellular processes can utilize the released energy by coupling with the ATP hydrolysis. The generation of ATP involves an electron transport series. Inevitably, electrons leak from the electron transport chain, producing free radicals, which have been implicated in the aging process.

Chloroplasts

Like mitochondria, chloroplasts have two membranes (inner and outer) and their own DNA. Yet most chloroplast proteins are still encoded by nuclear DNA. Inside the chloroplast, there are many thylakoids, each enclosed by a membrane. Chlorophylls, a group of photosynthetic pigments, are located on the thylakoid membrane and absorb light for photosynthesis. In the first step of photosynthesis, light energy is used to split water into hydrogen ions and oxygen molecules. The generated hydrogen ions create a concentration gradient across the thylakoid membrane. The movement of hydrogen ions through the membrane is coupled to ATP synthesis. The overall reactions can be written as

$$2H_2O \xrightarrow{\text{Light}} O_2 + 4H^+ + 4e^-$$

$$H^+ + ADP^{3-} + P_i^{2-} \rightarrow ATP^{4-} + H_2O$$

Viruses

Viruses are the smallest organisms, with diameters ranging from 20 nm to 300 nm. All viruses have a capsid enclosing genetic material and several enzymes. The genetic material and enzymes of a virus are enclosed by a surface structure called *capsid* (see Fig. A1.3). Some viruses also have an envelope surrounding the capsid. The shape of a viral capsid is helical or icosahedral.

Bacteriophage

Bacteriophage T4 is considered the 'Tyrannosaurus rex' of bacteriophages because it is one of the largest of the bacterial viruses. It is a tailed virus in that it has a tail with fibres that it uses to grip its host (see Fig. A1.4). Tailed viruses are very common; up to one billion phages can exist in a milliliter of freshwater.

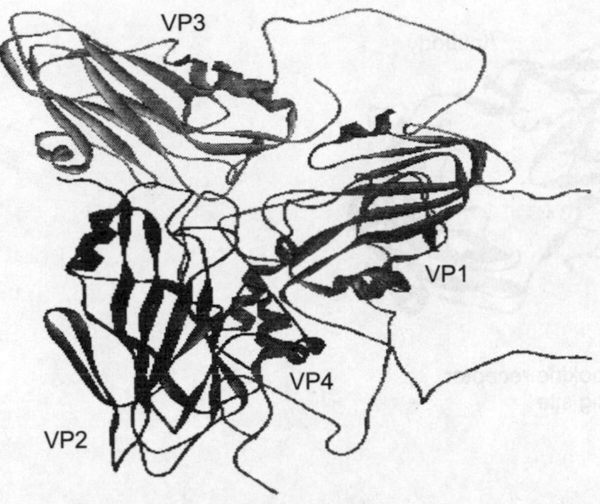

Fig. A1.3 A subunit in the capsid of the foot-and-mouth disease virus. The capsid is an icosahedron, comprising 60 subunits. Each subunit is made up of four proteins: VP1, VP2, VP3, and VP4.

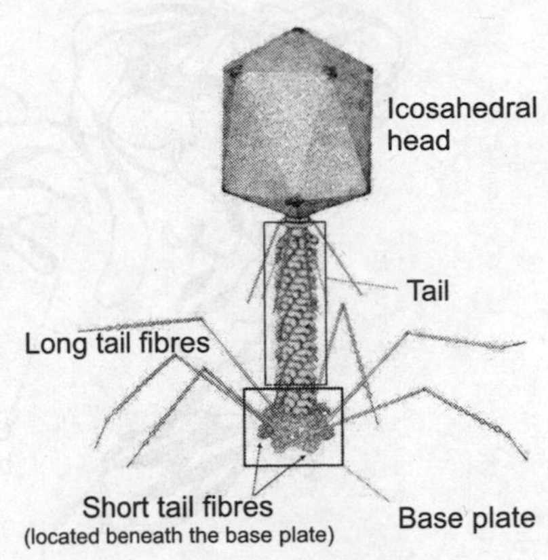

Fig. A1.4 Schematic figure of a T4 bacteriophage

Genetic contents

The genetic material in all types of cells is double-stranded DNA, but some viruses use RNA or single-stranded DNA to carry genetic information. The Baltimore classification system classifies viruses based on the type of genetic contents and replication strategies. It recognizes the following seven classes:

1. dsDNA viruses (Herpes virus)
2. ssDNA viruses
3. dsRNA viruses (common cold viruses)
4. (+)-sense ssRNA viruses (hepatitis A and C viruses)
5. (–)-sense ssRNA viruses (influenza virus)
6. RNA reverse transcribing viruses (HIV virus)
7. DNA reverse transcribing viruses (hepatitis B virus)

Here 'ds' denotes 'double strand' and 'ss' denotes 'single strand'.

Viral proteins

The proteins of viruses are of great interest to bioinformaticians especially those involved in unravelling protein structures and designing drug-protein interactions. One such example is depicted in Fig. A1.5. You may find it useful to visit the 'Big Pciture Book of Viruses' at http://www.virology.net/Big-Virology/BVHomepage.html

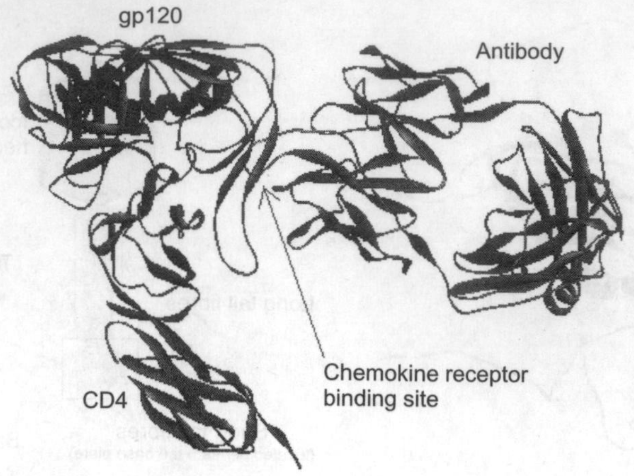

Fig. A1.5 Structure of the complex formed by HIV gp120, CD4 and an antibody against chemokine receptors (PDB ID = 1GC1)

Amino acid

An amino acid is defined as a molecule containing an amino group (NH_2), a carboxyl group (COOH), and an R group (see Fig. A1.6). It has the following general formula,

$$R\text{-}CH(NH_2)\text{-}COOH$$

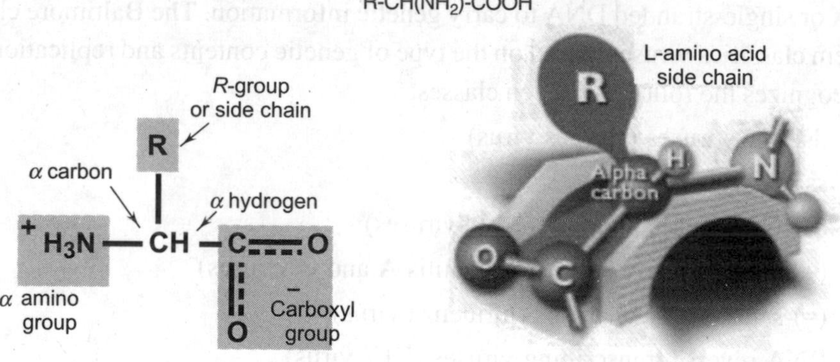

Fig. A1.6 Structure of of a typical amino acid showing the a carbon and R groups

The R group differs among various amino acids. In a protein, the R group is also called a *side chain*. Even though there are more than 300 naturally occurring amino acids, only 20 different amino acids are found in proteins. Based on the physicochemical properties of R groups, the 20 amino acids of proteins are classified as follows:

- ***Acidic*** In a neutral solution, the R group of an acidic amino acid may lose a proton and become negatively charged. Examples include aspartic acid (aspartate) and glutamatic acid (glutamate).

- ***Basic*** In a neutral solution, the R group of a basic amino acid may gain a proton and become positively charged. An interaction between positive and negative R groups may form a salt bridge, which is an important stabilizing force in proteins. Examples include lysine, arginine, and histidine.
- ***Aromatic*** Their R groups contain an aromatic ring. Examples include tyrosine, tryptophan, and phenylalanine.
- ***Sulphur containing*** Their R groups contain a sulphur atom (S). The disulphide bond formed between two cysteine residues provides a strong force for stabilizing the globular structure. A unique feature about methionine is that the synthesis of all peptide chains starts from methionine. Examples include cysteine and methionine.
- ***Uncharged hydrophilic*** Their R groups are hydrophilic and capable of forming hydrogen bonds. Examples include serine, threonine, asparagine, and glutamine.
- ***Inactive hydrophobic*** These amino acids are more likely to be buried in the protein interior. Their R groups do not form hydrogen bonds and rarely participate in chemical reactions. Examples include glycine, alanine, valine, leucine, and isoleucine.
- ***Special structure*** In most amino acids, the R group and the amino group are not directly connected. Proline is the only exception among 20 amino acids found in proteins. Due to this special feature, proline is often located at the turn of a peptide chain in the three-dimensional structure of a protein.

The chemistry of amino acids can be explored at http://www.biology.arizona.edu/biochemistry/problem-sets/aa/aa/html

*p*Ka

A molecule, or an atom group in a molcculc, may losc or gain a proton when it is placed in an aqueous solution. The exact probability that a molecule will be protonated or deprotonated depends on the *p*Ka of the molecule and the *p*H of the solution. If AH is an atom group in a molecule, it can be either neutral or charged. After AH loses a proton, it is denoted by A^-. The protonation/deprotonation reaction may be written as

$$AH \rightleftarrows H^+ + A^-$$

Define its equilibrium constant,

$$Ka = [H^+][A^-]/[AH]$$

where [...] represents the concentration. Taking logarithm on both sides,

$$\log ka = \log [H^+] + \log [A^-]/[AH]$$

Define

$$pKa = -\log Ka : pH = -\log [H^+]$$

we obtain

$$pKa - pH - \log [A^-]/[AH]$$

or

$$[A^-]/[AH] = 10^{(pH - pKa)}$$

It thus follows that half of the molecules will lose protons if they are in a solution where its $pH = pKa$. The higher the pH, the more the probability of a molecule to lose a proton. On the other hand, from the definition of pH, we find that the proton concentration of a solution becomes lower at a higher pH, thereby making it capable of accepting more protons.

Peptide bonds

Amino acids are linked together by peptide bonds to form polypeptides (see Fig. A1.7). This term is used to refer to long peptides, whereas *oligopeptides* are short peptides (less than 10 amino acids). Proteins are made up of one or more polypeptides with more than 50 amino acids. The primary structure of a protein refers to its amino acid sequence. The amino acid in a peptide is also called a *residue*.

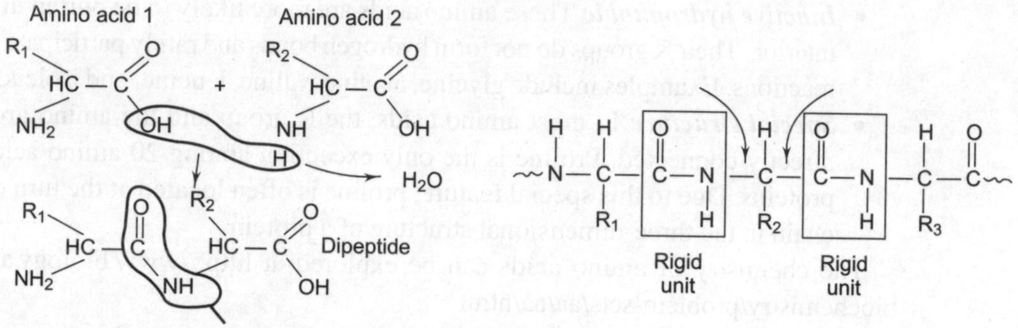

Fig. A1.7 Formation of peptide bond and rotational freedom in it

Phi-Psi angles

Due to the specific electronic structure of the peptide bond, the atoms on its two ends cannot rotate about the bond (see Fig. A1.7). Hence, the atoms of the group, O=C—N—H, are fixed on the same plane, known as the *peptide* plane (*see* Fig. A1.8). The whole plane may rotate about the N-Cα bond (ϕ angle) or C-Cα bond (ψ angle). Cα is the carbon atom connected to the R group.

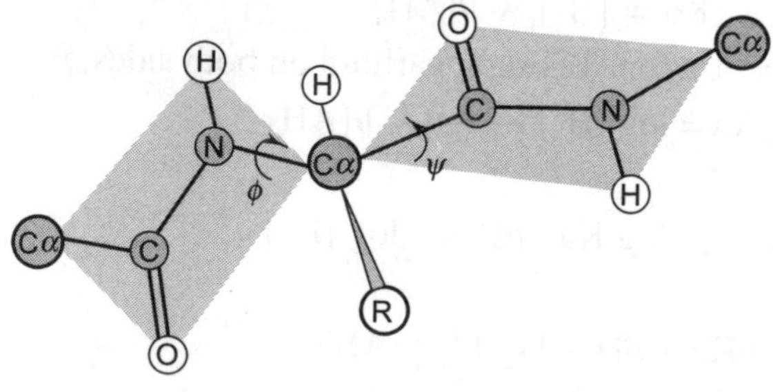

Fig. A1.8 Peptide plane (grey area) and ϕ-ψ angles. The red line formed by the repeating —Cα—C—N—Cα— is the backbone of the peptide chain

Phi (ϕ) and psi (ψ) are the dihedral angles (or torsional angles) between the point (e.g., Ca) at the end of a four-point sequence and the plane (e.g., the peptide plane) occupied by the other three points. In a peptide, phi-psi angles are restricted to certain ranges.

Ramachandran plot

A plot of the ϕ and ψ angles distribution is called the *Ramachandran plot*. The distance between two succession alpha carbon atoms in the backbone chain of a protein is approximately constant, as are the angles between the two bonds of such atoms. The proteins have only conformational freedom to rotate around the bonds in the backbone and in the side chain. The conformational angles show preferences for values that are expected based on simple energy considerations. Deviations from these angles may be used as indicators of any potential error in crystallographic projects. Phi and psi angles are also used in the classification of some secondary structure elements such as beta turns. In a Ramachandran plot, the core or allowed regions are the areas in the plot that show the preferred regions for ϕ-ψ angle pairs for residues in a protein. Presumably, if the determination of protein structure is reliable, most pairs will be in the favoured regions of the plot and only a few will be in 'disallowed' regions.

Tertiary structure of a protein

The three-dimensional (3D) structure is also called the tertiary structure. If a protein molecule consists of more than one polypeptide, it also has the quaternary structure, which specifies the relative positions among the polypeptides (subunits) in a protein. X-ray crystallography and nuclear magnetic resonance (NMR) are the major experimental techniques for determining the 3D structures of macromolecules. The Protein Data Bank (PDB) is the most important site for structural information, which can easily be accessed by the 'PDB ID' of a macromolecule (Fig. A1.5).

Nucleotides and nucleosides

Pentose, base, and phosphate groups make up a nucleotide. In DNA or RNA, a pentose has only one phosphate group, but within cells where free nucleotides such as ATP occur it may contain more than one phosphate group. If all phosphate groups are removed, a nucleotide becomes a nucleoside (see Fig. A1.9).

Nucleotide bases

Five different bases, each denoted by a single letter, are found in nucleic acids: adenine (A), cytosine (C), guanine (G), thymine (T), and uracil (U). Of these A, C, G, and T exist in DNA; and A, C, G, and U exist in RNA. Their chemical structures are shown in Fig. A1.10. A and G contain a pair of fused rings, classified as purines. C, T, and U contain only one ring, classified as pyrimidines.

At *p*H 7, all five bases are uncharged. Base pairing is a consequence of hydrogen bonding between acceptors and donors on the base molecules. There are three donor-acceptor pairs in the G≡C pairing, but only two in the A=T pairing. This means it is slightly more difficult to separate DNA strands containing a lot of G≡C pairs, and this property (the increased temperature required to 'melt', i.e., separate G≡C-rich DNA) has been used to classify bacteria. The *p*Ka of the groups in the bases has an effect on the H-bonding between complementary bases in nucleic acids. When any one of the three H-bonds between guanine and cytosine is disrupted by the addition of a proton (at low *p*H) to the middle N in the cytosine, it can potentially cause mutation.

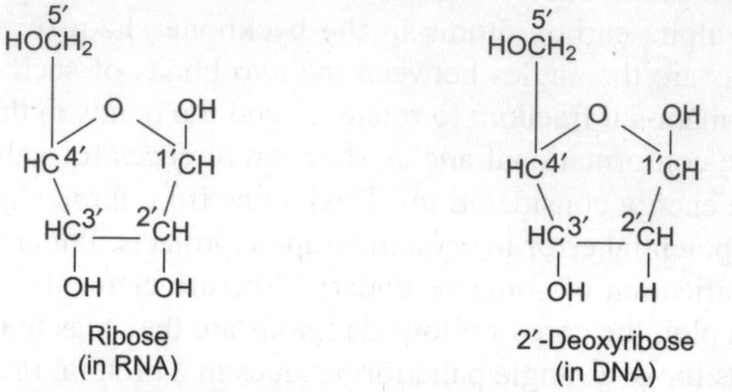

Ribose (in RNA)

2′-Deoxyribose (in DNA)

Fig. A1.9 Chemical structure of pentose containing five carbon atoms, labelled as C1′ to C5′. The pentose is called ribose in RNA and deoxyribose in DNA, because the DNA's pentose lacks an oxygen atom at C2′.

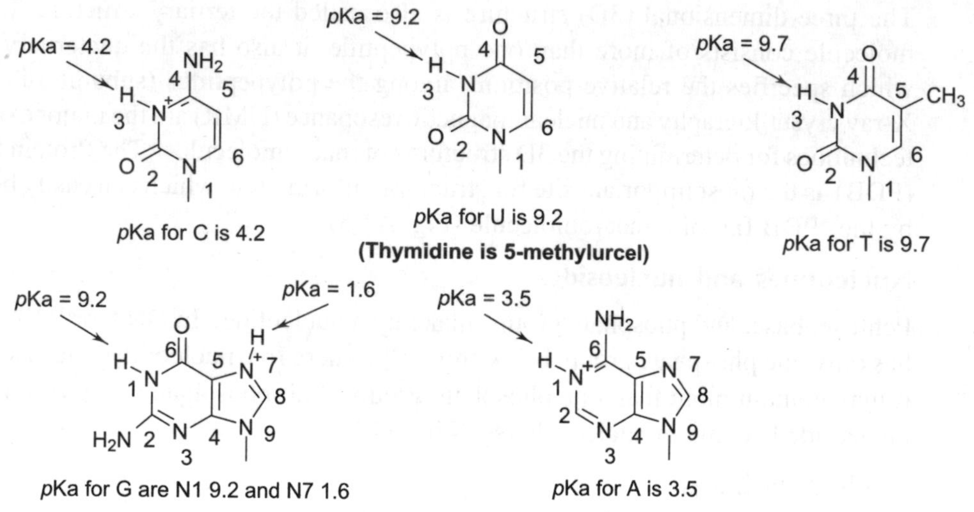

Fig. A1.10 Purines and pyrimidines with their *p*Ka

Isomery of nucleotide bases

Bases are prone to spontaneous isomery (tautomery), regardless of the pH. The minor tautomer is usually present at about 0.01% and will H-bond to the 'wrong' base. This is the main reason that DNA polymerase has a proofreading enzyme function so as to correct these errors.

Wobble

Nucleoside inosine (a derivative of the purine base hypoxanthine) is found in many tRNAs. It is capable of forming several sorts of hydrogen bonds and can consequently pair with A, C, and U in mRNA, leading to the phenomenon known as *wobble*: the ability of one tRNA to bond to more than one codon.

DNA

A DNA molecule has two strands, held together by the hydrogen bonding between their bases (see Fig. A3.11). Due to the specific base pairing, these two strands are complementary to each other. Hence, the nucleotide sequence of one strand determines the sequence on the other strand. Note that they obey the (A:T) and (C:G) pairing rule. Therefore, if we know the sequence on one strand, we can deduce the sequence on the other strand. *For this reason, a DNA database needs to store only the sequence of one strand.* By convention, the sequence in a DNA database refers to the sequence of the 5′ to 3′ strand (left to right).

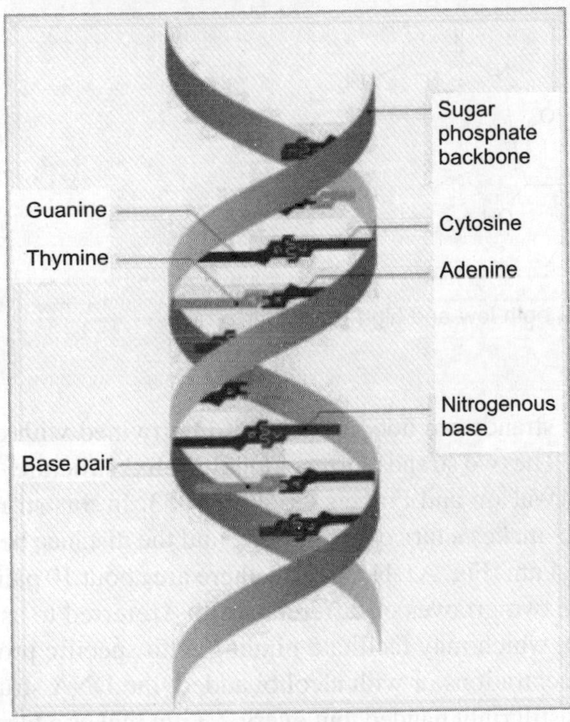

Fig. A1.11 Schematic drawing of the two strands of DNA

Stability of DNA and RNA

Purines take up protons at a low *p*H value, making them good leaving groups at low pH values. At low *p*H values, the bond between a base and ribose is readily hydrolysed.

Hence both DNA and RNA are unstable at a very low *p*H value. However, RNA has the additional disadvantage that the backbone of the RNA molecule is hydrolysed at high *p*H values (see Fig. A1.12) This is because the 2′ OH group is a good nucleophile and readily attacks the phosphate group on the 3′ carbon to form a cyclic product that is unstable to hydrolytic (hydroxide ion) attack. Thus when the phosphate group is swapped between the two carbons, the sugar phosphate backbone is broken. So RNA needs cosseting to stop it from falling apart in a way that DNA does not. RNA viruses are the only organisms whose genomes are written in RNA, and they have a notoriously high mutation rate.

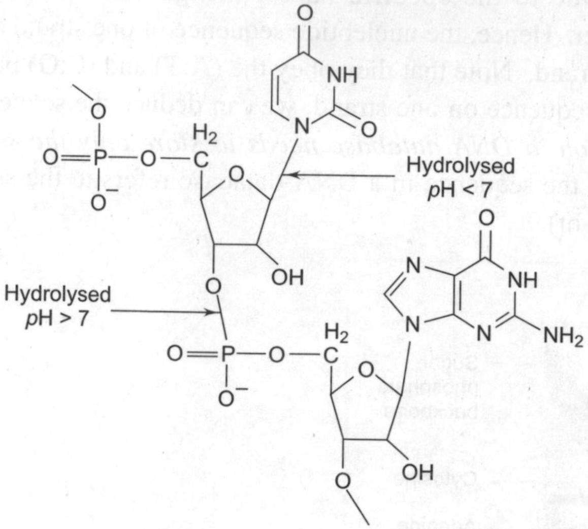

Fig. A1.12 RNA is inherently unstable at both low and high *p*H values

Forms of DNA

In a DNA molecule, the two strands are not parallel, but intertwined with each other. Each strand looks like a helix. The two strands form a 'double helix' structure, which was first discovered by James D. Watson and Francis Crick in 1953. In this structure, also known as the B form, the helix makes a turn every 3.4 nm, and the distance between two neighbouring base pairs is 0.34 nm (Fig. A1.11). Hence, there are about 10 pairs per turn. The intertwined strands make two grooves of different widths, referred to as the major groove and the minor groove, which may facilitate binding with specific proteins. In a solution with higher salt concentrations or with alcohol added, the DNA structure may change to an *A form*, which is still right handed, but every 2.3 nm makes a turn and there are 11 base pairs per turn. Another DNA structure is called the *Z form*, because its bases seem to zigzag. *Z* DNA is left-handed. One turn spans 4.6 nm and comprises 12 base pairs. The DNA molecule with alternating G-C sequences in alcohol or high salt solution tends to have such structure.

RNA

Most cellular RNA molecules are single stranded (see Fig. A1.13). The major role of RNA is to participate in protein synthesis. This requires three classes of RNA: messenger RNA (mRNA), transfer RNA (tRNA), and ribosomal RNA (rRNA). Other classes of RNA include ribozymes and small RNA molecules. They may form secondary structures such as stem loop and hairpin.

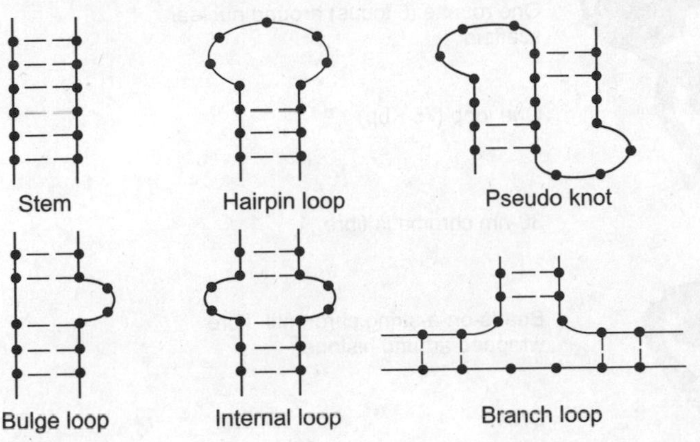

Fig. A1.13 Secondary structure of RNA

Chromatin

This is the substance which becomes visible chromosomes during cell division. Its basic unit is nucleosome, composed of 146 bp DNA and eight histone proteins. The structure of chromatin is dynamically changing, at least in part, depending on the need of transcription (see Fig. A1.14). In the metaphase of cell division, chromatin is condensed into the visible chromosome. At other times, the chromatin is less condensed, with some regions in a 'beads-on-a-string' conformation.

Histones

The basic repeating unit of chromatin, the nucleosome core particle, comprises approximately two turns of DNA wrapped around two molecules of each core histone protein: H2A, H2B, H3, and H4. The 30-nm chromatin fibre is associated with scaffold proteins (notably topoisomerase II) to form loops. Each loop contains about 75 kb DNA. Scaffold proteins are attached to DNA at specific regions, called *scaffold attachment regions* (SARs), which are rich in adenine and thymine. The chromatin fibre and associated scaffold proteins coil into a helical structure, which may be observed as a chromosome. G bands are rich in A-T nucleotide pairs while R bands are rich in G-C nucleotide pairs.

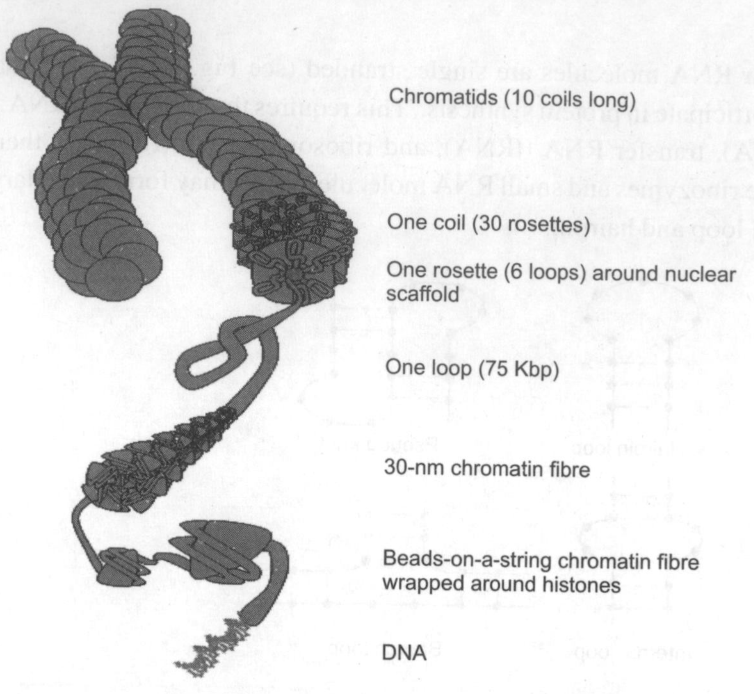

Fig. A1.14 Condensed structure of chromatin

Nucleosome

Nucleosomes are regularly spaced along eukaryotic DNA with a repeat length (i.e., centre-to-centre internucleosomal distance) of 180–200 bp, which probably reflects the most energetically favourable arrangement. The regular nucleosomal arrays comprise a 10-nm fibre of chromatin and fold into a 30-nm fibre upon the incorporation of a single molecule of linker histone, such as histone H1, per nucleosome. Higher levels of chromatin compaction involve additional proteins, many of which are unknown, in order to achieve the astounding degree of condensation required to yield mitotic chromosomes. Histone proteins are highly conserved among eukaryotic organisms. Consistent with this fact, the recently published crystal structure of the yeast nucleosome is surprisingly similar to the metazoan nucleosome structure. Even some archaebacteria contain nucleosomes comprised of 80 bp of DNA wrapped around two molecules each of the two archaeal histones. The similarity of nucleosome structures among eukaryotic organisms indicates that the mechanism of chromatin assembly is likely to be highly conserved across all eukaryotes. As such, we can combine insight provided from systems as diverse as yeast, *Drosophila*, *Xenopus*, and humans in order to build our understanding of the machinery and mechanism of chromatin assembly.

Genetic code

Protein synthesis is based on the sequence of mRNA, which is made up of nucleotides (proteins are made up of amino acids). There must be a specific relationship between the nucleotide sequence and the amino acid sequence. One of the approaches is illustrated in Table A1.2. It turns out that three nucleotides (a codon) code for one amino acid.

Table A1.2 Genetic code deciphered by Marshall Nirenberg and his colleagues in early 1960s

First position	Second position: U		C		A		G		Third position
U	UUU	Phe	UCU	Ser	UAU	Tyr	UGU	Cys	U
	UUC	Phe	UCC	Ser	UAC	Tyr	UGC	Cys	C
	UUA	Leu	UCA	Ser	UAA	*Stop*	UGA	*Stop*	A
	UUG	Leu	UCG	Ser	UAG	*Stop*	UGG	Trp	G
C	CUU	Leu	CCU	Pro	CAU	His	CGU	Arg	U
	CUC	Leu	CCC	Pro	CAC	His	CGC	Arg	C
	CUA	Leu	CCA	Pro	CAA	Gin	CGA	Arg	A
	CUG	Leu	CCG	Pro	CAG	Gin	CGG	Arg	G
A	AUU	Ile	ACU	Thr	AAU	Asn	AGU	Ser	U
	AUC	Ile	ACC	Thr	AAC	Asn	AGC	Ser	C
	AUA	Ile	ACA	Thr	AAA	Lys	AGA	Arg	A
	AUG	Met	ACG	Thr	AAG	Lys	AGG	Arg	G
G	GUU	Val	GCU	Ala	GAU	Asp	GGU	Gly	U
	GUC	Val	GCC	Ala	GAC	Asp	GGC	Gly	C
	GUA	Val	GCA	Ala	GAA	Glu	GGA	Gly	A
	GUG	Val	GCG	Ala	GAG	Glu	GGG	Gly	G

The synthesis of a peptide always starts from methionine (Met), coded by AUG. The stop codon (UAA, UAG, or UGA) signals the end of a peptide. This table applies to mRNA sequences. For DNA, U (uracil) should be replaced by T (thymine).

Effect of genetic code on mutation and replication errors

The genetic code is not randomly assigned. When an amino acid is coded by several codons, they often share the same sequence in the first two positions and differ in the third position. Such an assignment is accomplished by the design of *wobble position*. From Table A1.2, we see that the amino acids with close physical properties have similar codons. For example, both Asp and Glu are negatively charged. Their codons all contain 'GA' in the first two positions. The physical properties of Ser and Thr are also very close. They are coded by UCN and ACN (N = any), respectively. One advantage of this assignment is the reduction in the damage caused by replication errors or other mutations. For instance, if the third nucleotide of the Ser codon is mutated, the produced amino acid remains Ser. If the first nucleotide of the Ser codon is mutated to A, it will produce Thr, which is similar to Ser. Such mutation will not have significant effect on the protein structure and its function. Let us consider a hypothetical case that codons are randomly assigned. Suppose Ser were coded by GAG, GUA, GCU, and UGG; Thr were coded by AAA, AGU, UUU, and CCC. Then, any mutation would produce an amino acid different from Ser and Thr.

Open reading frame

In a DNA molecule, the sequence from an initiating codon (ATG) to a stop codon (TAA, TAG, or TGA) is called an *open reading frame* (ORF), which is likely (but not always) to encode a protein or polypeptide.

Codon usage bias

Biases in codon usage provide evidence for constraints on silent sites. If all the silent alternative codons were all functionally equivalent, we should expect only random variation in the frequency of those codons in a species. Table A1.3 shows that there are, in fact, consistent biases.

Table A1.3 Codon bias for a single amino acid arginine in humans, *Drosophila*, and *E. coli*

Codon	Human	Drosophila	E.coli
Arginine:			
AGA	22%	10%	1%
AGG	23%	6%	1%
CGA	10%	8%	4%
CGC	22%	49%	39%
CGG	14%	9%	4%
CGU	9%	18%	49%
Total number of arginine codons	2403	506	149
Total number of genes	195	46	149

How selection could discriminate among silent codons?

The nucleotide sequence controls the secondary structure of the DNA molecule; changes in nucleotides might then influence the molecular shape, which could make a difference to the organism's fitness. Silent substitutions would be as likely to influence the structure as replacements and selection would work on both for the same reason. The different silent codons use different tRNA molecules. It has been observed, in yeast and *E. coli*, that the frequency of the use of codons in a set of synonyms is correlated with the tRNA abundances in the cell. We should expect the relation if the frequencies of codons were set by some other factor: the tRNA abundances would then adjust (by natural selection) to the quantity needed. It has also been suggested that codon biases are caused by directional (or biased) mutation. Also it has been observed that genomes tend to avoid the usage of codons that are alike to the stop codons.

Gene

By definition, a gene includes the entire nucleic acid sequence necessary for the expression of its product (peptide or RNA). Such a sequence may be divided into a regulatory region and a transcriptional region.

Exons and introns

The regulatory region could be near to or far from the transcriptional region. The transcriptional region consists of *exons* and *introns*. Exons encode a peptide or functional RNA. Introns will be removed after transcription. As shown in Fig. A1.15, a typical DNA molecule consists of genes, pseudogenes, and an *extragenic region*.

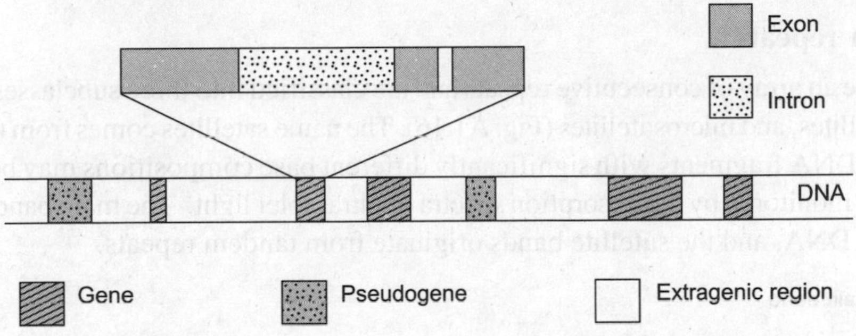

Fig. A1.15 General organization of exons and introns on the DNA sequence

Pseudogene

Pseudogenes are nonfunctional genes. They often originate from mutation of duplicated genes. Because duplicated genes have many copies, the organism can survive even if a couple of them become nonfunctional. An extreme case is that of the β-globin gene, which consists of three exons and two introns, with a total length of 1.6 kb.

Gene family

Gene family refers to a set of genes with homologous sequences. For example, H2A, H2B, H3, and H4 are in the same histone gene family. Their products have similar structures and functions. Another example is the β-globin gene family located on chromosome 11.

DNA repeats

A stretch of DNA sequence often repeats several times in the total DNA of a cell. For example, the entire telomeric ends of a chromosome contain 15 kb repeated sequences and reading as 'GGGTTA'. The number of repeated copies can be classified on the basis of DNA reassociation kinetics.

DNA reassociation kinetics

The total DNA is first randomly cleaved into fragments with an average size of about 1000 bp. Then, they are heated to separate the complementary strands of each fragment. Subsequently, the temperature is reduced to allow strand reassociation. If a fragment contains a sequence which is repeated many times in the total DNA, it will have greater chances to find a complementary strand and reassociate more quickly than other fragments with less repetitive sequences. Based on the reassociation rate, DNA sequences are divided into three classes:

Highly repetitive: About 10%–15% of mammalian DNA reassociates very rapidly. This class includes tandem repeats.

Moderately repetitive: Roughly 25%–40% of mammalian DNA reassociates at an intermediate rate. This class includes interspersed repeats.

Single copy (or a very low copy number): This class accounts for 50%–60% of mammalian DNA.

Tandem repeats

These are an array of consecutive repeats and are classified into three subclasses: satellites, minisatellites, and microsatellites (Fig. A1.16). The name satellites comes from their optical spectra. DNA fragments with significantly different base compositions may be separated and then monitored by the absorption spectra of ultraviolet light. The main band represents the bulk DNA, and the satellite bands originate from tandem repeats.

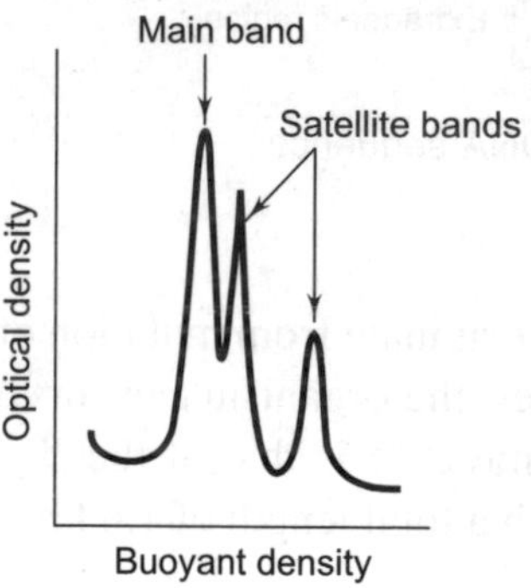

Fig. A1.16 Satellite bands using buoyant density gradient centrifugation

Satellites

The size of a satellite DNA ranges from 100 kb to over 1 Mb. In humans, a well known example is the alphoid DNA located at the centromere of all chromosomes. Its repeat unit is 171 bp and the repetitive region accounts for 3%–5% of the DNA in each chromosome. Other satellites have a shorter repeat unit. Most satellites in humans or in other organisms are located at the centromere.

Minisatellites

The size of a minisatellite ranges from 1 to 20 kb. One type of minisatellites is called *variable number of tandem repeats* (VNTR). Its repeat unit ranges from 9 to 80 bp. They are located in non-coding regions. The number of repeats for a given minisatellite may vary from individual to individual. This feature is the basis of DNA fingerprinting. Another type of minisatellites is the telomere. In a human germ cell, the size of a telomere is about 15 kb. In an aging somatic cell, the telomere is shorter. A telomere contains the tandemly repeated sequence GGGTTA.

Microsatellites

Microsatellites are also known as *short tandem repeats* (STR), because a repeat unit consists of only 1–6 bp and the whole repetitive region spans less than 150 bp. As is the case with minisatellites, the number of repeats for a given microsatellite may vary from individual to individual. Therefore, microsatellites can also be used for DNA fingerprinting. Both microsatellite and minisatellite patterns can provide information about paternity, such as the famous Thomas Jefferson case.

Polymorphism

This property describes the existence of different forms within a population, e.g., difference in the number of tandem repeats. All tandem repeat polymorphisms can result from DNA recombination during meiosis. The microsatellite polymorphism can also be caused by replication slippage.

Interspersed repeats

These are repeated DNA sequences located at dispersed regions in a genome. They are also known as mobile elements or transposable elements. A stretch of the DNA sequence may be copied to a different location through DNA recombination. After many generations, such a sequence (the repeat unit) could spread over various regions. Mobile elements were first discovered by Barbara McClintock in 1940s from the studies of corn. Subsequently, they were found in all kinds of organisms. In mammals, the most common mobile elements are *LINEs and SINEs*.

LINEs

LINEs stands for long interspersed nuclear elements. LINE elements code for two genes, one of which has known reverse transcriptase and integrase activity, enabling them to copy both themselves and other noncoding LINES such as Alu I elements (see Fig. A1.17). Because LINES move by copying themselves (instead of moving, like transposons do), they enlarge the genome. The human genome, for example, contains about 900,000 LINES, which is roughly 21% of the genome. LINES are used to generate genetic fingerprints. The most common LINEs in humans is the L1 family.

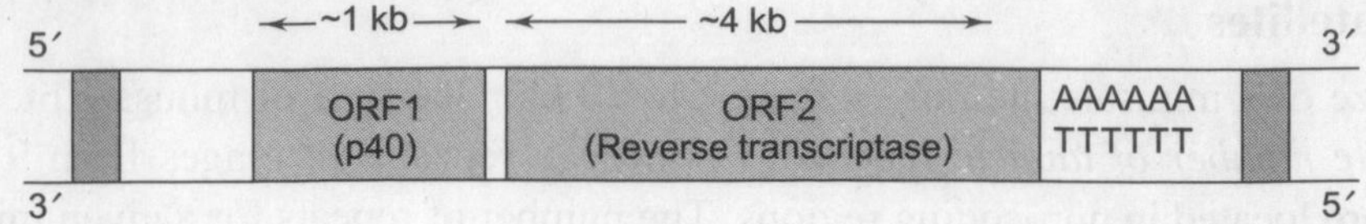

Fig. A1.17 Basic organization of LINEs

ORF1 and ORF2 are open reading frames. The protein product of ORF1 is called p40, with unknown function. ORF2 encodes a reverse transcriptase which is necessary for copying the element to other locations. The red color regions on both ends are direct repeats.

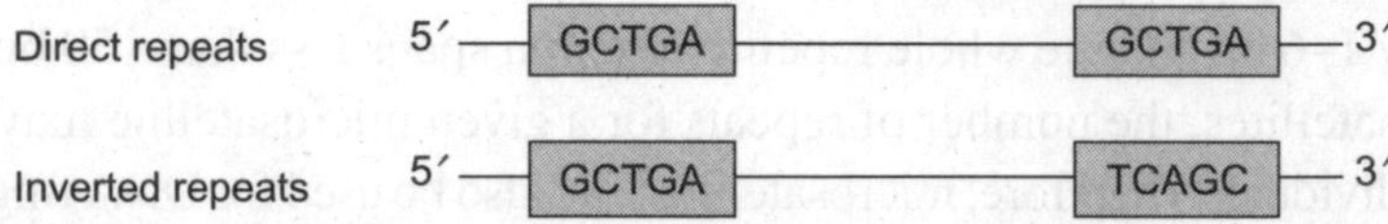

Fig. A1.18 Comparison between direct repeats and inverted repeats

Direct repeats have the same sequence along a given direction. Inverted repeats have complementary sequences along opposite directions (see Fig. A1.18).

SINEs

SINEs (short interspersed elements) are short DNA sequences (<500 bases) that represent reverse-transcribed RNA molecules originally transcribed by RNA polymerase III into tRNA, rRNA, and other small nuclear RNAs. SINEs do not encode a functional reverse transcriptase protein and rely on other mobile elements for transposition. The most common SINES in primates are called Alu sequences. Alu elements are about 300 bp long, do not contain any coding sequences, and can be recognized by the restriction enzyme Alu (thus the name). With about 1 million copies, SINEs make up about 11% of the human genome. While previously believed to be 'junk DNA', recent research suggests that both LINEs and SINEs have a significant role in gene evolution, structure, and transcription levels. The distribution of these elements has been implicated in some genetic diseases and cancers.

Cladistic markers using LINEs and SINEs

The analysis of SINEs and LINEs as molecular cladistic markers represents a particularly interesting complement to the DNA sequence and morphological data. The reason for this is that retrotransposons are assumed to represent powerful noise-free synapomorphies. The target sites are relatively unspecific so that the chance of an independent integration of exactly the same element into one specific site in different taxa is negligible even over evolutionary time scales. Retrotransposon integrations are assumed to be irreversible events since no biological mechanisms have yet been described for the precise re-excision of class I transposons. A clear differentiation between ancestral and derived character states at the respective loci thus becomes possible. In combination, the virtual lack of homoplasy together with a clear character polarity make retrotransposon integration markers ideal tools for determining the common ancestry of taxa by a shared, derived transpositional event. The presence of a given retrotransposon in related taxa implies their orthologue integration, a derived condition acquired via a common ancestry, while the absence of particular elements indicates the plesiomorphic condition prior to integration in more distant taxa. The use of presence/absence analyses to reconstruct the systematic biology of mammals depends on the availability of retrotransposons that were actively integrating before the divergence of a particular species.

Genome

Genome is the total genetic information of an organism. For most organisms, it is the complete DNA sequence. For RNA viruses, the genome is the complete RNA sequence, since their genetic information is encoded in RNA. Table A1.4 shows a comparison of genome sizes in various organisms.

Table A1.4 A comparison of genome sizes in various organisms

Organism	Genome Size (Mb)	Gene Number
Hepatitis D virus	0.0017	2
Hepatitis B virus	0.0032	4
HIV-1	0.0092	9
Bacteriophage l	0.0485	90
Escherichia coli	4.6	4437
S. cerevisiae (yeast)	12	6300
C. elegans (nematode)	97	19000
D. melanogaster (fruit fly)	137	14000
Mus musculus (mouse)	3000	20000–30000
Homo sapiens (human)	3000	20000–30000

Gene expression

All somatic cells in an organism have the same genome. Each organism may contain many types of somatic cells, each with a distinct function. The genes in a genome do not have any effect on cellular functions until they are expressed. Differential expression of various gene sets leads to a variety in size, shape, and function. Gene expression is the production of a protein or a functional RNA from its gene. Several steps are involved in this process:

Transcription: A DNA strand is used as the template to synthesize an RNA strand, which is called the primary transcript.

RNA processing: This step involves modifications in the primary transcript to generate a mature mRNA (for protein genes) or a functional tRNA or rRNA. For RNA genes (tRNA and rRNA), the expression is complete after a functional tRNA or rRNA is generated. However, protein genes require additional steps.

Nuclear transport: mRNA has to be transported from the nucleus to the cytoplasm for protein synthesis.

Protein synthesis

In the cytoplasm, mRNA binds to ribosomes, which can synthesize a polypeptide based on the sequence of mRNA. According to the gene expression process, the flow of genetic information is in the following direction: DNA → RNA → protein.

This rule was dubbed the 'central dogma', because it was thought that the same principle would apply to all organisms (see Fig. A1.19). However, we now know that for RNA viruses the flow of genetic information starts from RNA.

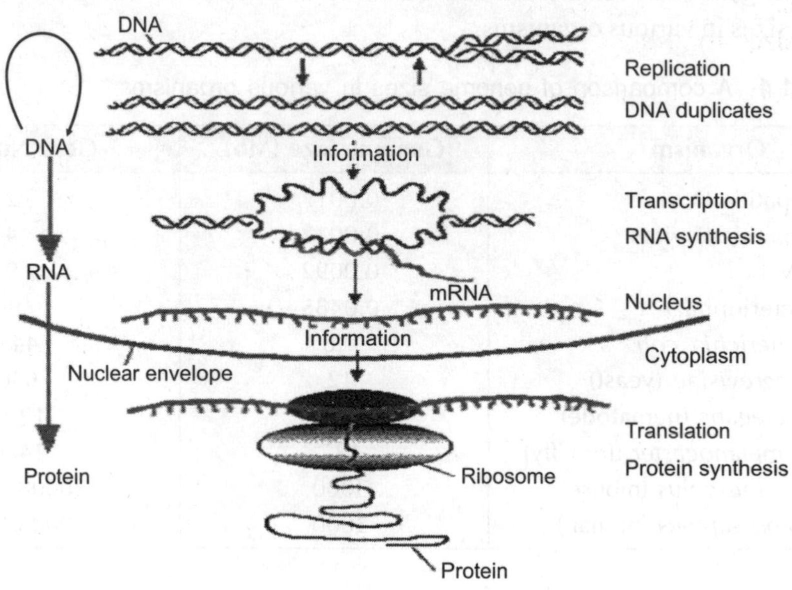

Fig. A1.19 Central dogma depicting steps involved in the expression of protein genes

Transcription

Transcription is a process in which one DNA strand is used as template to synthesize a complementary RNA. The following is an example:

```
5'ACATCGACGACGCGTTAATCCC -- 3'DNA coding strand (+)
3'TGTAGCTGCGCGTCAATTAAGGG -- 5'CNA template strand (-)
5'ACAUCGACGCGCAGUUAAUCCC -- 3'RNA (+)
```

Note that uracil (U) of RNA is paired with adenine (A) of DNA. There are a few different names for these nucleic acid strands. The DNA strand which serves as the template is called template strand, minus strand, or antisense strand. The other DNA strand is called the non-template strand, coding strand, plus strand, or sense strand. Since both the DNA coding and RNA strands are complementary to the template strand, they have the same sequences except that T in the DNA coding strand is replaced by U in the RNA strand (see Fig. A1.20). This concept allows the definition of the direction of any DNA sequence relative to a gene.

- Sense: same direction as the gene orientation.
- Antisense: opposite direction of the gene orientation. This concept is particularly important when the same sequence contains genes with different orientations.

Transcription of RNA from DNA

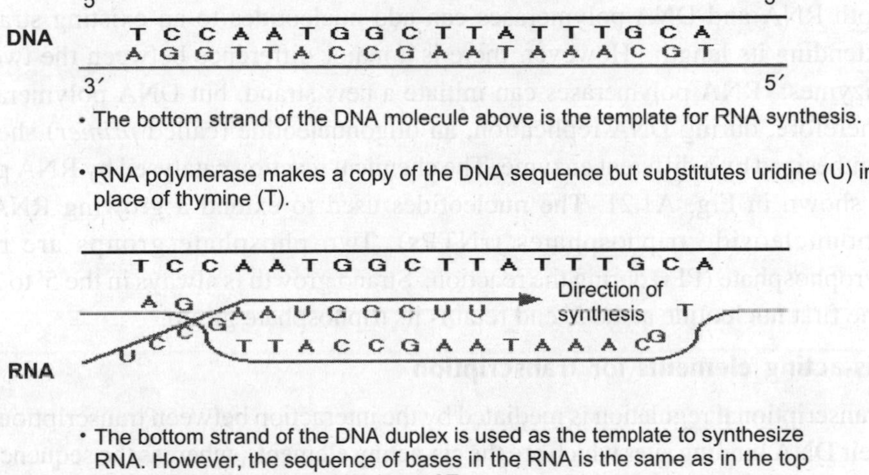

• The bottom strand of the DNA duplex is used as the template to synthesize RNA. However, the sequence of bases in the RNA is the same as in the top strand of the DNA, with U in place of T

RNA 5′ U C C A A U G G C U U A U U U G C A 3′

Fig. A1.20 Schematic illustration of transcription

The construction of a nucleic acid strand is always in the 5′ to 3′ direction. This is true not only for the synthesis of RNA during transcription, but also for the synthesis of DNA during replication. Polymerases, are used to catalyse the synthesis of nucleic acid strands. RNA strands are synthesized by RNA polymerases. DNA strands are synthesized by DNA polymerases. The entire transcription process involves the following essential steps: *Binding of polymerases to the initiation site*: The DNA sequence that signals the initiation of transcription is called the *promoter*. Prokaryotic polymerases can recognize the promoter and bind to it directly, but eukaryotic polymerases have to rely on other proteins called *transcription factors*.

Unwinding (melting) of the DNA double helix: The enzyme which can unwind the double helix is called *helicase*. Prokaryotic polymerases have the helicase activity, but eukaryotic polymerases do not. Unwinding of eukaryotic DNA is carried out by a specific transcription factor.
Synthesis of RNA (based on the sequence of the DNA template strand): RNA polymerases use nucleoside triphosphates (NTPs) to construct an RNA strand.
Termination of synthesis: Prokaryotes and eukaryotes use different signals to terminate transcription. (**Note:** The stop codon in the genetic code is a signal for the end of peptide synthesis, not the end of transcription.)

Transcription in eukaryotes is much more complicated than in prokaryotes, partly because eukaryotic DNA is associated with histones, which can hinder the access of polymerases to the promoter.

RNA polymerases

Both RNA and DNA polymerases can add nucleotides to an existing strand, thereby extending its length. However, there is a major difference between the two classes of enzymes: RNA polymerases can initiate a new strand, but DNA polymerases cannot. Therefore, during DNA replication, an oligonucleotide (called *primer*) should first be synthesized by a different enzyme. The chemical reaction catalysed by RNA polymerases is shown in Fig. A1.21. The nucleotides used to extend a growing RNA chain are ribonucleoside triphosphates (rNTPs). Two phosphate groups are released as pyrophosphate (PPi) during the reaction. Strand growth is always in the 5′ to 3′ direction. The first nucleotide at the 5′ end retains its triphosphate group.

Cis-acting elements for transcription

Transcriptional regulation is mediated by the interaction between transcription factors and their DNA binding sites which are the cis-acting elements, whereas the sequences encoding transcription factors are trans-acting elements. The cis-acting elements may be divided into the following four types:
Type I: Promoter This is the DNA region where the transcription initiation takes place. In prokaryotes, the sequence of a promoter is recognized by the sigma (σ) factor of the RNA polymerase. In eukaryotes, it is recognized by specific transcription factors. E. coli has σ70 (RpoD)—the 'housekeeping' sigma factor transcribes most genes in growing cells.

- σ38 (RpoS)—the starvation/stationary phase sigma factor
- σ28 (RpoF)—the flagellar sigma factor
- σ32 (RpoH)—the heat shock sigma factor
- σ24 (RpoE)—the extracytoplasmic stress sigma factor
- σ54 (RpoN)—the nitrogen-limitation sigma factor
- σ19 (FecI)—the ferric citrate sigma factor

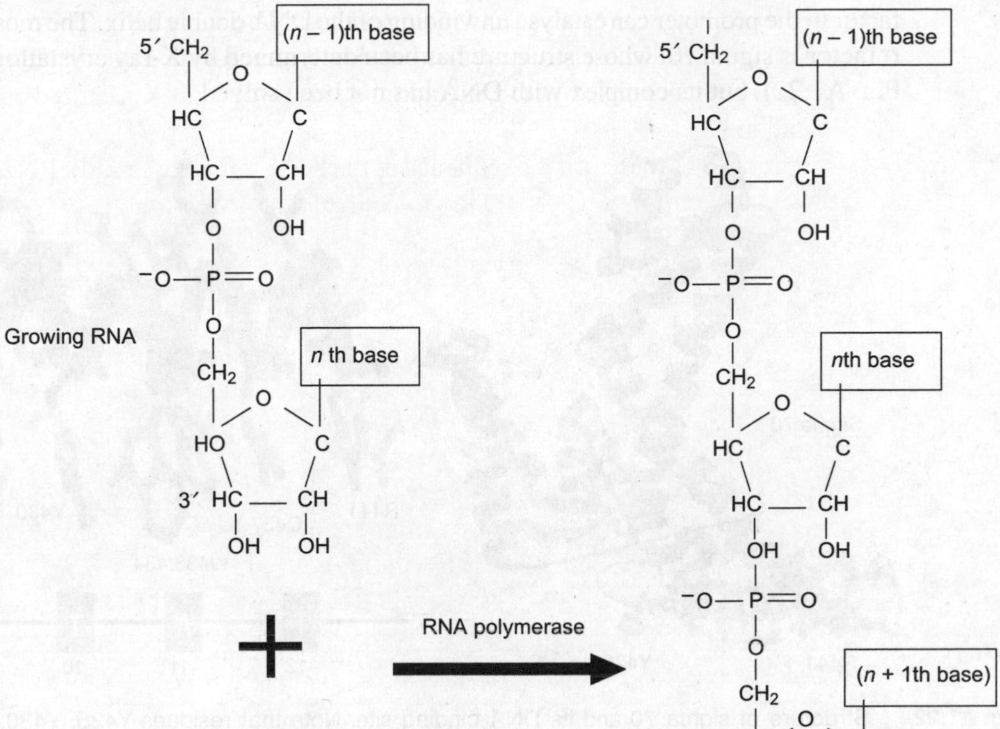

Growing RNA

5′ CH_2 | (n − 1)th base | O | HC | C | HC — CH | O | OH | ^-O—P=O | O | CH_2 | n th base | O | HO | C | 3′ HC — CH | OH | OH

NTP

^-O—P=O | ^-O—P=O | ^-O—P=O | O | CH_2 | (n + 1th base) | O | HC | C | HC — CH | OH | OH

RNA polymerase

5′ CH_2 | (n − 1)th base | O | HC | C | HC — CH | O | OH | ^-O—P=O | O | CH_2 | nth base | O | HC | C | HC — CH | OH | OH | ^-O—P=O | O | CH_2 | (n + 1th base) | O | HC | CH | 3′ HC | CH | OH | OH

PP_i

^-O | ^-O—P=O | O | ^-O—P=O | OH

Fig. A1.21 Chemical reaction catalysed by RNA polymerases

The consensus sequence is an ideal sequence for the interaction with its regulatory protein. A promoter such as the lac promoter should contain an element which is identical to or very close to the consensus sequence. In prokaryotes, binding of the polymerase's σ

factor to the promoter can catalyse unwinding of the DNA double helix. The most important σ factor is sigma 70, whose structure has been determined by X-ray crystallography (see Fig. A1.22), but its complex with DNA has not been solved.

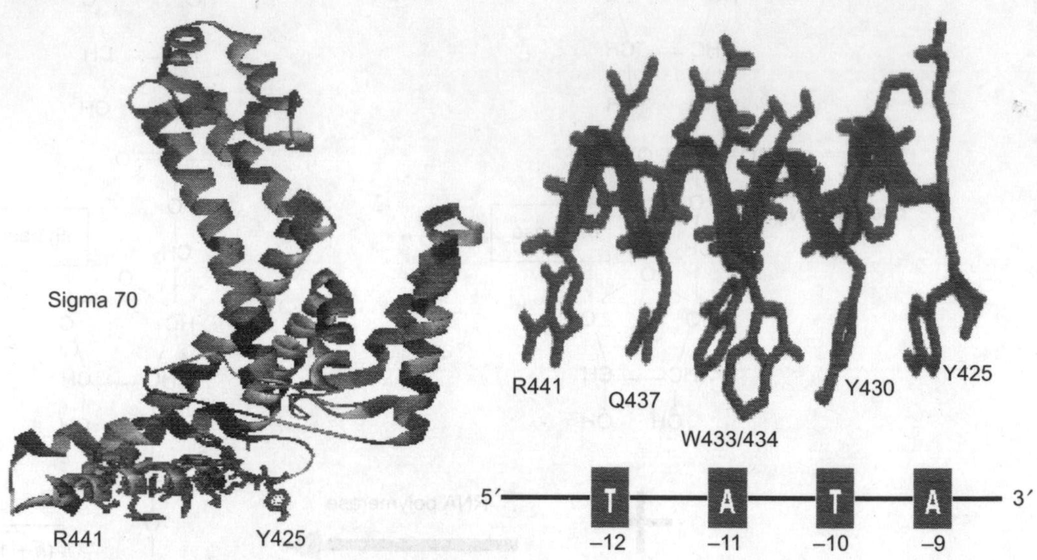

Fig. A1.22 Structure of sigma 70 and its DNA binding site. Note that residues Y425, Y430, W433, and W434 are directly involved in the unwinding (melting) of the double helix.

Type II: Enhancers These are the positive regulatory elements located either upstream or downstream of the transcriptional initiation site. However, most of them are located upstream. Downstream refers to the direction of transcription and upstream is opposite to the transcription direction. In prokaryotes, enhancers are quite close to the promoter, but eukaryotic enhancers could be far from the promoter. An enhancer region may contain one or more elements recognized by transcriptional activators.

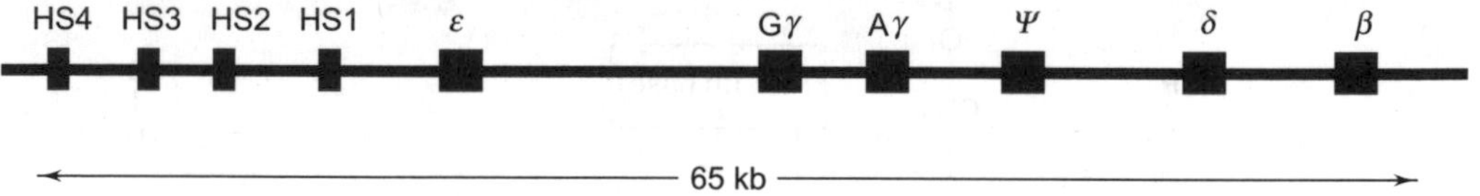

Fig. A1.23 Human β-globin gene cluster, where elements in blue are the enhancers

The human β-globin gene cluster is controlled by an enhancer region comprising HS1 to HS4, which contain the binding sites of GATA-1, NF-E2, AP-1, and other transcriptional activators (see Fig. A1.23). This region is known as the *locus control region* (LCR) and is responsible for regulating the expression of all five genes (e, Gg, Ag, d, and b), even though the distance between HS4 and the β gene is as far as 60 kb.

Type III: Silencers These are the DNA elements that interact with repressors (proteins) to inhibit transcription. Silencers are known as operators in prokaryotes and are found in many genes such as lac operon and trp operon. In eukaryotes, the human β-globin gene, human CD95(Fas/APO-1), and some others have been demonstrated to contain silencers. In a few cases, a DNA element may act either as an enhancer or a silencer, depending on the binding protein. For example, certain genes contain an element called *E box* (consensus CACGTG) which can bind either Max/Myc dimer or Max/Mad dimer. The Max/Myc dimer activates transcription, whereas the Max/Mad dimer suppresses transcription of these genes.

Type IV: Response elements These are the recognition sites of certain transcription factors (see Table A1.5). Most of them are located within 1 kb from the transcriptional start site.

Table A1.5 Eukaryotic response elements

Response Element	Transcription Factor	Consensus Sequence
CRE (cAMP response element)	CREB	TGACGTCA
ERE (Estrogen response element)	Estrogen receptor	AGGTCANNNTGACCT
GRE (Glucocorticoid response element)	Glucocorticoid receptor	AGAACANNNTGTTCT
HSE (Heat shock response element)	Heat shock factor	GAANNTTCNNGAA
SRE (Serum response element)	Serum response factor	CC(A/T)6GG

*(A/T)6 means six A or T; N = any

Gene organization in eukaryotes

The transcription region consists of exons and introns. The regulatory elements include promoter, response element, enhancer, and silencer. The numbering of base pairs in the promoter region is as follows. The number increases along the direction of transcription, with '+1' assigned for the initiation site. There is no '0' position. The base pair just upstream of +1 is numbered '–1', not '0' (see Fig. A1.24).

Promoter elements in eukaryotes

In eukaryotes, there is a significant difference between the transcription of protein genes and that of RNA genes. The most common promoter element in eukaryotic protein genes is the *TATA box*, located at –35 to –20. Its consensus sequence, TATAAA, is quite similar to the –10 region of the sigma 70 recognition site. Another promoter element is called the initiator (Inr). It has the consensus sequence PyPyAN(T/A)PyPy, where Py denotes pyrimidine (C or T), N = any, and (T/A) means T or A. The base A at the third position is located at +1 (the transcriptional start site). The TATA box and initiator are the core promoter elements. There are other elements often located within 200 bp of the transcriptional start site, such as the *CAAT* box and *GC box* which may be referred to as promoter-proximal elements (see Table A1.6).

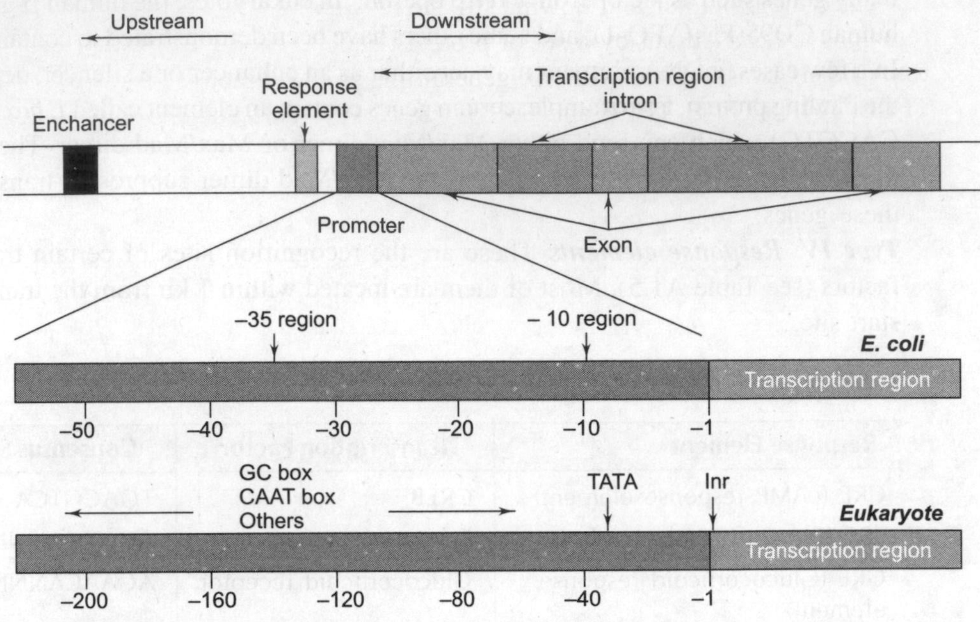

Fig. A1.24 Gene organization in eukaryotes

Table A1.6 Eukaryotic promoter elements

Promoter	Position	Transcription Factor	Consensus sequence
Inr	+ 1	TBP	pypya+1 N(T/A)pypy
Tata box	–35 ~ – 20	TBP	TATAA
CAAT box	–200 ~ –70	CBF, NF 1, C/EBP	CCAAT
GC box	–200 ~ – 70	SP1	GGGCGG

* Most, but not all, CAAT and GC boxes are located between –200 and –70.
† CBF = CAAT binding protein; C/EBP = CAAT/enhancer binding protein.
‡ N = any (A, T, C, pr G); Py = pyrimidine (C or T)

The protein that interacts with the initiator and TATA box is known as the TATA-box binding protein (TBP) (because the TATA box was discovered earlier than the initiator). TBP recognizes not only the core promoter of protein genes but also RNA promoters. It is a subunit of the general transcription factor TFIID. In eukaryotes, transcription requires several different general transcription factors and, in most cases, the regulatory transcription factors, for example, the promoter region of the IL-2 gene contains the TATA box and the regulatory elements of NFAT, Oct-1, NF-kB, and AP-1.

Initiation

RNA Pol II does not contain a subunit similar to the prokaryotic s factor which can recognize the promoter and unwind the DNA double helix. In eukaryotes, these two functions are carried out by a set of proteins, called general transcription factors. The RNA Pol II is associated with six general transcription factors, designated as TFIIA, TFIIB, TFIID, TFIIE, TFIIF, and TFIIH, where 'TF' stands for 'transcription factor' and 'II' for the RNA Pol II. TFIID consists of TBP (TATA-box binding protein) and TAFs (TBP associated factors). The role of TBP is to bind the core promoter. TAFs may assist TBP in this process. In human cells, TAFs are formed by 12 subunits. One of them, TAF250 (with molecular weight 250 kD), has the histone acetyltransferase activity, which can relieve the binding between DNA and histones in the nucleosome. The transcription factor which catalyses DNA melting is TFIIH. However, before TFIIH can unwind DNA, RNA Pol II and at least five general transcription factors (TFIIA is not absolutely necessary) have to form a pre-initiation complex (PIC).

Elongation

After the PIC is assembled at the promoter, TFIIH can use its helicase activity to unwind DNA. This requires energy released from ATP hydrolysis. The DNA melting starts from about -10 bp. Then, RNA Pol II uses nucleoside triphosphates (NTPs) to synthesize an RNA transcript. During RNA elongation, TFIIF remains attached to the RNA polymerase, but all of the other transcription factors dissociate from the PIC. The carboxyl-terminal domain (CTD) of the largest subunit of RNA Pol II is critical for elongation. In the initiation phase, CTD is unphosphorylated, but during elongation it has to be phosphorylated. This domain contains many proline, serine, and threonine residues.

Termination

Eukaryotic protein genes contain a poly-A signal located downstream of the last exon. This signal is used to add a series of adenylate residues during RNA processing. Transcription often terminates at 0.5–2 kb downstream of the poly-A signal, but the mechanism is unclear.

Role of regulatory transcription factors

It has been known for some time that binding of transcriptional activators to the enhancer region, in most cases, is not sufficient to stimulate transcription. Certain coactivators are also required. In 1996 a number of research groups discovered that certain transcriptional coactivators are histone acetyltransferases (HATs). Similarly, transcriptional repression often requires both repressor binding on the silencer element and the participation of corepressor proteins.

In eukaryotes, the association between DNA and histones prevents access of the polymerase and general transcription factors to the promoter. Histone acetylation catalysed by HATs can relieve the binding between DNA and histones. Although a subunit of

TFIID (TAF250 in human) has the HAT activity, participation of other HATs can make transcription more efficient. The following rules apply to most (but not all) cases:

- *Binding of activators to the enhancer element* recruits HATs to relieve association between histones and DNA, thereby enhancing transcription.
- *Binding of repressors to the silencer element* recruits histone deacetylases (denoted by HDs or HDACs) to tighten association between histones and DNA.

Histone acetylation

Acetylation of the lysine residues at the N terminus of histone proteins removes positive charges, thereby reducing the affinity between histones and DNA. This facilitates RNA polymerase and transcription factors' easy access the promoter region. Therefore, in most cases, histone acetylation enhances transcription, while histone deacetylation represses transcription. Histone acetylation is catalysed by HATs and histone deacetylation is catalysed by HDs or HDACs. Several different forms of HATs and HDs have been identified. Among them, CBP/p300 is probably the most important, since it can interact with numerous transcription regulators.

Lac operon

The lac operon of *E. coli* consists of three genes (Fig. A1.25) lacZ, lacY, and lacA, encoding β-galactosidase, lactose permease, and thiogalactoside transacetylase, respectively. Lactose permease is located on the cell membrane, which is capable of pumping lactose into the cell. β-galactosidase can convert lactose into glucose and galactose. Thiogalactoside transacetylase is responsible for degrading small molecules.

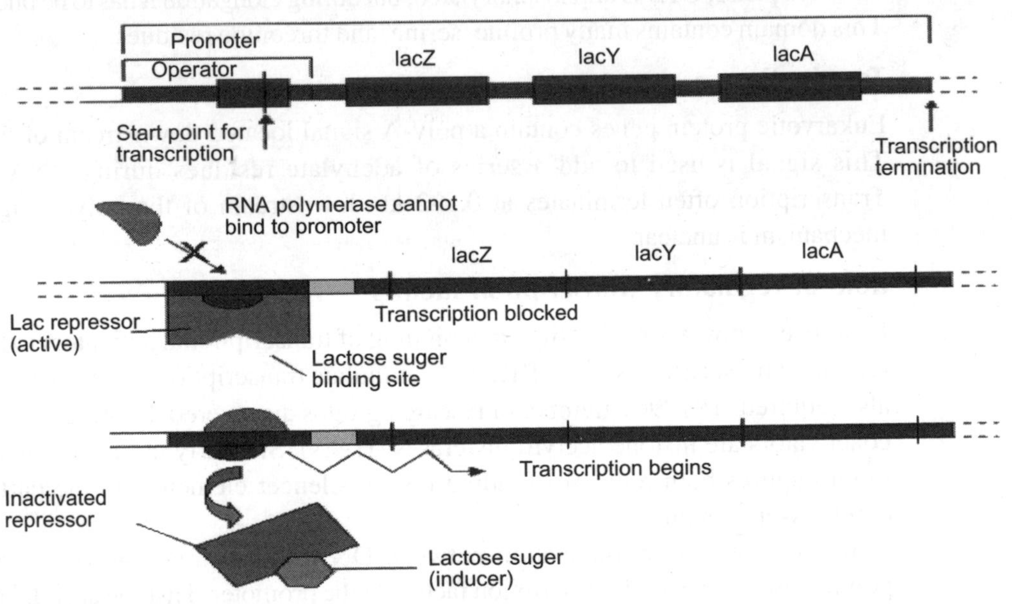

Fig. A1.25 Lac operon and its working

In the absence of lactose, transcription of the lac operon is inhibited by the lac repressor (Fig. A1.25). Lactose can bind to the lac repressor, preventing it from interacting with its DNA binding site. Hence, in a medium containing lactose, the lac operon is quickly transcribed, producing the enzymes to generate glucose, which is the major energy source for *E. coli*.

P53: A tumor suppressor protein

p53, also known as 'guardian of the genome', is a tumor suppressor protein,. It plays an important role in cell cycle control and apoptosis. Defective p53 can allow abnormal cells to proliferate, which can result in cancer. As many as 50% of all human tumors contain p53 mutants. In normal cells, the p53 protein level is low. DNA damage and other stress signals may trigger the increase of p53 proteins, which have three major functions: growth arrest, DNA repair and apoptosis (cell death). Growth arrest stops the progression of cell cycle, thus preventing replication of the damaged DNA. During growth arrest, p53 may activate the transcription of proteins involved in DNA repair. Apoptosis is the 'last resort' to avoid proliferation of cells containing abnormal DNA. The cellular concentration of p53 must be tightly regulated. While it can suppress tumors, high level of p53 may accelerate the aging process by excessive apoptosis. The major regulator of p53 is Mdm2, which can trigger the degradation of p53 by the ubiquitin system. p53 is a transcriptional activator and regulates the expression of Mdm2 (for its own regulation) and the genes involved in growth arrest, DNA repair, and apoptosis. Some important examples are as follows:

- Growth arrest: p21, Gadd45, and 14-3-3s
- DNA repair: p53R2
- Apoptosis: Bax, Apaf-1, PUMA, and NoxA

RNA processing

RNA processing is performed to generate a mature mRNA (for protein genes) or a functional tRNA or rRNA from the primary transcript. Processing of pre-mRNA is depicted in Fig. A1.26 and involves the following steps:

- Capping—adding 7-methylguanylate (m^7G) to the 5′ end.
- Polyadenylation—adding a poly-A tail to the 3′ end.
- Splicing—removing introns and joining exons.
- RNA editing is also involved in some cases

Capping

This occurs shortly after the transcription begins. The chemical structure of the cap is where m^7G is linked to the first nucleotide by a special 5′-5′ triphosphate linkage. In most organisms, the first nucleotide is methylated at the 2′-hydroxyl of the ribose. In vertebrates, the second nucleotide is also methylated.

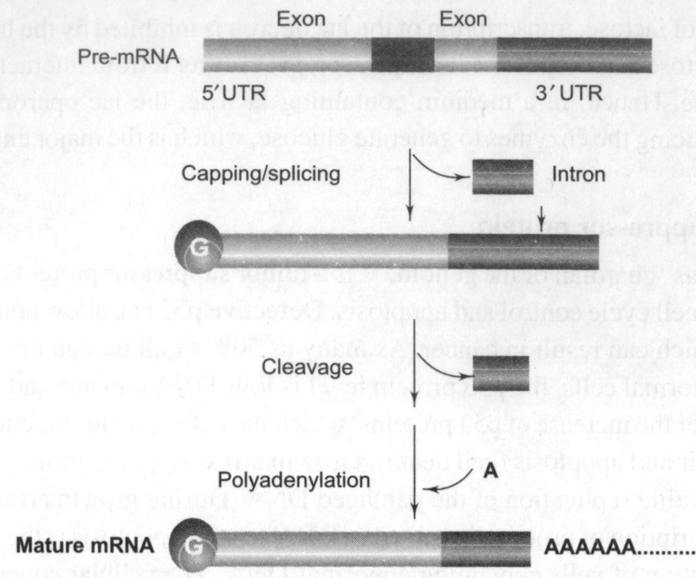

Fig. A1.26 RNA processing for protein genes

3′-Polyadenylation

A stretch of adenylate residues is added to the 3′ end. The poly-A tail contains ~250 A residues in mammals, and ~100 in yeasts. The major signal for the 3′ cleavage is the AAUAAA sequence. Cleavage occurs at 10–35 nucleotides downstream from the specific sequence. A second signal is located about 50 nucleotides downstream from the cleavage site. This signal is a GU-rich or U-rich region.

RNA splicing

This is the process that removes introns and joins exons in a primary transcript. An intron usually contains a clear signal for splicing (e.g., the β-globin gene). In some cases (e.g., the sex lethal gene of fruit fly), a splicing signal may be masked by a regulatory protein, thus resulting in alternative splicing. In rare cases (e.g., HIV genes), a pre-mRNA may contain several ambiguous splicing signals, which results in a few alternatively spliced mRNAs. Most introns start from the sequence GU and end with the sequence AG (in the 5′ to 3′ direction). They are referred to as the splice donor and splice acceptor site, respectively. However, the sequences at the two sites are not sufficient to signal the presence of an intron. Another important sequence is called the branch site located 20–50 bases upstream of the acceptor site. The consensus sequence of the branch site is 'CU(A/G)A(C/U)', where A is conserved in all genes (see Fig. A1.27). In over 60% of cases, the exon sequence is (A/C)AG at the donor site, and G at the acceptor site.

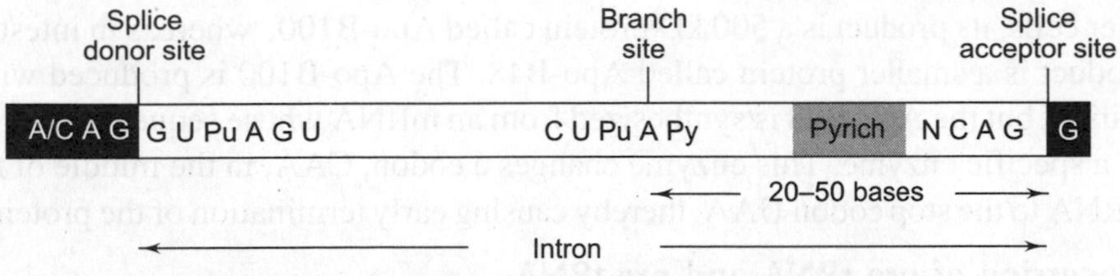

Fig. A1.27 Consensus sequence for splicing (Pu = A or G; Py = C or U)

The presence of GT at the 5′ splice site and AG at the 3′ splice site gives rise to the so called *GT-AG rule*.

Splicing

This rule holds in most cases, but exceptions have been found. For example, GC is occasionally found at the 5′ end of certain introns. GC-AG introns are processed by the same splicing pathway as conventional GT-AG introns. It had long been assumed that the removal of all introns from eukaryotic pre-mRNAs took place by the same splicing pathway until recent developments demonstrated the existence of a second pre-mRNA splicing pathway. The *AT-AC splicing pathway* was originally named after the distinctive sequences of the intron ends. An alternative designation for the two pathways is the *U2* dependent and *U12* dependent, reflecting their observed or expected requirements for one of the four snRNAs specific to each pathway. The splice signals have been used extensively by bioinformaticians in their quest to look for genes and in the development of gene finding algorithms.

The removal of introns from mRNA precursors (pre-mRNAs) involves two relatively straightforward chemical reactions. The recognition of intron–exon boundaries, the splice sites, however, requires the integration of information provided by many cis-acting elements and a complex splicing machinery. The cis-acting elements that define the borders between exons and introns are quite diverse and yet are recognized efficiently by the splicing machinery. This machinery is composed of general splicing factors (GSFs), which make up the spliceosome and its associated proteins, and of regulatory factors. The same machinery must also make cell-type-specific choices in cases in which pre-mRNAs are alternatively spliced. This is a monumental task given that it is estimated that transcripts from 30% of all genes in humans are alternatively spliced.

Spliceosome

The detailed splicing mechanism is quite complex. In short, it involves five snRNAs and their associated proteins. These ribonucleoproteins form a large (60S) complex, called spliceosome. A concise yet lucid introduction to splicing and spliceosome formation is available at http://www.neuro.wustl.edu/neuromuscular/pathwayspliceosome.htm

RNA editing

This step involves alteration of a base by specific enzymes. In mammals, the apo-B gene is expressed in both hepatocytes (liver cells) and intestinal epithelial cells. However, in liver cells, its product is a 500 kD protein called Apo-B100, whereas in intestine cells its product is a smaller protein called Apo-B48. The Apo-B100 is produced without RNA editing, but the Apo-B48 is synthesized from an mRNA whose sequence has been altered by a specific enzyme. This enzyme changes a codon, CAA, in the middle of the original mRNA to the stop codon UAA, thereby causing early termination of the protein synthesis.

Processing of pre-rRNA and pre-tRNA

The newly transcribed pre-rRNA is a cluster of three rRNAs: 18S, 5.8S, and 28S in mammals. These must be separated to become functional. Pre-rRNA is synthesized in the nucleolus. The U3 snRNA, other U-rich snRNAs, and their associated proteins in the nucleolus are involved in the cleavage of the pre-rRNA. 5S rRNA is synthesized in the nucleoplasm. It does not require any processing. After 5S rRNA is synthesized, it will enter the nulceolus to combine with 28S and 5.8S rRNAs and form the large subunit of the ribosome. Pre-tRNA requires extensive processing to become a functional tRNA. Four types of modifications are involved:

- Removing an extra segment (~16 nucleotides) at the 5′ end by RNase P.
- Removing an intron (~14 nucleotides) in the anticodon loop by splicing.
- Replacing two U residues at the 3′ end by CCA, which is found in all mature tRNAs.
- Modifying some residues to characteristic bases, e.g., inosine, dihydrouridine, and pseudouridine.

After RNA molecules (mRNA, tRNA, and rRNA) have been produced in the nucleus, these must be exported to the cytoplasm for protein synthesis. On the other hand, many proteins operating in the nucleus must be imported from the cytoplasm. The traffic through the nuclear envelope is mediated by a protein, family which can be divided into *exportins* and *importins*. Binding of a molecule (a 'cargo') to exportins facilitates its export to the cytoplasm. Importins facilitate import into the nucleus. The function of exportins and importins is regulated by a G protein called 'Ran'.

Proteins synthesis

Protein synthesis is carried out on ribosomes based on the sequence of mRNA. The whole process comprises three basic subprocesses.

- Initiation
- Translation by tRNA
- Protein Synthesis

Initiation

Synthesis of a peptide always starts from methionine. However, the first amino acid of a functional protein is not always methionine. After a peptide is synthesized, it may undergo a variety of modifications, including the cleavage of its N-terminus. Prokaryotes and eukaryotes use different mechanisms to recognize the initiating codon.

- ***In prokaryotes*** Many prokaryotic mRNAs are polycistronic, i.e., an mRNA encodes more than one peptide chain. The polycistronic mRNA should contain multiple initiating codons. However, a peptide may also contain several non-initiating methionine residues. In the standard genetic code, the codon for both initiating and non-initiating methionine is AUG. To distinguish them, prokaryotes use a specific sequence located about 5–10 bases upstream of the initiation AUG. The specific sequence, UAAGGAGG, is known as the *Shine–Dalgarno sequence*. The 16S rRNA of the ribosome contains a sequence which can pair with the Shine–Dalgarno sequence:

5′—UAAGGAGG(5-10 bases)AUG mRNA
3′—AUUCCUCC... 16S rRNA

Their association will recruit other parts of the ribosome to the initiation site for protein synthesis.

- ***In eukaryotes*** The mechanism used by eukaryotes to recognize the initiating AUG is not entirely clear. Evidence suggests that the eukaryotic ribosome may simply scan from the 5′ cap and identify the first AUG as the initiation site. This mechanism is reasonable because nearly all eukaryotic mRNAs are monocistronic (encode a single peptide). However, some viral mRNAs are polycistronic or lack the 5′ cap. It is not fully understood how the host eukaryotic cell recognizes the initiation sites of these viral mRNAs. Marilyn Kozak found that the following sequence may increase effectiveness as an initiation site:

  5′—ACCAUGG- mRNA

Translation by tRNA

Translation is a process by which the nucleotide sequence of mRNA is converted into the amino acid sequence of a peptide. It starts from the initiation codon, and then follows the mRNA sequence in a strictly 'three nucleotides for one amino acid' manner. Therefore, a minor change in the mRNA sequence can produce a very different peptide. For instance, if a codon which codes for an amino acid is changed to a 'stop' codon, the subsequent sequence will not be translated. Another example is frameshift. In most cases, frameshift involves the insertion or deletion of a single nucleotide in mRNA (see Fig. A1.28). Theoretically, it can involve more than one nucleotide, as long as the number is not a multiple of 3. When a nucleotide is added to or deleted from the mRNA, the subsequent sequence produces an entirely different peptide.

Translation is carried out by tRNA through the relationship between its anticodon and the associated amino acid. When a tRNA is brought to the ribosome by pairing between its anticodon and the mRNA's codon, the amino acid attached at its 3′ end is added to the growing peptide. In bacteria, there are 30–40 tRNAs with different anticodons. In animal and plant cells, about 50 different tRNAs are found. However, there are 61 codons coded for amino acids. Suppose each codon can pair with only a unique anticodon, then 61 tRNAs would be needed.

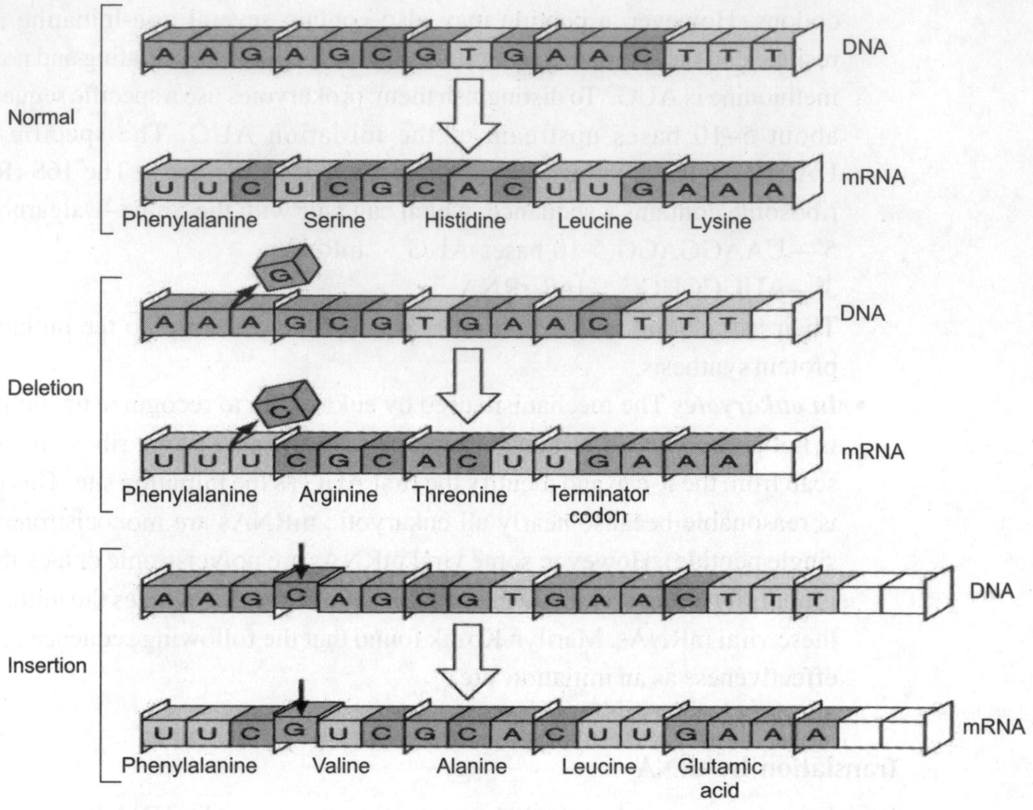

Fig. A1.28 Frameshift indels. Note that the translated amino acids are entirely different after the insertion point

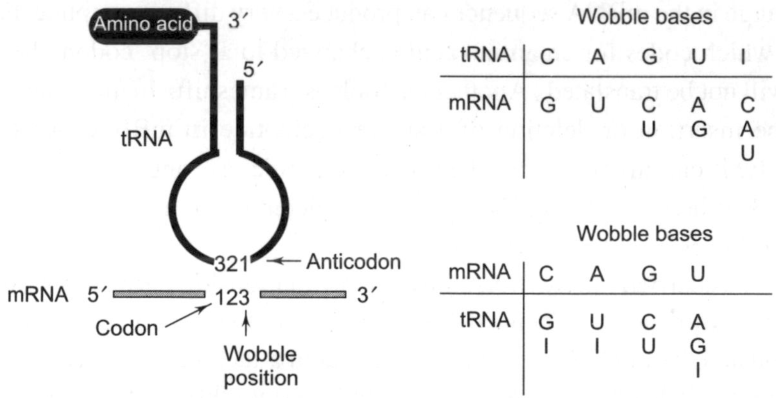

Fig. A1.29 Pairing between (a) tRNA's anticodon and mRNA's codon, the wobble position where base pairing does not obey the standard rule; (b) All possible base pairings at the wobble position

Note in Fig. A1.29 guanine (G) can pair with both cytosine (C) and uracil (U); inosine (I) can pair with cytosine, adenine, and uracil.

Gene overlapping

This translation mechanism allows for gene overlapping. From the phenomenon of frameshift, we see that the same nucleotide sequence can encode different amino acid sequences. This allows two genes with entirely different protein products to share the same nucleotide sequence. Gene overlapping is often found in a compact genome such as HBV and HIV. In human mitochondrial DNA, there are two overlapping genes: ATPase subunits 6 and 8. The gene of ATPase subunit 8 is located between 8366 and 8569 of the human mitochondrial genome. The gene of ATPase subunit 6 is located between 8527 and 9204. ATPase 8 contains 68 amino acids, while ATPase 6 has 226 amino acids. Their amino acid sequences are different even in the region where they share the same DNA sequence. It is interesting to note that the end of the ATPase 6 gene is not a stop codon, but two nucleotides 'TA'. It becomes the stop codon 'TAA' only after the poly-A tail is added during RNA processing.

Methylation during protein synthesis

Peptide synthesis always starts from methionine (Met); therefore, the initial aminoacyl-tRNA is Met-tRNAiMet, where the subscript 'i' means 'initiation'. In bacteria, the methionine of the initial aminoacyl-tRNA has been modified by the addition of a formyl group (HCO) to its amino group. The modified methionine is called formylmethionine (fMet), which is unique for bacteria. Thus, fMet is an obvious foreign substance in eukaryotes. It can elicit a strong immune response. A ribosome contains two major tRNA-binding sites: A site and P site. After the large subunit joins the initiation complex, the initial Met-tRNAiMet enters the P site and the newly arrived aminoacyl-tRNA is always placed at the A site ('A' for 'aminoacyl'). Then, methionine is transferred to the new aminoacyl-tRNA, forming a 'peptidyl-tRNA' where a peptide is attached to the tRNA. Subsequently, the empty tRNA at the P site is ejected from the ribosome and the peptidyl-tRNA jumps to the P site ('P' for 'peptidyl'). During this translocation step, the ribosome also moves one codon down the mRNA chain. Similar steps are repeated in the next cycles of elongation.

Termination of protein synthesis

Protein synthesis is terminated when the ribosome arrives at one of three stop codons. The termination process is assisted by special proteins called termination factors which recognize the stop codons. Their association stimulates the release of the peptidyl-tRNA from the ribosome. Subsequently, the released peptidyl-tRNA divides into tRNA and a newly synthesized peptide chain. The ribosome also divides into the large and small subunits, which are ready for synthesizing another peptide. In the absence of mRNA, the large and small subunits of a ribosome are separated. At the beginning of peptide synthesis, initiation factors (IF) first assist the assembly of the small subunit, mRNA, and the initial aminoacyl-tRNA. Then, the large subunit is recruited to join the complex. In the elongation process, one cycle involves the following steps:

- ***New entry*** A new aminoacyl-tRNA with a correct anticodon is brought to the A site. This step is catalysed by elongation factors Tu and Ts in prokaryotes, and by elongation factors EF1 and EF1b in eukaryotes.
- ***Peptide synthesis*** The peptide attached to the peptidyl-tRNA at the P site is transferred to the new aminoacyl-tRNA at the A site, generating a peptidyl-tRNA with a longer peptide. This step is catalysed by peptidyl transferase.
- ***Translocation*** The empty tRNA at the P site is ejected from the ribosome and the peptidyl-tRNA generated at the A site takes over the vacant P site. In the mean time, the ribosome moves one codon down the mRNA chain. The A to P switch is catalysed by the elongation factor G in bacteria, and by the elongation factor EF2 in eukaryotes.

Sorting of proteins

Most proteins are synthesized on ribosomes which are located mainly in the cytosol. Only a small number of ribosomes are located in mitochondria and chloroplasts. Proteins synthesized on these ribosomes can be directly incorporated into the compartments within these organelles. However, most mitochondrial and chloroplast proteins are encoded by nuclear DNA and synthesized on cytosolic ribosomes. These and all other proteins synthesized in cytosol must be transported to appropriate locations in the cell. This is made possible by the specific signal sequence in the newly synthesized peptide.

Cell signalling

A cell needs to communicate with its environment so that it can make appropriate responses. The external signal may enter a cell via four major pathways(see Fig. A1.30):

- Hydrophobic molecules
- Ion channels
- G-protein-coupled receptors
- Enzymes

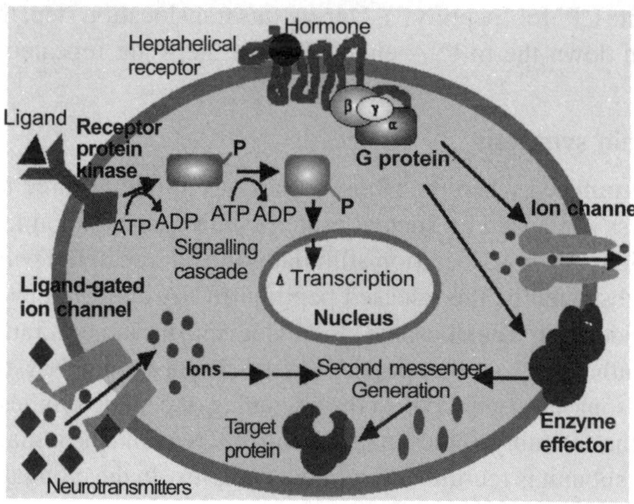

Fig. A1.30 Major pathways for signal interaction in a cell

Hydrophobic molecules

Such molecules can move in and out of cells by passing through lipid bilayers. Nitric oxide, arachidonic acid, and steroids have been shown to play important roles in cell signalling.

Ion channels

Ion channels are membrane proteins that allow ions to pass through (see Fig. A1.31). In terms of ion selectivity, these channels are classified as calcium channels, sodium channels, potassium channels, etc. In terms of gating (how channels are opened), these may be classified as voltage-gated channels, ligand-gated channels, etc.

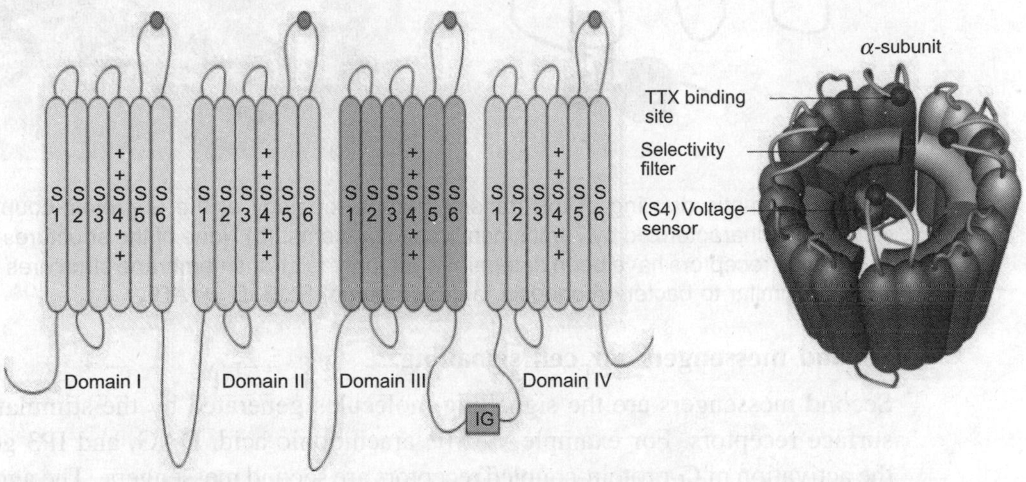

Fig. A1.31 Sodium ion channel is a voltage-gated channel. (a) The protein has four homologous domains containing multiple potential a-helical transmembrane segments. The fourth transmembrane segment (S4) of each domain is highly positively charged, and thought to be a voltage sensor. (b) The ionic pore formed from the protein has a large aqueous cavity, with a gate close to the interior and a selectivity filter on the outer vestibule.

G-protein-coupled receptors

The major role of G-protein-coupled receptors is to transmit signals into the cell. They are characterized by seven transmembrane segments (see Fig. A1.32). This class of membrane proteins can respond to a wide range of agonists, including photons, amines, hormones, neurotransmitters, and proteins. Some agonists bind to the extracellular loops of the receptor; others may penetrate into the transmembrane region.

Effectors are the target molecules of G protein a or bg subunit. The major effectors of the α subunit. The βy subunits may act on adenylyl cyclase, phospholipase A2, phospholipase C, ion channels, calcium ATPase, etc.

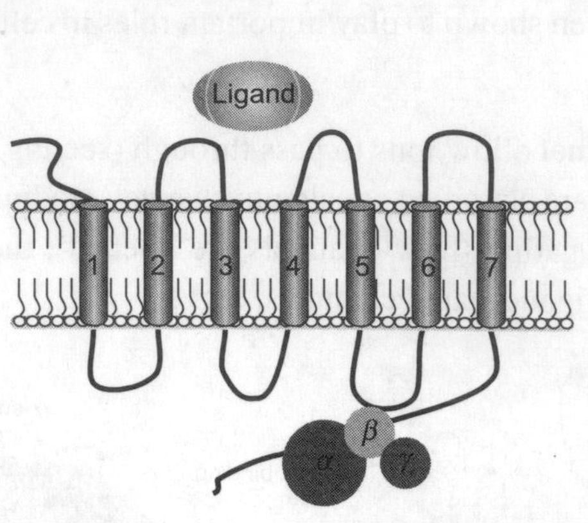

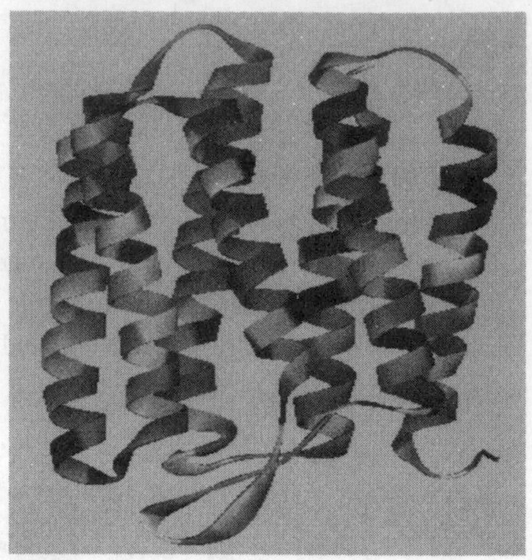

Fig. A1.32 (a) Schematic drawing of the transmembrane topology of the G-protein-coupled receptor, which is characterized by 7 transmembrane segments. (b) None of the structures of G-protein-coupled receptors have been determined yet, but their transmembrane structures are expected to be similar to bacteriorhodopsin (a proton pump) PDB ID = 1AT9.

Second messengers for cell signalling

Second messengers are the signalling molecules generated by the stimulation of cell-surface receptors. For example, cAMP, arachidonic acid, DAG, and IP3 generated by the activation of G-protein-coupled receptors are second messengers. The agonists which activate G-protein-coupled receptors are first messengers. Table A1.7 shows mammalian G protein α subunits

Table A1.7 Mammalian G protein α subunits

Class	Gene Variant	Effector	Toxin Sensitivity
G_s	$a_{s(s)}$	(+) Adenylyl cyclase	
	$a_{s(L)}$	(+) Ca^{2+} channel	
		(–) Na^+ channel	Cholera
	a_{olf}	(+) Adenylyl cyclase	Cholera
G_i	a_{i1}	(–) Adenylyl cyclase	
	a_{i2}	(–) Ca^{2+} channel(+)	
	a_{i3}	K^+ channel	Pertussis
	a_{oa}	(–) Ca^{2+} channel	
	a_{ob}	(+) Phospholipase C	
		(+) Phospholipase A_2	Pertussis
	a_{t1}	(+) cGMP phosphodiesterase	Cholera and
	a_{t2}		pertussis

(contd)

Table A1.7 *(contd)*

Class	Gene Variant	Effector	Toxin Sensitivity
	a_2	(+) Phospholipase C	pertussis
	a_2	(–) Adenylyl cyclase	
G_q	a_q	(+) Phospholipase C	
	a_{11}		
	a_{14}		
	a_{15}		
	a_{16}		

Note: (+) indicates activation and (–) indicates inhibition.

The targets of cAMP include the following:

- cAMP-dependent protein kinase (PKA). PKa is its principal target.
- Cyclic nucleotide gated ion channels.
- Guanine exchanging factors Epac1 and Epac2 that regulate the activity of Rap1.

Apoptosis

This is a tightly regulated form of cell death, also called the programmed cell death. Morphologically, it is characterized by chromatin condensation and cell shrinkage in the early stage. Then the nucleus and cytoplasm fragment to form membrane-bound apoptotic bodies which can be engulfed by phagocytes. In contrast, cells undergo another form of cell death, called necrosis, and swell and rupture. The released intracellular contents can damage the surrounding cells and often cause inflammation.

Apoptosis is an important process during normal development. It is also involved in aging and various diseases such as cancer, AIDS, Alzheimer's disease, and Parkinson's disease.

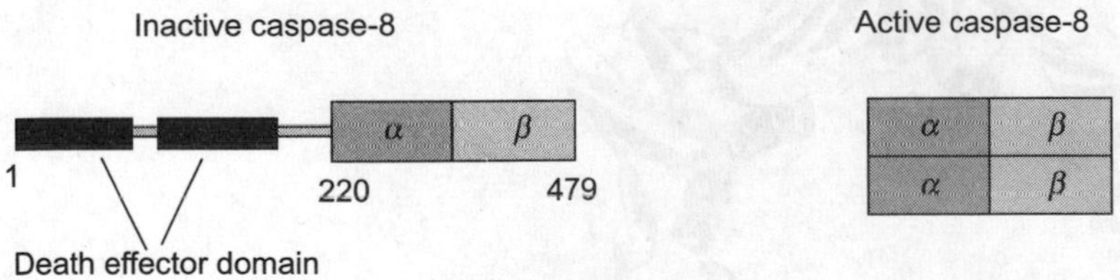

Fig. A1.33 Active and inactive forms of caspases

During apoptosis, the cell is killed by a class of proteases called caspases. More than 10 caspases have been identified. Some of them (e.g., caspase 8 and 10) are involved in the initiation of apoptosis, others (caspase 3, 6, and 7) execute the death order by destroying essential proteins in the cell (Fig. A1.33). Newly produced caspases are inactive. The apoptotic process can be summarized as follows:

- Activation of initiating caspases by specific signals.

- Activation of executing caspases by the initiating caspases which can cleave inactive caspases at specific sites.
- Degradation of essential cellular proteins by the executing caspases with their protease activity.

Specifically cleaved caspases dimerize and become active. A variety of death ligands (FasL/CD95L, TRAIL, APO-3L, and TNF) can induce apoptosis. It is natural to see if they can kill cancer cells without affecting normal cells. TNF was first investigated in the 1980s for cancer therapy, but with disappointing results. Then CD95L (FasL) was tested in the 1990s. The results were still not satisfactory. Recently, TRAIL has been demonstrated to be highly selective for transformed cells, with minimal effects on normal cells. It could be an effective drug for both cancer and AIDS.

DNA polymerases

In prokaryotes Three types of DNA polymerases exist in *E. coli*: types I, II, and III. DNA polymerase I is used to fill the gap between DNA fragments of the lagging strand. It is also the major enzyme for gap filling during DNA repair. DNA polymerase II is encoded by the PolB gene, which is involved in the SOS response to DNA damage. DNA replication is mainly carried out by DNA polymerase III. DNA polymerase III consists of several subunits, with a total molecular weight exceeding 600 kD. Among them, α, ε, and θ subunits constitute the core polymerase. The major role of other subunits is to keep the enzyme from falling off the template strand. As shown in Fig. A1.34 two β subunits can form a doughnut-shaped structure to clamp a DNA molecule in its centre, and slide with the core polymerase along the DNA molecule. This allows continuous polymerization of up to 5×10^5 nucleotides. In the absence of β subunits, the core polymerase would fall off the template strand after synthesizing 10–50 nucleotides.

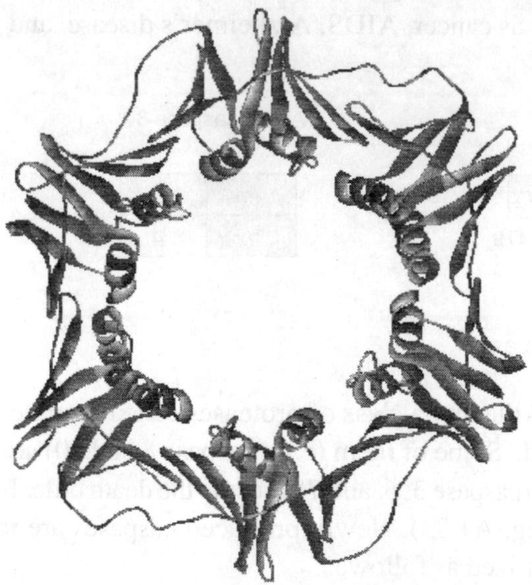

Fig. A1.34 Doughnut shaped structure formed by two β subunits of the *E. coli* DNA polymerase III

In eukaryotes There are five types of DNA polymerases in mammalian cells: α, β, γ, δ, and ε. The g subunit is located in the mitochondria and is responsible for the replication of mtDNA. Other subunits are located in the nucleus. Their major roles are given below:

- α: synthesis of lagging strand
- β: DNA repair
- δ: synthesis of leading strand
- ε: DNA repair

The synthesis of the lagging strand requires a short primer which will be removed. At the extreme end of a chromosome, there is no way to synthesize this region when the last primer is removed. Therefore, the lagging strand is always shorter than its template by at least the length of the primer. This is the so called *end-replication problem* (see Fig. A1.35).

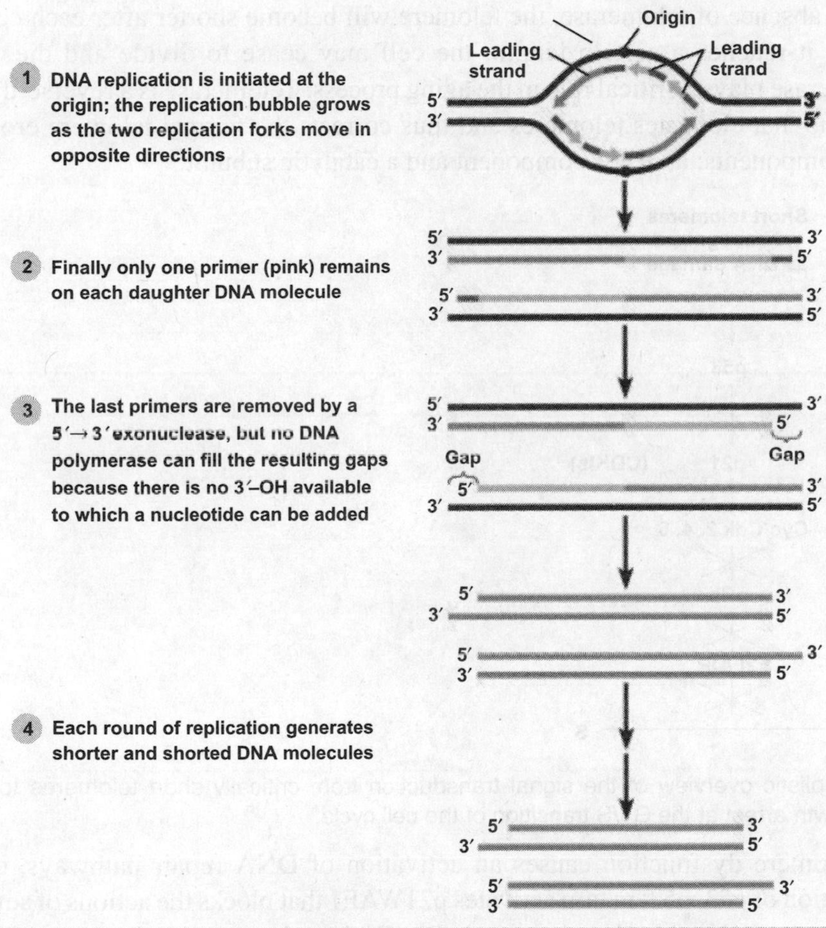

Origin
Leading strand
Leading strand
1 DNA replication is initiated at the origin; the replication bubble grows as the two replication forks move in opposite directions
2 Finally only one primer (pink) remains on each daughter DNA molecule
3 The last primers are removed by a 5′→3′ exonuclease, but no DNA polymerase can fill the resulting gaps because there is no 3′-OH available to which a nucleotide can be added
Gap
Gap
4 Each round of replication generates shorter and shorted DNA molecules
5′
3′

Fig. A1.35 The end-replication problem

Bacteria do not have the end-replication problem, because their DNA is circular. In eukaryotes, the chromosome ends are called telomeres, which have at least two functions:

- to protect chromosomes from fusing with each other.
- to solve the end-replication problem. To extend the length of a telomere, telomerase first extends its longer strand. Then, using the same mechanism as synthesizing the lagging strand, the shorter strand is extended. In a human chromosome, the telomere is about 10 to 15 kb in length and is composed of the tandem repeat sequence TTAGGG. The telomerase contains an essential RNA component which is complementary to the telomere repeat sequence. Hence, internal RNA can serve as the template for synthesizing DNA. Through telomerase translocation, a telomere may be extended by many repeats.

Telomeres

In the absence of telomerase, the telomere will become shorter after each cell division. When it reaches a certain length, the cell may cease to divide and die. Therefore, telomerase plays a critical role in the aging process. Telomerase is a reverse-transcriptase enzyme that elongates telomeres and thus corrects the normal telomere erosion. It has two components: an RNA component and a catalytic subunit.

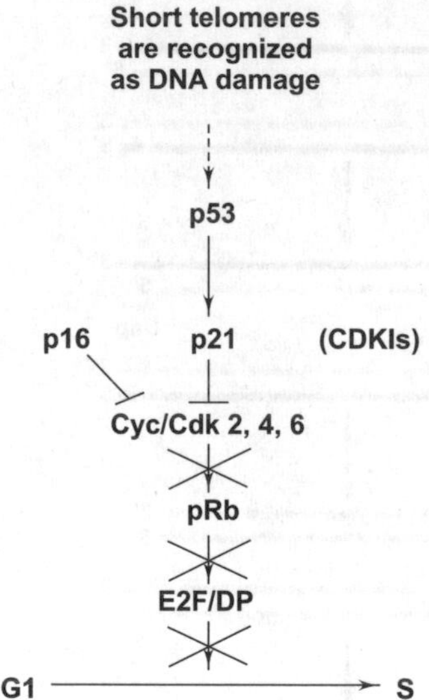

Fig. A1.36 Simplistic overview of the signal transduction from critically short telomeres to irreversible growth arrest at the G1/S transition of the cell cycle.

Telomere dysfunction causes an activation of DNA repair pathways, e.g., the an activation of p53. p53, in turn activates p21WAF1 that blocks the actions of several CDKs and prevents the phosphorylation of pRb. Without hyper-phosphorylated pRb, several critical genes in the G1/S transition are not transcribed, thus blocking the cell cycle. Immortalization with viral proteins is not as simple as it may seem at first.

CpG islands

The CpG island is a short stretch of DNA in which the frequency of the CG sequence is higher than in other regions. It is also called the CG island. p simply indicates that C and G are connected by a phosphodiester bond. These islands are often located around the promoters of housekeeping genes (which are essential for general cell functions) or other genes frequently expressed in a cell. At these locations, the CG sequence is not methylated. By contrast, the CG sequences in inactive genes are usually methylated to suppress their expression. The methylated cytosine may be converted into thymine by accidental deamination. Unlike the cytosine to uracil mutation (which is efficiently repaired), the cytosine to thymine mutation can be corrected only by the mismatch repair, which is very inefficient. Hence, over evolutionary time scales, the methylated CG sequence will be converted into the TG sequence. This explains the deficiency of the CG sequence in inactive genes. A good FAQ on CpG islands is available at http://data.microarrays.ca/cpg/faq.htm

Inheritance of the DNA methylation pattern

All types of cells have their own methylation patterns so that a unique set of proteins may be expressed to perform functions specific for a particular cell type. Thus, during cell division, the methylation pattern should also pass over to the daughter cell. This is achieved by a specific enzyme called the maintenance methylase. This enzyme can methylate only the CG sequence paired with methylated CG.

DNA cloning

DNA cloning is a technique for reproducing DNA fragments. It can be achieved by two different approaches: (1) cell based approach and (2) polymerase chain reaction (PCR) approach. In the cell-based approach, a vector is required to carry the DNA fragment of interest into the host cell. Figure A1.37 shows the typical procedure by using plasmids as the cloning vector.

DNA cloning involves the following steps:

- DNA recombination
- Transformation
- Selective amplification
- Isolation of desired DNA clones

Preparation of a DNA library

DNA library is is a collection of cloned DNA fragments. There are two types of DNA libraries:

- The genomic library contains DNA fragments representing the entire genome of an organism.
- The cDNA library contains only complementary DNA molecules synthesized from mRNA molecules in a cell.

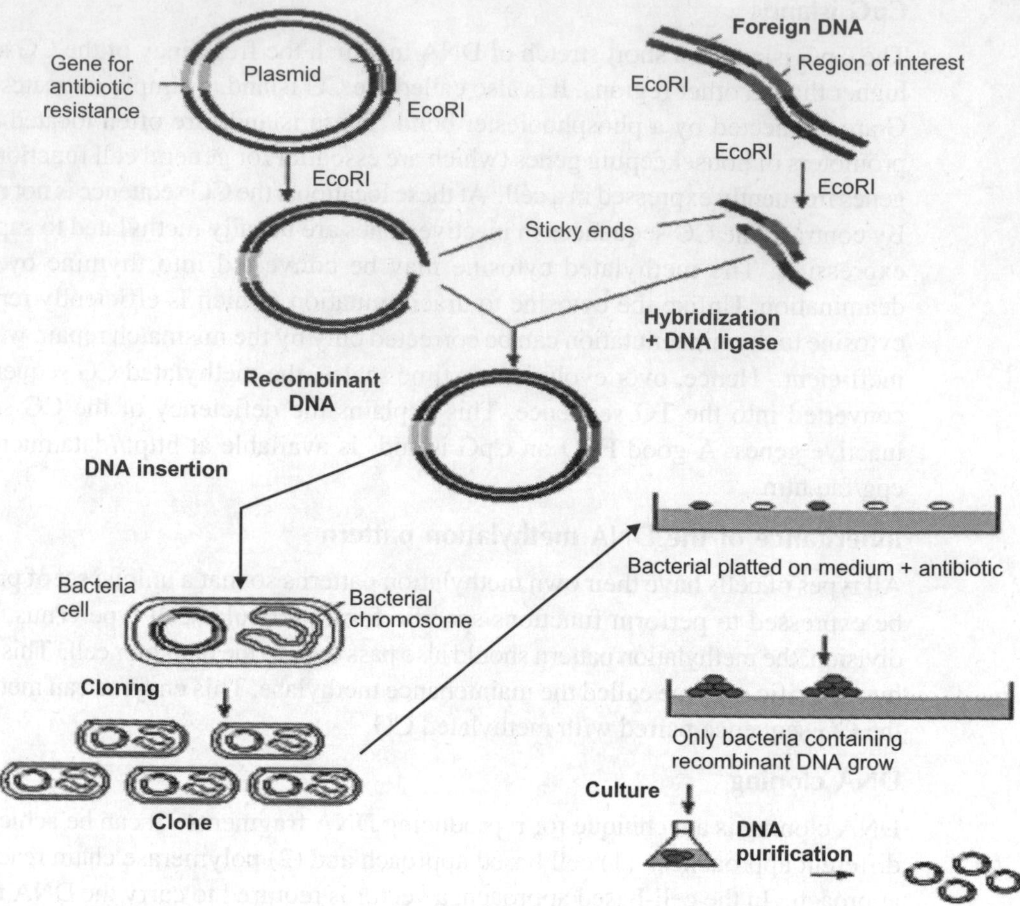

Fig. A1.37 Essential steps in DNA cloning using plasmids as vectors

Genomic library

It is basically the cloning of all DNA fragments representing the entire genome. The genomic library is normally made by l phage vectors, instead of plasmid vectors, for the following reasons: The entire human genome is about 3 × 109 bp long while a plasmid or l phage vector may carry up to 20 kb fragments. This would require 1.5 × 105 recombinant plasmids or l phages. When plating *E. coli* colonies on a 3" Petri dish, the maximum number to allow isolation of individual colonies is about 200 colonies per dish. Thus, at least 700 Petri dishes are required to construct a human genomic library. By contrast, as many as 5 × 104 l phage plaques can be screened on a typical Petri dish. This requires only 30 Petri dishes to construct a human genomic library. Another advantage of the l phage vector is that its transformation efficiency is about 1000 times higher than the plasmid vector.

cDNA library

The advantage of the cDNA library is that it contains only the coding region of a genome. To prepare a cDNA library, the first step is to isolate the total mRNA from the cell type of interest. Since eukaryotic mRNAs consist of a poly-A tail, they can easily be separated. The enzyme reverse transcriptase is used to synthesize a DNA strand complementary to each mRNA mlecule. After the single-stranded DNA molecules are converted into double-stranded DNA molecules by DNA polymerase, they are inserted into vectors and cloned.

Probes

A probe is a piece of DNA or RNA used to detect specific nucleic acid sequences by hybridization (binding of two nucleic acid chains by base pairing). Probes are radioactively labelled so that the hybridized nucleic acid can be identified by autoradiography. The size of probes ranges from a few nucleotides to hundreds of kilobases. Long probes are usually made by cloning. Originally they may be double stranded, but the working probes must be single stranded. Short probes (oligonucleotide probes) can be made by chemical synthesis. They are single stranded.

Suppose we have cloned a specific gene in yeast and want to find its homologous gene in human, then we may use the specific yeast gene as a probe to detect its homologous gene from the human genomic library. On the other hand, if we know the conserved sequence in the specific gene between yeast and human, we may use oligonucleotide probes containing only the conserved sequence. Typically, an oligonucleotide about 20 nucleotides long is sufficient to screen a library.

In some cases, we have known the partial sequence of a protein and want to detect its gene in the library. Then we may synthesize oligonucleotide probes based on the known peptide sequence. Since an amino acid may be encoded by several DNA triplets, many different oligonucleotide probes are often needed.

Screening

Once a particular DNA fragment is identified, it can be isolated and amplified to determine its sequence. If we know the partial sequence of a gene and want to determine its entire sequence, the probe should contain the known sequence so that the detected DNA fragment may contain the gene of interest. After recombinant virions form plaques on the lawn of *E. coli*, a nitrocellulose filter (membrane) is placed on the surface of the Petri dish to pick up l phages from each plaque. Then, the filter is incubated in an alkaline solution to disrupt the virions and release the encapsulated DNA, which is subsequently denatured. Next, the probe is added to hybridize with the target DNA fragment, whose position may be displayed by autoradiography.

Following gel electrophoresis, probes are often used to detect specific molecules from the mixture. However, probes cannot be applied directly to a gel. The problem can be solved by three types of blotting methods: *Southern blotting, Northern blotting, and Western blotting*.

Southern blotting

This is a technique for detecting specific DNA fragments in a complex mixture (Fig. A1.38). This technique was invented in mid 1970s by Edward Southern. It has been applied to detect restriction fragment length polymorphism (RFLP) and variable number of tandem repeat polymorphism (VNTR). The latter is the basis of DNA fingerprinting.

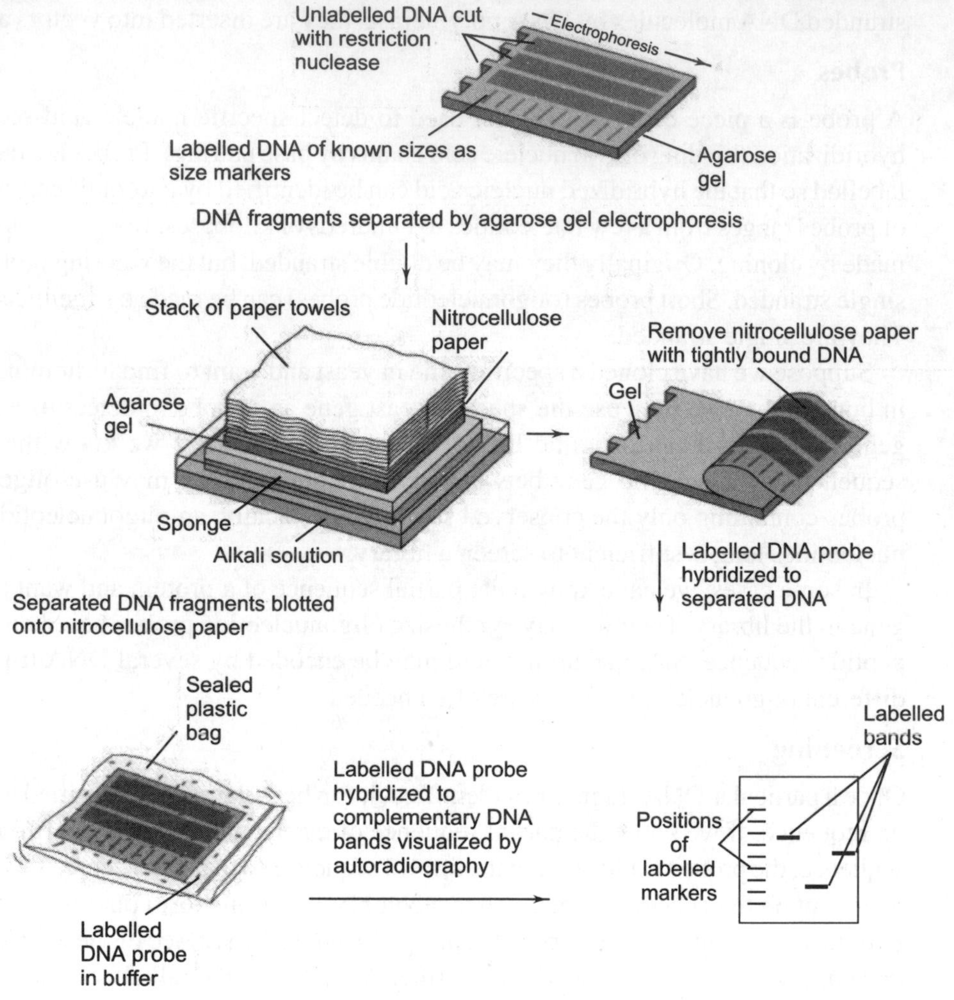

Fig. A1.38 Southern blotting method

The DNA to be analysed is digested with restriction enzymes and then separated by agarose gel electrophoresis. The DNA fragments in the gel are denatured with an alkaline solution and transferred onto a nitrocellulose filter or nylon membrane by blotting, preserving the distribution of the DNA fragments in the gel. The nitrocellulose filter is incubated with a specific probe. The location of the DNA fragment that hybridizes with the probe can be displayed by autoradiography.

Northern blotting

This technique is used for detecting RNA fragments. The technique is called 'Northern' not because it was invented by a person named 'Northern' but simply because the word 'Northern' rhymes with 'Sourthern'. In the Southern blotting, DNA fragments are denatured with an alkaline solution. In this technique, RNA fragments are treated with formaldehyde to ensure linear conformation.

Western blotting

Western blotting is used to detect a particular protein in a mixture. The probe used is therefore not DNA or RNA, but some antibody. The technique is also called immunoblotting.

Appendix

A2 PERL for Bioinformatics

Assuming that you already have some knowledge of computer programming, here we will look at some PERL examples that deal with some elementary problems of bioinformatics.

Q1. Write a program that will allow the user to input two DNA sequences and concatenate the sequences. Display the input sequences as well as the concatenated sequence.

Program

```
#!/usr/bin/perl -w

print "Enter the first DNA Sequence:  ";
$seq1=<STDIN>;
print "Enter the second DNA Sequence:  ";
$seq2=<STDIN>;

chomp($seq1);
chomp($seq2);
$concat=$seq1.$seq2;

print "The sequence1 is $seq1 \n";
print "The sequence2 is $seq2 \n";
print "The concatenated sequence is $concat \n";
```

Q2. Write a program to count the number of residues for A, T, G, C of a given sequence. Display the number of GC rich region of the sequence.

Program

```
#!/usr/bin/perl -w

print "Enter the DNA sequence: ";
chomp($dna=<STDIN>);
```

```
counter($dna);
gcrich($dna);
sub counter
{ $seq=$_[0];
$len=length($seq);
for($i=0;$i<$len;$i++)
  {  $arr[$i]=substr($_[0],$i,1);
  }
foreach $x(@arr)
    {   if( $x eq 'A')
        { $a++;
        }
  elsif( $x eq 'T')
        { $t++;
        }
  elsif( $x eq 'G')
        { $g++;
        }
  elsif( $x eq 'C')
        { $c++;
        }
  }
  print "The number of A residues is $a \n";
  print "The number of T residues is $t \n";
  print "The number of G residues is $g \n";
  print "The number of C residues is $c \n";
}
sub gcrich
{ $seq=$_[0];
$len=length($seq);
for($i=0;$i<$len;$i++)
  {  $arr[$i]=substr($_[0],$i,2);
  }
foreach $x(@arr)
  {   if( $x eq 'GC')
        {$gc++;
        }
  }
 print "The number of GC rich region $gc \n";
}
```

Q3. Write a program to calculate and display the AT content and GC content of a given sequence.

Program

```
#!/usr/bin/perl -w

@array=('a','c','t','g','g','t','t','g','a','g','c','g','t','a','t','g',
'c','g','g','g','g','a','a','a','t','g');
$len=@array;

foreach $x(@array)
{
if(($x eq 'a')|($x eq 't'))
     {  $at++;
     }
     elsif (($x eq 'g')|($x eq 'c'))
     {  $gc++;
     }
}

$at=($at*100)/$len;
$gc=($gc*100)/$len;

print "The AT content: $at \n";
print "The GC content: $gc \n";
```

Q4. Write a program to display the complement of the given DNA sequence.

Program

```
#!/usr/bin/perl -w

@array=qw(A C T G G T T G A G C G T A T G C G G G G A A A T G);
$i=0;
foreach $x(@array)
{   if( $x eq 'A')
    {  $dna[$i]= 'T';
    }
    elsif( $x eq 'T')
    {  $dna[$i]= 'A';
    }
    elsif( $x eq 'G')
    {  $dna[$i]= 'C';
    }
    elsif( $x eq 'C')
    {  $dna[$i]= 'G';
    }
$i++;
}
print " The complement is:@dna \n";
```

Q5. Write a program that will allow the user to input the gene name, nucleotide sequence, species name, and sequencing date. Store these data in a file. Display the file content.

Program

```
#!/usr/bin/perl -w

print "Enter the Gene name  : ";
chomp($name=<STDIN>);
print "Enter the nucleotide sequence  : ";
chomp($seq=<STDIN>);
print "Enter the species  : ";
chomp($spec=<STDIN>);
print "Enter the sequencing date  : ";
chomp($dat=<STDIN>);

open(gen,">>DNAdata.out");
print gen "Gene name : $name \n";
print gen "Nucleotide sequence :$seq \n";
print gen "Species :$spec \n";
print gen "Date : $dat \n";
close(gen);

open(gen,"DNAdata.out");
while($con=<gen>)
      {  print $con;
      }
close(gen);
```

Q6. Write a program that will allow the user to enter a DNA sequence and then display the complement as well as the reverse complement of the sequence.

Program

```
#!/usr/bin/perl -w

print "Enter the DNA sequence : ";
chomp($rev=<STDIN>);
$seq=$rev;
$rev=~ tr/ATGC/TACG/;
$comp=reverse($rev);

print "The DNA sequence is : $seq \n";
print "The complement of DNA sequence is : $rev \n";
print "The reverse complement of DNA sequence is : $comp \n";
```

Q7. Write a program to find out the TATA box and its position in a given sequence.

Program

```
#!/usr/bin/perl -w

@DNA=('A','T','A','G','A','G','A','G','A','T','A','T','A','G',
'A','G','A','A','A','A','G','A','A','G','A');
```

```
$i=0;
$flag=0;
$len=@DNA;
for($k=0;$k<$len-3;$k++)
     {   $con="";
     $count=4;
     for($j=$i;$j<$count+$i;$j++)
            {   $con.=$DNA[$j];
             }
            if ($con eq "TATA")
                   {   $flag=1;
                   print "The position of TATA box in the given array $i \n";
                   }
            $i++;
     }
if($flag==0)
{   print "There is no TATA box in the given array \n";
}
```

Q8. Write a program to calculate the number of promoters in each input sequence and then find out the sequence that will transcribe.

Program

```
#!/usr/bin/perl -w

@Euk1=('A','G','A','G','A','G','C','A','A','T','G','A','G','A','G',
'A','G','A','T','A','T','A','T','G');
@Euk2=('A','T','A','C','A','G','A','G','A','T','T','G','T',
'A','T','A','G','A','G','A','G','A','G','A');

$i=0;
$flag1=0;
$len=@Euk1;

for($k=0;$k<$len-3;$k++)
     {   $con="";
     $count=4;
     for($j=$i;$j<$count+$i;$j++)
            {   $con.=$Euk1[$j];
            }
     if (($con eq "TATA")|($con eq "CAAT"))
            {   $flag=1;
                $seq1++;
            }
     $i++;
     }
```

```
$i=0;
$flag2=0;
$len=@Euk2;

for($k=0;$k<$len-3;$k++)
    {   $con="";
    $count=4;
    for($j=$i;$j<$count+$i;$j++)
         {   $con.=$Euk2[$j];
         }
    if (($con eq "TATA")|($con eq "CAAT"))
         {   $flag=1;
         $seq2++;
         }
    $i++;
    }

if ($flag1==0)
    { print "The Euk1 sequence  will not transcribe as it does not have TATA
or
             CAAT box";
    }
elsif ($flag2==0)
    {print "The Euk2 sequence  will not transcribe as it does not have TATA
or
             CAAT box";
    }
else
    { print "The number of promoters in Euk1:  $seq1 \n";
     print "The number of promoters in Euk2:  $seq2 \n";

    if ($seq1 gt $seq2)
         {   print " Euk1 is better transcribed than Euk2 \n";
         }
    else
         {   print " Euk2 is better transcribed than Euk1 \n";
         }
```

Q9. Write a program that will allow the user to input two sequences and then find out the frame in which the similarity between the two sequences is the greatest.

Program

```
#!/usr/bin/perl  -w

print "Enter the sequence1 :  ";
chomp($seq1=<STDIN>);

print "Enter the sequence2 :  ";
chomp($seq2=<STDIN>);
```

```
$len1=length($seq1);
$len2=length($seq2);
&compare($seq1,$seq2);

sub compare
{
  for($i=0;$i<$len1;$i++)
  {  $arr1[$i]=substr($_[0],$i,1);
   }
  for($i=0;$i<$len2;$i++)
  {  $arr2[$i]=substr($_[1],$i,1);
   }

 print(@arr1);
 print("\n");
 print(@arr2);
 print("\n");

 $k=0;
 $greatest=0;

 if($len1>$len2)
  {
    for($i=$k;$i<$len1-$len2+1;$i++,$k++)
     { $count=0;
       for($j=0;$j<$len2;$j++)
       {  if($arr1[$i+$j] eq $arr2[$j])
           {   $count++;
            }
        }

       if($count gt $greatest)
       {  $greatest=$count;
          $frame=$k;
        }
     }
  }
 else
 {  for($i=$k;$i<$len2-$len1+1;$i++,$k++)
     { $count=0;
       for($j=0;$j<$len1;$j++)
       {  if($arr2[$i+$j] eq $arr1[$j])
           {   $count++;
            }
        }

       if($count gt $greatest)
```

```
        {   $greatest=$count;
             $frame=$k;
          }
      }
   }
   print "The similarity between the two sequences is $greatest at frame
$frame\n";
}
```

Q10. Write a program that will allow the user to input two sequences and write in a file the sequences, number of similarities and the percentage similarity of the two sequences.

Program

```
#!/usr/bin/perl -w

print "Enter the sequence1 : ";
chomp($seq1=<STDIN>);
print "Enter the sequence2 : ";
chomp($seq2=<STDIN>);
$count=0;
$len1=length($seq1);
$len2=length($seq2);

for($j=0;$j<$len1;$j++)
{  $arr1[$j]=substr($seq1,$j,1);
}
for($j=0;$j<$len2;$j++)
{  $arr2[$j]=substr($seq2,$j,1);
}

if($len1 le $len2)
     {   for($i=0;$i<$len1;$i++)
           {   if($arr1[$i] eq $arr2[$i])
                 {   $count++;
                 }
           }
     }
else
     {
     for($i=0;$i<$len2;$i++)
           {   if($arr1[$i] eq $arr2[$i])
                 {   $count++;
                 }
           }
     }
$per=($count*100)/$len;
```

```
open (Hom," > Similarity.out");
print Hom "DNA sequence1 \n";
print Hom "$seq1 \n";
print Hom "DNA sequence2 \n";
print Hom "$seq2 \n";
print Hom "The homology similarity between two sequences is $count \n";
print Hom "The similarity percentage between two sequences is $per;
close Hom;
open (Hom,"Similarity.out");
while($con=<Hom>)
{  print $con;
}
close Hom;
```

Q11. Write a program to read a DNA sequence from a file and then ask the user to enter the query sequence and then look for the query in the DNA sequence file.

Program

```
#!/usr/bin/perl -w

if (open(DNA,"dna.txt"))
{  $seq=<DNA>;
   print "Enter the query to be searched in nucleotide sequence : ";
   chomp($query=<STDIN>);
   if ($seq=~ m/$query/)
   {      $x = pos($seq);
     print "The query $query exists in the nucleotide sequence at $x
position \n";
    }
   else
   { print "The query $query does not exist in the nucleotide sequence\n";
    }
}
else
{  print "File does not exist in current directory \n";
}
```

Q12. Write a program to enter a three letter codon and display its corresponding amino acid. When a user gives a wrong input display an appropriate error message.

Program

```
#!/usr/bin/perl -w

%arr=("UCA"=>"S","UUA"=>"L","CCA"=>"P","AUA"=>"I","AGA"=>"R","AAG"=>"K");
print "Enter the three letter code : ";
chomp($code=<STDIN>);
$flag=0;
```

```
if(length($code) eq 3)
    {   while(($cc,$amino)=each(%arr))
            {  if($code eq $cc)
                    {   $am=$amino;
            $flag=1;
            last;
             }
    }
if($flag eq 1)
    {   print "The Translation of code $code is $am \n";
    }
else
    {   print "No code for Translation \n";
    }
else
{   print "The length of code is not equal to 3 \n";  }
```

Appendix

A3

LINUX for Bioinformatics

The basic commands required to work in Linux environment are as follows:

Command	Example	Description
cat		To send the file contents to standard output
cd		To change directory
	cd /home	To change the current working directory to /home
	cd work	To change the current working directory to work, relative to the current location which is "/home". The full path of the new working directory is "/home/work"
	cd ..	To move to the parent directory of the current directory. This command will make the current working directory "/home
	cd ~	To move to the user's home directory "/ home/username". The '~' sign indicates the users home directory.
cp		To copy files
	cp file1 file2	To copy the files "file1" to the file "file2" in the current working directory. This command will create the file "file2" if it does not exist. It will normally overwrite it without warning if it exists.

Command	Example	Description
	cp i file1 file2	To copy the file where if "file2" file exists, you will be prompted before it is overwritten.
	cp i /data/ file1	To copy the file "/data/file1" to the current working directory and name it "file1". Prompt before overwriting the file.
	cp -dpr dir1 dir2	To copy all files from the directory "dir1" to directory "dir2" preserving links (-p option), file attributes (-p option), and copy recursively (-r option). With these options, a directory and all it contents can be copied to another directory.
dd	dd if=/dev/hdb1of=/backup/	To duplicate the contents of the disk. The "if" parameter means input file, "of" means output file.
df		To show the amount of disk space used on each mounted file system.
less	less textfile	To display the contents of a text file. This is similar to the more command, but the user can page up and down through the file.
ln		To create a symbolic link to a file.
	ln -s test symlink	To create a symbolic link named symlink that points to the file test. Typing "ls -i test symlink" will show the two files are different with different inodes. Typing "ls -l test symlink" will show that symlink points to the file test.
locate		A fast database driven file locator.
	slocate -u	To build the slocate database. It takes several minutes to complete this command.
	locate whereis	To list all files whose names contain the string "whereis".
logout		To log off the current user from the system.
Is		To list files
	Is	To list files in the current working directory
	Is -al	To list all files in the current working directory in a long listing format showing permissions, ownership, size, and time, and date stamp.

Command	Example	Description
more		To allow file contents or piped output to be sent to the screen one page at a time.
	more/etc /profile	To list the contents of the "/etc/profile" file to the screen one page at a time.
	ls -al lmore	To perform a directory listing of all files and pipes the output of the listing through more. If the directory listing is longer than a page, it will be listed one page at a time.
mv		To move or rename file.
	mv -i file1 file2	To move a file from "file1" to "file2". This effectively changes the name of "file1" to "file2".
	mv -i /data /myfile	Move the file from "myfile" from the directory "/data" to the current working directory.
pwd		To show the name of the current working directory
	more /etc /profile	To list the contents of the "/etc/profile" file to the screen one page at a time.
shutdown		To shut down the system.
	shutdown-h now	To shut down the system to halt immediately.
	shutdown-r now	To shut down the system immediately and reboot it.
whereis		To show where the binary, source, and manual page files are for a command.
	whereis is	To locate binaries and manual pages for the ls command.

Here, let us look at few Linux Shell Scripts that you may need for bioinformatics tasks.

1. Write a script to create a directory structure as below:

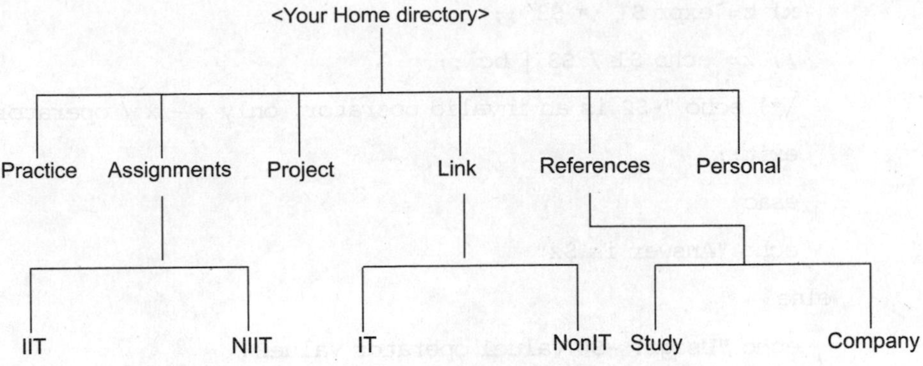


```
# Shell script to create a directory structure

mkdir practice
mkdir assignments
mkdir project
mkdir links
mkdir references
mkdir personal
cd assignments
cd IIT
cd NIIT
cd ..
cd links
cd IT
cd nonIT
cd ..
cd references
mkdir study
mkdir company
```

2. Write script, using case statement to perform basic real number math operations such as addition, subtraction, multiplication, and division. The script should accept the numbers and operator as command line arguments.

```
# Shell script to perform basic maths operation using case statements
clear
if [ $# -eq 3 ]
then
   case $2 in
   +) z=`expr $1 + $3`;;
   -) z=`expr $1 - $3`;;
```

```
    x) z=`expr $1 \* $3`;;
    /) z=`echo $1 / $3 | bc`;;
    \*) echo "-$2 is an invalid operator, only + - x / operator allowed"
    exit;;
    esac
    echo "Answer is $z"
else
    echo "Usage: -$0 value1 operator value2"
    echo "Value1 and Value2 are numeric values"
    echo "Operator can be + - / x"
fi
```

3. Write script to determine whether a given file exists or not; the file name is supplied as command line argument. If the file does not exist, the script should display an appropriate message.

```
# Shell script to check whether a file exists or not.
if [ $# -ne 1 ]
then
    echo "Usage - $0  file-name"
    exit 1
fi
if [ -f $1 ]
then
    echo "$1 file exist"
else
    echo "Sorry, $1 file does not exist"
fi
```

4. Write script to greet the user and print the following messages: Good Morning, Good Afternoon or Good Evening, according to the system time. Incorporate this script into your startup file called .bash_profile so that it gets automatically executed as soon as you log on to your system.

```
# Shell script to greet the user as per time
clear
templdate=`date | cut -c12-c13`
if [ $templdate -le 12 ]
then
```

```
   echo "Hello $LOGNAME! Good Morning"
elif [ $temp1date -gt 12 ] && [ $temp1date -le 16 ]
then
   echo "Hello $USER! Good Afternoon"
elif [ $temp1date -gt 16 ] && [ $temp1date -le 18 ]
then
   echo "Hello `whoami`! Good Evening"
fi
```

5. Write shell script to show various system configurations,e.g.,
(a) currently logged number of users and logged users
(b) your login id
(c) your current shell
(d) your home directory
(e) your current working directory
(f) your operating system type
(g) your current path setting
(h) about your os and version, release number , kernel version
(i) show computer cpu information such as the processor type, speed, etc.
(j) show memory information and hard disk information such as the size of hard disk, cache memory, model, etc

```
# Shell script to display system configuration
clear
temp1date=`date +"%A, %dth %B %Y"`
echo "Today is &temp1date"
echo "Currently logged number of user(s) $nouser"
echo "Currently logged users are `who`"
echo "Your login id `whoami`"
echo -e "Your Home Directory: $HOME"
echo -e "Your current working directory: `pwd`"
echo -e "OS Type: $OSTYPE"
echo -e "Current Shell: $SHELL"
echo -e "PATH: $PATH"
```

/*a. About your os and version, release number, kernel version: Show computer CPU information like processor type, speed etc

b. Show memory information and hard disk information like size of hard-disk, cache memory, model etc*/

6. Write a shell script to remove blank spaces, special characters, and new line characters from a text file containing DNA sequence.

```
# Shell script to remove blank spaces, special characters and new line
  characters from a text file containing DNA sequence
# Consider the input file name is myinput and output file name is
  myoutput
  tr -s ' ' '' <myinput.txt>myoutput
  tr -s '@' '' <myinput.txt>myoutput
  tr -s ' ' '' <myinput.txt>myoutput
```

Glossary of Bioinformatics Terms

Accession number

An identifier assigned by the curators of major biological databases upon submission of a novel entry that uniquely identifies that sequence (or other) entry.

Agent

Independent, autonomous, software modules that can search the Internet for data or content pertinent to a particular application, such as a gene, protein, or biological system.

Algorithm

A series of steps, which can be coded into a programming language and executed, defining a procedure or formula for solving a problem,. Bioinformatics algorithms typically are used to process, store, analyse, visualize, and make predictions from biological data.

Alignment

The result of a comparison of two or more gene or protein sequences in order to determine their degree of base or amino acid similarity. Sequence alignments are used to determine the similarity, homology, function, or other degree of relatedness between two or more genes or gene products.

Analogy

Reasoning by which the function of a novel gene or protein sequence may be deduced from comparisons with other gene or protein sequences of known function. Identifying analogous or homologous genes via similarity searching and alignment is one of the chief uses of bioinformatics. (*See also* alignment, similarity search.)

Annotation

A combination of comments, notations, references, and citations, either in free format or utilizing a controlled vocabulary, that together describe all the experimental and inferred information about a gene or protein. Annotations can also be applied to the description of other biological systems. Batch, automated annotation of bulk biological sequences is one of the key uses of bioinformatics tools.

Assembly

Compilation of overlapping sequences from one or more related genes that have been clustered together based on their degree of sequence identity or similarity. Sequence assembly may be used to piece together 'shotgun' sequencing fragments (see shotgun sequencing) based upon overlapping restriction enzyme digests. It can also be used to identify and index novel genes from 'single-pass' cDNA sequencing efforts.

Bacterial artificial chromosome (BAC)

It is a cloning vector that can incorporate large fragments of DNA.

Beta sheet

A three-dimensional arrangement taken up by polypeptide chains which consists of alternating strands linked by hydrogen bonds. The alternating strands together form a frequently twisted sheet. It is one of the secondary structural elements characteristic of proteins.

Chromat

This is data file output from most popular DNA sequencers. Chromat files consist of the fluorescent traces generated by the sequencer for each of the four chemical bases,(A, C, G, and T) together with the sequence and measures of the error in the traces at each sequence position.

Cluster

The grouping of similar objects in a multidimensional space results in a cluster. Clustering is used for constructing new features, which are abstractions of the existing features of those objects. The quality of clustering depends crucially on the distance metric in the space. In bioinformatics, clustering is performed on sequences, high-throughput expressiona, and other experimental data. Clusters of partial or complete gene sequences can be used to identify a complete (contiguous) sequence and to better identify its function. Clustering expression data enables the researcher to discern patterns of co-regulation in groups of genes.

Combinatorial chemistry

This is he use of chemical methods to generate all possible combinations of chemicals starting with a subset of compounds. The building blocks may be peptides, nucleic acids, or small molecules. The libraries of compounds formed by this methodology are used to probe for new pharmaceutical reagents (see high-throughput screening).

Complexity (of a gene sequence)

The term 'low complexity sequence' may be thought of as synonymous with regions of locally biased amino acid composition. In these regions, the sequence composition deviates from the random model that underlies the calculation of the statistical significance (P-value) of an alignment. Such alignments among low complexity sequences are statistically, but not biologically, significant, i.e., one cannot infer homology (common ancestry) or functional similarity.

Configuration

The complete ordering and description of all parts of a software or database system is termed its configuration. Configuration management is the use of software to identify, inventory, and maintain the component modules that together comprise one or more systems or products.

Conformation

It is the precise three-dimensional arrangement of atoms and bonds in a molecule describing its geometry and hence its molecular function.

Consensus sequence

A single sequence delineated from an alignment of multiple constituent sequences which represents a 'best fit' for all those sequences. A 'voting' or other selection procedure is used to determine which residue (nucleotide or amino acid) is placed at a given position in the event that not all of the constituent sequences have the identical residue at that position.

Contig

A length of contiguous sequence assembled from partial, overlapping sequences, generated from a 'shotgun' sequencing project. Contigs are typically created computationally, by comparing the overlapping ends of several sequencing reads generated by restriction enzyme digestion of a segment of genomic DNA. The creation of contigs in the presence of sequencing errors, ambiguities, and the presence of repeats is one of the most computationally challenging aspects of the role of bioinformatics in genome analysis.

Convergence

The end-point of any algorithm that uses iteration or recursion to guide a series of data processing steps. An algorithm is usually said to have reached convergence when the difference between the computed and observed steps falls below a predefined threshold.

Cosmid

DNA vector that allow the insertion of long fragments of DNA (up to 50 kbases).

Crystal structure

It is the term used to describe the high resolution molecular structure derived by X-ray crytallographic analysis of protein or other biomolecular crystals.

Data cleaning

A process whereby automated or semi-automated algorithms are used to process experimental data, including noise, experimental errors, and other artifacts, in order to generate and store high-quality data for use in subsequent analysis. Data cleaning is typically required in high-throughput sequencing where compression or other experimental artifacts limit the amount of sequence data generated from each sequencing run or 'read.'

Data mining

The ability to query very large databases in order to satisfy a hypothesis ('top-down' data mining); or to interrogate a database in order to generate new hypotheses based on rigorous statistical correlations ('bottom-up' data mining).

Data processing

Data processing is defined as the systematic performance of operations upon data, e.g., handling, merging, sorting, and computing. The semantic content of the original data should not be changed, but the semantic content of the processed data may be changed.

Data warehouse

Vast arrays of heterogeneous (biological) data, stored within a single logical data repository, which are accessible to different querying and manipulation methods.

Database

Any file system by which data gets stored following a logical process. (See also relational database)

Dendrogram

A graphical procedure for representing the output of a hierarchical clustering method. A dendrogram is strictly defined as a binary tree with a distinguished root, which has all the data items at its leaves. Conventionally, all the leaves are shown at the same level of the drawing. The ordering of the leaves is arbitrary, as is their horizontal position. The heights of the internal nodes may be arbitrary, or may be related to the metric information used to form clustering.

DNA microarray

It is the deposition of oligonucleotides or cDNAs onto an inert substrate such as glass or silicon. Thousands of molecules may be organized spatially into a high-density matrix. These DNA chips may be probed to allow expression monitoring of many thousands of genes simultaneously. They are used to study polymorphisms in genes, de novo sequencing, or molecular diagnosis of disease.

Domain (protein)

It is a region of special biological interest within a single protein sequence. A domain may also be defined as a region within the three-dimensional structure of a protein that may encompass regions of several distinct protein sequences that accomplish a specific function. A domain class is a group of domains that share a common set of well-defined properties or characteristics.

Drug

A drug is an agent that affects a biological process. It is a molecule whose molecular structure can be correlated with its pharmacological activity.

Expressed sequence tag (EST)

An EST is a small sequence from an expressed gene that can be amplified by PCR. ESTs act as physical markers for cloning and full length sequencing of cDNAs of expressed genes. ESTs are typically identified by purifying mRNAs, converting to cDNAs, and then sequencing a portion of the cDNAs.

Expression (gene or protein)

It is a measure of the presence, amount, and time-course of one or more gene products in a particular cell or tissue. Expression studies are typically performed at the RNA (mRNA) or protein level in order to determine the number, type, and level of genes that may be up-regulated or down-regulated during a cellular process, in response to an external stimulus, or in sickness or disease. Gene chips and proteomics now allow the study of expression profiles of sets of genes or even entire genomes.

Expression profile

The level and duration of expression of one or more genes, selected from a particular cell or tissue type, generally obtained by a variety of high-throughput methods, such as sample sequencing, serial analysis, or microarray-based detection.

Functional genomics

It is the use of genomic information to delineate protein structure, function, pathways, and networks. Function may be determined by 'knocking out' or 'knocking in' expressed genes in model organisms such as worm, fruitfly, yeast or mouse.

Gap (affine gap)

A gap is defined as any maximal, consecutive run of spaces in a single string of a given alignment. Gaps help create alignments that better conform to underlying biological models and more closely fit patterns that one expects to find in a meaningful alignment. The idea is to take into account the number of continuous gaps and not only the number of spaces when calculating an alignment. Affine gaps contain a component each for gap insertion and gap extension, where the extension penalty is usually much lower than the insertion penalty. This mimics biological reality as multiple gaps would imply multiple mutations; but a single mutation can lead to a long gap quite easily.

Gap penalty

It is the penalty applied to a similarity score for the introduction of an insertion or deletion gap, the extension of a gap, or both. Gap penalties are usually subtracted from a cumulative score being determined for the comparison of two or more sequences via an optimization algorithm that attempts to maximize that score.

Gene index

It is a listing of the number, type, label, and sequence of all the genes identified within the genome of a given organism. Gene indices are usually created by assembling overlapping EST sequences into clusters, and then determining whether each cluster corresponds to a unique gene. Methods by which a cluster can be identified as representing a unique gene include identification of long open reading frames (ORFs), comparison to a genomic sequence,, and detection of SNPs or other features in the cluster that are known to exist in the gene.

Gene family

Gene families are subsets of genes containing homologous sequences which usually correlate with a common function.

Genome

It is the complete genetic content of an organism.

Genomic DNA (sequence)

The DNA sequence typically obtained from mammalian or other higher-order species, which includes both intron and exon sequences (coding sequence), as well as non-coding regulatory sequences such as promoter and enhancer sequences.

Genomics

Genomics is the analysis of the entire genome of a chosen organism.

Hidden Markov model (HMM)

HMM is a joint statistical model for an ordered sequence of variables. It is the result of stochastically perturbing the variables in a Markov chain (the original variables are thus 'hidden'), where the Markov chain has discrete variables which select the 'state' of the HMM at each step. The perturbed values can be continuous and are the 'outputs' of the HMM. A hidden Markov model is equivalently a coupled mixture model where the joint distribution over states is a Markov chain. Hidden Markov models are valuable in bioinformatics because they allow a search or alignment algorithm to be trained using unaligned or unweighted input sequences. They are useful also because they allow position-dependent scoring parameters such as gap penalties, thus more accurately modeling the consequences of evolutionary events on sequence families.

High-throughput screening

It is the method by which very large numbers of compounds are screened against a putative drug target in either cell-free or whole-cell assays. Typically, these screenings are carried out in 96 well plates using automated, robotic station based technologies or in higher-density array ('chip') formats.

Homology

Strict: Two or more biological species, systems or molecules that share a common evolutionary ancestor. General: Two or more gene or protein sequences that share a significant degree of similarity, typically measured by the amount of identity (in the case of DNA), or conservative replacements (in the case of protein), that they register along their lengths. Sequence 'homology' searches are typically performed with a query DNA or protein sequence to identify known genes or gene products that share significant similarity and hence might inform on the ancestry, heritage, and possible function of the query gene.

in silico

(Lit. computer mediated) The use of computers to simulate, process, or analyse a biological experiment.

Intron

Nucleotide sequences found in the structural genes of eukaryotes that are non-coding and interrupt the sequences containing information that codes for polypeptide chains. Intron sequences are spliced out of their RNA transcripts before maturation and protein synthesis. (cf. Exon)

Iteration

A series of steps in an algorithm whereby the processing of data is performed repetitively until the result exceeds a particular threshold. Iteration is often used in multiple sequence alignments whereby each set of pairwise alignments is compared with every other, starting with the most similar pairs and progressing to the least similar, until there are no longer any sequence pairs remaining to be aligned.

Lead compound

A candidate compound identified as the best 'hit' (tight binder) after screening of a combinatorial (or other) compound library, which is then taken into further rounds of screening to determine its suitability as a drug.

Lead optimization

The process of converting a putative lead compound ('hit') into a therapeutic drug with maximal activity and minimal side effects, typically using a combination of computer-based drug design, medicinal chemistry, and pharmacology.

Markov chain

Any multivariate probability density whose independence diagram is a chain. The variables are ordered, and each variable 'depends' only on its neighbours in the sense of being conditionally independent of others. Markov chains are an integral component of hidden Markov models.

Microarray

A two-dimensional array, typically on a glass, filter, or silicon wafer, upon which genes or gene fragments are deposited or synthesized in a predetermined spatial order, allowing them to be made available as probes in a high-throughput, parallel manner.

Microfluidics

The miniaturization of chemical reactions or pharmacological assays into microscopic tubes or vessels in order to greatly increase their throughput, by placing many of them side by side in an array.

Modelling

In bioinformatics, modelling usually refers to molecular modelling, a process whereby the three-dimensional architecture of biological molecules is interpreted (or predicted), visually represented, and manipulated in order to determine their molecular properties. General: A series of mathematical equations or procedures which simulate a real-life process, given a set of assumptions, boundary parameters, and initial conditions.

Motif

A conserved element of a protein sequence alignment that usually correlates with a particular function. Motifs are generated from a local multiple protein sequence alignment corresponding to a region whose function or structure is known. It is sufficient that it is conserved, and is hence likely to be predictive of any subsequent occurrence of such a structural/functional region in any other novel protein sequence.

Multigene family

A set of genes derived by duplication of an ancestral gene, followed by independent mutational events resulting in a series of independent genes either clustered together on a chromosome or dispersed throughout the genome.

Multiple (sequence) alignment

A multiple alignment of k sequences is a rectangular array, consisting of characters taken from the alphabet A, which satisfies the following conditions: There are exactly k rows; ignoring the gap character, row number i is exactly the sequence sI; and each column contains at least one character different from '-'. In practice, multiple sequence alignments include a cost/weight function, which defines the penalty for the insertion of gaps (the '-' character) and weights identities and conservative substitutions accordingly. Multiple alignment algorithms attempt to create the optimal alignment defined as the one with the lowest cost/weight score.

Multiplex sequencing

Approach to high-throughput sequencing that uses several pooled DNA samples run through gels simultaneously and then separated and analysed.

Neural net

A neural net is an interconnected assembly of simple processing elements, units or nodes, whose functionality is loosely based on the animal brain. The processing ability of the network is stored in the inter-unit connection strengths, or weights, obtained by a process of adaptation to, or learning from, a set of training patterns. Neural nets are used in bioinformatics to map data and make predictions such as taking a multiple alignment of a protein family as a training set in order to identify novel members of the family from their sequence data alone.

Object-relational database

Object databases combine the elements of object orientation and object-oriented programming languages with database capabilities. They provide more than persistent storage of programming language objects. Object databases extend the functionality of object programming languages (e.g., C++, Smalltalk, or Java) to provide full-featured database programming capability. The result is a high level of congruence between the data model for the application and the data model of the database. Object-relational databases are used in bioinformatics to map molecular biological objects (such as sequences, structures, maps, and pathways) to their underlying representations (typically within the rows and columns of relational database tables.) This enables the user to deal with the biological objects in a more intuitive manner, as they would in the laboratory, without having to worry about the underlying data model of their representation.

Open reading frame (ORF)

It is any stretch of DNA that potentially encodes a protein. Open reading frames start with a start codon, and end with a termination codon. No termination codons may be present internally. The identification of an ORF is the first indication that a segment of DNA may be part of a functional gene.

Ortholog (or Orthologue)

Orthologs are genes in different species that evolved from a common ancestral gene by speciation. Normally, orthologs retain the same function in the course of evolution. Identification of orthologs is critical for reliable prediction of gene function in newly sequenced genomes. (See also Paralogs)

Pattern

Molecular biological patterns usually occur at the level of the characters making up the gene or protein sequence. A pattern language must be defined in order to apply different criteria to different positions of a sequence. In order to have position-specific comparison done by a computer, a pattern-matching algorithm must allow alternative residues at a given position, repetitions of a residue, exclusion of alternative residues, weighting, and, ideally, combinatorial representation.

Pathway

Bioinformatics strives to define representations of key biological data types, algorithms and inference procedures, including sequences, structures, biological pathways, and reactions. Representing and computing with biological pathways requires ontologies for representing pathway knowledge, user interfaces to these databases, physico-chemical properties of enzymes and their substrates in pathways, and pathway analysis of whole genomes including identifying common patterns across species and species differences.

Paralog (or Paralogue)

Paralogs are genes related by duplication within a genome. Orthologs retain the same function in the course of evolution, whereas paralogs evolve new functions, even if the new functions are related to the original one.

Pharmacogenomics

This is the use of (DNA-based) genotyping in order to target pharmaceutical agents to specific patient populations. Genetic differences are known to affect responses to many types of drug therapy, and pharmacogenomics analysis serves to customize the use of pharmaceuticals for specific subgroups of patients. The rationale for this approach is that observed gene expression differences may correlate with, and explain, the differences in side effects and efficacy of drugs in humans.

Pharmacophore

The three-dimensional spatial arrangement of atoms, substituents, functional groups, or chemical features that together are sufficient to describe the pharmacologically active components of a drug molecule or molecule series.

Profile

Sequence profiles are usually derived from multiple alignments of sequences with a known relationship, and consist of tables of position-specific scores and gap penalties. Each position in the profile contains scores for all of the possible amino acids, as well as one penalty score for opening gap and one for continuing a gap at the specified position. Attempts have been made to further improve the sensitivity of the profile by refining the procedures to construct a profile starting from a given multiple alignment. Other representations for sequence domains or motifs do not necessarily require the presence of a correct and complete multiple alignment, e.g., hidden Markov models.

Protein families

These are sets of proteins that share a common evolutionary origin reflected by their relatedness in function which is usually reflected by similarities in sequence, or in primary, secondary, or tertiary structure. Subsets of proteins with related structure and function.

Proteome

Proteome is the entire protein complement of a given organism.

Query (sequence)

A DNA, RNA of protein sequence used to search a sequence database in order to identify close or remote family members (homologs) of known function, or sequences with similar active sites or regions (analogs), from whom the function of the query may be deduced.

Rational drug design (Structure based drug design)

The development of drugs based on the three-dimensional molecular structure of a particular target.

Recursion

An algorithmic procedure whereby an algorithm calls on itself to perform a calculation until the result exceeds a threshold, in which case the algorithm exits. Recursion is a powerful data processing procedure and is computationally quite efficient.

Reading frame

Each of the three bases in the RNA sequence codes for one amino acid. Thus there are six possible predicted protein sequences resulting from such a piece of code. For example, the RNA sequence given below can be translated as

```
CAAUGGCUAGGUACUAUGUAUGAGAUCAUGAUCUU

   □ Forward frames
 CAA UGG CUA GGU ACU AUG  UAU GAG AUC AUG   RNA
  Q   W   L   G   T   M    Y   E   I   M  a.a.
 C AAU GGC UAG GUA CUA UGU  AUG AGA UCA UGA  RNA
  N   G   *   V   L   C    M   R   S   *  a.a.
 CA AUG GCU AGG UAC UAU GUA UGA GAU CAU GAU  RNA
  M   A   R   Y   Y   V    *   D   H   D  a.a.
   □ Reverse frames
 GAG CCU AAA CAU UUC UAG UAC UAG AGU AUG UAU  RNA
  E   P   K   H   F   *    Y   *   S   M     a.a.
```

Relational database

It is a database that follows E.F. Codd's 11 rules, a series of mathematical and logical steps for the organization and systemization of data into a software system that allows easy retrieval, updating, and expansion. A relational database management system (RDBMS) stores data in a database consisting of one or more tables of rows and columns. The rows correspond to a record (tuple); the columns correspond to attributes (fields) in the record. In an RDBMS, a view, defined as a subset of the database that is the result of the evaluation of a query, is a table. RDBMSs use structured query language (SQL) for data definition, data management, and data access and retrieval. Relational and object-relational databases are used extensively in bioinformatics to store sequence and other biological data.

Repeats (Repeat sequences)

Repeat sequences and approximate repeats occur throughout the DNA of higher organisms (mammals). For example, the Alu sequences of length about 300 characters appear hundreds of thousands of times in human DNA with about 87% homology to a consensus Alu string. Some short substrings such as TATA-boxes, poly-A and (TG)* also appear more often than by chance. Repeat sequences may also occur within genes as mutations or alterations to those genes. Repetitive sequences, especially mobile elements, have many applications in genetic research. DNA transposons and retroposons are routinely used for insertional mutagenesis, gene mapping, gene tagging, and gene transfer in several model systems.

Secondary structure (protein)

The organization of the peptide backbone of a protein that occurs as a result of hydrogen bonds, e.g., alpha helix, beta pleated sheet.

Selectivity

Selectivity of bioinformatics similarity search algorithms is defined as the significance threshold for reporting database sequence matches. As an example, for BLAST searches the parameter E is interpreted as the upper bound on the expected frequency of the chance occurrence of a match within the context of the entire database search. E may be thought of as the number of matches one expects to observe by chance alone during database search.

Sensitivity

Sensitivity of bioinformatics similarity search algorithms centers around two areas: First, how well can the method detect biologically meaningful relationships between two related sequences in the presence of mutations and sequencing errors. Second, how does the heuristic nature of the algorithm affect the probability that a matching sequence will not be detected. At the user's discretion, the speed of most similarity search programs can be sacrificed in exchange for greater sensitivity--with an emphasis on detecting lower scoring matches.

Sequence tagged site (STS)

A unique sequence from a known chromosomal location that can be amplified by PCR. STSs act as physical markers for genomic mapping and cloning.

Similarity (homology) search

Given a newly sequenced gene, there are two main approaches to the prediction of structure and function from the amino acid sequence. Homology methods are the most powerful and are based on the detection of significant extended sequence similarity to a protein of known structure, or of a sequence pattern characteristic of a protein family. Statistical methods are less successful but more general and are based on the derivation of structural preference values for single residues, pairs of residues, short oligopeptides, or short sequence patterns. The transfer of structure/function information to a potentially homologous protein is straightforward when the sequence similarity is high and extended in length, but the assessment of the structural significance of sequence similarity can be difficult when sequence similarity is weak or restricted to a short region.

Signal sequence (leader sequence)

A short sequence added to the amino-terminal end of a polypeptide chain that forms an amphipathic helix allowing the nascent polypeptide to migrate through membranes such as the endoplasmic reticulum or the cell membrane. It is cleaved from the polypeptide after the protein has crossed the membrane.

Single nucleotide polymorphism (SNP)

Variations of single base pairs scattered throughout the human genome that serve as measures of the genetic diversity in humans. About 1 million SNPs are estimated to be present in the human genome. SNPs are useful markers for gene mapping studies.

Single-pass sequencing

Rapid sequencing of large segments of the genome of an organism by isolating as many expressed (cDNA) sequences as possible and performing single sequencer runs on their 5` or 3` ends. Single-pass sequencing typically results in individual, error-prone sequencing reads of 400-700 bases, depending on the type of sequencer used. However, if many of these are generated from numerous clones from different tissues, they may be overlapped and assembled to remove the errors and generate a contiguous sequence for the entire expressed gene.

Site

Sites in sequences can be located either in DNA (e.g., binding sites, cleavage sites) or in proteins. In order to identify a site in DNA, ambiguity symbols are used to allow several different symbols at one position. Proteins, however, need a different mechanism (see Pattern). Restriction enzyme cleavage sites, for instance, have the following properties: limited length (typically, less than 20 base pairs), definition of the cleavage site and its appearance (3', 5' overhang or blunt), definition of the binding site.

Structure prediction

Algorithms that predict the secondary, tertiary, and sometimes even quarternary structure of proteins from their sequences. Determining protein structure from sequence has been dubbed 'the second half of the genetic code' since it is the folded tertiary structure of a protein that governs how it functions as a gene product. As yet most structure prediction methods are only partially successful, and typically work best for certain well-defined classes of proteins.

Substitution matrix

A model of protein evolution at the sequence level resulting in the development of a set of widely used substitution matrices. These are frequently called Dayhoff, MDM (mutation data matrix), BLOSUM, or PAM (percent accepted mutation) matrices. They are derived from global alignments of closely related sequences. Matrices for greater evolutionary distances are extrapolated from those for lesser ones.

Tertiary structure

Folding of a protein chain via interactions of its side chain molecules including formation of disulphide bonds between cysteine residues.

Virtual library

The creation and storage of vast collections of molecular structures in an electronic database. These databases may be queried for subsets that exhibit specific physico-chemical features, or may be 'virtually screened' for their ability to bind a drug target. This process may be performed prior to the synthesis and testing of the molecules themselves.

Visualization

Visualization is the process of representing abstract scientific data as images that can aid in understanding the meaning of the data.

Weight matrix

The density of binding sites in a gene or sequence can be used to derive a ratio of density for each element in a pattern of interest. The combined individual density ratios of all elements are then collectively used to build a scoring profile known as a weight matrix. This profile can be used to test the prediction of the identification of the selected pattern and the ability of the algorithm to discriminate them from non-pattern sequences.

References

Arthur M. Lesk 2002, *Introduction to Bioinformatics*, Oxford University Press, India

Attwood, Teresa K. and David J. Parry-Smith 1999, *Introduction to Bioinformatics*, Longman Higher Education.

Baxevanis, Andreas and B.F. Francis Ouellette (eds) 2001, *Bioinformatics: A Practical Guide to the Analysis of Genes and Proteins*, 2nd edn, John Wiley.

Berg, Jeremy M., John L. Tymoczko, and Lubert Stryer 2002, *Biochemistry*, W.H. Freeman and Co.

Brinkman, Fiona S.L. and Detlef D. Leipe 2001, 'Phylogenetic Analysis' in *Bioinformatics: A Practical Guide to the Analysis of Genes and Proteins*, Andreas D. Baxevanis and B.F. Francis Ouellette (eds), 2nd edn, John Wiley & Sons.

Benson, D.A., M.S. Boguski, D.J. Lipman, J. Ostell, B.F. Ouellette, B.A. Rapp, and D.L. Wheeler 1999 *Nucleic Acids Research*, vol. pp 27, (1), pp 12–17.

Gardiner-Garden, M. and M. Frommer 1987, J. *Mol. Biol.*, vol. 196 (2), pp. 261–82.

Gibas, Cynthia and Per Jambeck 2001, *Developing Bioinformatics Computer Skill*, Shroff Publishers and Distributors Pvt Ltd, India.

Higgins, Des and Willie Taylor, 2000, *Bioinformatics: Sequence and Genome Analysis*, Cold Spring Harbour Laboratory Press, New York.

I, Georgeiv et al. 2006, *Bioinformatics*, vol. 24 (14), pp. 174–83.

Jones, Neil C. and Pavel A. Pevzner 2004, *An Introduction to Bioinformatics Algorithms*, Ane Books, India.

Krane, Dan E. and Michael L. Raymer 2003, *Fundamental Concepts of Bioinformatics*, Pearson Education, India.

Lesk, Arthur M. 2002, *Introduction to Bioinformatics*, Oxford University Press.

Lewin, Benjamin 2003, *Genes III*, Prentice Hall.

Lodish, H. et al. 2000, *Molecular Cell Biology*, W.H. Freeman & Co.

Marilyn Kozak 1999, *Gene*, vol. 234, pp. 187–208.

Min-Hao Kuo and C. David Allis 1998, *BioEssays*, vol. 20(8), pp. 615–26.

Mount, David W. 2001, *Bioinformatics: Sequence and Genome Analysis*, Cold Spring Harbor Laboratory.

Murrey, R.K. et al. 2002, *Harper's Biochemistry*, 25th edn, Appleton et Lange.

Nelson, D.L. and M.M. Cox 2000, *Principles of Biochemistry*, Lehninger 3rd edn, Worth.

R, Giegrick 2000, *Bioinformatics*, vol. 16(8), pp. 665–77.

Rastogi, S.C., Namita Mendiratta, and Parag Rastogi 2003, *Bioinformatics: Concepts, Skills and Applications*, CBS Publishers & Distributors, India.

Westhead, David R. and J. Howard Parish 2003, *Instant Notes: Bioinformatics*, Viva Books Pvt Ltd, India.

http://www.mindbranch.com/products/R2-909.html (last visited on 12th November 2006)

Note: All other links that appear in different chapters, especially in *Database and Tools*, were last visited in June 2006.

Index

A

acceptor sites 374
accession number 37
agriculture 11
algorithm 431–439
  implementation 439
  biological 441
  end state 431
  genetic 437
  heuristic 438
  initial state 431
  intermediate state 440
  internal state 440
  iteration 431
  logic 431
  parallel 440
  population 437
  probabilistic 437
  randomized 437
  recursive algorithm 439
  serial 440
  series 431
  simulistic approach 437
  design 433
  efficiency 432
  big-O notation 432
  binary search 433
  BLOSUM matrix construction 456
  Character state methods 489
  Chou–Fasman method 493
  data analysis 10, 44
  distance matrix methods 488
  divide and conquer 433
  dynamic programming 434
  gene prediction 477
  GOR 497
  graph exploration 436
  greedy 435
  linear programming 436
  machine learning 443
  mathematical programming 436
  maximum likelihood 491
  maximum parsimony 489
  multiple alignment 443
  multiple sequence heuristic alignment
  optimal substructures 434
  PAM matrix construction 453
  phylogenetic prediction 482
  prediction 10, 443, 476
  probabilistic and heuristic 437
  process of reduction 433
  protein structure prediction 491
  search and enumeration 436
  search-and-backtracking 436
  sequence alignment heuristic 443
  sequence alignment optimal 442, 443, 458
  sequence comparison 442, 443, 444

Substitution matrices 442, 443, 452
Needleman–Wunsch 459
Smith–Waterman 466
alibee 247
alignments 450
global sequence 459
local sequence alignment 465
multiple 467
pairwise 459
amino acids 5, 6
anthropology 11
AutoDep 247

B

bankit 247
bases 4
basic local alignment search tool 246
bioarchaeology 11
biodiversity 11
bioinformatics 8, 11, 12
bioprocessing 11
BLAST 246, 297, 470
bl2seq 297
BLASTall 297
BLASTclust 297
BLASTn 297, 298
BLASTp 297, 324
BLASTpgp 297
BLASTx 297, 306
fastacmd 297
formatdb 297
Mega Blast 297
PSI-Blast 330
RPS-BLAST 337
RPS-Blast 297
tBLASTn 297
tBLASTx 297, 342
Boolean operators 40
BRENDA 167
Braunschweig enzyme database 167

C

Catalytic Site Atlas 173
CATH 119
data formats 123
sample file 122
structure classification databases 124
CD search 246
CDART 246, 347
chimeric proteins 152
Chou–Fasman 248
clan 148, 150
clancards 150
clans 152
Cn3D 246
coding region 372
codon 5
codon bias 374
COGs 246
comparative genomics 11
compositional complexity 299
conserved domain architecture retrieval tool 347
contiguous words 313
CpG islands 287, 294
CpGPlot 287
CpGReport 287
CSA 173

D

DALI 128
data analysis tools 283
nucleotide sequence analysis 283
protein sequence analysis 324
Transeq 284
data submission tools 10, 251
nucleotide sequence submission 252
protein submission 267
databases 10, 19, 21, 23
ASN.1 29
categories 22
characteristics 22
flat file 25
information retrieval systems 37
navigating 35
object-oriented/object relational 29
others 10
relational 27
retrieving information 39
sequence 10
storage system 38
structure 10

XML 27
DDBJ 56
abbreviation key 57
DDBJ entry 56
sample DDBJ format 57
Deep View 414
discontiguous words 313
DNA 4
DNA forensics 11
DomPred 248
dot plots 445
dotmatcher 449
dottup 449

E

E-value 300, 303
EC system 147
edges 436
electronic PCR 246
EMBL 47
abbreviation key 50
EMBL entry 48
sample EMBL format 48
endopeptidases 152
ensemble 226
entrez 81
enzyme databases 145
enzymes 145, 146
classes 146
ePCR 246
eukaryotic promoter database 59
evolution 11
exons 372
exopeptidases 153
expect 300

F

FamCards 150
families 151
family 148, 150
FASTA 468
filtering 299
fingerprint 79
forward ePCR 311
FSSP 127

G

gamma distribution 370
Garnier–Osguthorpe-Robson 248
GCUA 294
GenBank 52
abbreviation key 54
entry 52
sample format 53
gene expression 11
gene expression omnibus 247
gene map 179
gene prediction 372
gene regulation 11
GeneCards 186
genefinder 248
GENSCAN 374
GenScan 248
GEO 247
GI numbers 36
GOR 248
GrailEXP 377
graph 436
graphical codon usage analyser 294

H

HIV sequence database 59
HMMER 249
human repeats 299
hyperchem 249

I

identifier 36
informatics 14
information retrieval system 81, 93
cross-database search system 81
information retrieval 86, 91
information storage system 83, 90
intergenic region 372
International Protein Sequence Database Collaboration 60
introns 372

K

knowledge discovery 242

L

LITDB 196
livestock breeding 11

M

map viewer 246
masking 299
MedlinePlus 189
MERNUM 149
MEROPS 147
metadata 219
microbial genomics 11
migration 11
MMDB 117
  abbreviation key 117
  MMDB entry 117
  Sample format 117
modeler 249
modelling tools 400
  for 2D protein modelling 400
  for 3D protein modelling 414
molecular medicine 11
morbid map 179
motif 79

N

noise 446
nucleic acid identifier 37
nucleotide sequence analysis
  transeq 284
nucleotide sequence databases 45

O

oligomers 446
omega-peptidases 152
OMIM 177
open reading frames 373
ORF 373
ORF finder 246, 315
Orthologous groups 80
  cluster 246
other databases 145
  chemical databases 197
  disease databases 177
  expression databases 209
  genome database and genome browser 226
  microarray database 222

P

PAM matrix construction algorithm 453
paralogs 80
PDB 108
  abbreviation key 115
  entry 109
  sample format 111
PepCards 147
peptidase 147–152
domain 150
identifier 149
inhibitors 150
unit 150, 152
peptidases 147, 148, 150, 151, 152
Pfam 248
PHI pattern 332
Phylip 247, 356
  CLIQUE 363
  CONSENSE 363
  CONTML 361
  CONTRAST 362
  DNACOMP 358
  DNADIST 359
  DNAINVAR 359
  DNAML 359
  DNAMLK 359
  DNAMOVE 358
  DNAPARS 358
  DNAPENNY 358
  DOLLOP 362
  DOLMOVE 363
  DOLPENNY 363
  DRAWGRAM and DRAWTREE 363
  FACTOR 363
  FITCH 361
  GENDIST 362
  KITSCH 361
  MIX 362
  MOVE 362
  NEIGHBOR 361
  PARS 362

PENNY 362
PROML 359
PROMLK 359
PROTDIST 359
PROTPARS 358
RESTDIST 360
RESTML 360
RETREE 363
SEQBOOT 360
TREEDIST 363
phylogenetic analysis 354
phylogenetic trees 354
bootstrap 355
branch 355
branch length 355
characters 355
clade 355
distance scale 355
long branch attraction 355
monophyletic 355
node 355
outgroup 355
polyphyletic 356
root 355
topology 355
Phyml 368
PIR-PSD 66
PIR-PSD entry 66
sample PIR format 67
polydot 450
PP 248
Pred TMR 249
predict protein 248
prediction algorithm 443
prediction tools 354
probabilistic algorithms 437
promoter region 372
PROSITE 77, 248, 387
InterProScan 390
MotifScan 390
pftools 390
PRATT 390
ps_scan 390
ScanProsite 390
protein data bank 68
entry 68
sample format 69
protein inhibitors 150
protein sequence databases 60
protein structure database collaboration 108
proteins 5
PSI-Blast 330
PubMed 193

R

Rasmol 249, 401
REBASE 59
ribosomal database project 59
risk assessment 11
RNA 4
codon 5
mRNA 5
rRNA 5
tRNA 5
uracil 5

S

Sakura 247
scissile bond 152
SCOP 131
linking SCOP 139
parsing SCOP 139
sCOP entry 132
search 133
score 303
secondary and specialized protein sequence database 75
CATH 77
COG 80
ENZYME 77
GOA 76
GPCRDb 76
HGVbase 80
InterPro 79
MEROPS 76
PRINTS 78
PROSITE 77
SWISS-2DPAGE 79
YPD 77
secondary nucleotide sequence databases 58
sequence databases 44
sequin 247
siftware 241
signal 446

similarity based approaches to gene prediction 480
SNPs 81
SPDV 249
specialized databases 197
spin 247
splice site 372
spliced 372
splicing 372
SQL or structured (english) query language 40
  complicated conditions 40
  create tables 41
  drop tables 41
  insert data 42
  joints 40
  simple condition 40
  update data 42
  view 41
SRS 93
STACK 59
start codon 372
statistical approaches 478
stop codon 372
stringency 447
structure databases 106
structure file formats 107
  chemistry rule approach 108
  explicit bonding approach 108
Swiss model 249
Swiss-Prot 62
  entry 63
  sample format 63

T

Tmpred 249
tools 10, 241
  data submission 247
  data analysis 10
  data mining 245
  genome analysis and gene expression
  modelling 10, 249
  prediction 10, 247
  protein sequence analysis and proteomics 246
  sequence analysis 246
  structure analysis 246
transcribed region 372
transeq 284
  attributes 285
  result page 286
TrEMBL 62

U

UniGene 58
UniProt 61
untranslated region (UTR) 372

V

vecscreen 246, 319
version numbers 36
vertices 436

W

webin 247
windows technique 447